AF356167

Modeling and Analysis of Damage and Failure of Concrete-Like, Brittle and Quasi-brittle Materials

Editors

Dan Huang
Lisheng Liu
Zhanqi Cheng
Liwei Wu

Basel • Beijing • Wuhan • Barcelona • Belgrade • Novi Sad • Cluj • Manchester

Editors

Dan Huang
College of Mechanics and
Engineering Science
Hohai University
Nanjing
China

Lisheng Liu
School of Science
Wuhan University
of Technology
Wuhan
China

Zhanqi Cheng
School of Civil Engineering
Zhengzhou University
Zhengzhou
China

Liwei Wu
College of Mechanics and
Engineering Science
Hohai University
Nanjing
China

Editorial Office
MDPI
St. Alban-Anlage 66
4052 Basel, Switzerland

This is a reprint of articles from the Special Issue published online in the open access journal *Materials* (ISSN 1996-1944) (available at: www.mdpi.com/journal/materials/special_issues/M39UYZM8IC).

For citation purposes, cite each article independently as indicated on the article page online and as indicated below:

Lastname, A.A.; Lastname, B.B. Article Title. *Journal Name* **Year**, *Volume Number*, Page Range.

ISBN 978-3-7258-0494-8 (Hbk)
ISBN 978-3-7258-0493-1 (PDF)
doi.org/10.3390/books978-3-7258-0493-1

Contents

About the Editors

Dan Huang

Dr. Huang Dan is a Professor and Doctoral Supervisor at Hohai University. He is primarily involved in research on computational mechanics and engineering. He has published over 150 papers, authored 5 books, and holds more than 40 authorized patents and software copyrights. In addition to his research contributions, he serves as a Director of the International Association for Computational Mechanics (IACM), an Executive Council Member of the International Chinese Association for Computational Mechanics (ICACM), Vice Chairman of the Computational Mechanics Committee of the Chinese Society of Theoretical and Applied Mechanics, Vice Chairman of the Southern Computational Mechanics Liaison Committee, and Chairman of the Computational Mechanics Committee of the Jiangsu Society of Theoretical and Applied Mechanics. He also plays a role as an editorial board member or reviewer for over 50 journals, including prestigious ones such as the *Journal of the Mechanics and Physics of Solids*, *Computer Methods in Applied Mechanics and Engineering*, the *International Journal of Solids and Structures*, the *International Journal of Mechanical Sciences*, and *Computational Mechanics*.

Lisheng Liu

Dr. Lisheng Liu is a Professor at Wuhan University of Technology (P.R. China). He is the Vice Dean of the School of Science, the Vice Director of the Hubei Key Laboratory of Theory and Application of Advanced Materials Mechanics, and the Chairman of the Hubei Society for Composite Materials. He became a faculty member at Wuhan University of Technology in 1992 and received his Ph.D. degree from this university in 2004. His current research interests include computational mechanics, multiscale mechanics, and nanomechanics. He has published more than 80 journal articles.

Zhanqi Cheng

Dr. Cheng Zhanqi, a professor and doctoral supervisor at the School of Civil Engineering, Zhengzhou University, mainly engages in research in the field of dynamic failure of advanced composite materials and computational solid mechanics. He obtained bachelor's, master's, and doctoral degrees in solid mechanics from Huazhong University of Science and Technology and Tongji University in 1994, 1996, and 2006, respectively. Additionally, he has hosted three National Natural Science Foundation projects and one national major science and technology sub-project, and received a second prize for scientific and technological progress in Henan Province. He also has published over 40 academic papers in international authoritative journals.

Liwei Wu

Dr. Liwei Wu is a postdoctor at the College of Mechanics and Engineering Science, Hohai University. He is mainly engaged in computational mechanics and engineering research. He has led one project from the National Natural Science Foundation of China (NSFC), and was selected for the Jiangsu Funding Program for Excellent Postdoctoral Talent. He has published over 20 papers, with 12 of them as the first author, and holds two software copyrights.

Preface

The modeling and analysis of the damage and failure of materials and structures is an active and persistent challenge in computational mechanics, materials, and various scientific and industrial fields. Understanding the mechanisms that lead to the degradation and eventual failure of materials is crucial for the design, maintenance, and safety of structures and components. This reprint provides an informative and stimulating forum to enhance academic communications on this challenging topic, focusing on the development and applications of computational theories, numerical and experimental methods, models, and algorithms for modeling and analyzing the damage and failure of concrete-like, brittle, and quasi-brittle materials and structures. The primary objective of this reprint is to provide a valuable resource for researchers, engineers, and students interested in the fundamental principles and advanced techniques of modeling and analyzing damage and failure in materials and structures. It offers insights into the fundamental principles of damage and failure in concrete-like, brittle, and quasi-brittle materials, as well as the latest advancements in relevant modeling and analysis techniques. We hope that this reprint will be a useful reference for those seeking to deepen their knowledge and will inspire further research and innovation in the field of damage and failure analysis.

Dan Huang, Lisheng Liu, Zhanqi Cheng, and Liwei Wu
Editors

Article

Performance of Eco-Friendly Cement Mortars Incorporating Ceramic Molds Shells and Paraffin Wax

Sandra Cunha [1], Raphael Silva [1], José Aguiar [1,*] and Fernando Castro [2]

1   Centre for Territory, Environment and Construction (CTAC), Department of Civil Engineering, Campus de Azurém, University of Minho, 4800-058 Guimarães, Portugal; sandracunha@civil.uminho.pt (S.C.); raphael.rezende@live.com (R.S.)
2   Mechanical Engineering and Resource Sustainability Center (MEtRICs), Department of Mechanical Engineering, Campus de Azurém, University of Minho, 4800-058 Guimarães, Portugal; fcastro@dem.uminho.pt
*   Correspondence: aguiar@civil.uminho.pt

Abstract: The lost wax foundry industry has been rapidly expanding in recent years, generating a large amount of waste due to the fact that most of the durable goods include castings and the need for dimensional precision castings for specific purposes, such as the automotive and aeronautics sectors. The waste produced by this industry is currently being deposited in landfills because practical applications are not known and cannot be reused in a new production process, and recycling is also a challenge because of the economics of the process. Thus, the main objective of this study consists in the incorporation of the produced wastes by the lost wax casting foundry industry (ceramic molds shells and paraffin wax) as substitutes for natural aggregate in exterior coatings mortars, evaluating their behavior under normal operating conditions and against freeze–thaw actions. The obtained results revealed porosity, flexural strength, and compressive strength adequate under normal operating conditions. The freeze–thaw performance of the mortars with waste incorporation was similar to the mortars developed with natural aggregates. Thus, the potential of the ceramic mold shells and paraffinic waxes utilization in cementitious mortars for the construction sector was demonstrated.

Keywords: foundry wastes; ceramic mold shells; paraffin wax; eco-friendly mortars; freeze–thaw performance

Citation: Cunha, S.; Silva, R.; Aguiar, J.; Castro, F. Performance of Eco-Friendly Cement Mortars Incorporating Ceramic Molds Shells and Paraffin Wax. *Materials* **2023**, *16*, 5764. https://doi.org/10.3390/ma16175764

Academic Editors: Dan Huang, Lisheng Liu, Zhanqi Cheng and Liwei Wu

Received: 31 July 2023
Revised: 14 August 2023
Accepted: 21 August 2023
Published: 23 August 2023

1. Introduction

Presently, the world witnesses economic growth propelled by its industrial and urban advancements [1]. Consequently, an increasing number of industries are generating waste [2–4], which needs to be recovered to prevent landfill deposition and mitigate adverse environmental impacts. The waste production from foundry industries is escalating on a global scale [5], given that castings are an integral part of about 90% of durable goods [6]. Simultaneously, the rapid progress in global infrastructure development has resulted in a significant acceleration of construction rates [2,7]. Thus, the construction industry, a huge consumer of raw materials for the production of construction materials, namely concrete and mortars [8], needs to find sustainable and alternative solutions that allow to reduce the use of natural raw materials, namely aggregates.

In 2020, the foundry industry produced about 103,300 thousand tons of castings [9], produced by about 35,000 active foundry companies [10]. The main world producers are China, India, and the United States, representing about 50.3%, 10.9%, and 9.4% of world casting production, respectively [9]. At this time, it is estimated that around 90,000 thousand tons of foundry waste are generated worldwide annually, with only between 15% and 30% being reused, with the remainder being landfilled [11–14].

In recent times, the precision casting or lost wax casting sector has experienced remarkable growth due to its significant advantages, including high dimensional accuracy, excellent surface quality, and the ability to accommodate highly complex designs. Moreover, it offers versatile applications across various industries such as automotive, valves and accessories, domestic equipment, railway construction, naval engineering, aeronautics, electrical components, machine construction, and even civil construction [15]. As evidence of its flourishing, the lost wax casting output value in China reached USD 3.3 billion in 2021, representing a remarkable increase of 17.5% compared with 2020 [16]. During the production process, the lost wax casting industry produces mainly two different types of wastes: ceramic mold shells and paraffin wax. In the lost wax casting process, the initial step involves fabricating sacrificial pieces made of paraffin wax, which are assembled onto wax support to create a cluster. Subsequently, a ceramic mold is built by repeatedly dip-coating the cluster in a ceramic slurry to form multiple ceramic layers. Once all the ceramic layers have dried, the ceramic mold and the paraffin are subjected to an oven, removing the paraffin wax by its evaporation, obtaining in this way the first waste, the paraffin wax, after condensation. The next step consists of pouring the molten metal into the empty ceramic mold and allowing it to cool and solidify. The final step consists in breaking the ceramic mold shell and separating the individual castings, obtaining the second waste type, the ceramic mold shells [17,18]. Until now, these wastes are deposited in landfills because any practical applications are not known or economic, and during the production process, their reincorporation is not possible. Thus, the raw materials used during the lost wax casting process are used only a single time. According to data from the producers, for each ton of casting, around one ton of ceramic mold shells and about 150 kg of paraffinic wax are produced.

In the construction sector, around 48 billion tons of aggregates are consumed annually [19]. The aggregates represent about 80% of the total weight of the concrete and mortars, being its main constituent [20]. Thus, the exploitation of natural resources for making aggregates is mandatory. However, this leads to serious problems related to the high energy consumption for their extraction and also to the possibility of their depletion. In the future, natural aggregates will become less available, increasing their price and difficulty in fulfilling worldwide needs. Thus, it is essential to look for alternative materials that allow replacing the aggregate in part or totally.

Due to the large number of foundry wastes available, in recent years, many researchers have developed works searching solutions for foundry wastes, essentially foundry sand [21–28] and iron and steel foundry slags [1,17,29–32], but few studies have been conducted with wastes from the precision or lost wax casting industry. Guney et al. [23] conducted an assessment of the feasibility of reusing foundry sand waste in high-performance concrete. They replaced natural sand with foundry sand in different contents (5%, 10%, and 15%). The outcomes of their study revealed that the inclusion of foundry waste resulted in a decrease in the compression and tensile strengths, as well as in the elasticity modulus of the concrete. However, the concrete containing 10% of foundry sand exhibited similar results compared to the reference concrete. Thiruvenkitam et al. [25] also evaluated the possibility of partially replacing natural sand with foundry sand in concrete, with contents varying from 0% to 25%. The results allowed us to observe that the concrete compressive strength increases until a foundry sand content of 20%. However, the incorporation of higher foundry sand content by natural aggregate replacement leads to a slight reduction in compressive strength. The concrete with the incorporation of 20% of foundry sand presents the same compressive strength classification as the concrete with 100% of natural sand (M30). In their research, Devi et al. [30] conducted a study where they utilized steel slag as an aggregate for concrete production. They assessed the durability of the concrete incorporating steel slag under different exposure conditions. Remarkably, the concrete with steel slag demonstrated satisfactory performance, even when compared with conventional concrete, presenting a pulse velocity higher than 3.5 mm/s in various exposure conditions and a weight loss 60% lower than the traditional concrete in the acid resistance tests. Keertan et al. [20] evaluated the mechanical performance of concrete mixtures produced by replacing coarse aggregate with 40%, 45%, and 50% of steel slag, observing a better performance for a steel slag content of 50% in terms of compressive strength and denser microstructure when compared with the other mixtures.

However, even though there is already research into some foundry wastes, the application of foundry wastes from the lost wax casting process is still an area with great research needs and is underdeveloped. Taking into account the expansion of the lost wax casting industry, it is also important to develop solutions that allow the reuse of ceramic mold shells and paraffin wax waste. This research team has been making efforts in an attempt to reuse these wastes [10,12]. Thus, during these preliminary works, the focus of the work was the valorization of the ceramic mold shells and paraffin wax. The presence of these wastes leads to the development of some microcracks in the mortar microstructure, especially the presence of ceramic mold shells, due to their chemical composition. However, the incorporation of 20% of paraffin wax led to a decrease of about 17%, 27%, and 64% in the flexural strength, compression strength, and adhesion of the mortars, respectively [10]. This situation was evaluated and solved by Cunha et al. [12] in a study in which they evaluated different methods to minimize the effect of the incorporation of ceramic mold shell wastes. Three different methods were analyzed: ceramic mold shell washing, polymeric fibers incorporation, and simultaneous utilization of ceramic mold shell washing and polymeric fibers incorporation. It was possible to verify that a single washing treatment was efficient in eliminating the presence of the chemical elements that leads to the alkali-aggregate reaction.

In recent years, we have observed climate changes that arise from the significant emissions of greenhouse gases, atmospheric pollution, and excessive energy consumption [33,34]. These factors are interconnected and present cumulative effects on the climate. While the term "global warming" is commonly linked to rising temperatures, climate change also influences weather patterns worldwide, resulting in extremes of both heat and cold. Consequently, issues concerning buildings in cold regions have become critical and require comprehensive attention, particularly regarding their durability, especially the durability of exterior coating mortars [35]. Thus, several researchers have explored the freeze–thaw behavior of different composite materials. Zhao et al. [36] studied the geopolymers' freeze–thaw behavior when subjected to 50 freeze–thaw cycles, observing a decrease in their compressive strength and destruction of the specimens after the 44th freeze–thaw cycle. Cunha et al. [37] developed a study evaluating the durability of coating mortars with the incorporation of commercial paraffin. The test results allowed us to observe that cement-based mortars with higher paraffin contents reveal a higher resistance to the freeze–thaw action and lower mass loss during the tests. Guney et al. [23] studied the behavior of cementitious materials with the incorporation of foundry sand, observing a decrease in the concrete mechanical and physical properties after freeze–thaw cycles. These studies allowed us to observe the influence of freeze–thaw actions on geopolymer concrete and cementitious materials incorporating commercial paraffin and foundry wastes. However, the evaluation of the behavior against the freeze–thaw tests of cementitious materials with the incorporation of precision foundry industry wastes (ceramic mold shells and paraffin wax) remains an underdeveloped area.

According to the previously mentioned, the main objectives of this study were the following:

- Evaluate the possibility of reusing ceramic mold shells and paraffin wax, from the lost wax foundry process, as a substitute for natural aggregate in cement mortars;
- Evaluate the physical and mechanical performance of the developed mortars in normal conditions;
- Evaluate the performance of the developed mortars subjected to freeze–thaw actions.

This is the first work carried out by the research team with the simultaneous incorporation of washed ceramic mold shells and paraffinic wax. On the other hand, the evaluation of the behavior against freeze–thaw actions in mortars with the incorporation of ceramic mold shells and paraffin wax, wastes from the precision foundry industry, or lost wax foundry industry has not yet been carried out. Thus, the novelty associated with this study is based on the following:

- Utilization of waste from the lost wax foundry industry to replace natural aggregates in mortars;
- Simultaneous incorporation of washed ceramic mold shells and paraffin wax to replace natural aggregate in mortars;
- Replacement of high contents of natural aggregate by lost wax foundry industry waste;
- Evaluation of the behavior of the developed mortars under normal operation conditions;
- Evaluation of the behavior of the developed mortars against freeze–thaw actions.

2. Materials and Methods

2.1. Foundry Wastes Preparation

The wastes were collected without any treatment, so they did not present ideal conditions to be used, being necessary to give them some methods of treatment and preparation.

2.1.1. Ceramic Mold Shells

The ceramic mold shells were initially subjected to a process of reducing their dimensions through the use of a crusher, in which they were subjected to several grinding cycles. Subsequently, they were subjected to a sieving process in order to eliminate particles with dimensions superior to 4 mm. Finally, the ground ceramic mold shells were subjected to a washing process in order to remove the presence of some constituents that originate alkali-aggregate reactions in cementitious mixtures [12]. The washing was carried out manually using two sieves, the square mesh sieve with an opening of 0.125 mm and the sieve with an opening of 0.063 mm (Figure 1). For the ceramic mold shells with dimensions between 0.125 mm and 4 mm, the 0.125 mm sieve was used, and for ceramic mold shells with dimensions inferior to 0.125 mm, the 0.063 mm sieve was used. The washing process was carried out in accordance with the European Standard NP EN 933-1 [38].

(a)   (b)

Figure 1. Ceramic mold shells washing process: (**a**) Washing of the particles with dimensions between 0.125 mm and 4 mm; (**b**) Washing of the particles with dimensions lower than 0.125 mm.

The importance of the ceramic shell-washing process was developed in a preliminary study by the research team, and observing that this process allowed for the removal of the presence of sodium, sulfur, aluminum, potassium, magnesium, chlorides, and sulfates. Thus, the utilization of washed ceramic mold shells allows the elimination of the expansibility problems and, consequently, microcracking in the cement mortars [12]. Table 1 shows the chemical composition of the obtained crystals, resulting from a leaching procedure according to DIN EN 12457-4:2003-01 [39], obtained in a scanning electron microscope by electron dispersive spectrometry.

Table 1. Chemical composition of obtained crystals from the water resulting from the ceramic mold shells washing [10].

Element	wt (%)
Oxygen (O)	59.4
Silicon (Si)	15.9
Calcium (Ca)	11.8
Chlorine (Cl)	3.9
Sodium (Na)	3.3
Sulfur (S)	2.0
Potassium (K)	1.8
Phosphorus (P)	1.1
Magnesium (Mg)	0.4
Aluminum (Al)	0.4

Figure 2 presents the granulometry of the ceramic mold shells after grinding and washing, obtained according to the European Standard NP EN 933-1 [38]. Based on the granulometric distribution, it can be observed that a minimum dimension of 0.063 mm, a maximum dimension of 4 mm, a D10 of 0.19 mm, a D50 of 1.25 mm, and a D90 of 3.5 mm. The ceramic mold shells presented a water absorption of 6.6%.

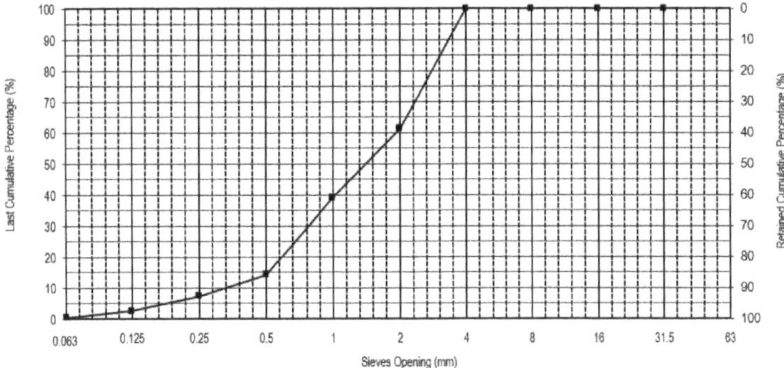

Figure 2. Particle size distribution of the ceramic mold shells.

Figure 3 presents the surface and geometry of the ceramic mold shells through scanning electron microscopy observations. It was possible to observe the existence of particles with different dimensions and angular shapes, typical of processed aggregates.

(**a**) (**b**)

Figure 3. Scanning electron microscopy observations of ceramic mold shells: (**a**) magnification of 200× and (**b**) magnification of 500×.

2.1.2. Paraffin Wax

The paraffin wax wastes also need to be processed in order to obtain smaller particles, again using a crusher. Finally, after grinding, particles with dimensions higher than 4 mm are separated using a sieve with a 4 mm of mesh opening. These particles are rejected and inserted again in a new grinding cycle until the desired particle size is obtained.

Figure 4 presents the granulometry of the paraffin wax after grinding, obtained according to the European Standard NP EN 933-1 [38]. Figure 5 shows the paraffin wax morphology based on scanning electron microscopic observations. According to the granulometric distribution, it was possible to observe a minimum dimension of 0.063 mm, a maximum dimension of 4 mm, a D10 of 0.355 mm, a D50 of 1.8 mm, and a D90 of 3.6 mm. The paraffin wax presented a water absorption of 1.6%.

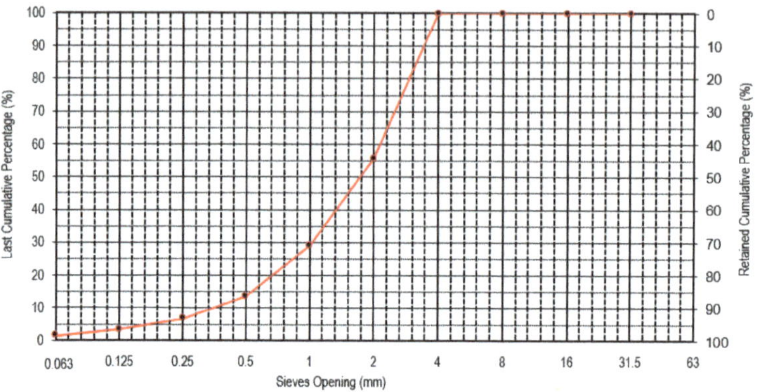

Figure 4. Particle size distribution of the paraffin wax [10].

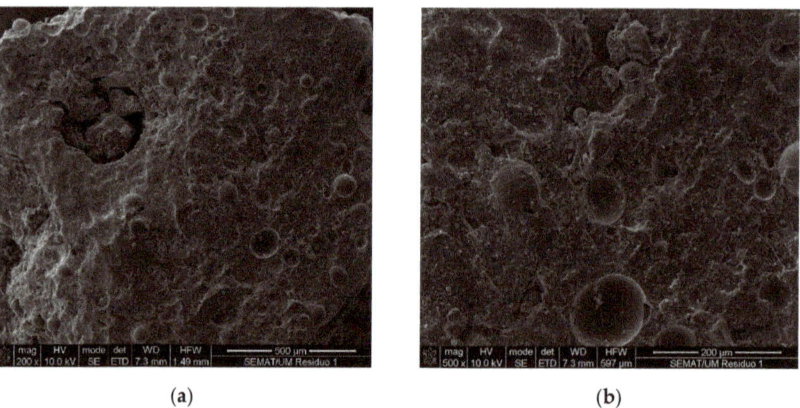

(**a**)   (**b**)

Figure 5. Scanning electron microscopy observations of paraffin wax waste: (**a**) magnification of 200× and (**b**) magnification of 500×.

Table 2 shows the minor elements of the paraffin wax. The major elements are carbon and hydrogen. The presence of these metals in the wax may be explained by contamination of the wax during the process.

Table 2. Chemical composition of paraffin wax (minor elements) [10].

Element	Weight (%)
Sodium (Na)	0.4
Aluminum (Al)	0.2
Calcium (Ca)	0.9

2.2. Materials Characterization

The materials selected for the development of mortars were based on previous work by the research team [10,12]. The Portland cement used was a CEM I 42.5R, produced by a Portuguese company (Secil, Lisbon, Portugal). The cement chemical composition is shown in Table 3, presenting a loss ignition value of 2.82%. CaO was the major chemical component, followed by SiO_2, Al_2O_3, SO_3, and Fe_2O_3.

Table 3. Chemical composition of cement CEM I 42.5 R.

Chemical Component	wt (%)
Loss on ignition	2.82
Insoluble residue	1.70
Silicon oxide (SiO_2)	19.65
Aluminum oxide (Al_2O_3)	4.28
Iron oxide (Fe_2O_3)	3.35
Calcium oxide (CaO)	61.35
Magnesium oxide (MgO)	1.70
Sulfates (SO_3)	3.36
Potassium oxide (K_2O)	0.89
Sodium oxide (Na_2O)	0.19
Chlorides (Cl^-)	0.04

The superplasticizer selected was based on a polyacrylate. In order to evaluate the replacement of natural aggregate, natural sand was selected with a minimum dimension of 0.063 mm and a maximum dimension of 4 mm (Figure 6). The natural sand presents a D10 of 0.16 mm, a D50 of 0.68 mm, and a D90 of 3 mm, based in their particle size distribution, obtained according to the European Standard NP EN 933-1 [38] and a water absorption of 1.19%.

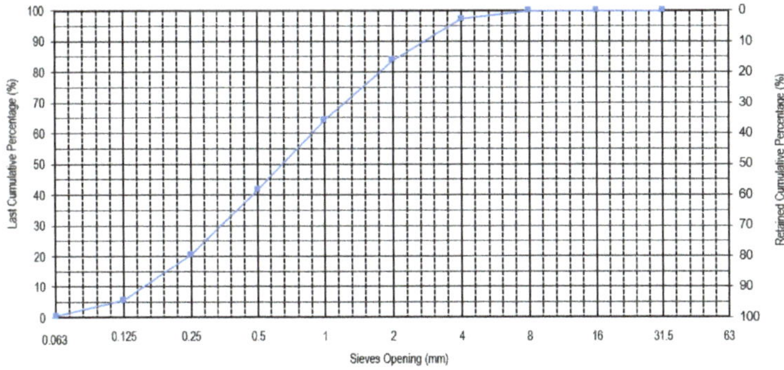

Figure 6. Particle size distribution of the natural sand [12].

Table 4 shows the density of the raw materials used to produce the different mortars.

Table 4. Materials densities (kg/m^3).

Material	Density
CEM I 42.5R	3049
Superplasticizer	1041
Ceramic mold shells	2630
Paraffin wax	1013
Natural sand	2569

2.3. Mortars and Specimens Production

Taking into account the high water absorption capacity of the used aggregates, they were initially saturated. The saturation water was mixed with the dry aggregates for 60 s, and a further 4 min were waited before continuing the production process. Then, the remaining solid materials were added to the mixture, having been mixed for 60 s.

Finally, the mixing water and the superplasticizer were added to the mixture, and the mortar was homogenized for 120 s. The different specimens were molded.

The mortar's water absorption by capillarity, water absorption by immersion, flexural strength, and compressive strength were determined using 3 prismatic specimens for each composition and each test with dimensions of 40 mm × 40 mm × 160 mm. The behavior of mortar's face to freeze–thaw cycles was determined using 5 specimens for each composition with dimensions of 50 mm × 50 mm × 50 mm. Regarding the microstructure observations, 2 cylindrical specimens with a diameter and height of approximately 1 cm were prepared for each mortar composition. A total of 88 specimens were produced. After the preparation, all the specimens were stored for 7 days in polyethylene bags (2 days in the mold and 5 days out of the mold). Posteriorly, the specimens were placed in a laboratory room with a controlled and constant temperature of 20 °C and humidity of 65% for 21 days.

2.4. Formulations

Six different mortars were developed with different materials constituting the aggregate mixture (Table 5). The cement and superplasticizer contents were fixed at 750 kg/m^3 and 7.5 kg/m^3, respectively. The superplasticizer content adopted was 1% of the binder weight. The volume of the aggregate and water varied according to the mortar dosage for 1 m^3 and taking into account the maintenance of the same workability for all mortars. A reference composition with 100% aggregate constituted by natural sand (NS100) and also composition with 100% aggregate constituted by ceramic mold shells (CMS100) were developed. Four different compositions were developed with different contents of ceramic mold shells and paraffin wax (20%, 40%, 60%, and 80%).

Table 5. Mortars formulation (kg/m^3).

Composition	Cement	Ceramic Mold Shells	Paraffin Wax	Natural Sand	Superplasticizer	Water
NS100	750	0	0	1225	7.5	273.3
CMS100	750	1136	0	0	7.5	315.2
CMS80PW20	750	699	175	0	7.5	305
CMS60PW40	750	427	284	0	7.5	303
CMS40PW60	750	234	351	0	7.5	285
CMS20PW80	750	103	412	0	7.5	280

2.5. Methods

The mortar's performance was tested in the fresh state and in the hardened state. For the fresh behavior, the workability was determined according to the standard EN 1015-3 [40]. The hardened behavior was determined based on the water absorption by capillarity [41], water absorption by immersion [42], flexural strength [43], compressive strength [43], and freeze–thaw tests [44].

The microstructure observations were performed using a scanning electron microscope (EDAX-Pegasus X4M, AMETEK, Unterschleissheim, Germany).

The workability tests were performed in order to determine the mortar spreading diameter. The test procedure consists in using the spreading table standardized by the EN 1015-3 standard [40]. After producing the mortar, the mold was filled in 2 layers, compacting each one with 10 strokes. Subsequently, the excess mortar was retained from the mold, and it was vertically removed. Finally, 15 strokes were applied to the mortar. The spreading diameter was obtained through the average of two measurements perpendicular to each other of the resulting spreading mortar.

The water absorption by capillarity tests were performed following the standard EN 1015-18 [41]. The test procedure starts by drying the specimens until constant mass in an oven at 60 °C. The specimen's lateral surfaces were coated with silicone to ensure that contact with the water happened only on the specimen's inferior face. After drying the side waterproofing, the first contact of the specimens with a 6 mm thick sheet of water is carried out. A weighing plan was established (Table 6), starting with the dry mass (Measurement 0) and interrupted after 7 h (Measurement 8). The coefficient of water absorption by capillarity was obtained based on the slope of a linear regression obtained from the line constructed with the different measurements.

Table 6. Weighing plan of the water absorption by capillarity tests.

Measurement	Time (Hours)	Time (Minutes)
1	0.0	0
2	0.2	10
3	0.7	40
4	1.3	77
5	2.1	125
6	2.7	159
7	4.0	241
8	7.0	420

The water absorption by immersion was performed based on the Portuguese specification LNEC E 394 [42]. Initially, the samples were dried in an oven at 60 °C until constant mass (m_3). Then, the specimens were immersed in water at a temperature of about 20 °C, at atmospheric pressure, obtaining the saturated mass (m_1). Finally, the hydrostatic mass was determined by weighing the sample in water (m_2). The water absorption by immersion was determined according to Equation (1) [42]:

$$A = ((m_1 - m_3)/(m_1 - m_2)) \times 100 \tag{1}$$

where

A—water absorption by immersion (%);
m_1—saturated sample mass (g);
m_2—hydrostatic sample mass (g);
m_3—dry sample mass (g).

The flexural and compressive strengths of the mortars were performed in accordance with the European Standard EN 1015-11 [43]. The flexural and compressive tests were performed with load control at a speed of 50 N/s and 500 N/s, respectively. The strengths can be determined according to Equations (2) and (3) [43].

$$FS = 1.5 \times ((F \times l)/(b \times d^2)) \tag{2}$$

where

FS—flexural strength (MPa);
F—maximum load (N);
l—the distance between supports (mm);
b—specimen width (mm);

d—specimen height (mm).

$$CS = F/A \tag{3}$$

where

CS—compressive strength (MPa);
F—maximum load (N);
A—area (mm^2).

The freeze–thaw tests were carried out based on the European specification CEN/TS 12390-9 [44]. The equipment used to carry out the tests was programmed with a temperature law in which each freeze–thaw cycle has a duration of 24 h, with a total of 56 cycles having been carried out. During each, the temperature varies between 24 °C and −18 °C. The specimens were saturated and subsequently subjected to temperature cycles. During the tests, each specimen was placed in contact with a layer of water, with the aim of the specimen absorbing the water lost by evaporation and also by ventilation of the equipment, thus ensuring that the pores of the specimens would always be saturated. The equipment was also programmed with a constant relative humidity of 90% in order to avoid weight loss due to water evaporation. The measurement of mortar degradation took into consideration the mass losses suffered in each cycle.

3. Results and Discussion

3.1. Workability

The workability tests were developed in order to maintain the spreading diameter between 195 and 205 mm. The mortar's water/binder ratio was iteratively determined, considering the mortar's constituent material's density and a mortar volume of 1 m^3.

According to Figure 7, it was possible to observe that the mortar with 100% of the natural aggregate (NS100) presents a lower water/binder ratio compared with the mortars with the incorporation of ceramic mold shells and paraffin wax. The replacement of 100% of natural sand with 100% of ceramic mold shell leads to an increase in the water/binder ratio of about 16.7%. The presence of paraffin wax in the mortar's constituents causes an increase in the water/binder ratio higher than 2.8% compared with the NS100 mortar. This behavior can be justified by the lower water absorption capacity of the natural aggregate compared with the recycled aggregates, about 1.19%.

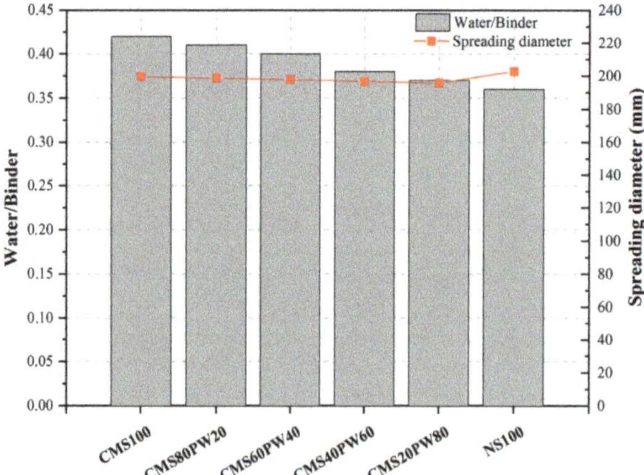

Figure 7. Water/binder ratio of the developed mortars.

The mortar produced with 100% of ceramic mold shells (CMS100) shows a higher water-binder ratio, which can be justified by the more angular shape, roughness, and higher water absorption capacity of the aggregate, about 6.6%. The ceramic mold shells replacement by paraffin wax waste leads to a decrease in the water/binder ratio. The incorporation of 20% of paraffin wax causes a reduction in the water/binder ratio higher than 2.4%. This behavior can be justified by the higher average particle size of the paraffin wax particle compared with the ceramic mold shells (Figure 4) and also due to their lower water absorption capacity compared with the ceramic mold shells, about 1.6%.

3.2. Water Absorption by Capillarity

The water absorption by capillarity was evaluated based on the water absorption capacity (Figure 8) and on the coefficient of water absorption by capillarity (Figure 9).

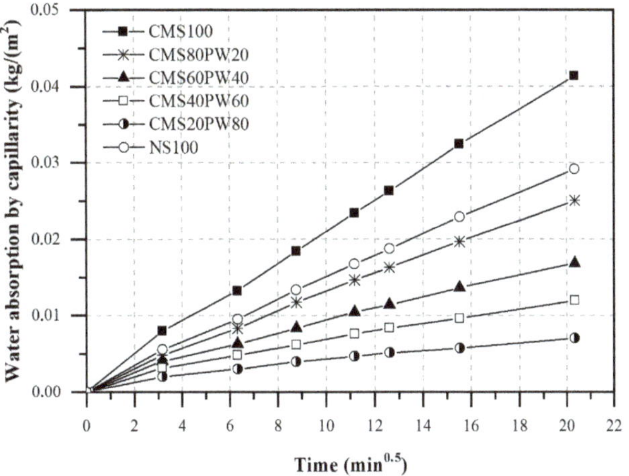

Figure 8. Water absorption by capillarity of the developed mortars.

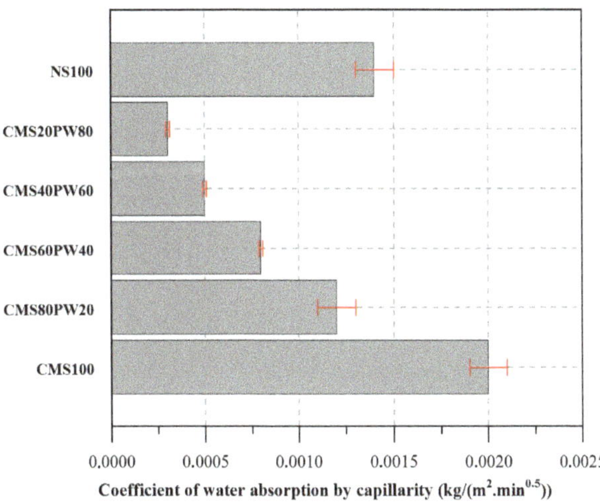

Figure 9. Coefficient of water absorption by capillarity of the developed mortars.

According to Figure 8, it is possible to observe that the composition with the incorporation of 100% aggregate constituted by ceramic mold shells (CMS100) shows the greatest capacity to absorb water by capillarity during the first seven hours of contact with water. This behavior can be associated with the presence of a higher water/binder ratio of these mortars (Figure 7) and also due to the higher capacity of water absorption characteristic of the waste material. The replacement of ceramic mold shells with paraffinic wax leads to a decrease in water absorption by capillarity. The incorporation of 20% paraffinic wax results in a decrease in water absorption by capillarity of about 39%. The observed behavior can be explained by the lower ratio of water/binder present in mortars with simultaneous incorporation of the two foundry wastes (CMS80PW20, CMS60PW40, CMS40PW60, and CMS20PW80), as shown in Figure 7. The lower water absorption capacity of the paraffin wax waste is also due the higher particle dimensions of the paraffin wax compared with the ceramic mold shells average particle size. Finally, the mortars developed with the aggregate constituted by 100% of natural sand (NS100) showed a water absorption by capillarity higher than the mortars with simultaneous incorporation of ceramic mold shells and paraffin wax, even presented the lower water/binder ratio of all developed mortars. This behavior can be related to different aspects, namely the lower water absorption capacity of the natural raw material and the fact that natural sand has a smaller particle size, allowing a greater adjustment in the microstructure of the mortar.

In Figure 9, it was possible to observe the coefficient of water absorption by capillarity of the developed mortars. These results are in line with the results presented for the capillary water absorption capacity (Figure 8), with the CMS100 mortar having the highest capillary water absorption coefficient, followed by the NS100 mortar and then the mortars with simultaneous incorporation of ceramic mold shells and paraffinic wax (CMS80PW20, CMS60PW40, CMS40PW60, and CMS20PW80). The replacement of 100% of natural sand with 100% of ceramic mold shell leads to an increase in the capillary water absorption coefficient of about 42.8%. The presence of paraffin wax causes a decrease in the capillary water absorption coefficient higher than 14.3% compared with the reference mortar (NS100).

According to Figure 10, it was possible to observe a more compact microstructure for the mortar with the incorporation of 40% paraffin, replacing the ceramic mold shells (CMS60PW40), proving its lower microporosity, reflected by the smaller size and quantity of pores, comparatively to the mortar with the incorporation of 100% of ceramic mold shells (CMS100). The obtained results are also in accordance with previous works developed by the research team [10,12].

(a)

(b)

Figure 10. Mortar matrix scanning electron microscopy observations, magnification of 25,000×: (**a**) mortar with aggregate constituted by 100% of ceramic mold shells (CMS100); and (**b**) mortar with aggregate constituted by 60% of ceramic mold shells and 40% of paraffin wax (CMS60PW40).

3.3. Water Absorption by Immersion

In Figure 11, it was possible to observe the water absorption by immersion of the developed mortars.

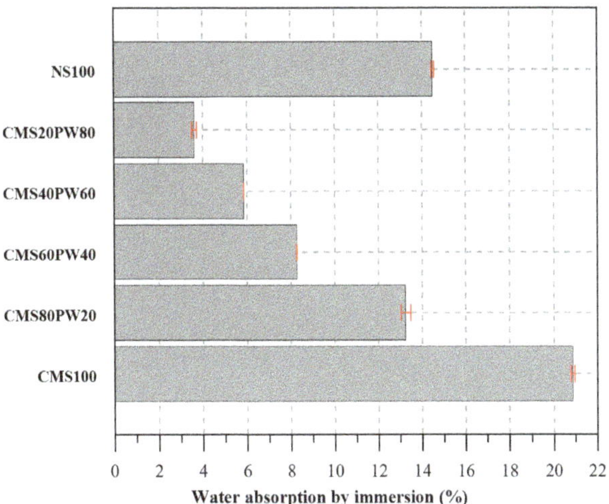

Figure 11. Water absorption by immersion of the developed mortars.

The mortar with incorporation of 100% of aggregate constituted by ceramic mold shells (CMS100) presents the highest water absorption by immersion. Once again, this behavior can be associated with the higher water content present in the mortars (Table 5), which leads to a more porous mortar due to the evaporation of chemically uncombined water. However, in this case, the characteristics of the recycled crushed aggregate can also contribute to greater absorption of water by immersion because this type of aggregate has a higher water absorption capacity, which, even with its initial saturation during the mortar production process, may still have some water absorption capacity. On the other hand, the ceramic mold shells have an average particle size greater than the natural sand and, also, due to their manufacturing process, a more angular shape, which provides the appearance of pores at higher dimensions. The replacement of natural sand (NS100) with ceramic mold shells (CMS100) leads to an increase in water absorption by immersion of about 44%.

The replacement of ceramic mold shells with paraffin wax translated into a decrease in water absorption by immersion. The incorporation of 20%, 40%, 60%, and 80% of paraffinic wax results in a decrease of about 37%, 60%, 72%, and 83% in water absorption by immersion, respectively. This behavior is not only associated with the presence of a smaller amount of water in these mortars (Table 5 and Figure 7) but also with the lower water absorption capacity of the paraffinic wax, around 5% inferior when compared with the water capacity of the ceramic mold shells. The obtained results are also in accordance with previous works developed by the research team [10,12].

3.4. Flexural and Compressive Strength

Regarding the mechanical behavior (Figure 12), it was possible to observe that the mortar with the incorporation of 100% of natural sand (NS100) exhibits a higher performance, presenting high flexural and compressive strengths when compared with the mortars with the incorporation of waste materials (ceramic mold shells and paraffin wax), which can be related with the lower water/binder ratio present in these mortars.

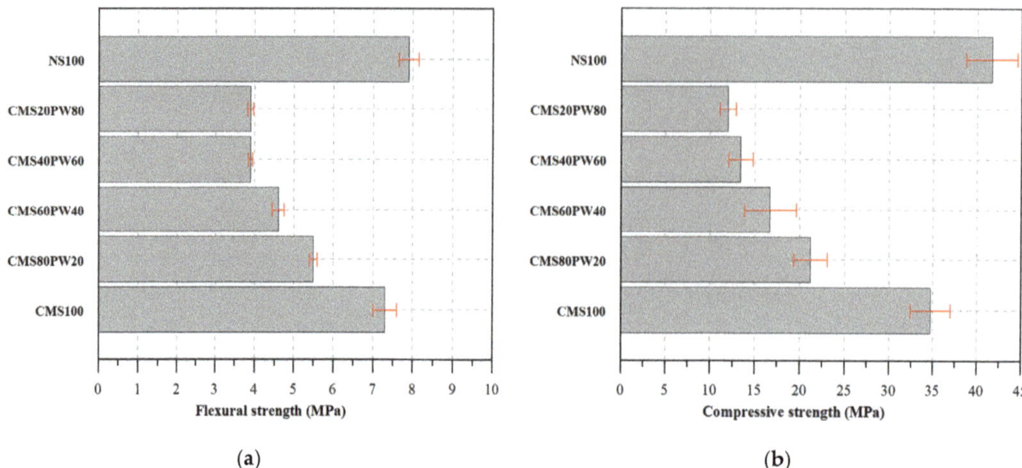

Figure 12. Mechanical behavior of the developed mortars: (**a**) flexural strength and (**b**) compressive strength.

The total replacement of the natural aggregate (NS100) by ceramic mold shells (CMS100) results in a decrease in the flexural and compressive strength of about 8% and 17%, respectively. Once again, this behavior can be justified by the higher water content present in the mortar's composition (Table 5) and consequent higher water absorption by capillarity and immersion (Figures 8, 9 and 11), resulting in a greater mortar porosity.

The paraffin wax incorporation causes a decrease in flexural and compressive strength. The addition of 20% of paraffin leads to a decrease in the flexural and compressive strength of more than 24% and 39%, respectively. In this case, the observed behavior is associated with the lower adhesion of the paraffin wax to the cementitious matrix and the lower resistance of the paraffin wax particles compared with the ceramic mold shell particles [10,12]. Some studies report that the incorporation of free and pure paraffin in cementitious mixtures also indicates a delay in the cement hydration process, which may also explain the lower mechanical performance observed for mortars with higher content of paraffinic wax incorporation [45]. Even having verified a decrease in the flexural and compressive strengths of the developed mortars, it is important to note that its application in the construction industry was not compromised. According to European Standardization NP EN 998-1 [46], which provides the classification of coating mortars according to their compressive strength, the developed mortars presented the maximum classification provided, CSIV, which is obtained with a compressive strength equal to or higher than 6 MPa.

3.5. Freeze–Thaw Tests

The tests started in cycle 0, in which the specimens presented their mass unaltered, 100% of the mass.

In Figure 13, it was possible to observe that the mortars constituted by 100% of natural sand (NS100) showed lower mass loss and faced freeze–thaw cycles due to the higher compressive strength (Figure 12). The mortar with the replacement of natural sand by 100% of ceramic mold shells (CMS100) exhibited a mass loss slightly higher than the reference mortar, which is related to its higher porosity (Figures 9 and 11) and lower compressive strength (Figure 12). The higher porosity of the mortars will favor a higher water storage capacity, which, consequently, with the volume variation relative to the freezing and thawing cycles, will cause internal tensions in the microstructure of the mortar, originating the appearance of microcracks, which allow the specimen's degradation, resulting in a higher mass loss.

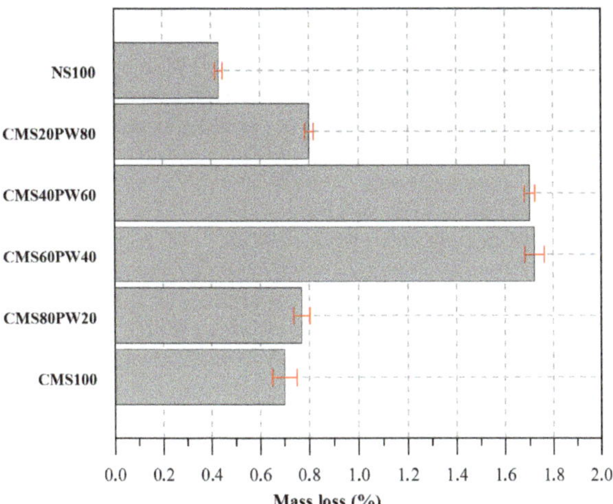

Figure 13. Total mass loss during the freeze–thaw tests of the developed mortars.

The incorporation of paraffin wax in higher contents leads to a higher mass loss during the freeze–thaw tests, which is directly connected to the lower mechanical performance of the mortars due to the lower resistance of the paraffin wax compared with the ceramic mold shells (Figure 13), even presenting lower porosity. However, it is important to observe that all developed mortars exhibit an appropriate behavior to freeze–thaw action because, for all developed mortars, the mass loss is lower than 1.7%, and even the developed mortars present high levels of wastes incorporation (ceramic mold shells and paraffin wax). Another study evaluated the incorporation of paraffin directly into cementitious mixtures against freeze–thaw actions, revealing higher mass losses [37], which is justified by the lower mechanical performance of the developed mortars compared with the mechanical performance of the mortars presented in this study. The embrittlement of cementitious materials with the incorporation of wastes from the foundry industry, in particular, foundry sands subjected to freeze–thaw cycles, was also reported in another study [23].

4. Conclusions

This study allowed the evaluation of the influence of the incorporation of ceramic mold shells and paraffinic wax wastes in cement mortars, replacing the natural aggregate and evaluating their physical, mechanical, and freeze–thaw behavior.

Taking into account the high amount of precision foundry wastes currently generated around the world and the continuously growing prospection of this industry, it will be expected that this study has enormous applicability worldwide due to the need to reuse these wastes, but also to the need of the construction industry in reduce the natural raw materials consumption.

The most relevant conclusions achieved were the following:

- The water content in the mortars increases with the total replacement of natural sand by ceramic mold shells. However, the replacement of ceramic mold shells with paraffinic wax leads to a decrease in the water/binder ratio. These behaviors were directly connected with the particle size and water absorption capacity of the different aggregates.
- The water absorption by capillarity and immersion of the mortars with simultaneous incorporation of ceramic mold shells and paraffinic wax were lower compared with the mortar with 100% of natural sand and 100% of ceramic mold shells due to the lower ratio of water/binder of these mortars and lower water capacity absorption of the paraffinic wax. The higher mortar porosity was presented by the mortars with

- 100% of aggregate constituted by ceramic mold shells due to the higher water/binder ratio and water capacity absorption of the ceramic mold shells.
- The higher mechanical performance was presented by the mortar with 100% aggregate constituted by natural sand. The presence of ceramic mold shells and paraffinic wax leads to a decrease in the flexural and compressive strengths due to the lower adhesion of the paraffin wax to the cementitious matrix and the delay in the cement hydration process.
- The freeze–thaw behavior of the developed mortars was very satisfactory because all the mortars presented mass losses very reduced and close to the mortar with 100% of natural aggregate.
- The replacement of natural sand by foundry waste (peels from ceramic molds and paraffinic wax) in the compositions CMS100, CMS80PW20, CMS60PW40, CMS40PW60, and CMS20PW80 will lead to a 100% saving in the aggregate acquisition cost for these mortars.

Considering the mechanical performance (flexural and compressive behavior) and the performance against freeze–thaw actions, it was possible to observe that the mortars with the incorporation of 100% of ceramic mold shells (CMS100) and 80% of ceramic mold shells and 20% of paraffin wax (CMS80PW20) showed the more interesting behavior.

It can be concluded that the reutilization of these wastes (paraffin waxes and ceramic mold shells) as a replacement for natural aggregate in mortars can be seen as a viable solution for decreasing the consumption of natural raw materials. On the other hand, the reuse of these wastes, which so far do not have any type of application and cannot be reused during a new foundry production process, will result in a substantial reduction in their deposit in landfills.

Author Contributions: Conceptualization, S.C., J.A. and F.C.; methodology, S.C., R.S., J.A. and F.C.; validation, S.C., J.A. and F.C.; formal analysis, S.C., R.S. and J.A.; investigation, S.C., R.S. and J.A.; resources, S.C., J.A. and F.C.; data curation, S.C., R.S. and J.A.; writing—original draft preparation, S.C.; writing—review and editing, J.A. and F.C.; visualization, S.C., J.A. and F.C.; supervision, S.C. and J.A. All authors have read and agreed to the published version of the manuscript.

Funding: This research was funded by FUNDAÇÃO PARA A CIÊNCIA E TECNOLOGIA (FCT), grant number UIDB/04047/2020.

Institutional Review Board Statement: Not applicable.

Informed Consent Statement: Not applicable.

Data Availability Statement: Not applicable.

Acknowledgments: This work was supported by FCT/MCTES through national funds (PIDDAC) under the R&D Unit Centre for Territory, Environment and Construction (CTAC) under reference UIDB/04047/2020.

Conflicts of Interest: The authors declare no conflict of interest.

References

1. Murmu, M.; Mohanta, N.R.; Bapure, N. Study on the fresh and hardened properties of concrete with steel slag as partial replacement for natural aggregates. *Mater. Today Proc.* **2023**, *in press*. [CrossRef]
2. Zhang, M.; Li, W.; Wang, Z.; Liu, H. Urbanization and production: Heterogeneous effects on construction and demolition waste. *Habitat Int.* **2023**, *13*, e102778.
3. Juvan, E.; Grün, B.; Dolnicar, S. Waste production patterns in hotels and restaurants: An intra-sectoral segmentation approach. *Ann. Tour. Res. Empir. Insights.* **2023**, *4*, e100090.
4. Sadh, P.K.; Chawla, P.; Kumar, S.; Das, A.; Kumar, R.; Bains, A.; Sridhar, K.; Duhan, J.S.; Sharma, M. Recovery of agricultural waste biomass: A path for circular bioeconomy. *Sci. Total Environ.* **2023**, *870*, e161904.
5. Qasrawi, H. Fresh properties of green SCC made with recycled steel slag coarse aggregate under normal and hot weather. *J. Clean. Prod.* **2018**, *204*, 980–991. [CrossRef]
6. American Foundry Society. Available online: https://www.afsinc.org/industry-statistics (accessed on 1 June 2023).

7. Baalamurugan, J.; Kumar, V.G.; Chandrasekaran, S.; Balasundar, S.; Venkatraman, B.; Padmapriya, R.; Raja, V.K.B. Recycling of steel slag aggregates for the development of high density concrete: Alternative & environment-friendly radiation shielding composite. *Compos. Part B Eng.* **2021**, *216*, e108885.
8. Tang, Y.; Xiao, J.; Zhang, H.; Duan, Z.; Xia, B. Mechanical properties and uniaxial compressive stress-strain behavior of fully recycled aggregate concrete. *Constr. Build. Mater.* **2022**, *323*, e126546.
9. The European Foundry Association. Available online: https://www.caef.eu/downloads-links/#statistics (accessed on 1 June 2023).
10. Cunha, S.; Tavares, A.; Aguiar, J.; Castro, F. Cement mortars with ceramic molds shells and paraffin waxes wastes: Physical and mechanical behavior. *Constr. Build. Mater.* **2022**, *342*, e127949. [CrossRef]
11. Torres, A.; Bartlett, L.; Pilgrim, C. Effect of foundry waste on the mechanical properties of Portland Cement Concrete. *Constr. Build. Mater.* **2017**, *135*, 674–681.
12. Cunha, S.; Costa, D.; Aguiar, J.; Castro, F. Mortars with the incorporation of treated ceramic molds shells wastes. *Constr. Build. Mater.* **2023**, *365*, e130074.
13. Sithole, N.T.; Tsotetsi, N.T.; Mashifana, T.; Sillanpää, M. Alternative cleaner production of sustainable concrete from waste foundry sand and slag. *J. Clean. Prod.* **2022**, *336*, e130399. [CrossRef]
14. Soares, H.; Lessa, C. Replacement of the gravel by ceramic waste from the lost wax casting process in the high resistence concretes. *Tecnol. Metal Mater. Min.* **2021**, *18*, e2140. [CrossRef]
15. Portuguese Association of Foundry. Available online: https://apf.com.pt/numeros-do-setor/ (accessed on 1 June 2023).
16. Jia, Z.; Aguiar, J.; Jesus, C.; Castro, F.; Cunha, S. Physical and mechanical properties of lightweight concrete with incorporation of ceramic mold casting waste. *Materialia* **2023**, *28*, e101765. [CrossRef]
17. Wang, J.; Sama, S.; Lynch, P.; Manogharan, G. Design and Topology Optimization of 3D-Printed Wax Patterns for Rapid Investment Casting. *Procedia Manuf.* **2019**, *34*, 683–694. [CrossRef]
18. Pattnaik, S.; Karunakar, D.B.; Jha, P.K. Developments in investment casting process—A review. *J. Mater. Process. Technol.* **2012**, *212*, 2332–2348. [CrossRef]
19. Shmlls, M.; Abed, M.; Horvath, T.; Bozsaky, D. Multicriteria based optimization of second generation recycled aggregate concrete. *Case Stud. Constr. Mater.* **2022**, *17*, e01447. [CrossRef]
20. Keertan, T.S.; Kumar, V.P.; Bommisetty, J.; Ramanjaneyulu, B.; Kumar, M.A. High strength fiber reinforced concrete with steel slag as partial replacement of coarse aggregate: Overview on mechanical and microstructure analysis. *Mater. Today Proc.* **2023**, *in press*. [CrossRef]
21. Khatib, J.M.; Herki, B.A.; Kenai, S. Capillarity of concrete incorporating waste foundry sand. *Constr. Build. Mater.* **2013**, *47*, 867–871. [CrossRef]
22. Behnood, A.; Golafshani, E.M. Machine learning study of the mechanical properties of concretes containing waste foundry sand. *Constr. Build. Mater.* **2020**, *243*, e118152. [CrossRef]
23. Guney, Y.; Sari, Y.D.; Yalcin, M.; Tuncan, A.; Donmez, S. Re-usage of waste foundry sand in high-strength concrete. *Waste Manag.* **2010**, *30*, 1705–1713. [CrossRef]
24. Matos, P.R.; Pilar, R.; Bromerchenkel, L.H.; Schankoski, R.A.; Gleize, P.J.P.; Brito, J. Self-compacting mortars produced with fine fraction of calcined waste foundry sand (WFS) as alternative filler: Fresh-state, hydration and hardened-state properties. *J. Clean. Prod.* **2020**, *252*, e119871. [CrossRef]
25. Thiruvenkitam, M.; Pandian, S.; Santra, M.; Subramanian, D. Use of waste foundry sand as a partial replacement to produce green concrete: Mechanical properties, durability attributes and its economical assessment. *Environ. Technol. Innov.* **2020**, *19*, e101022. [CrossRef]
26. Chalangaran, N.; Farzampour, A.; Paslar, N.; Fatemi, H. Experimental investigation of sound transmission loss in concrete containing recycled rubber crumbs. *Adv. Concr. Constr.* **2021**, *6*, 447–454.
27. Matos, P.R.; Marcon, M.F.; Schankoski, R.A.; Prudêncio, L.R., Jr. Novel applications of waste foundry sand in conventional and dry-mix concretes. *J. Environ. Manag.* **2019**, *244*, 294–303. [CrossRef] [PubMed]
28. Siddique, R.; Singh, G.; Belarbi, R.; Ait-Mokhtar, K. Comparative investigation on the influence of spent foundry sand as partial replacement of fine aggregates on the properties of two grades of concrete. *Constr. Build. Mater.* **2015**, *83*, 216–222. [CrossRef]
29. Papachristoforou, M.; Anastasiou, E.K.; Papayianni, I. Durability of steel fiber reinforced concrete with coarse steel slag aggregates including performance at elevated temperatures. *Constr. Build. Mater.* **2020**, *262*, e120569. [CrossRef]
30. Devi, V.S.; Kumar, M.M.; Iswarya, N.; Gnanavel, B.K. Durability of Steel Slag Concrete under Various Exposure Conditions. *Mater. Today Proc.* **2019**, *22*, 2764–2771. [CrossRef]
31. Lai, M.H.; Zou, J.; Yao, J.C.M.B.; Ho, X.Z.; Wang, Q. Improving mechanical behavior and microstructure of concrete by using BOF steel slag aggregate. *Constr. Build. Mater.* **2021**, *277*, e122269. [CrossRef]
32. Dhivya, K.; Anusha, G.; Vinoth, S.; Shanjai, P.R.; Sreedharan, A.; Rahuma, V.M. Steel slag's effect on concrete mechanical properties and durability. *Mater. Today Proc.* **2023**, *in press*. [CrossRef]
33. Salimi, M.; Al-Ghamdi, S.G. Climate change impacts on critical urban infrastructure and urban resiliency strategies for the Middle East. *Sustain. Cities Soc.* **2020**, *54*, e101948.
34. Tahir, F.; Al-Ghamdi, S.G. Climatic change impacts on the energy requirements for the built environment sector. *Energy Rep.* **2023**, *9*, 670–676. [CrossRef]

35. Mermerdaş, K.; İpek, S.; Mahmood, Z. Visual inspection and mechanical testing of fly ash-based fibrous geopolymer composites under freeze-thaw cycles. *Constr. Build. Mater.* **2021**, *283*, e122756. [CrossRef]
36. Zhao, M.; Zhang, G.; Htet, K.W.; Kwon, M.; Liu, C.; Xu, Y.; Tao, M. Freeze-thaw durability of red mud slurry-class F fly ash-based geopolymer: Effect of curing conditions. *Constr. Build. Mater.* **2019**, *215*, 381–390.
37. Cunha, S.; Silva, M.; Aguiar, J. Behavior of cementitious mortars with direct incorporation of non-encapsulated phase change material after severe temperature exposure. *Constr. Build. Mater.* **2020**, *230*, e117011. [CrossRef]
38. *NP EN 933-1*; Tests for Geometrical Properties of Aggregates—Parte 1: Determination of Particle Size Distribution—Sieving Method. Portuguese Institute for Quality (IPQ): Caparica, Portugal, 2000. (In Portuguese)
39. *DIN EN 12457-4:2003-01*; Characterization of Waste—Leaching—Compliance Test for Leaching of Granular Waste Materials and Sludges—Part 4: One Stage Batch Test at a Liquid to Solid Ratio of 10 L/kg for Materials with Particle Size below 10 mm (without or with Limited Size Reduction). German Institute for Standardisation (DIN): Berlin, Germany, 2003.
40. *EN 1015-3*; Methods of Test for Mortar for Masonry—Part 3: Determination of Consistence of Fresh Mortar (by Flow Table). European Committee for Standardization (CEN): Brussels, Belgium, 1999.
41. *EN 1015-18*; Methods of Test for Masonry—Part 18: Determination of Water Absorption Coefficient due to Capillary Action of Hardened Mortar. European Committee for Standardization (CEN): Brussels, Belgium, 2002.
42. *E394:1993*; Concrete—Determination of Water Absorption by Immersion. National Laboratory for Civil Engineering (LNEC): Lisbon, Portugal, 1993. (In Portuguese)
43. *EN 1015-11*; Methods of Test for Mortar for Masonry—Part 11: Determination of Flexural and Compressive Strength of Hardened Mortar. European Committee for Standardization (CEN): Brussels, Belgium, 1999.
44. *CEN/TS 12390-9*; Testing Hardened Concrete—Part 9: Freeze-Thaw Resistance. European Committee for Standardization (CEN): Brussels, Belgium, 2006.
45. Kheradmand, M.; Vicente, R.; Azenha, M.; Aguiar, J. Influence of the incorporation of phase change materials on temperature development in mortar at early ages: Experiments and numerical simulation. *Constr. Build. Mater.* **2019**, *225*, 1036–1051. [CrossRef]
46. *NP EN 998-1*; Specification for Masonry Mortars—Part 1: Mortars for Interior Exteriors. Portuguese Institute for Quality (IPQ): Caparica, Portugal, 2010. (In Portuguese)

Article

Rapid-Hardening and High-Strength Steel-Fiber-Reinforced Concrete: Effects of Curing Ages and Strain Rates on Compressive Performance

Fan Mo [1], Boxiang Li [1], Mingyi Li [1], Zhuangcheng Fang [2,3,*], Shu Fang [2,3] and Haibo Jiang [1,*]

[1] School of Civil and Transportation Engineering, Guangdong University of Technology, Guangzhou 510006, China; 13018556303@163.com (F.M.); lpx3199@163.com (B.L.); m15915841513@163.com (M.L.)

[2] Earthquake Engineering Research & Test Center, Guangzhou University, Guangzhou 510006, China; shufang@gzhu.edu.cn

[3] Guangdong Key Laboratory of Earthquake Engineering & Applied Technique, Guangzhou 510006, China

* Correspondence: fangzc@gzhu.edu.cn (Z.F.); hbjiang@gdut.edu.cn (H.J.)

Abstract: High-strength steel-fiber-reinforced concrete (HSFRC) has become increasingly popular as a cast-in-place jointing material in precast concrete bridges and buildings due to its excellent tensile strength and crack resistance. However, working conditions such as emergency repairs and low-temperature constructions require higher demands on the workability and mechanical properties of HSFRC. To this end, a novel rapid-hardening HSFRC has been proposed, which is produced using sulphoaluminate cement (SC) instead of ordinary Portland cement. In this study, quasi-static and dynamic tests were carried out to compare the compressive behavior of conventional and rapid-hardening HSFRCs. The key test variables included SC replacement ratios, concrete curing ages, and strain rates. Test results showed: (1) Rapid-hardening HSFRC exhibited high early strengths of up to 33.14 and 44.9 MPa at the curing age of 4 h, respectively, but its compressive strength and elastic modulus were generally inferior to those of conventional HSFRC. (2) The strain rate sensitivity of rapid-hardening HSFRC was more significant compared to its conventional counterpart and increased with increasing curing ages and strain rates. This study highlights the great potential of rapid-hardening HSFRC in rapid bridge construction.

Keywords: high-strength steel-fiber-reinforced concrete; rapid hardening; compressive performance; strain rate; curing age

Citation: Mo, F.; Li, B.; Li, M.; Fang, Z.; Fang, S.; Jiang, H. Rapid-Hardening and High-Strength Steel-Fiber-Reinforced Concrete: Effects of Curing Ages and Strain Rates on Compressive Performance. *Materials* **2023**, *16*, 4947. https://doi.org/10.3390/ma16144947

Academic Editor: Carlos Leiva

Received: 11 June 2023
Revised: 3 July 2023
Accepted: 4 July 2023
Published: 11 July 2023

1. Introduction

As the service life of highway bridges increases and traffic flows become more dramatic, bridge deck pavement inevitably suffers from varying degrees of damage. A common typical disease is the breakage of expansion joints (Figure 1), which regulate the connection and displacement of the upper bridge deck structure to ensure smooth vehicle movement. It is noted that the infilled concrete is subjected to external factors such as structural shrinkage, impact fatigue, natural disasters, and vehicle overloading, which can lead to cracking, spalling of the concrete cover, and even outright failure [1,2]. High-strength steel-fiber-reinforced concrete (HSFRC) has become increasingly popular as a cast-in-place jointing material in precast assembled concrete bridges and buildings due to its excellent tensile strength and resistance to crack development [3,4]. However, the Highway Agency often requires repairs to be completed within six hours during the night so that lanes can be reopened the next morning to avoid disruption to road users [5]. However, the replacement of the existing expansion joint concrete requires a number of operations, such as cutting and chiseling of the old concrete and casting of the new concrete, which often takes a long time and can cause extensive traffic congestion. Moreover, the poor stability of the construction quality and the extremely high rework rate reduce the efficiency of the highway. The search for a rapid-hardening concrete with high early strength that can be used to repair bridge decks quickly and with minimum disruption to traffic has therefore become a high priority. For this reason, a new rapid-hardening HSFRC has been proposed, which is produced using sulphoaluminate cement (SC) instead of ordinary Portland cement [6,7]. This type of concrete is feasible in the actual construction site using rapid-paced construction and is suitable for the rapid repair of various bridge pavements. It can effectively solve the current problems of prolonged traffic closure, low traffic throughput, and economic loss due to the insufficient life of repair materials during bridge deck pavement repair.

Figure 1. Expansion joints and corresponding repair work in bridge engineering.

With the increasing demand for better behavior and lower costs, mounting research has been devoted to upgrading or exploiting the structural mechanical performance by introducing innovative materials, structures, and techniques [8–17]. For conventional HSFRC, the presence of silica-fume reactive admixtures and nanoparticles is used to improve defective and porous structures within the concrete, achieving increased strength while reducing the permeability of the concrete material to chloride ions, thus protecting the internal steel fibers from corrosion [18]. In addition, fine quartz sand was used instead of river sand, and the maximum particle size was limited by not mixing coarse aggregates to improve the compactness and uniformity of the aggregate [19]. In addition, the presence of steel fibers reduces the rate of damage development in the concrete and allows for a longer yielding phase, further improving the static and dynamic strength, ductility, and toughness of HSFRC [20]. Rapid-hardening HSFRC not only retains the excellent properties of its conventional counterparts but also has four improvements [6,7]: (1) The introduction of SC with micro-expansion characteristics can completely fill the bridge deck pavement, whose early hydration can achieve rapid construction. (2) The incorporation of gypsum helps to regulate the initial setting time of concrete, which can form an encapsulation layer on the surface of particles to prevent further hydration of SC, thus enabling rapid-hardening HSFRC to achieve slow setting. (3) The addition of water-reducing agents ensures the fluidity of the fresh concrete and solves the problem that the elements cannot be formed densely due to the short setting time of SC. (4) High-temperature steam curing at 90 °C for 3 days is used to reduce material shrinkage and improve the microstructure of the material. The above improvements of rapid-hardening HSFRC ultimately result in an improvement in the three performance indicators of HSFRC, namely, rapid hardening, slow setting, and high early strength. In recent years, rapid-hardening HSFRC has been successfully adopted to repair the deteriorated pavement of a roadway in China, as shown in Figure 2. Within two hours of hardening, the lanes were reopened. After serving for three years, no visible damage was observed on the pavement (Figure 2c). This phenomenon demonstrated that the rapid-hardening HSFRC as investigated in this study exhibited favorable durability. However, the mechanical properties of rapid-hardening HSFRC still remain unclear, which hinders the application of this innovative type of concrete.

(a)   (b)   (c)

Figure 2. Applications of rapid-hardening HSFRC in practices. (**a**) Casting of concrete. (**b**) Hardening of concrete. (**c**) Operation situation for 3 years.

Static mechanical properties are the main performance indicators of concrete, which determine the load-bearing capacity of structures such as bridges and pavements. They are also an important basis for structural design and construction and are an essential part of the in-depth study of the dynamic mechanical properties of concrete [21,22]. Luo et al. [23] showed that the incorporation of polypropylene fibers had a significant effect on the frost resistance of HSFRC after hundreds of freeze–thaw (F–T) cycles, with a greater effect on the splitting tensile strength than on the compressive strength. Lancellotti et al. [24] used alkali-activated materials instead of Portland cement in fiber-reinforced concrete and found that the presence of fibers neither facilitated nor hindered the ground polymerization process, even though there was an increase in ionic conductivity in the samples containing fibers. As a result, a hypothesis was obtained that the samples containing fibers were less consolidated or that the dissolution of fibers contributed to the conductivity values. Vaitkevicius et al. [25] have found that a large number of microsteel fibers (up to 147 kg/m^3) were incorporated into HSFRC to obtain excellent salt-scale resistance and favorable mechanical properties. Rady et al. [26] investigated the bond mechanism of high-strength lightweight concrete containing steel fibers with different geometries. The tested results showed that the steel fibers enhanced the internal bond strength and prevented crack extension, regardless of the geometries of the steel fibers. Furthermore, most of the existing equations for predicting tensile and bond strengths need to be modified for the case of high-strength lightweight concrete. Yu et al. [6] conducted an experimental study on HSFRC with rapid-hardening characteristics. The experimental results indicated that the 3 h strength developed fastest at 55 MPa when the gypsum substitution rate was 15%, while the later strength could be continuously increased to 81 MPa after 7 days. In addition, 2.5% (by volume) of steel fibers could increase the 3 h and 28 d compressive strength of the newly designed UHPC by 157.0% and 46.1%, respectively, compared to the reference group without fibers. Despite the above-mentioned investigations of HSFRC, few studies have focused on the static mechanical properties of rapid-hardening HSFRC.

In addition, pavements and bridges are mainly subjected to dynamic loads during services, for example, the impact friction caused by vehicles coming into direct contact with pavement, which can cause the concrete to deform and crack. Then there are the more destructive earthquakes, which can even cause bridge structures to collapse, resulting in huge losses of life and road safety. Note that the strain rates generated by impacts and earthquakes are generally higher than 100 s^{-1}. When concrete is subjected to short and strong loads, its failure is considerably different from that under quasi-static loads, and the brittleness of concrete under high strain rates of impact loading is more obvious. For this reason, it is also crucial to study the dynamic mechanical properties of HSFRC. It is worth noting that in recent years there have been a large number of studies on the mechanical properties of conventional HSFRC under quasi-static loading. Murali and Vinodha [27] carried out an experimental campaign to assess the impact failure strength of steel hybrid fiber reinforced concrete (SHFRC) subjected to freezing–thawing cycles in water containing

4.0% solution of NaCl. The experimental results revealed that when the number of freezing–thawing cycles was increased, the loss in weight of SHFRC specimens was increased, and the impact failure strength of SHFRC specimens was decreased. The impact failure strength of SHFRC incorporating a higher amount of long fibers was higher compared to short fibers, which implies that long fiber played a predominant role in enhancing its impact failure strength. Li et al. [28] investigated the dynamic mechanical properties of HSFRC under the influence of freeze–thaw cycles and found that the compressive strength and energy absorption capacity of concrete gradually decreased with the increase in F–T cycles. During F–T cycles, the mechanical properties of concrete increased with the addition of steel fibers, and the optimum amount of steel fibers to enhance the resistance to F–T cycles was 1% within the evaluated range. Moein et al. [29] conducted a systematic experimental study on conventional types of HSFRC concrete cured under wet and dry conditions by drop-weight impact assessment and revealed that hooked steel fibers were more effective than crimped steel fibers in improving the impact strength, even though the length diameter was smaller. The compressive strength of concrete containing hybrid fibers (hooked + crimped) was also lower than other fibers. In addition, the moisture-cured samples had higher compressive strength (up to 12%) and tensile strength (up to 21%). Sharma et al. [30] concluded that HSFRC was effective against abrasion erosion and cavitation erosion. A systematic experimental study showed that SFRC with the addition of 1.25–1.5% steel fibers showed significant improvements in impact resistance, toughness, and energy absorption. Sun et al. [31] demonstrated that as the steel-fiber content increased, the peak stress, energy absorption, and multiple impact compressive resistance of the specimens were greatly improved. When the steel-fiber content was 6%, the dynamic impact peak strain, dynamic impact compressive strength ratio, and energy absorption capacity of the specimens were 3.09, 1.45, and 4.1 times higher than those of the reference group, respectively. Dalvand et al. [32] used zeolite material to partially replace ordinary silicate cement, which could effectively enhance the bond strength in the interfacial transition zone between fine aggregate and cement paste, thus improving the toughness and postpeak behavior. In recent years, a number of researchers have studied the mechanical properties of rapid-hardening HSFRC under quasi-static loading; however, there is limited research on its dynamic mechanical properties under impact loading [29,33]. Previous studies have shown that rapid-hardening HSFRC has good dynamic mechanical properties and its use in repairing bridge deck pavement allows a structure to dissipate more energy at the material level, reduce amplitude and stress, and improve the overall structural damping [33].

Against the above background, this paper aimed to investigate the effects of SC replacement ratios and concrete curing ages on the failure mode, ultimate condition, and stress–strain response of HSFRC under static and dynamic compression tests. It is noted that the Split Hopkinson pressure bar (SHPB) test can skillfully decouple the inertia effect in the structure and the strain rate effect in the material, which is the typical experimental technique for obtaining the dynamic compressive behavior of concrete at relatively high strain rates ($10^2 \sim 10^4$ s^{-1}). Therefore, the dynamic compression tests in this study are carried out based on the SHPB experimental technique.

2. Experimental Program

2.1. Specimen Design

A number of 35 groups of (or 112) specimens were prepared and tested, including 7 groups of (or 21) cubic specimens with 100 mm sides, 7 groups (or 21) cylindrical specimens with 100 mm diameter and 200 mm height for quasi-static compression tests, and 35 groups of (or 70) flattened Brazilian discs (FBD) with 100 mm diameter and 50 mm height for dynamic compression tests. Each group of quasi-static compression specimens consisted of three nominally identical specimens and two dynamic counterparts. Two series were designed for each type of compression test. Specifically, for the quasi-static compression tests, three groups of cubic specimens and three groups of cylindrical specimens made from conventional HSFRC were tested together as control specimens to form Series I, while Series II comprised the remaining cubic and cylindrical specimens infilled with rapid-hardening HSFRC. It is noted that conventional HSFRC was introduced to better investigate the effect of SC replacement ratios on HSFRC. For the dynamic compression tests, Series I consisted of 15 groups of FBD specimens made of conventional HSFRC, while Series II consisted of 20 groups of FBD specimens made of rapid-hardening HSFRC. Each series (i.e., Series I and II in the quasi-static and dynamic compression tests) examined the influence of test parameters on the concrete curing age, using 3, 7, and 28 days for the conventional HSFRC and 4 h, 3 days, 7 days, and 28 days for the rapid-hardening HSFRC, to cover the emergency repair work opening time (4 h) and typical concrete curing ages (3, 7, and 28 days). In addition, the effect of strain rate was investigated in both Series I and II of the dynamic compression test. It should be mentioned here that the strain rate test range was determined by designing the gas pressure for the SHPB technique in the dynamic compression test. Five SHPB gas pressures were used for the dynamic compression tests, 0.5, 0.6, 0.7, 0.8, and 0.9, representing strain rates in the range of 50 to 134 s^{-1}, covering a wide range of strain rates to which concrete is subjected when used in structural members. The detailed specimen arrangement for the quasi-static and dynamic compression tests is shown in Table 1. It is worth noting that instead of the quasi-static compression test where the stress and strain are directly output by the instrument and strain gauge, the average strain rate, stress, and its corresponding strain at each step (or time t) of the dynamic compression test can be obtained based on the one-dimensional stress wave propagation theory. Specifically, the average strain rate ($\dot{\varepsilon}_s$), compressive strength (σ_s), and its corresponding strain (referred to as the ultimate strain ε_s) of all FBD specimens under dynamic compressive loading are expressed in the following equations [34,35]:

$$\dot{\varepsilon}_s(t) = \frac{2C_0}{l_s}\varepsilon_R(t) \tag{1}$$

$$\sigma_s(t) = \frac{E_b A_b}{A_s}\varepsilon_T(t) \tag{2}$$

$$\varepsilon_s(t) = \frac{2C_0}{l_s}\int_0^t \varepsilon_R(t)dt \tag{3}$$

where C_0 is the speed of propagation of a one-dimensional stress wave in the SHPB; l_s and A_s are the thickness and cross-sectional area of FBD specimens, respectively; ε_R is the strain of the reflected wave; and E_b and A_b are the elastic modulus and cross-sectional area of the SHPB, respectively.

Each specimen is given a name consisting of 3 or 4 sets of letters and/or numbers (3 sets for the quasi-static compression test and 4 sets for the dynamic compression test), plus a numeral "1, 2, or 3" to distinguish between 2 or 3 nominally identical specimens. The first set consists of the two or three byte letters "CQC, QC, or DC", representing cubic and cylindrical specimens for quasi-static compression tests (i.e., "cubes under quasi-static compression" is shortened to "CQC" and "cylinders under quasi-static compression" is shortened to "QC"), and FBD specimens for dynamic compression tests (i.e., "FBD specimens under dynamic compression" is shortened to "DC"), respectively. The second set consists of the number "0 or 60" and the unit "%", representing the type of infilled concrete as conventional or rapid-hardening HSFRC, respectively. The third set consists of the numbers "4, 3, 7, or 28" and the letters "h or d", representing the curing age of the specimen as 4 h, 3 days, 7 days, or 28 days, respectively. The fourth set of specimens for dynamic compression tests only represents the gas pressure of the SHPB technique in dynamic compression tests. For example, specimen DC-60%-4h-0.5-1 represents one of two identical dynamic compressive specimens infilled with rapid-hardening HSFRC, with a curing age of 4 h and a target SHPB gas pressure of 0.5 MPa.

Table 1. Detailed design of test specimens.

Specimen ID	Test Type	Specimen Type	SC Replacement Ratio	Curing Age	Diameter × Height (Side Length)	SHPB Gas Pressure
CQC-0%-3d-1,-2,-3	Quasi-static	Cube	0%	3 days	100 mm	--
CQC-0%-7d-1,-2,-3				7 days	100 mm	
CQC-0%-28d-1,-2,-3				28 days	100 mm	
CQC-60%-4h-1,-2,-3			60%	4 h	100 mm	
CQC-60%-3d-1,-2,-3				3 days	100 mm	
CQC-60%-7d-1,-2,-3				7 days	100 mm	
CQC-60%-28d-1,-2,-3				28 days	100 mm	
QC-0%-3d-1,-2,-3		Cylinder	0%	3 days	100 mm × 200 mm	
QC-0%-7d-1,-2,-3				7 days	100 mm × 200 mm	
QC-0%-28d-1,-2,-3				28 days	100 mm × 200 mm	
QC-60%-4h-1,-2,-3			60%	4 h	100 mm × 200 mm	
QC-60%-3d-1,-2,-3				3 days	100 mm × 200 mm	
QC-60%-7d-1,-2,-3				7 days	100 mm × 200 mm	
QC-60%-28d-1,-2,-3				28 days	100 mm × 200 mm	
DC-0%-3d-0.5-1,-2	Dynamic	FBD	0%	3 days	100 mm × 50 mm	0.5
DC-0%-3d-0.6-1,-2					100 mm × 50 mm	0.6
DC-0%-3d-0.7-1,-2					100 mm × 50 mm	0.7
DC-0%-3d-0.8-1,-2					100 mm × 50 mm	0.8
DC-0%-3d-0.9-1,-2					100 mm × 50 mm	0.9
DC-0%-7d-0.5-1,-2				7 days	100 mm × 50 mm	0.5
DC-0%-7d-0.6-1,-2					100 mm × 50 mm	0.6
DC-0%-7d-0.7-1,-2					100 mm × 50 mm	0.7
DC-0%-7d-0.8-1,-2					100 mm × 50 mm	0.8
DC-0%-7d-0.9-1,-2					100 mm × 50 mm	0.9
DC-0%-28d-0.5-1,-2				28 days	100 mm × 50 mm	0.5
DC-0%-28d-0.6-1,-2					100 mm × 50 mm	0.6
DC-0%-28d-0.7-1,-2					100 mm × 50 mm	0.7
DC-0%-28d-0.8-1,-2					100 mm × 50 mm	0.8
DC-0%-28d-0.9-1,-2					100 mm × 50 mm	0.9
DC-60%-4h-0.5-1,-2			60%	4 h	100 mm × 50 mm	0.5
DC-60%-4h-0.6-1,-2					100 mm × 50 mm	0.6
DC-60%-4h-0.7-1,-2					100 mm × 50 mm	0.7
DC-60%-4h-0.8-1,-2					100 mm × 50 mm	0.8
DC-60%-4h-0.9-1,-2					100 mm × 50 mm	0.9
DC-60%-3d-0.5-1,-2				3 days	100 mm × 50 mm	0.5
DC-60%-3d-0.6-1,-2					100 mm × 50 mm	0.6
DC-60%-3d-0.7-1,-2					100 mm × 50 mm	0.7
DC-60%-3d-0.8-1,-2					100 mm × 50 mm	0.8
DC-60%-3d-0.9-1,-2					100 mm × 50 mm	0.9
DC-60%-7d-0.5-1,-2				7 days	100 mm × 50 mm	0.5
DC-60%-7d-0.6-1,-2					100 mm × 50 mm	0.6
DC-60%-7d-0.7-1,-2					100 mm × 50 mm	0.7
DC-60%-7d-0.8-1,-2					100 mm × 50 mm	0.8
DC-60%-7d-0.9-1,-2					100 mm × 50 mm	0.9
DC-60%-28d-0.5-1,-2				28 days	100 mm × 50 mm	0.5
DC-60%-28d-0.6-1,-2					100 mm × 50 mm	0.6
DC-60%-28d-0.7-1,-2					100 mm × 50 mm	0.7
DC-60%-28d-0.8-1,-2					100 mm × 50 mm	0.8
DC-60%-28d-0.9-1,-2					100 mm × 50 mm	0.9

2.2. Raw Materials

The material mass proportion for both types of HSFRC was cementitious material: fine aggregate: steel fiber: water: water reducer = 42:40:6:10:2. Note that the two types of concrete differed only in the mass proportion of cementitious material, with the conventional type using ordinary Portland cement: microsilica fume: Nano-$CaCO_3$ = 77:20:3, and the rapid-hardening type using sulphoaluminate concrete: ordinary Portland cement: gypsum: microsilica fume: Nano-$CaCO_3$ = 45:16:16:20:3. A close-up view of the raw materials is shown in Figure 3.

2.3. Specimen Preparation

The same preparation procedure was used for all specimens, and the key steps are given in Figure 4. It is noted that the same type of concrete was cured in the same environment using the same methods. In particular, conventional HSFRC specimens were demolded after one day of hardening, and then placed outdoors and covered with plastic molds for one week with three dripping treatments per day. Furthermore, rapid-hardening HSFRC specimens were demolded after 2 h of hardening and subsequently steamed at 90 °C for 3 days, followed by 4 days of outdoor conditioning as with their conventional counterparts. In addition, to ensure smooth surfaces during loading to avoid stress concentrations at the specimen ends, the top and bottom surfaces of the specimens for the static compression test were covered with a high-strength plaster, while the top and bottom surfaces of the specimens for the dynamic compression test were polished with an MY259 grinder. Note that the nonparallel depth of the top and bottom surfaces of the specimens was kept below 0.02 mm [36,37].

Figure 3. Close-ups of raw materials.

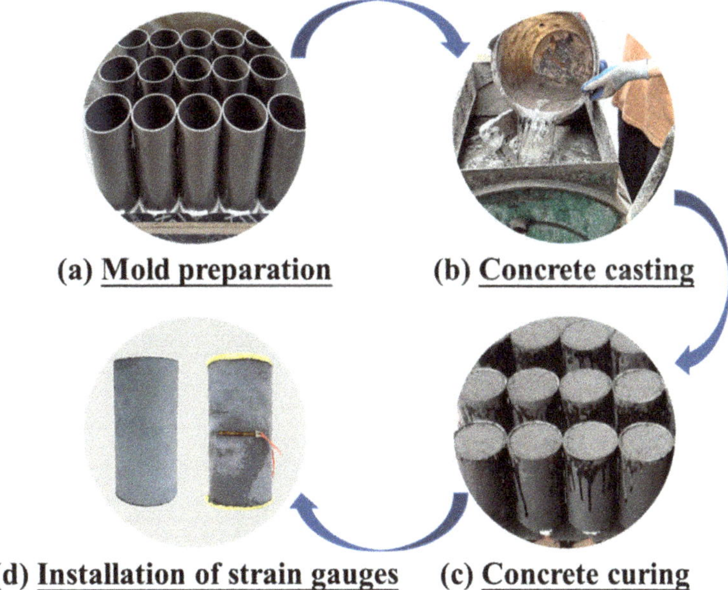

Figure 4. Schematic diagram of specimen preparation.

2.4. Test Setup and Instrumentation

Quasi-Static Compression Tests

All quasi-static compression tests were carried out in accordance with the American Specification ASTM C469/C469M-14 [38]. The test setup and instrumentation are shown in Figure 5. Specifically, a displacement-controlled loading mode was used with a constant rate of 0.18 mm/min, corresponding to 10^{-5} s^{-1}. All specimens were loaded in two steps: (1) Each specimen was preloaded to 10% of the expected peak load to verify loading axis alignment and proper instrument operation. (2) After preloading, all specimens were formally loaded until the load dropped to 40% of the measured peak load. In addition, the same linear displacement transducers (LVDTs) and strain gauge arrangement were used for all specimens: (1) two LVDTs installed symmetrically to cover the middle two-fifths of the column height to measure axial shortenings; (2) two strain gauges installed symmetrically in the axial direction with a gauge length of 50 mm to measure axial deformations of the midheight section in the specimen; and (3) two strain gauges installed symmetrically in the hoop direction with a gauge length of 50 mm to measure hoop deformations of the midheight section in the specimen. All test data were automatically collected every second using a data acquisition system.

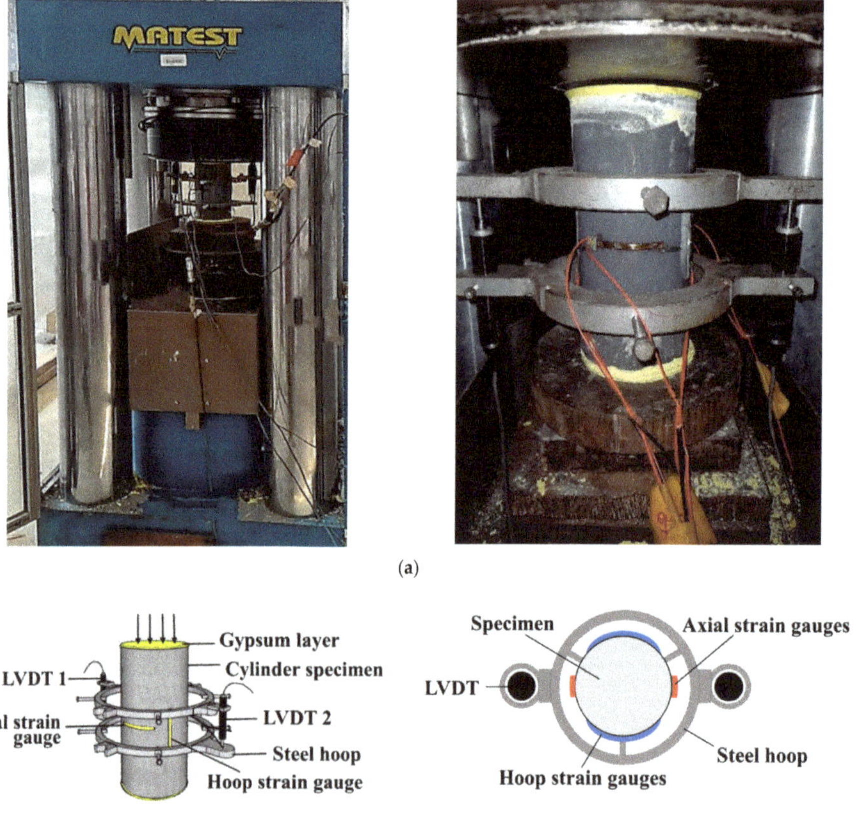

Figure 5. Setup and instruments of quasi-static compression tests. (**a**) Test setup. (**b**) Layout of strain gauges and LVDTs.

All dynamic compression tests were performed in a separate SHPB apparatus in the Mechanics Laboratory of Guangdong University of Technology, China, following the Chinese Specifications GB/T 7314-2017 and GB/T 34108-2017 [39,40]. The test setup and instrumentation is shown in Figure 6. Specifically, the striker, incident, transmitted, and absorbent bars in the SHPB technique were fabricated from 60Si2Mn, which had an elastic modulus, Poisson's ratio, mass density, and yield strength of 206 GPa, 0.3, 7740 kg/m^3, and 1180 MPa, respectively. According to the definition of the elastic wave velocity of a bar as the square ratio of the elastic modulus to its mass density, the above-mentioned elastic wave velocity of the bar in the used SHPB technique was 5169 m/s. It was noted that wave fluctuations and dispersion would greatly affect the dynamic performance of the specimens during SHPB impact tests. Therefore, two additional measures were routinely taken to ensure uniform forces at both ends of the concrete specimen: (1) the contact surfaces of the incident bar, specimen, and transmitted bar were covered with Vaseline to reduce the frictional resistance of the contact surfaces; and (2) brass pulse shapers of 2 mm thickness and 20 mm diameter were applied between the incident bar and the specimen to retard the slope of the rising phase of the incident wave and increase the reflection time of the gravitational wave inside the specimen. In addition, four strain gauges with a gauge length of 50 mm were installed equally on the incident and transmitted bars on the SHPB technique, which were used to monitor the signals of the incident (ε_I) and reflected (ε_R) waves from the incident bar, and the transmitted (ε_T) waves from the transmitted bar, respectively.

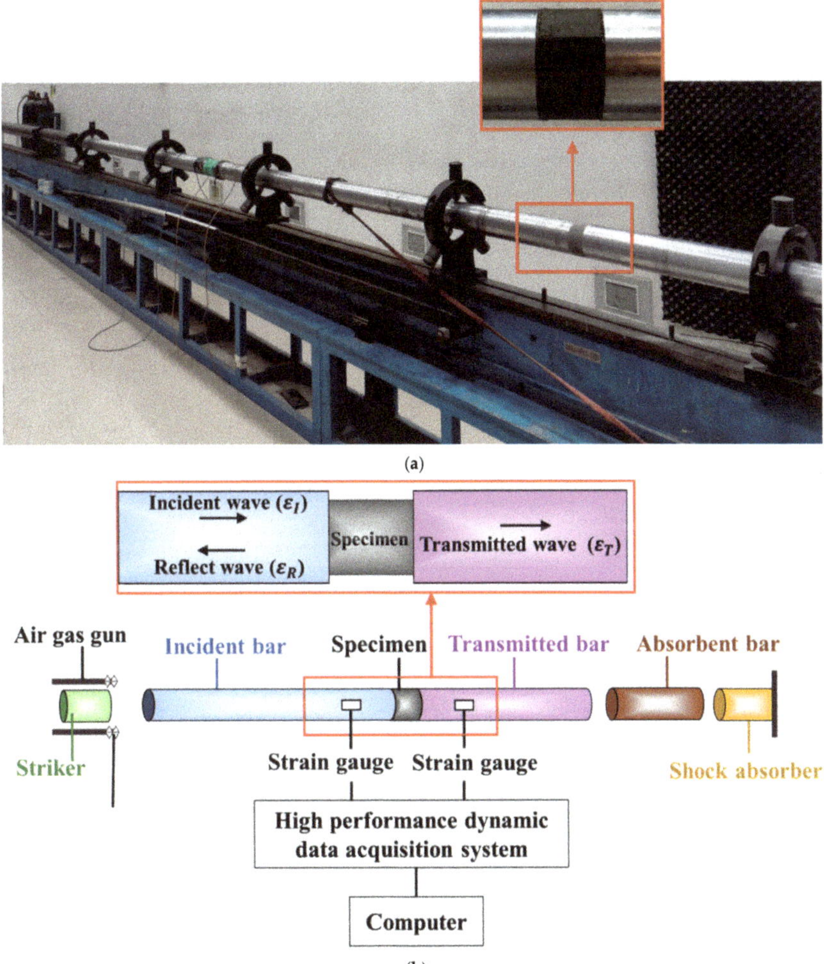

Figure 6. Setup and instruments of dynamic compression tests. (**a**) Test setup. (**b**) Layout of strain gauges.

3. Results and Discussion of Quasi-Static Compression Test

3.1. Failure Modes and Ultimate Conditions

Figure 7 shows typical failure modes of cubic and cylindrical specimens tested under quasi-static compression loading. All of the quasi-static compressive specimens failed by early crack initiation followed by rapid crack propagation and even concrete crushing. A conical final failure mode was observed in the midheight region of all specimens. It is noteworthy that as the load approached its peak, all specimens showed surface cracking of the concrete protective layer, followed by the formation of major vertical cracks and the development of multiple diagonal cracks along the vertical cracks. Furthermore, both types of HSFRC specimens showed a gradual decrease in the crack numbers as the concrete curing age increased. Furthermore, the rapid-hardening HSFRC specimens all had higher crack numbers than their conventional counterparts at the same curing ages. Note that with earlier curing ages or higher SC replacement ratios, the cubic and cylindrical specimens exhibited more significant diagonal splitting cracks. This relationship twisted with the increasing curing age or decreasing SC replacement ratio, i.e., the specimens developed a dominant failure with the vertical cracking. This finding may be related to the strength development level of the base material, with the diagonal-crack-dominated failure occurring when their diagonal shear capacity is lower than the corresponding vertical splitting capacity and, conversely, with vertical-crack-dominated failure.

Table 2 summarizes the average key test results for cubic and cylindrical specimens under quasi-static compressive loading, which include the compressive strength and its corresponding axial strain (corresponding to the ultimate axial strain), elastic modulus, and Poisson's ratio. Specifically, the axial strain was determined by averaging the two LVDT readings installed in the midheight region of the specimen.

Table 2. Key results of quasi-static compression tests.

Specimen ID	Compressive Strength (MPa)	Ultimate Axial Strain	Elastic Modulus (GPa)	Poisson's Ratio
CQC-0%-3d-1,-2,-3	72.62	--	--	--
CQC-0%-7d-1,-2,-3	90.65	--	--	--
CQC-0%-28d-1,-2,-3	124.89	--	--	--
CQC-60%-4h-1,-2,-3	45.16	--	--	--
CQC-60%-3d-1,-2,-3	52.56	--	--	--
CQC-60%-7d-1,-2,-3	62.55	--	--	--
CQC-60%-28d-1,-2,-3	80.94	--	--	--
QC-0%-3d-1,-2,-3	49.52	0.0022	33.1	0.15
QC-0%-7d-1,-2,-3	56.17	0.0020	37.0	0.19
QC-0%-28d-1,-2,-3	90.95	0.0018	43.6	0.21
QC-60%-4h-1,-2,-3	33.14	0.0046	26.5	0.19
QC-60%-3d-1,-2,-3	39.14	0.0030	27.6	0.20
QC-60%-7d-1,-2,-3	54.06	0.0028	34.1	0.21
QC-60%-28d-1,-2,-3	70.62	0.0023	38.3	0.22

Figures 8 and 9 show the effect of the examined parameters on the ultimate condition of quasi-static compression tests for cubic and cylindrical specimens, where all results are represented by the average of the test results for three nominally identical specimens in each test case. Note that in addition to the test results for cubic specimens shown in Figure 7, the test results for cylindrical specimens are shown in Figure 8. Figures 8 and 9 show the effects of SC replacement ratios and concrete curing ages, respectively. Generally, the curing age of both types of HSFRC is proportional to the compressive strength of cubic and cylindrical specimens and inversely proportional to the cylindrical ultimate axial strain. Conversely, as the SC replacement ratio increased, the compressive strength of cylindrical HSFRC specimens at the same curing age gradually decreased while the corresponding ultimate axial strain gradually increased. It can be seen from Figure 9 that the relationship curve between the concrete curing age and compressive strength or ultimate axial strain is generally higher for specimens with conventional HSFRC than for their rapid-hardening counterparts. However, it is worth noting that the rate of strength enhancement (i.e., the slope of the curve) was significantly greater for rapid-hardening HSFRC than for conventional counterparts at the early curing age. Such a finding is understandable, as the low elastic modulus of SC makes the matrix more rigid than ordinary Portland cement, and thus exhibits a lower compressive strength at the same stress state. Nevertheless, Table 2 shows that cubic and cylindrical specimens of conventional HSFRC cured for 3 days reached 58% and 54% of those cured for 28 days, respectively, and similar comparative results were found for rapid-hardening HSFRC counterparts cured for 4 h and 28 days.

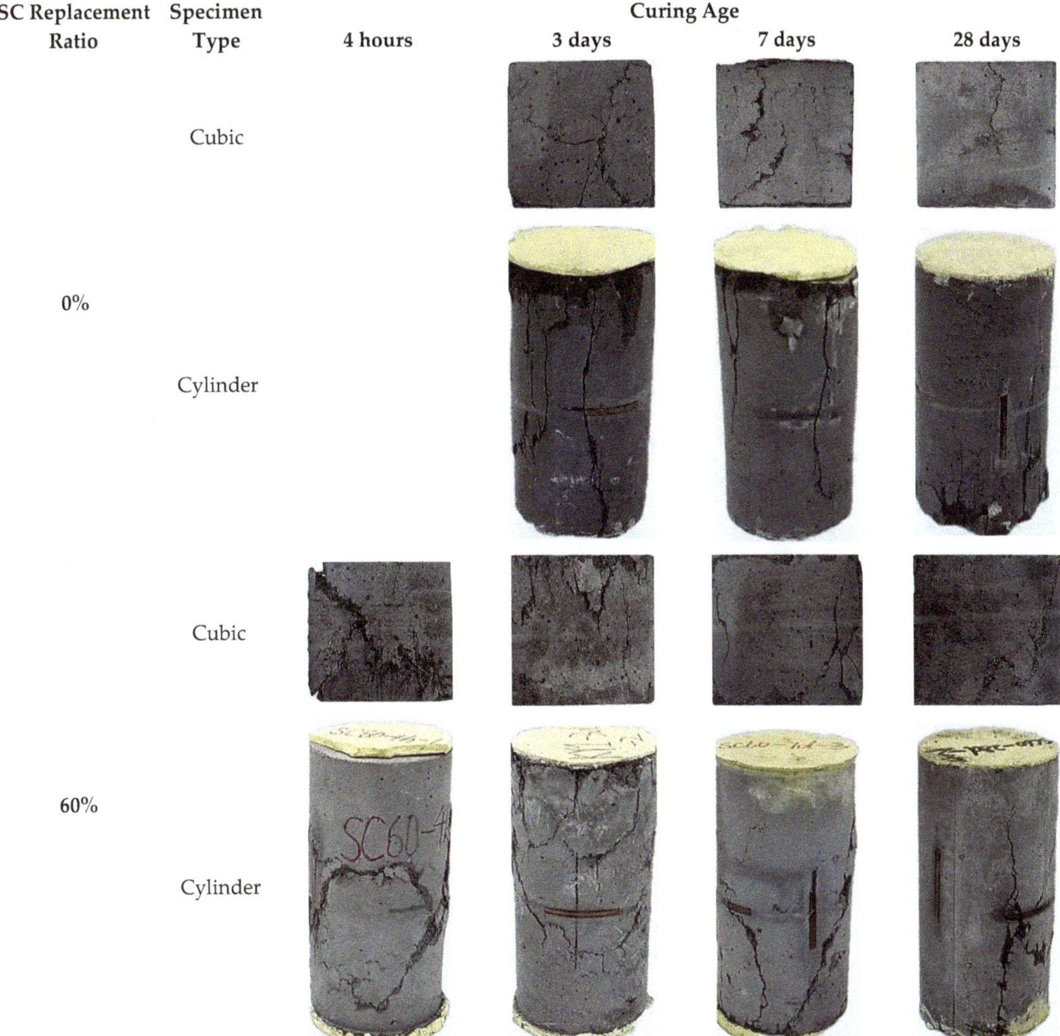

Figure 7. Typical failure modes of HSFRC specimens under quasi-static compression tests.

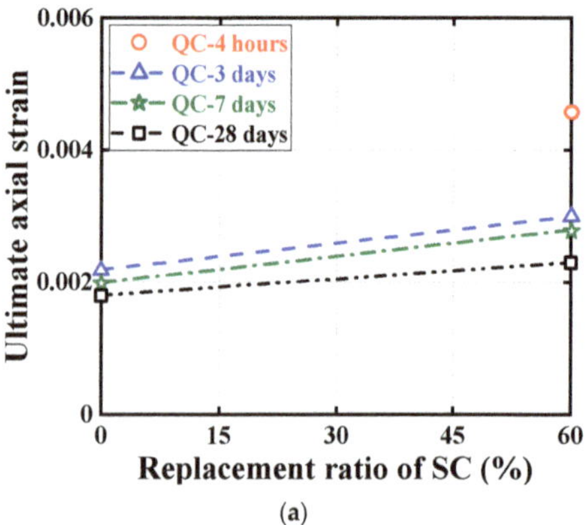

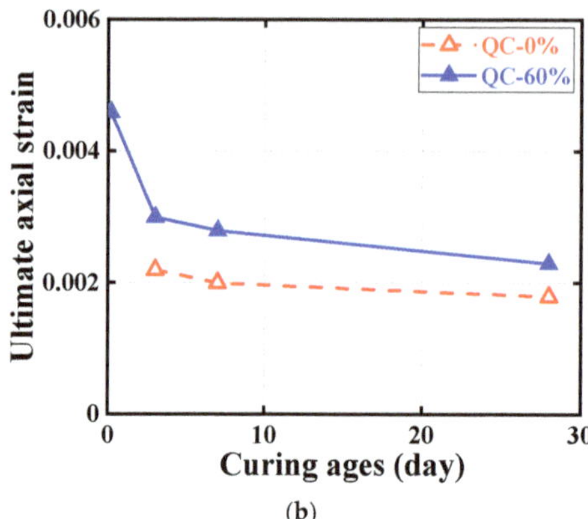

Figure 8. Effect of SC replacement ratios on ultimate conditions of quasi-static compression tests. (**a**) Ultimate axial stress. (**b**) Ultimate axial strain.

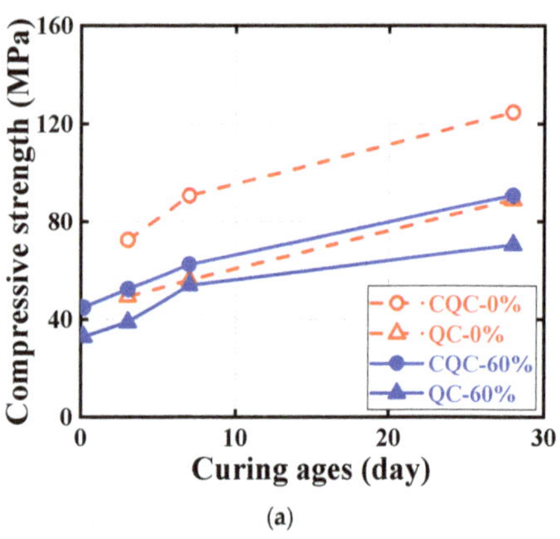

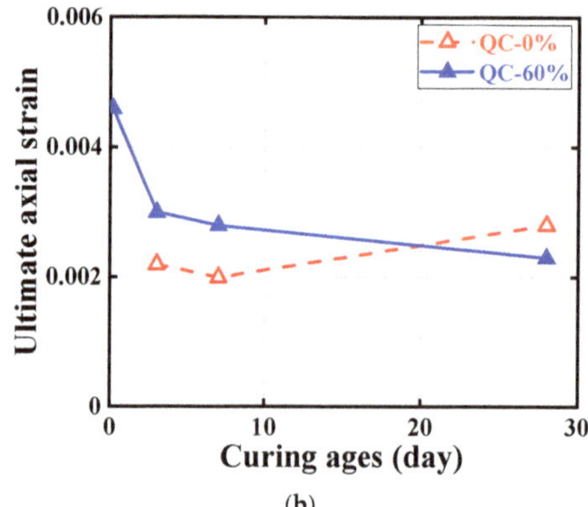

Figure 9. Effect of concrete curing ages on ultimate conditions of quasi-static compression tests. (**a**) Ultimate axial stress. (**b**) Ultimate axial strain.

3.2. Stress–Strain Responses

Quasi-static compressive stress–strain curves for cylindrical specimens are shown in Figures 10 and 11, each examining the effect of one test parameter. The test results show that all three nominally identical specimens have close stress–strain curves. Therefore, for ease of comparison, only one of the test results for one of the repeated specimens is given in both Figures 10 and 11. Note that the positive strains in Figures 10 and 11 represent the axial strains of the specimens measured by the LVDTs, while the negative strains refer to the hoop strains of the specimens measured by the strain gauges. In addition, to further illustrate the quasi-static compressive stress–strain curves for the cylindrical specimens, Figures 12 and 13 each depict the effect of one test parameter on the quasi-static compressive index (including the elastic modulus and Poisson's ratio).

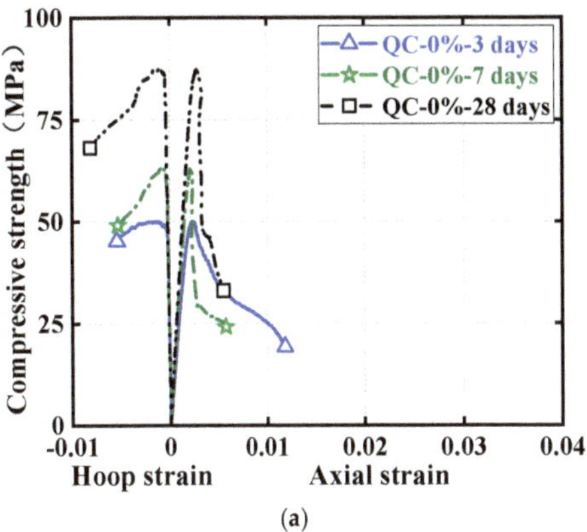
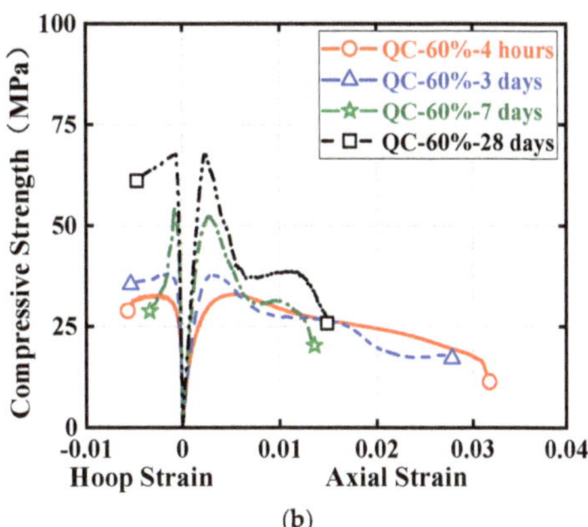

Figure 10. Effect of SC replacement ratios on stress–strain curves of quasi-static compression tests. (**a**) Conventional HSFRC. (**b**) Rapid-hardening HSFRC.

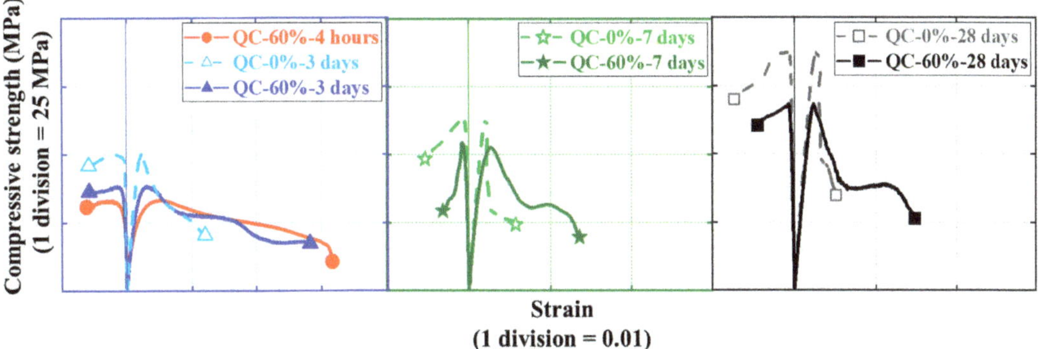

Figure 11. Effect of concrete curing ages on stress–strain curves of quasi-static compression tests.

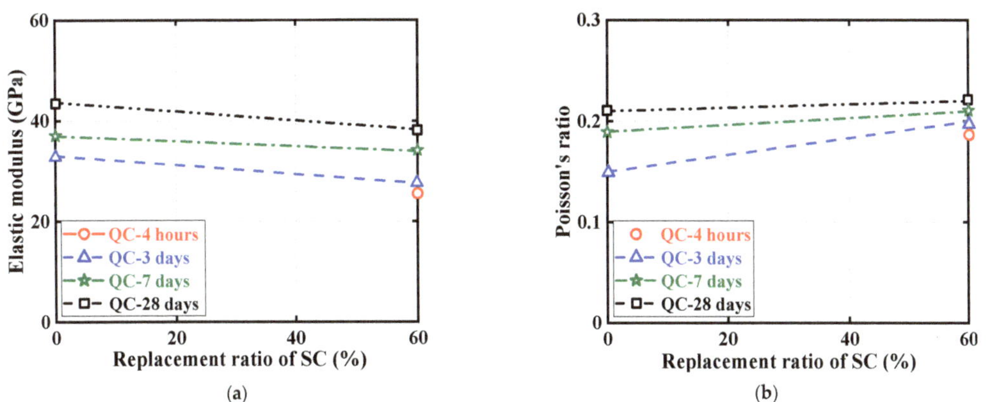

Figure 12. Effect of SC replacement ratios on ultimate conditions of quasi-static compression tests. (**a**) Elastic modulus. (**b**) Poisson's ratio.

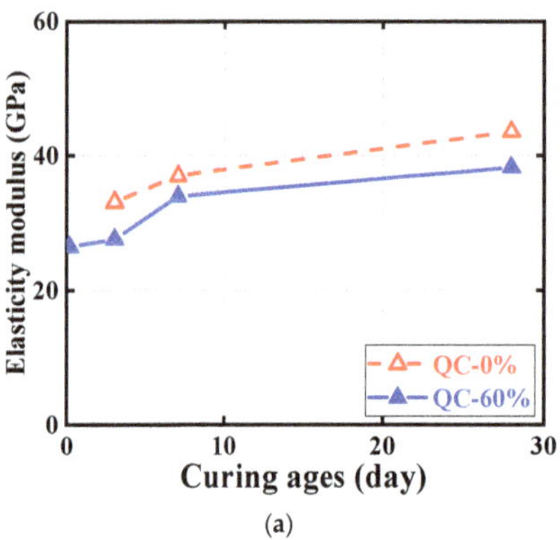 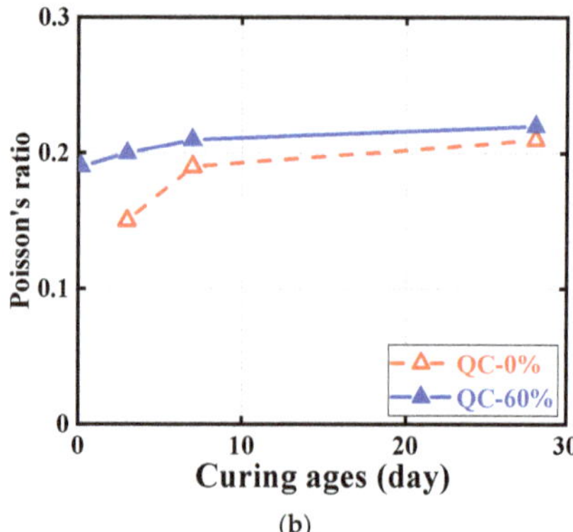

(a)   (b)

Figure 13. Effect of concrete curing ages on ultimate conditions of quasi-static compression tests. (**a**) Elastic modulus. (**b**) Poisson's ratio.

Figure 10 regroups and compares the quasi-static compressive stress–strain responses of cylindrical specimens using the two different types of HSFRC. Three subplots are included in Figure 12 to illustrate the difference in quasi-static compressive performance of cylindrical specimens with two types of HSFRC for ages of 3 days, 7 days, and 28 days, respectively. It can be seen that the nonlinear growth phase of the quasi-static compressive stress–strain response was generally higher for conventional HSFRC cylindrical specimens than for their rapid-hardening HSFRC counterparts. The difference gradually increased with the increasing concrete curing age. Furthermore, Figure 12 clearly shows that the elastic modulus of HSFRC cylindrical specimens under quasi-static compression decreases with increasing SC replacement ratio; the opposite relationship was found for Poisson's ratio and SC replacement ratio. Additionally, it can be found that rapid-hardening HSFRC cylindrical specimens had a more stable development of quasi-static compressive elastic modulus and Poisson's ratio with the changing concrete curing age compared to their conventional counterparts.

Figure 11 illustrate the effect of concrete curing ages on the quasi-static compressive stress–strain response of cylinders with conventional and rapid-hardening HSFRC, respectively. The effect of concrete curing ages is similar to previous studies [36]: as the concrete curing age increased, the compressive strength of the cylindrical specimens under quasi-static loading increased; by contrast, the corresponding ultimate strain decreased. Furthermore, Figure 13 clearly shows that the elastic modulus and Poisson's ratio of both types of HSFRC cylindrical specimens generally increased with increasing curing age. Interestingly, the effect of concrete curing ages on the quasi-static compressive elastic modulus and Poisson's ratio of HSFRC cylindrical specimens was significant until the curing age was less than 7 days; however, the effect of curing age became less pronounced when the curing age was greater than 7 days. This phenomenon demonstrates that an early curing age is crucial for the development of the quasi-static compressive properties of HSFRC.

4. Results and Discussion of Dynamic Compression Tests
Failure Modes and Ultimate Conditions

Figure 14 shows the typical failure modes of the FBD specimens tested under dynamic compression loading. All of the dynamic compressive specimens experienced either macroscopic cracking or crushing damage. Specifically, the concrete cracking first occurred on the sides of the FBD specimens and subsequently developed in both two-specimen planes along the height direction. Such crack propagations triggered both the spalling of the concrete protective layer and the successive appearance of crushing in the specimen core. Additionally, it was noteworthy that the specimens with both types of HSFRC exhibited more significant cracking and even crushing as the strain rate (or SHPB gas pressure) increased or the concrete curing age decreased, i.e., the integrity of the specimens after failure became increasingly degraded. Specifically, with the increasing strain rate (or SHPB gas pressure) or the decreasing concrete curing age, the specimens eventually failed in more fragments with smaller volumes. It is noted that there may be a threshold between the strain rate (or SHPB gas pressure) and specimen cracking (or crushing): when the strain rate (or SHPB gas pressure) is greater than a certain threshold, HSFRC specimens will crack or crush. Specifically, when the SHPB gas pressure is greater than 0.5 MPa (corresponding to a strain rate of 43 to 90 s^{-1}), all HSFRC specimens will experience macroscopic cracking; when the SHPB gas pressure is greater than 0.8 MPa (corresponding to a strain rate of 81 to 135 s^{-1}), all HSFRC specimens will experience almost complete separation of mortar, aggregate, and steel fibers. Further observations of the specimen's failure modes were made after testing. Note that a large number of steel fibers were observed to be pulled out as a whole but not broken. This phenomenon indicates that the anchoring capacity of the steel fibers used in this study was poor, i.e., better mechanical properties of HSFRC could be achieved if steel fibers with stronger anchoring capacity were used. In addition, FBD specimens with rapid-hardening HSFRC had significantly fewer cracks than their conventional counterparts, maintaining a higher degree of postfailure integrity. This was more pronounced at the early curing age, suggesting that the introduction of SC favored the dynamic performance of the HSFRC, which may be attributed to the higher early strength.

Table 3 summarizes the key test results for all FBD specimens under dynamic compressive loading, which include the average strain rate ($\dot{\varepsilon}_s$), compressive strength (σ_s), and ultimate strain (ε_s).

Table 3. Key results of dynamic compression tests.

Specimen ID	Average Strain Rate (s^{-1})	Compressive Strength (MPa)	Ultimate Axial Strain
DC-0%-3d-0.5-1	43	35.56	0.013
DC-0%-3d-0.5-2	49	37.82	0.021
DC-0%-3d-0.6-1	66	45.32	0.012
DC-0%-3d-0.6-2	67	47.99	0.015
DC-0%-3d-0.7-1	71	45.84	0.019
DC-0%-3d-0.7-2	76	46.34	0.017
DC-0%-3d-0.8-1	119	50.98	0.024
DC-0%-3d-0.8-2	122	50.63	0.030
DC-0%-3d-0.9-1	127	54.20	0.022
DC-0%-3d-0.9-2	132	54.79	0.030
DC-0%-7d-0.5-1	51	40.35	0.012
DC-0%-7d-0.5-2	59	43.21	0.022
DC-0%-7d-0.6-1	72	46.72	0.023
DC-0%-7d-0.6-2	79	46.79	0.017
DC-0%-7d-0.7-1	89	47.76	0.016
DC-0%-7d-0.7-2	97	49.63	0.037
DC-0%-7d-0.8-1	102	49.08	0.036
DC-0%-7d-0.8-2	104	51.89	0.016
DC-0%-7d-0.9-1	110	57.85	0.019
DC-0%-7d-0.9-2	123	62.19	0.027
DC-0%-28d-0.5-1	80	63.68	0.027
DC-0%-28d-0.5-2	90	65.74	0.029
DC-0%-28d-0.6-1	97	72.84	0.024
DC-0%-28d-0.6-2	100	74.13	0.018
DC-0%-28d-0.7-1	125	78.06	0.043
DC-0%-28d-0.7-2	126	80.89	0.032
DC-0%-28d-0.8-1	124	81.34	0.040
DC-0%-28d-0.8-2	135	82.06	0.039
DC-0%-28d-0.9-1	128	93.81	0.047
DC-0%-28d-0.9-2	133	95.34	0.032

Table 3. *Cont.*

Specimen ID	Average Strain Rate (s^{-1})	Compressive Strength (MPa)	Ultimate Axial Strain
DC-60%-4h-0.5-1	65	33.42	0.022
DC-60%-4h-0.5-2	70	34.41	0.029
DC-60%-4h-0.6-1	83	37.07	0.026
DC-60%-4h-0.6-2	96	39.18	0.036
DC-60%-4h-0.7-1	97	39.43	0.038
DC-60%-4h-0.7-2	98	41.08	0.024
DC-60%-4h-0.8-1	98	39.52	0.031
DC-60%-4h-0.8-2	99	40.99	0.033
DC-60%-4h-0.9-1	100	42.32	0.029
DC-60%-4h-0.9-2	104	43.97	0.020
DC-60%-3d-0.5-1	70	39.04	0.023
DC-60%-3d-0.5-2	71	40.82	0.039
DC-60%-3d-0.6-1	93	44.85	0.032
DC-60%-3d-0.6-2	95	46.73	0.032
DC-60%-3d-0.7-1	98	51.41	0.026
DC-60%-3d-0.7-2	100	55.89	0.022
DC-60%-3d-0.8-1	100	53.92	0.032
DC-60%-3d-0.8-2	103	54.86	0.027
DC-60%-3d-0.9-1	105	61.82	0.017
DC-60%-3d-0.9-2	122	64.53	0.028
DC-60%-7d-0.5-1	63	43.33	0.029
DC-60%-7d-0.5-2	66	45.83	0.024
DC-60%-7d-0.6-1	80	52.75	0.029
DC-60%-7d-0.6-2	83	52.91	0.030
DC-60%-7d-0.7-1	98	61.82	0.017
DC-60%-7d-0.7-2	103	61.93	0.032
DC-60%-7d-0.8-1	107	67.01	0.030
DC-60%-7d-0.8-2	107	67.08	0.015
DC-60%-7d-0.9-1	134	67.82	0.030
DC-60%-7d-0.9-2	111	69.21	0.021
DC-60%-28d-0.5-1	80	60.13	0.029
DC-60%-28d-0.5-2	82	60.40	0.030
DC-60%-28d-0.6-1	98	68.43	0.032
DC-60%-28d-0.6-2	99	69.36	0.033
DC-60%-28d-0.7-1	106	70.03	0.036
DC-60%-28d-0.7-2	110	71.63	0.038
DC-60%-28d-0.8-1	108	88.54	0.038
DC-60%-28d-0.8-2	110	93.89	0.042
DC-60%-28d-0.9-1	123	99.06	0.043
DC-60%-28d-0.9-2	126	99.86	0.044

Figures 15–17 show the effects of the SC replacement ratio, concrete curing age, and strain rate (or SHPB gas pressure) on the average strain rate, compressive strength, and ultimate axial strain of FBD specimens in dynamic compression tests, respectively, where all ultimate conditions are represented by the average of the test results of two nominally identical specimens in each test case. As expected, Figures 15 and 17 demonstrate that for FBD specimens with different types of HSFRC, the average strain rate, dynamic compressive strength, and ultimate axial strain generally increased with increasing concrete curing age or strain rate (or SHPB gas pressure). Furthermore, Figure 15 shows that the introduction of SC enhanced the early curing age dynamic compressive performance of HSFRC specimens to some extent. Specifically, when the SHPB gas pressure was increased from 0.5 to 0.9 MPa, the mean strain rate of the specimens with conventional HSFRC increased from 43 to 132 s^{-1} and 51 to 123 s^{-1} at the concrete curing ages of 3 and 7 days, respectively. Their corresponding peak stresses increased from 35.56 to 54.79 MPa and 40.35 to 62.19 MPa,

respectively, and their corresponding ultimate axial strains increased from 0.013 to 0.030 and 0.012 to 0.027, respectively. Meanwhile, the average strain rate of the specimens with rapid-hardening HSFRC increased from 70 to 122 s^{-1} and 63 to 111 s^{-1} at the concrete curing ages of 3 and 7 days, respectively. Their corresponding peak stresses increased from 39.04 to 64.53 MPa and 43.33 to 69.21 MPa, respectively, and their corresponding ultimate axial strains increased from 0.023 to 0.028 and 0.029 to 0.030, respectively. However, when the SHPB gas pressure was increased from 0.5 to 0.9 MPa at the concrete curing age of 28 days, the average strain rate of the specimens with conventional and rapid-hardening HSFRC increased from 80 to 133 s^{-1} and 80 to 126 s^{-1}, respectively, and their corresponding peak stresses increased from 63.68 to 95.34 MPa and 60.13 to 99.86 MPa, respectively. In addition, their corresponding ultimate axial strains increased from 0.027 to 0.047 and 0.029 to 0.044, respectively. The above phenomenon shows that there were some differences in the dynamic compressive performance between the two types of HSFRC at the concrete curing age of 28 days, i.e., when the material properties had stabilized, but that both were generally acceptable. At the same time, rapid-hardening HSFRC has a clear advantage over its conventional counterparts in terms of dynamic compressive performance at early curing ages. These two aspects confirm the potential of rapid-hardening HSFRC for rapid bridge construction.

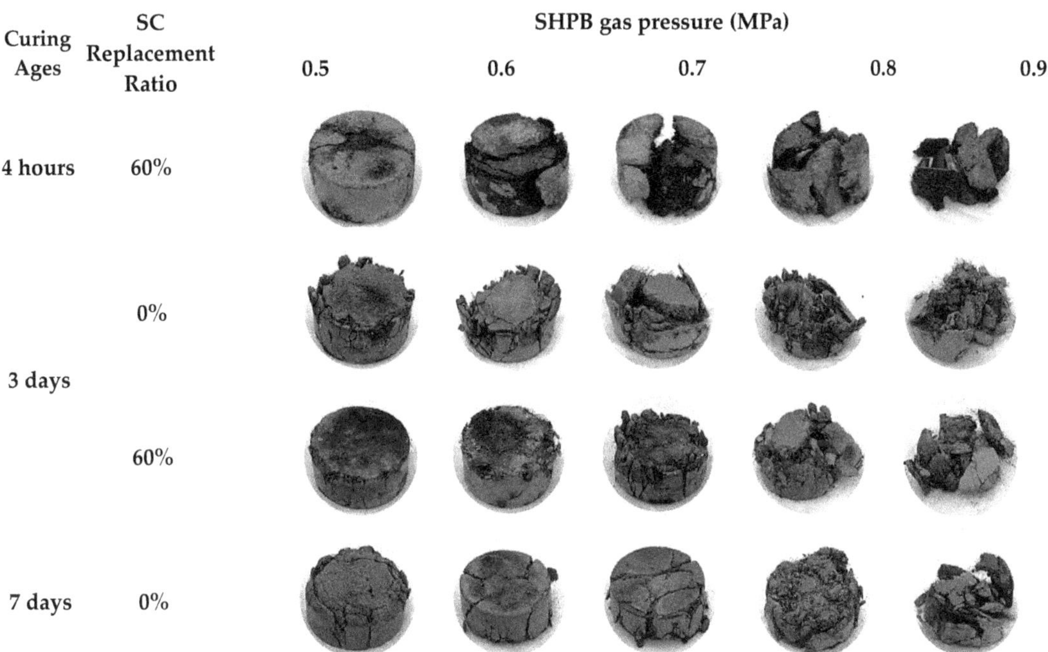

Figure 14. *Cont.*

Figure 14. Typical failure modes of HSFRC specimens under dynamic compression tests.

The dynamic compression stress–strain curve for the FBD specimen is shown in Figure 18. Note that the plot consists of 20 subplots in 5 rows and 4 pillars, where the subplots in the same row can be used to examine the effect of concrete curing ages on the dynamic compression stress–strain responses of HSFRC specimens, while their counterparts in the same pillar are used to investigate the effect of strain rate (or SHPB gas pressure). In addition, each subplot gives a comparison of specimens differing only in SC replacement ratio while other parameters remain the same, for investigating the effect of SC replacement ratio on the dynamic compression behavior of HSFRC specimens. The test results showed that both nominally identical specimens had close stress–strain curves, indicating the good repeatability of the majority of the test results. The following observations can be made from Figure 18: (1) All specimens presented comparable stress–strain responses under dynamic compression, which can be divided into four main components: a linear elastic phase, a nonlinear ascent phase, a yielding phase, and a softening phase where the specimen is completely destroyed. (2) With other parameters being the same, the longer the concrete curing age, the higher the peak point of dynamic compressive stress–strain responses of FBD specimens, but at the same time leading to a shorter yielding stage (i.e., with a constant stress but large change in strain). (3) When other parameters remained consistent, the dynamic compressive stress–strain response of specimens with rapid-hardening HSFRC was generally higher than that of their conventional counterparts when the concrete curing age was less than 7 days, but the opposite relationship existed when the curing age was greater than 7 days. (4) The dynamic compressive stress–strain response of specimens with rapid-hardening HSFRC exhibited an overall shorter yielding phase compared to their conventional counterparts. (5) The higher the strain rate (or SHPB gas pressure), the higher the dynamic compression stress–strain response of the FBD specimen when all other parameters are kept consistent.

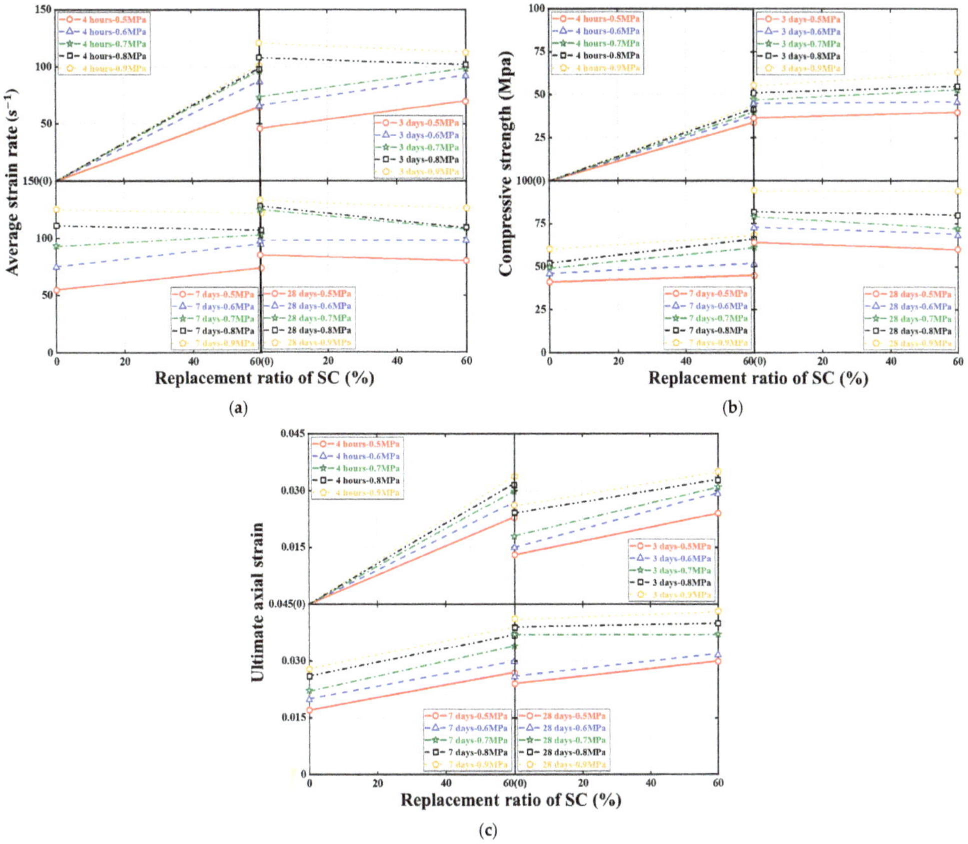

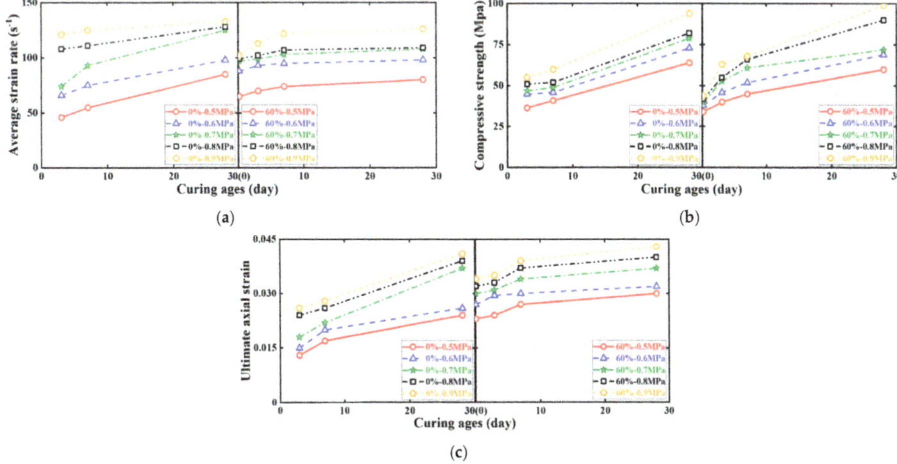

Figure 15. Effect of SC replacement ratios on ultimate conditions of dynamic compression tests. (**a**) Average strain rate. (**b**) Compressive strength. (**c**) Ultimate axial strain.

Figure 16. Effect of concrete curing ages on ultimate conditions of dynamic compression tests. (**a**) Average strain rate. (**b**) Compressive strength. (**c**) Ultimate axial strain.

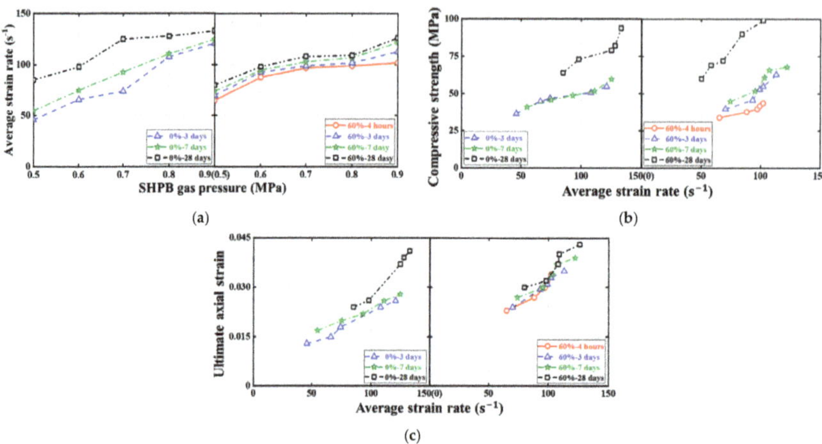

Figure 17. Effect of strain rates on ultimate conditions of dynamic compression tests. (**a**) Average strain rate. (**b**) Compressive strength. (**c**) Ultimate axial strain.

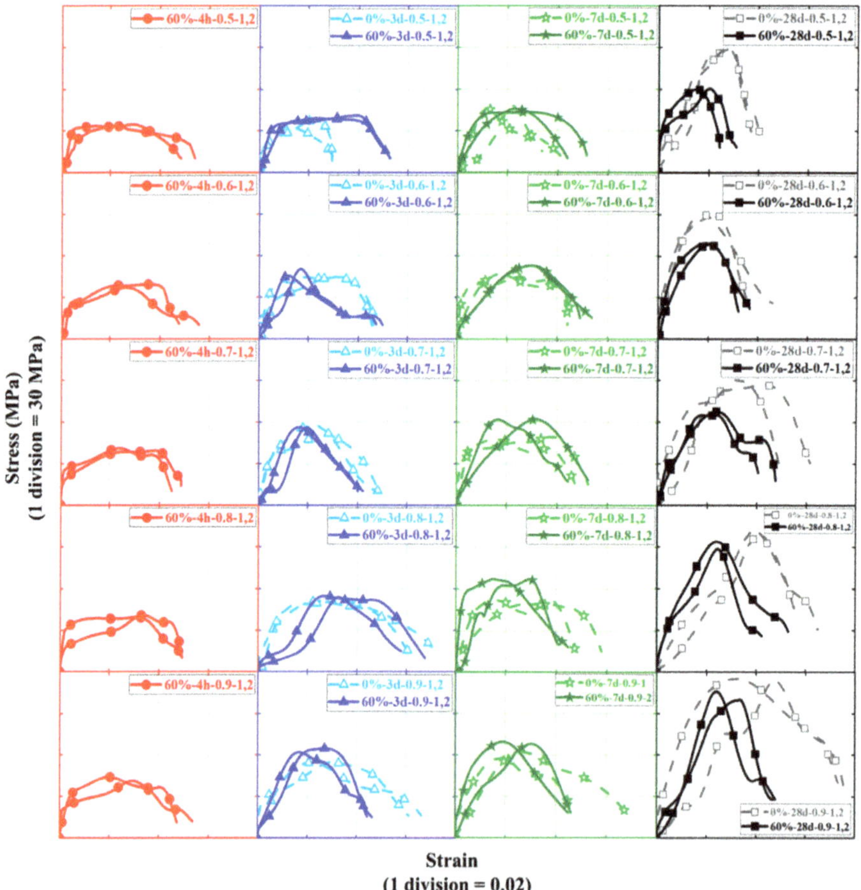

Figure 18. Stress–strain curves of dynamic compression tests.

5. Conclusions

This paper presents an experimental study of the quasi-static and dynamic compressive behavior of conventional and rapid-hardening HSFRCs. The key parameters examined in this study include concrete curing ages (from 4 h to 28 days) and strain rates (from 43 to 133 s^{-1}). Based on the experimental results and discussion presented in this paper, the following main highlights, main findings, limitations, and areas of future research can be drawn.

Main highlights and findings include the following:

(1) Rapid-hardening HSFRC exhibited high early strength characteristics: the quasi-static compressive strengths of the cubes and cylinders at 4 h of curing age reached 45.16 and 33.14 MPa, respectively, which were 55 and 48% of their corresponding 28-day age strengths; in addition, the FBD specimen at 4 days of curing age had a dynamic compressive strength of up to 43.97 MPa, roughly 0.4 times their corresponding 28-day age strength.

(2) The introduction of SC resulted in an increase in the strain rate sensitivity of HSFRC, where the strain rate sensitivity of rapid-hardening HSFRC increased with the increasing curing age and strain rate.

(3) The compressive strength, elastic modulus, and Poisson's ratio of rapid-hardening HSFRC were inferior to those of their conventional counterparts in both quasi-static and dynamic compressive loading, but rapid-hardening HSFRC's postpeak deformation capacity was significantly better than that of the conventional counterparts.

(4) The elastic modulus and Poisson's ratio of HSFRC with different cement types under quasi-static compression increased with increasing curing age; the negative relationship between the elastic modulus and curing ages or strain rates was found under dynamic counterparts.

Limitations and areas of future research include the following: Overall, rapid-hardening HSFRC has great potential for rapid bridge construction. In addition, the objective of this research was to verify the application feasibility of rapid-hardening HSFRC in rapid bridge construction through experimental investigations. To this end, this research only investigated the short-term compressive performance of a specific rapid-hardening HSFRC under normal conditions, lacking consideration of the effects of different material compositions, fiber types, and environmental conditions. Furthermore, to fully assess the properties of this innovative type of HSFRC, additional investigations should also be supplemented to determine the skin, the durability under the action of water and frost, the alkaline reactivity of HSFRC concrete with new cement, and so on. Moreover, the quasi-static and dynamic compressive constitutive models of this new type of concrete were unavailable due to the lack of test data on their compressive performance. Therefore, further research in this area is needed to investigate more systematically the compressive performance and mechanical modelling for the engineering application of such novel concrete.

Author Contributions: Conceptualization, Z.F., S.F. and H.J.; methodology, F.M., B.L. and M.L.; software, F.M., B.L. and M.L.; validation, Z.F., S.F. and H.J.; formal analysis, Z.F. and S.F.; investigation, F.M., B.L. and M.L.; resources, Z.F.; data curation, F.M., B.L. and M.L.; writing—original draft preparation, F.M.; writing—review and editing, Z.F., S.F. and H.J.; visualization, F.M. and S.F.; supervision, Z.F. and H.J.; project administration, Z.F. and H.J.; funding acquisition, Z.F. and H.J. All authors have read and agreed to the published version of the manuscript.

Funding: This research was funded by National Natural Science Foundation of China with the grant number of 52208156, the National Natural Science Foundation of China with the grant number of 51778150, the Postdoctoral Science Foundation of China with the grant number of 2022M720874 and 2023M730769, and the Basic and Applied Basic Research Foundation of Guangdong Province with the grant number of 2022A1515110394.

Institutional Review Board Statement: Not applicable.

Informed Consent Statement: Not applicable.

Data Availability Statement: Not applicable.

Acknowledgments: The authors would like to thank all the anonymous referees for their constructive comments and suggestions.

Conflicts of Interest: The authors declare no conflict of interest.

Nomenclature

The following abbreviations and symbols are used in this paper (sorted by first appearance).

Abbreviations
HSFRC	High-strength steel-fiber-reinforced concrete
SC	Sulphoaluminate cement
SHPB	Split Hopkinson pressure bar
FBD	Flattened Brazilian discs
CQC	Cubes under quasi-static compression
QC	Cylinders under quasi-static compression
DC	FBD specimens under dynamic compression
h	Hour
d	Day
LVDT	Linear displacement transducer

Symbols
$\dot{\varepsilon}_s$	Average strain rate of FBD specimens
σ_s	Compressive strength of FBD specimens
ε_s	Strain corresponding to σ_s
t	Time
C_0	Propagation speed of one-dimensional stress wave in SHPB
l_s	Thickness of FBD specimens
A_s	Cross-sectional area of FBD specimens
ε_R	Strain of reflected wave from incident bar of SHPB
E_b	Elastic modulus of SHPB
A_b	Cross-sectional area of the SHPB
ε_I	Strain of incident wave from incident bar of SHPB
ε_T	Strain of transmitted wave from transmitted bar of SHPB

References

1. Deng, L.; Yan, W.C.; Zhu, Q.J. Vehicle impact on the deck slab of concrete box-girder bridges due to damaged expansion joints. *J. Bridge Eng.* **2016**, *21*, 2. [CrossRef]
2. Ding, Y.; Zhang, W.; Au, F.T. Effect of dynamic impact at modular bridge expansion joints on bridge design. *J. Eng. Struct.* **2016**, *127*, 645–662. [CrossRef]
3. Gong, J.H.; Ma, Y.W.; Fu, J.Y.; Hu, J.; Ouyang, X.W.; Zhang, Z.H.; Wang, H. Utilization of fibers in ultra-high performance concrete: A review. *J. Compos. Part B Eng.* **2022**, *241*, 109995. [CrossRef]
4. Zhang, J.H.; Wu, Z.Y.; Yu, H.F.; Ma, H.Y.; Da, B. Mesoscopic modeling approach and application for steel fiber reinforced concrete under dynamic loading: A review. *J. Eng.* **2022**, *16*, 220–238. [CrossRef]
5. Yeo, S.H.; Mo, K.H.; Hosen, M.A.; Mahmud, H.B. Properties of cementitious repair materials for concrete pavement. *Adv. Mater. Sci. Eng.* **2022**, *2022*, 3057801. [CrossRef]
6. Yu, R.; Zhang, X.Y.; Hu, Y.W.; Li, J.H.; Zhou, F.J.; Liu, K.N.; Zhang, J.J.; Wang, J.N. Development of a rapid hardening ultra-high performance concrete (R-UHPC): From macro properties to micro structure. *Constr. Build. Mater.* **2022**, *329*, 127188. [CrossRef]
7. Al-musawi, H.; Huang, H.D.; Di Benedetti, M.; Guadagnini, M.; Pilakoutas, K. Effect of shrinkage on rapid hardening plain and recycled steel fiber concrete overlays. *Cem. Con. Comp.* **2021**, *125*, 104246. [CrossRef]
8. Cao, X.Y.; Feng, D.C.; Li, Y. Assessment of various seismic fragility analysis approaches for structures ex-cited by non-stationary stochastic ground motions. *Mech. Syst. Signal Process.* **2023**, *186*, 109838. [CrossRef]
9. Zhao, B.D.; Liu, C.Q.; Wu, H.D.; Ge, Y.H.; Yang, J.J.; Yi, Q. Study on out-of-plane flexural stiffness of unstiffened multi-planar CHS X-joints. *Eng. Struct.* **2019**, *188*, 137–146. [CrossRef]

10. Xiong, Z.; Lin, L.H.; Qiao, S.H.; Li, L.J.; Li, Y.L.; He, S.H.; Li, Z.W.; Liu, F.; Chen, Y.L. Axial performance of seawater sea-sand concrete columns reinforced with basalt fiber-reinforced polymer bars under concentric compressive load. *J. Build. Eng.* **2022**, *47*, 103828. [CrossRef]

11. Sun, J.B.; Lin, S.; Zhang, G.B.; Sun, Y.T.; Zhang, J.F.; Chen, C.F. The effect of graphite and slag on electrical and mechanical properties of electrically conductive cementitious composites. *Con. Build. Mat.* **2021**, *281*, 122606. [CrossRef]

12. Zhao, B.D.; Chen, Y.; Liu, C.Q.; Wu, H.D.; Wang, T.; Wei, X.D. An axial semi-rigid connection model for cross-type transverse branch plate-to-CHS joints. *Eng. Struct.* **2019**, *181*, 413–426. [CrossRef]

13. Feng, W.H.; Tang, Y.C.; He, W.M.; Wei, W.B.; Yang, Y.M. Model dynamic fracture toughness of rubberised concrete using a drop hammer device and split Hopkinson pressure bar. *J. Build. Eng.* **2022**, *48*, 103995. [CrossRef]

14. Cao, X.Y.; Feng, D.C.; Beer, M. Consistent seismic hazard and fragility analysis considering combined capacity-demand uncertainties via probability density evolution method. *Struct. Saf.* **2023**, *103*, 102330. [CrossRef]

15. Fang, S.; Li, L.J.; Luo, Z.P.; Fang, Z.C.; Li, Z.W.; Liu, F.; Wang, H.L.; Xiong, Z. Novel FRP interlocking multi-spiral rein-forced-seawater sea-sand concrete square columns with longitudinal hybrid FRP–steel bars: Monotonic and cyclic axial compressive behaviors. *Compos. Struct.* **2023**, *305*, 116487. [CrossRef]

16. Zhao, B.D.; Ke, K.; Liu, C.Q.; Hong, L. Computational model for the flexural capacity and stiffness of eccentric RHS X-connections under brace out-of-plane bending moment. *J. Struct. Eng.* **2020**, *146*, 04019227. [CrossRef]

17. Cao, X.Y.; Feng, D.C.; Wang, C.L.; Shen, D.J.; Wu, G. A stochastic CSM-based displacement-oriented design strategy for the novel precast SRC-UHPC composite braced-frame in the externally attached seismic retrofitting. *Compos. Struct.* **2023**, *321*, 117308. [CrossRef]

18. Seok, S.H.; So, S. An experimental study on the strength development of ultra-high strength steel fiber reinforced cementitious composites. *J. Archit. Inst. Korea Struct.* **2009**, *25*, 63–70.

19. Rong, Q.; Hou, X.M.; Ge, C. Quantifying curing and composition effects on compressive and tensile strength of 160-250 MPa RPC. *Con. Build. Mat.* **2020**, *241*, 117987. [CrossRef]

20. Song, P.S.; Hwang, S. Mechanical properties of high-strength steel fiber-reinforced concrete. *Con. Build. Mat.* **2004**, *18*, 669–673. [CrossRef]

21. Kizilirmak, C.; Aydin, S.; Yardimci, M.Y. Effect of the steel fiber hook geometry on the flexural properties of high strength steel fiber reinforced concretes under static and impact loading. *J. Fac. Eng. Archit. Gazi Univ.* **2019**, *34*, 1610–1627.

22. Xu, P.; Ma, J.Y.; Ding, Y.H.; Zhang, M.X. Influences of steel fiber content on size effect of the fracture energy of high-strength concrete. *J. Civil Eng.* **2021**, *25*, 948–959. [CrossRef]

23. Luo, D.M.; Wang, Y.; Niu, D.T. Evaluation of the performance degradation of hybrid steel-polypropylene fiber reinforced concrete under freezing-thawing conditions. *Adv. Civ. Eng.* **2020**, *2020*, 8863047. [CrossRef]

24. Lancellotti, I.; Piccolo, F.; Nguyen, H.; Mastali, M.; Alzeer, M.; Illikainen, M.; Leonelli, C. The effect of fibrous reinforcement on the polycondensation degree of slag-based alkali activated composites. *Polymers* **2021**, *13*, 2664. [CrossRef]

25. Vaitkevicius, V.; Serelis, E.; Vaiciukyniene, D.; Raudonis, V.; Rudzionis, Z. Advanced mechanical properties and frost damage resistance of ultra-high performance fiber reinforced concrete. *Constr. Build. Mater.* **2016**, *126*, 26–31. [CrossRef]

26. Rady, S.; Al-Sibahy, A. Bond strength behavior of high strength structural lightweight concrete containing steel fibers with different geometries. *Innov. Infrastruct. Solut.* **2023**, *8*, 184. [CrossRef]

27. Murali, G.; Vinodha, E. Experimental and analytical study of impact failure strength of steel hybrid fibre reinforced concrete subjected to freezing and thawing cycles. *J. Sci. Eng.* **2018**, *43*, 5487–5497. [CrossRef]

28. Li, Y.; Zhang, Q.R.; Wang, R.J.; Xiong, X.B.; Li, Y.; Wang, J.Y. Experimental investigation on the dynamic mechanical properties and microstructure deterioration of steel fiber reinforced concrete subjected to freeze-thaw cycles. *Buildings* **2022**, *12*, 2170. [CrossRef]

29. Moein, M.M.; Saradar, A.; Rahmati, K.; Shirkouh, A.H.; Sadrinejad, I.; Aramali, V.; Karakouzian, M. Investigation of impact resistance of high-strength portland cement concrete containing steel fibers. *Materials* **2022**, *15*, 7157. [CrossRef]

30. Sharma, S.; Arora, V.V.; Kumar, S.; Daniel, Y.N.; Sharma, A. Durability study of high-strength steel fiber-reinforced concrete. *ACI Mater. J.* **2018**, *115*, 219–225. [CrossRef]

31. Sun, K.W.; Wu, Y.; Li, S.L.; Feng, Y.; Feng, L.H. Study on dynamic impact mechanical properties of UHPC with high-content and directional reinforced steel fiber. *Appl. Sci.* **2023**, *13*, 3753. [CrossRef]

32. Dalvand, A.; Ahmadi, M. Impact failure mechanism and mechanical characteristics of steel fiber reinforced self-compacting cementitious composites containing silica fume. *Eng. Sci. Technol. Int. J. Jestech* **2021**, *24*, 736–748. [CrossRef]

33. Li, J.; Shi, S.Q.; He, Q.L.; Chen, S. Split-Hopkinson pressure bar test and numerical simulation of steel fiber-reinforced high-strength concrete. *J. Compos. Mater.* **2019**, *29*, 109–117. [CrossRef]

34. Zhang, P.; Han, X.; Zheng, Y.X.; Wan, J.Y.; Hui, D.V. Effect of PVA fiber on mechanical properties of fly ash-based geopolymer concrete. *J. Compos. Mater.* **2021**, *60*, 418–437. [CrossRef]

35. Cheng, Z.J.; Li, S.; Lu, Y.Y.; Li, W.T.; Liu, Z.Z. Dynamic compressive properties of lightweight engineered geopolymer composites containing ceramsite (LW-EGC). *Con. Build. Mat.* **2023**, *370*, 130717. [CrossRef]

36. Chen, D.; Liu, F.; Yang, F.; Jing, L.; Feng, W.H.; Lv, J.B.; Luo, Q.Z. Dynamic compressive and splitting tensile response of unsaturated polyester polymer concrete material at different curing ages. *Con. Build. Mat.* **2018**, *177*, 477–498. [CrossRef]

37. Guo, Y.C.; Xiao, S.H.; Zeng, J.J.; Zheng, Y.; Li, X.; Liu, F. Fiber reinforced polymer-confined concrete under high strain rate compression: Behavior and a unified dynamic strength model. *Con. Build. Mat.* **2020**, *260*, 120460. [CrossRef]

38. *C469/C469M-22*; Standard Test Method for Static Modulus of Elasticity and Poisson Ratio of Concrete in Compression. American Society of Testing Materials (ASTM): West Conshohocken, PA, USA, 2022.
39. *GB/T 7314-2017*; Metallic Materials—Compression Test Method at Room Temperature. State General Administration of the People's Republic of China for Quality Supervision and Inspection and Quarantine: Beijing, China, 2017. (In Chinese)
40. *GB/T 34108-2017*; Metallic Materials—High Strain Rate Compression Test Method at Ambient Temperature. State General Administration of the People's Republic of China for Quality Supervision and Inspection and Quarantine: Beijing, China, 2017. (In Chinese)

Article

Modeling and Simulation of the Hysteretic Behavior of Concrete under Cyclic Tension–Compression Using the Smeared Crack Approach

Pei Zhang [1,*], Shenshen Wang [2] and Luying He [2]

[1] College of Mechanical Engineering and Mechanics, Xiangtan University, Xiangtan 411105, China
[2] College of Civil Engineering, Xiangtan University, Xiangtan 411105, China;
202121572230@smail.xtu.edu.cn (S.W.); 202121572183@smail.xtu.edu.cn (L.H.)
* Correspondence: zhangpei@xtu.edu.cn

Abstract: Concrete structures under wind and earthquake loads will experience tensile and compressive stress reversals. It is very important to accurately reproduce the hysteretic behavior and energy dissipation of concrete materials under cyclic tension–compression for the safety evaluation of concrete structures. A hysteretic model for concrete under cyclic tension–compression is proposed in the framework of smeared crack theory. Based on the crack surface opening–closing mechanism, the relationship between crack surface stress and cracking strain is constructed in a local coordinate system. Linear loading–unloading paths are used and the partial unloading–reloading condition is considered. The hysteretic curves in the model are controlled by two parameters: the initial closing stress and the complete closing stress, which can be determined by the test results. Comparison with several experimental results shows that the model is capable of simulating the cracking process and hysteretic behavior of concrete. In addition, the model is proven to be able to reproduce the damage evolution, energy dissipation, and stiffness recovery caused by crack closure during the cyclic tension–compression. The proposed model can be applied to the nonlinear analysis of real concrete structures under complex cyclic loads.

Keywords: concrete; hysteretic behavior; energy dissipation; cyclic loading; smeared crack model; numerical simulation; dynamic behavior

Citation: Zhang, P.; Wang, S.; He, L. Modeling and Simulation of the Hysteretic Behavior of Concrete under Cyclic Tension–Compression Using the Smeared Crack Approach. *Materials* **2023**, *16*, 4442. https://doi.org/10.3390/ma16124442

Academic Editor: Gwenn Le Saout

Received: 16 May 2023
Revised: 13 June 2023
Accepted: 15 June 2023
Published: 17 June 2023

1. Introduction

The safety assessment of concrete structures subjected to cyclic loading such as seismic excitation requires realistic constitutive models to reproduce the real behavior of the materials. Due to the low tensile strength, the concrete subjected to seismic load usually presents softening behavior in tension and hysteretic behavior in tension–compression reversals. As a result, the hysteretic model for concrete plays a significant role in determining the seismic responses of concrete structures including the deformation and energy evolution. However, due to the lack of experimental data, studies on modeling the hysteresis behavior of concrete under cyclic tension and tension–compression reversal are scarce compared to those on cyclic compression [1–5].

At present, most concrete constitutive models considering the cyclic tension assumed that the unloading path coincides with the reloading path, neglecting the hysteretic behavior and energy dissipation caused by the crack opening and closing [6–10]. Several researchers [11–14] have suggested various modeling strategies to replicate the hysteretic stress–strain curves of concrete under cyclic loading, including the complete unloading–reloading path and the partial unloading–reloading path.

Yankelevsky and Reinhardt [11] proposed a focal model to reproduce the hysteretic curves through a series of focal points. These predefined focal points govern the unloading and reloading paths either by the rays from themselves or by their stress level. When all

43

the focal points are determined, the complete unloading–reloading curves can be plotted through a simple graphical process. Although the focal model performed well in replicating the test results, it is thought that the concept of a focal point, which is solely derived from the graphic feature, lacks physical significance and that the process for determining unloading and reloading paths is excessively complex.

Chang and Mander [12] proposed a rule-based model to simulate the hysteretic behavior of confined and unconfined concrete in both cyclic compression and tension. Fifteen unloading and reloading paths governed by fifteen different rules are prescribed in the model. The unloading and reloading curves in the model are characterized by the polynomial equation. The rule-based model is capable to simulate the hysteretic curve of concrete under various loading conditions. However, due to the overabundance of rules and paths in the model, the parameters of the unloading and reloading paths are difficult to calibrate, and the calculation is very time-consuming.

Aslani and Jowkarmeimandi [13] proposed a constitutive model for concrete under cyclic loading based on the findings of previous experimental and analytical studies. The model for concrete subjected to monotonic and cyclic loading comprises four components: an envelope curve, an unloading curve, a reloading curve, and a transition curve. A crack-closing model is suggested, in which the crack-closure mechanism is governed by the crack-closure stress. The unloading curve in compression is described by a power equation, whereas the reloading curve is described by a linear equation. However, the model fails to account for hysteretic behavior and energy dissipation during cyclic tension due to the use of the same path for both unloading and reloading processes in tension.

The most popular model for modeling the mechanical behavior of concrete is the plastic damage model, which can replicate concrete's irreversible deformation and stiffness degradation under monotonic and cyclic loads with confinement [8–10]. The unloading–reloading law in the plastic damage model is usually controlled by the damage modulus and the residual plastic strain. Chen et al. [14] proposed analytical expressions introducing damage index to describe the response of concrete to cyclic loadings. A power type function is selected to model the unloading and reloading curve and the power is expressed as a function of the damage index and the strain rate. McCall and Guyer [15] suggested that cracks inside nonlinear mesoscopic elastic material such as sandstone were denoted as hysteretic mesoscopic elastic units. Based on the assumption, Preisach–Mayergoyz (P-M) model was used to simulate the phenomenon of the hysteretic loop in concrete under cyclic tension and alternating tensile–compressive loading [16,17]. However, the P-M model can only reproduce the hysteretic curves and cannot simulate the process of concrete crack propagation under cyclic tension.

Based on the continuum damage theory, Long et al. [18,19] developed an improved anisotropic damage model to simulate the nonlinear behavior of concrete by proposing the independent tensile and compressive damage evolution laws. The hysteretic behavior is described by a nonlinear unloading path and a linear reloading path. A stiffness recovery coefficient is defined to model the stiffness recovery phenomenon caused by crack-closing. However, the simulated results of concrete hysteretic behavior under tension–compression reversals are not satisfactory.

Liu et al. [20] combined the loading and unloading characteristic points and the loading and unloading paths in the hysteretic rules to construct a four-parameter plastic damage model considering the hysteretic effect under cyclic loading. The reloading curve is simplified to linear and the unloading curve is represented by an exponential equation reported in the literature [5]. Nonlinear characteristics of concrete such as stiffness degradation, strength softening, irreversible plastic deformation, and hysteresis effects under cyclic loading were simulated by the model. However, the model assumes that damage accumulates during the unloading process and remains unaltered during the reloading process, which is inconsistent with the concrete damage mechanism.

In this paper, an efficient model capable of predicting the hysteretic behavior of concrete under cyclic tension and tension–compression reversals is proposed within the framework of smeared crack theory. By decomposing the total strain into the elastic strain and crack strain, it is possible to directly model the crack surface behavior that causes the hysteretic phenomenon. Straight lines are adopted to describe the unloading and reloading paths, and several necessary hysteretic rules are proposed based on the closing–opening mechanism of the crack surface. Furthermore, the crack surface-closure effect during the tension–compression reversals is considered by the definition of crack-closure stress parameters. Compared to the existing hysteretic models, only two model parameters are needed in the proposed model to simulate the complete and partial unloading–reloading curves. The model has been validated by comparison with available test results under different loading conditions.

2. Theoretical Framework of Smeared Crack Model

2.1. Overview of the Smeared Crack Model

The nonlinear behavior of concrete is mainly caused by the initiation and propagation of cracks. In the finite element simulation, the description of concrete cracks generally has two ways: one is the discrete crack method [21–24] and the second is the smeared crack method [25–29]. The discrete crack model assumes that cracks appear on the element boundaries. Once the crack generates, new nodes are added and the mesh is re-divided to produce new element boundaries. In this way, the geometric discontinuity caused by the crack is directly characterized, and the position, shape, and width of the crack can be clearly described.

The smeared crack method assumes that local discontinuities caused by cracks are uniformly distributed in the fracture zone of the finite element (Figure 1). Based on this assumption, the relative displacement between the crack surfaces can be characterized by the crack strain, and the constitutive behavior of the cracked concrete can be directly modeled in terms of the stress–strain relations. Contrary to the discrete crack concept, the smeared crack concept fits the nature of the finite element displacement method, as the continuity of the displacement field remains intact [30]. The main ingredients of the smeared crack model are introduced here, and a full description can be found in the literature [29].

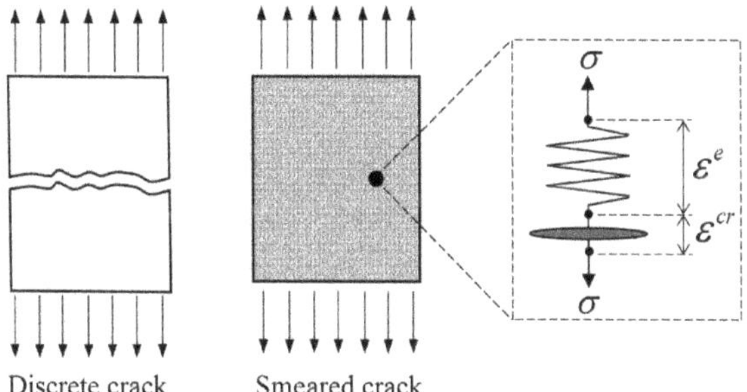

Figure 1. A schematic diagram of the discrete crack and smeared crack.

The smeared crack model decomposes the total strain in the fracture zone into two parts: elastic strain and crack strain. The elastic strain corresponds to the deformation yielded by the uncracked portion, and the crack strain corresponds to the equivalent deformation generated by the crack opening. In the one-dimensional case, the schematic diagram of strain decomposition is illustrated in Figure 1, and the corresponding formula can be written as:

$$\varepsilon = \varepsilon^e + \varepsilon^{cr} \tag{1}$$

where ε^e represents the elastic strain and ε^{cr} represents the crack strain. The major advantage of this decomposition is that it allows the crack surface behavior to be treated separately from the constitutive behavior of the uncracked concrete [30]. As a result, the stress–strain relationship of the fracture zone in concrete can be treated as the superposition of the elastic part and the cracked part, as illustrated in Figure 2.

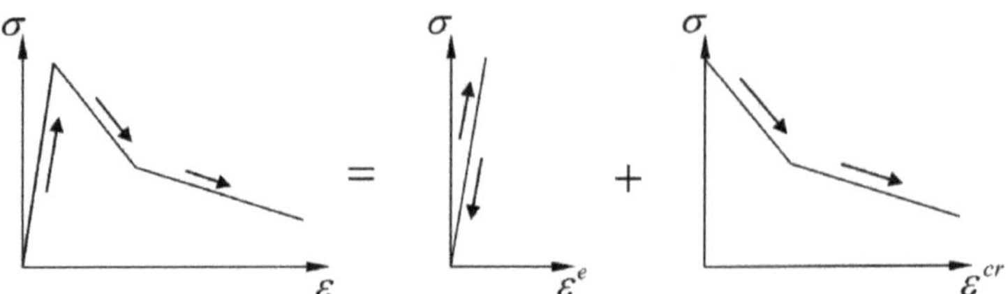

Figure 2. Decomposition of the stress–strain relationship for cracked concrete.

2.2. Constitutive Relations in Local Coordinate System

In the multiaxial stress state, the strain decomposition can be written in a tensor form as:

$$\varepsilon = \varepsilon^e + \varepsilon^{cr} \tag{2}$$

The corresponding incremental form is:

$$\Delta\varepsilon = \Delta\varepsilon^e + \Delta\varepsilon^{cr} \tag{3}$$

When the first principal stress exceeds the tensile strength, the crack generates and the crack surface is perpendicular to the direction of the first principal stress. In the two-dimensional case, the behavior of the crack surface can be clearly described in the local coordinate system as illustrated in Figure 3, where the subscript n represents the direction normal to the crack surface, and the subscript t represents the direction tangential to the crack surface. Projecting the stress on the crack surface σ_n^{cr} onto the local coordinate system, the corresponding normal stress component σ_{nn}^{cr} and tangential stress component σ_{nt}^{cr} can be obtained. Correspondingly, the cracking strain on the crack surface ε_n^{cr} can be decomposed into the normal cracking strain ε_{nn}^{cr} and the tangential cracking strain γ_{nn}^{cr} in the local coordinate system. In the smeared crack model, the normal cracking strain ε_{nn}^{cr} corresponds to the open displacement of the crack surface, and the tangential cracking strain γ_{nn}^{cr} corresponds to relative slip displacement on the crack surface.

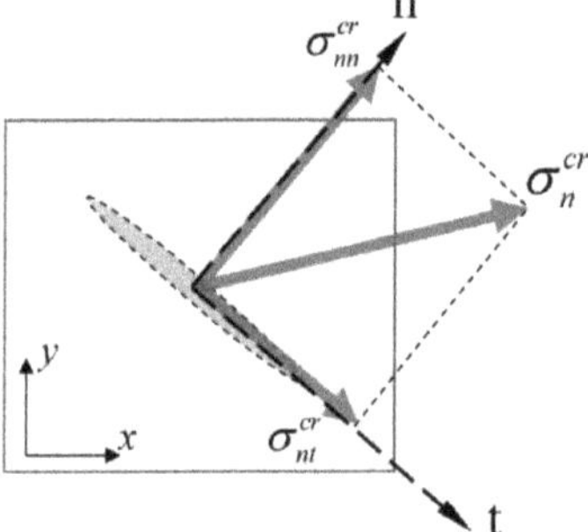

Figure 3. Local coordinates system normal to the crack surface.

Assuming that the stress–strain relationships between the normal and tangential directions of the crack surface are decoupled from each other, the relationship between the stress and cracked strain in the local coordinate system can be written as follows:

$$\left\{ \begin{array}{c} \Delta\sigma_{nn}^{cr} \\ \Delta\sigma_{nt}^{cr} \end{array} \right\} = \left[\begin{array}{cc} D_c & 0 \\ 0 & G_c \end{array} \right] \left\{ \begin{array}{c} \Delta\varepsilon_{nn}^{cr} \\ \Delta\gamma_{nt}^{cr} \end{array} \right\} \tag{4}$$

where D_c is the modulus governing the crack surface opening and closing behavior, and G_c is the modulus governing the relative slip between the crack surfaces. Considering the different loading cases, the modulus D_c can be further divided into the strain-softening modulus D_c^s, which determines the tensile strain-softening behavior, and the hysteretic modulus D_c^h, which determines the unloading–reloading behavior. The physical meaning of each modulus is illustrated in Figure 4.

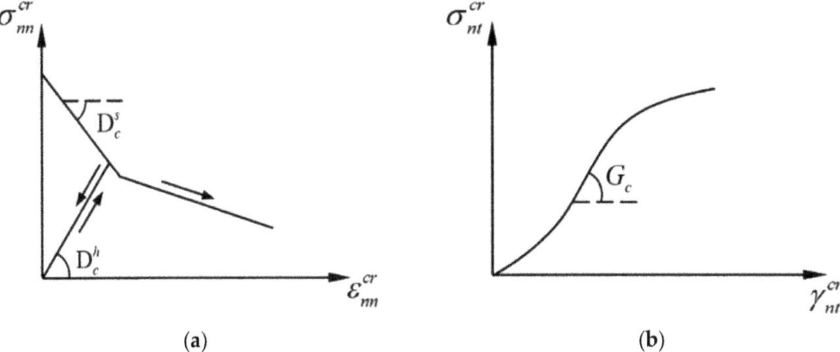

Figure 4. A schematic diagram of the cracking modulus: (**a**) the strain-softening modulus and hysteretic modulus and (**b**) the shear modulus on the crack interface.

Equation (4) can be written in a tensor form as:

$$\Delta \mathbf{s}^{cr} = \mathbf{D}^{cr} \Delta \mathbf{e}^{cr} \tag{5}$$

where $\Delta \mathbf{s}^{cr}$ and $\Delta \mathbf{e}^{cr}$ are the increment of crack surface stress and strain in the local coordinate system, respectively. $\mathbf{D}^{cr}$ is the cracking modulus matrix controlling the complex behavior of the crack surface.

2.3. Constitutive Relations of Cracked Concrete in Global Coordinate System

The crack surface stress and crack strain increments in the local coordinate system can be converted into the stress and crack strain increments in global coordinates through coordinate transformation:

$$\begin{Bmatrix} \Delta \varepsilon_{xx}^{cr} \\ \Delta \varepsilon_{yy}^{cr} \\ \Delta \gamma_{xy}^{cr} \end{Bmatrix} = \begin{bmatrix} \cos^2 \theta & -\sin \theta \cos \theta \\ \sin^2 \theta & \sin \theta \cos \theta \\ 2 \sin \theta \cos \theta & \cos^2 \theta - \cos^2 \theta \end{bmatrix} \begin{Bmatrix} \Delta \varepsilon_{nn}^{cr} \\ \Delta \gamma_{nt}^{cr} \end{Bmatrix} \tag{6}$$

$$\begin{Bmatrix} \Delta \sigma_{xx} \\ \Delta \sigma_{yy} \\ \Delta \sigma_{xy} \end{Bmatrix} = \begin{bmatrix} \cos^2 \theta & -\sin \theta \cos \theta \\ \sin^2 \theta & \sin \theta \cos \theta \\ 2 \sin \theta \cos \theta & \cos^2 \theta - \cos^2 \theta \end{bmatrix} \begin{Bmatrix} \Delta \sigma_{nn}^{cr} \\ \Delta \sigma_{nt}^{cr} \end{Bmatrix} \tag{7}$$

The formulas (6) and (7) in a tensor form are as follows:

$$\Delta \boldsymbol{\varepsilon}^{cr} = \mathbf{N} \Delta \mathbf{e}^{cr} \tag{8}$$

$$\Delta \boldsymbol{\sigma} = \mathbf{N} \Delta \mathbf{s}^{cr} \tag{9}$$

where $\Delta \boldsymbol{\sigma}$ and $\Delta \boldsymbol{\varepsilon}^{cr}$ are the stress increment and crack strain increment in the global coordinate system, respectively. $\mathbf{N}$ is the coordinate transformation matrix, and θ is the angle between the normal direction of the crack surface and the x-axis in the global coordinate system.

The relationship between the stress increment and the total strain increment in the global coordinate system is derived as follows. In the initial undamaged state, the relationship between the stress tensor increment $\Delta\sigma$ and the elastic strain tensor increment $\Delta\varepsilon^e$ can be written as follows:

$$\Delta\sigma = \mathbf{D}^e \Delta\varepsilon^e \tag{10}$$

where $\mathbf{D}^e$ is the elastic modulus tensor.

Substituting Equations (3) and (8) into (10):

$$\Delta\sigma = \mathbf{D}^e[\Delta\varepsilon - \mathbf{N}\Delta\mathbf{e}^{cr}] \tag{11}$$

Combining Equations (5) and (9):

$$\Delta\sigma = \mathbf{N}\mathbf{D}^{cr}\Delta\mathbf{e}^{cr} \tag{12}$$

Substituting Equation (12) into (11):

$$\Delta\mathbf{e}^{cr} = [\mathbf{D}^{cr} + \mathbf{N}^{\mathbf{T}}\mathbf{D}^e\mathbf{N}]^{-1}\mathbf{N}^{\mathbf{T}}\mathbf{D}^{\mathbf{e}}\Delta\varepsilon \tag{13}$$

Substituting Equation (13) back into Equation (11):

$$\Delta\sigma = [\mathbf{D}^{\mathbf{e}} - \mathbf{D}^{\mathbf{e}}\mathbf{N}[\mathbf{D}^{cr} + \mathbf{N}^{\mathbf{T}}\mathbf{D}^e\mathbf{N}]^{-1}\mathbf{N}^{\mathbf{T}}\mathbf{D}^{\mathbf{e}}]\Delta\varepsilon \tag{14}$$

Equation (14) represents the constitutive equation of the cracked concrete. It can be seen that the constitutive relationship of the cracked concrete is mainly determined by the cracking modulus matrix $\mathbf{D}^{cr}$.

2.4. Determination of Cracking Modulus

The crack surface strain-softening modulus D_c^s is determined based on the crack band theory [31–33]. The crack band theory assumes that the microcracks are uniformly distributed in a crack band with a width of h. According to the principle of conservation of energy, the relationship between the area g_f enclosed by the constitutive curve $\sigma_{nn}^{cr} - \varepsilon_{nn}^{cr}$ and the material fracture energy G_f satisfies the following equation:

$$G_f = hg_f = h\int \sigma_{nn}d\varepsilon_{nn}^{cr} \tag{15}$$

It can be seen that the strain-softening modulus D_c^s of the crack surface is determined by the parameters G_f and h. The parameter h is known as the character length, which is mesh-dependent and affected by the shape, size, type, and integration scheme of the finite element. In the case of a bilinear softening model as shown in Figure 5, the corresponding strain-softening modulus can be calculated as:

$$D_c^s = \begin{cases} -\dfrac{5}{6}\dfrac{f_t^2 h}{G_f} & \left(0 < \varepsilon_{nn}^{cr} < \dfrac{2}{9}\varepsilon_u\right) \\[2ex] -\dfrac{5}{42}\dfrac{f_t^2 h}{G_f} & \left(\dfrac{2}{9}\varepsilon_u < \varepsilon_{nn}^{cr} < \varepsilon_u\right) \end{cases} \tag{16}$$

where ε_u is the ultimate strain. It can be seen that the strain-softening modulus D_c^s and the ultimate strain ε_u of the softening curve have to be adjusted to the crack band width h.

The hysteretic modulus D_c^h of the crack surface is determined by the unloading and reloading paths as well as the loading history. In the smeared crack approach, the unloading and reloading paths of concrete under cyclic load can be governed by proposing different hysteretic rules without modifying the constitutive equation. The crack shear modulus G_c is usually assumed to be constant, and its value can be determined according to the method in the literature [30].

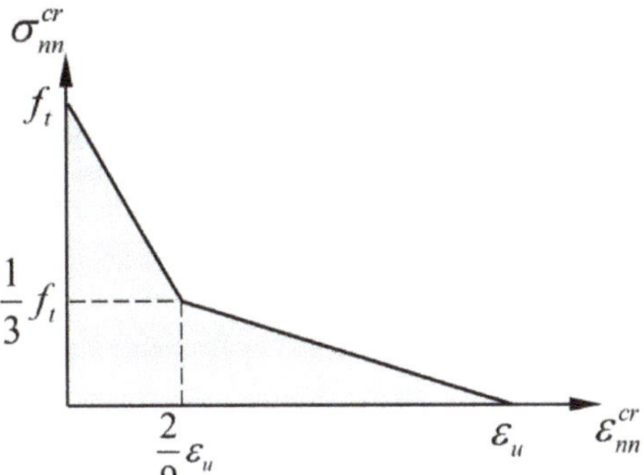

Figure 5. A schematic diagram of the bilinear softening model.

3. Hysteretic Rules Based on Crack Opening–Closing Mechanism

3.1. Hysteretic Characteristics of Concrete under Tension–Compression Reversals

The unloading and reloading behavior of concrete under tension–compression reversals usually presents obvious hysteretic characteristics, accompanied by a large amount of energy dissipation [34,35]. Figure 6 shows the characteristic hysteretic curve of concrete under uniaxial tension–compression reversals, where the horizontal axis represents the tensile strain and the vertical axis represents the tensile stress. In this paper, the loading procedure is defined as the process of the tensile strain increase, and the unloading procedure is defined as that of the tensile strain decrease.

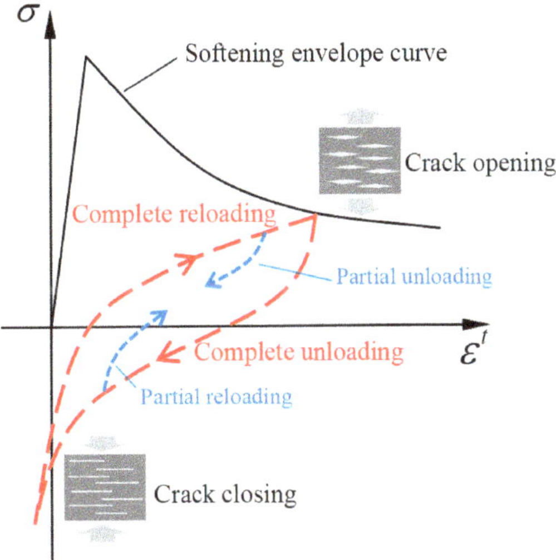

Figure 6. The characteristic hysteretic curve of concrete under tension–compression reversals.

The complete and partial unloading–reloading process under tension–compression reversals is shown as the dotted line in Figure 6. It can be seen that in the process of complete unloading, the stiffness gradually decreases with the reduction in tensile stress, and the inelastic strain is observed when the tensile stress is unloaded to zero. When the stress is reversed, the inelastic strain decreases gradually, while the stiffness increases with the increase in compressive stress. The phenomenon of stiffness recovery under compressive stress is called the "unilateral effect" [36], which is believed to be caused by the complete closure of an open crack under compressive stress [37]. Nouailletas et al. [38]

According to the above provisions, the partial reloading and unloading modulus under the cyclic tension can be calculated as follows:

$$D_c^{hp} = \begin{cases} \infty & (\Delta\varepsilon^{cr} \leq 0) \\ \dfrac{\sigma_{un}^{cr} - \sigma_p^{cr}}{\varepsilon_{un}^{cr} - \varepsilon_p^{cr}} & (\Delta\varepsilon^{cr} > 0) \end{cases} \tag{20}$$

where the condition $\Delta\varepsilon^{cr} \leq 0$ indicates the unloading process, and the condition $\Delta\varepsilon^{cr} > 0$ indicates the reloading process.

Under the condition of tension–compression reversals, all of the possible partial unloading–reloading paths are presented as the dotted line ①, ②, ③, and ④ with an arrow shown in Figure 7. The partial unloading and reloading paths in this condition are specified only dependent on the stress state (tension or compression). According to the geometric relationship, the partial loading and unloading modulus under tension–compression reversals can be calculated as follows:

$$D_c^{hp} = \begin{cases} \infty & (\sigma^{cr} \leq 0) \\ \dfrac{\sigma_{un}^{cr}}{\varepsilon_{un}^{cr} - \varepsilon_{re}^{cr}} & (\sigma^{cr} > 0) \end{cases} \tag{21}$$

3.4. Determination of the Model Parameters

Two model parameters of the proposed unloading–reloading paths are determined in this section. Six stress-deformation curves from the uniaxial cyclic loading tests in the literature [40] were selected to determine the value of the initial closing stress f_0^{cl}. The value of f_0^{cl} corresponding to different unloading points was identified according to the procedure demonstrated in Figure 9. Figure 10 shows the relationship between the value of f_0^{cl} and the corresponding unloading deformations. It can be seen that the result appears very scattered and f_0^{cl} decreases nonlinearly with the increase in unloading deformation. Figure 11 shows the relationship between the value of f_0^{cl} and the corresponding unloading stress σ_{un}^{cr}. A clear linear feature can be observed in the figure.

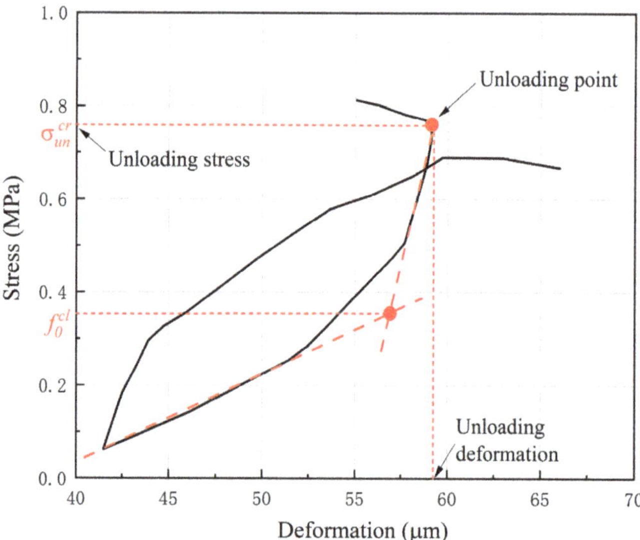

Figure 9. Identification of the initial closing stress from the test curve.

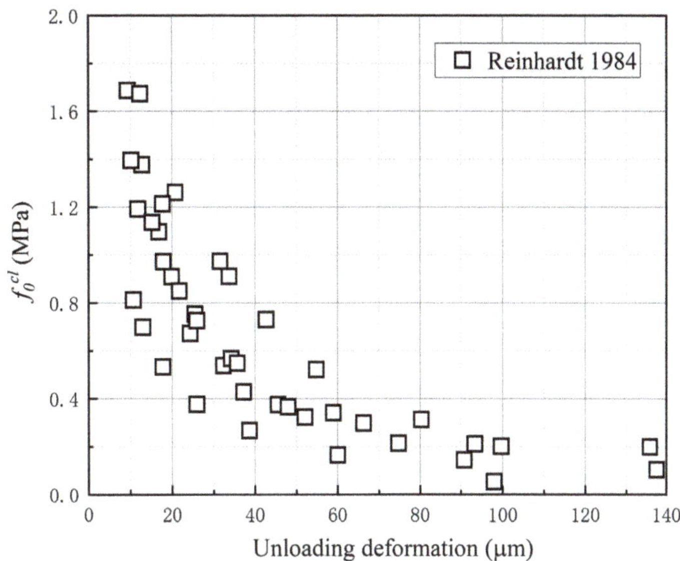

Figure 10. The relationship between the value of f_0^{cl} and the corresponding unloading deformations [40].

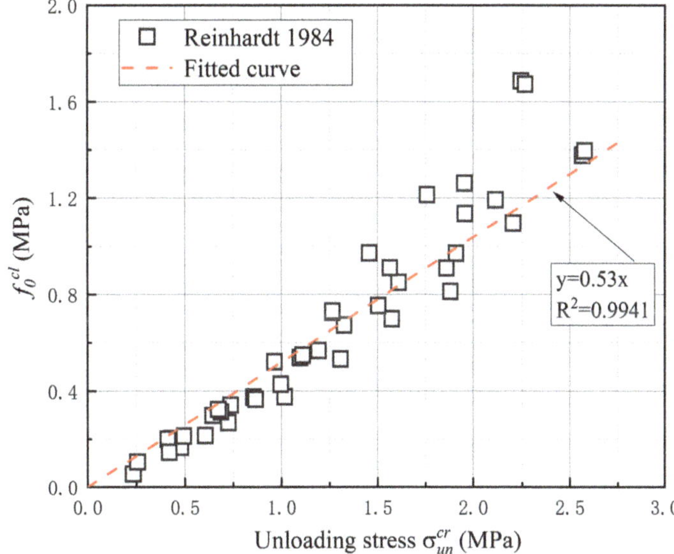

Figure 11. The relationship between the value of f_0^{cl} and the corresponding unloading stress [40].

Based on the comparison between Figures 10 and 11, it is more reasonable to relate the value of f_0^{cl} to the corresponding unloading stress. A fitted linear equation was obtained as shown in Figure 11 and the R-squared of the fitting is equal to 0.9941. The fitted line passes through the origin and the slope is defined as the initial closing coefficient $\alpha = f_0^{cl}/\sigma_{un}^{cr}$, which represents the ratio of the initial closing stress to the unloading stress. In the proposed model, the dimensionless parameter α is assumed to be a constant and consequently the value of f_0^{cl} is determined by the corresponding unloading stress σ_{un}^{cr}.

For the determination of the complete closing stress f_1^{cl}, there is no unified conclusion at present. In the literature [39], the value of f_1^{cl} was set as one-third of tensile strength based on the experimental results. In the literature [40], the value of f_1^{cl} was considered related to the degree of compressive damage in concrete.

3.5. Numerical Implementation of the Proposed Model

According to the hysteretic rules proposed in this paper, the unloading and reloading paths of the model can be implemented by modifying the cracking modulus within the framework of smeared crack theory. The numerical algorithm of the proposed model is implemented in a special finite element program using Matlab. The flow chart for updating the crack surface modulus is shown in Figure 12.

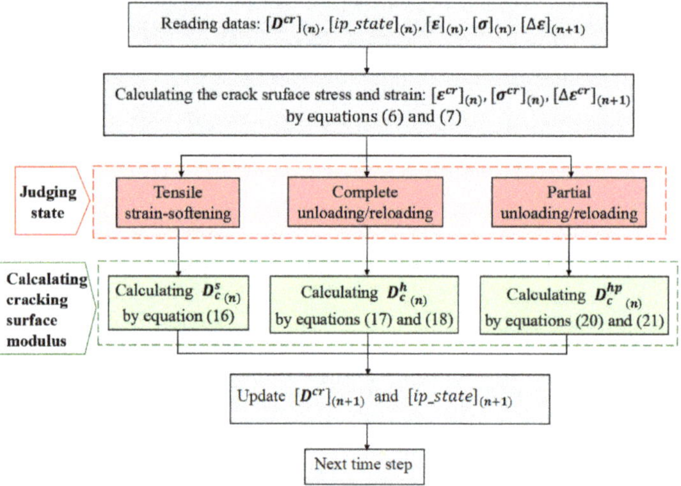

Figure 12. A flow chart of the computational implementation of the proposed model.

4. Simulation Results and Discussion

4.1. Cyclic Tests of Concrete

The concrete cyclic tests conducted by Reinhardt [40] were chosen in this study to validate the proposed model. The average tensile strength of concrete was 3.2 MPa, the average elastic modulus was 39.3 GPa and the fracture energy was 132 N/m. The specimens for cyclic tension and tension–compression tests were concrete prisms with dimensions of 240 mm × 220 mm × 50 mm. Two grooves with a depth of 20 mm were precut on both sides of the specimen along the thickness direction to control the initial position of cracking. The schematic diagram of the specimen is shown in Figure 13.

The cyclic loading procedure was displacement-controlled, and three cyclic loading programs T1, T2, and T3 with different stress ranges were adopted. The lowest stress in cyclic loading program T1 was the tensile stress with an amount equal to 5% of the tensile strength. The lowest stress in loading program T2 was the compressive stress with an amount equal to 15% of the tensile strength. The lowest stress in loading program T3 was the compressive stress with the amount equal to the tensile strength. Four extensometers with a gauge length of 35 mm were installed on the front and back of the specimen to measure the deformation during the cracking process. The arrangement of the measuring device is shown in Figure 13. Based on the measured test data, the complete stress-deformation curves during cyclic loading were obtained.

Figure 14 shows the history of deformation measured by the extensometers in the cyclic loading programs T1, T2, and T3. The decrease in deformation is defined as unloading and the increase in deformation is defined as loading. The extreme value points of the stress history are also plotted in the figure. It can be seen that under the three loading programs, the maximum value of stress decreases with cyclic loading, while the minimum value of stress is stable at the design value.

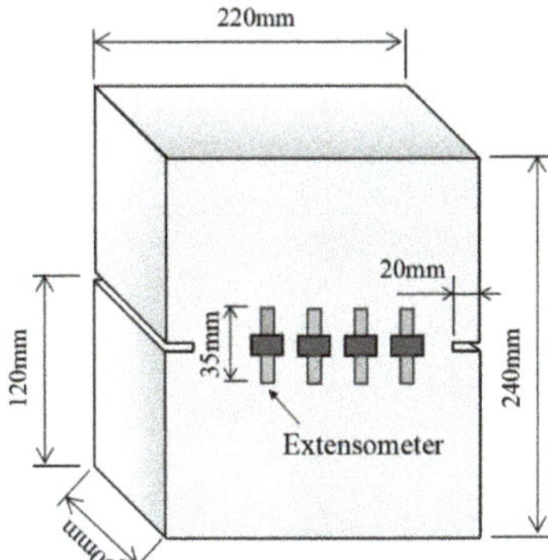

Figure 13. The schematic diagram of the concrete specimen.

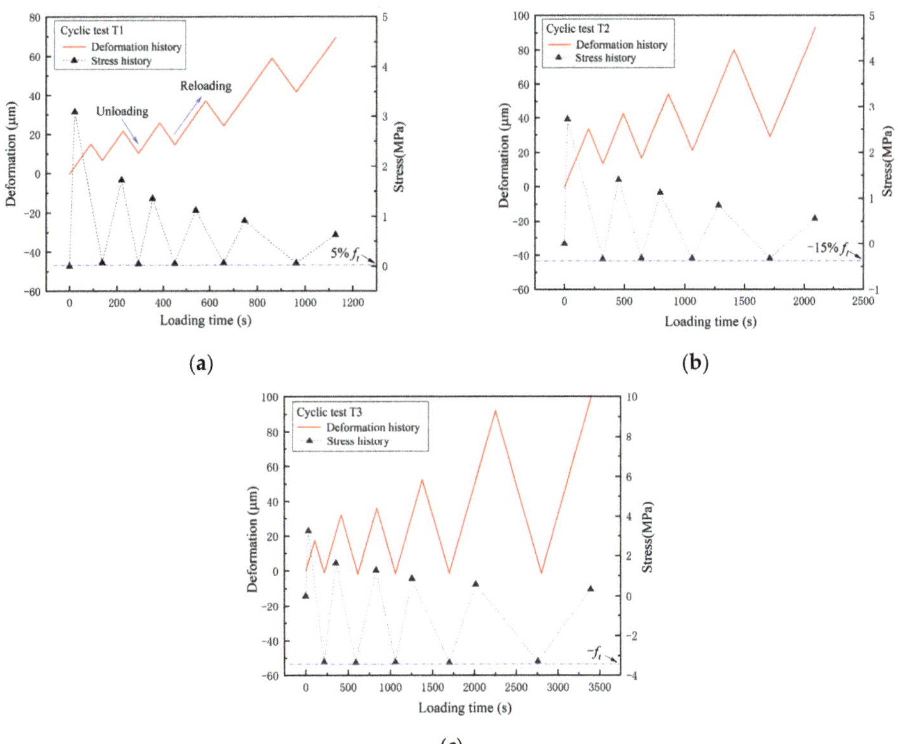

Figure 14. The cyclic loading history of programs (**a**) T1, (**b**) T2, and (**c**) T3.

4.2. Numerical Model and Parameters

According to the constraints and loading conditions, the specimen is always in a plane stress state during the cyclic loads, so the two-dimensional finite element model is adopted, as shown in Figure 15. The finite element mesh of the notched specimen is shown in the figure. The meshes consist of 528 nodes and 479 four-node linear elements, numerically integrated by means of the four-point Gaussian scheme. There is no constraint on both sides of the specimen. The bottom surface of the specimen is fixed, and the top surface is fixed in the x direction and moves in the y direction according to the prescribed displacement history.

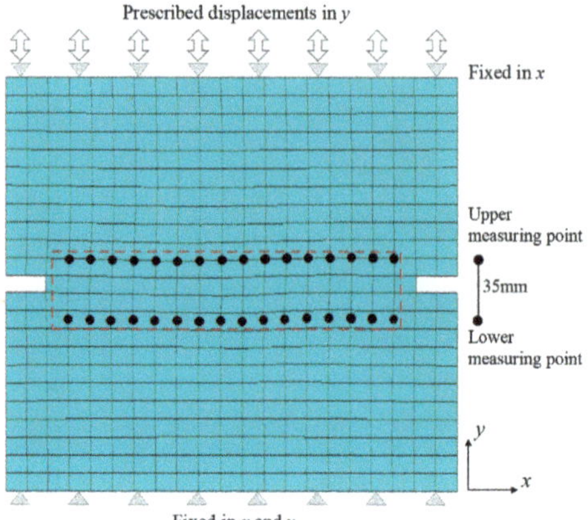

Figure 15. The finite element mesh of the notched specimen.

The following material properties are adopted for concrete: Young's modulus E_0 = 39.3 GPa, Poisson's ratio v = 0.2, tensile strength f_t =3.2 MPa, and fracture energy G_f = 132 N/m. The hysteretic model is introduced into the concrete constitutive relationship. Based on the experimental data of Reinhardt's test, the model parameters α and f_1^{cl} are set to be 0.53 and $f_t/3$, respectively.

The crack band theory is used to guarantee the proper energy dissipation during the fracturing process. A bilinear softening relationship is adopted in this paper. In order to reduce the dependency of results on finite element mesh sizes, the width of crack band h is calculated for each finite element as the element length projected into the normal crack direction [41], as depicted in Figure 16.

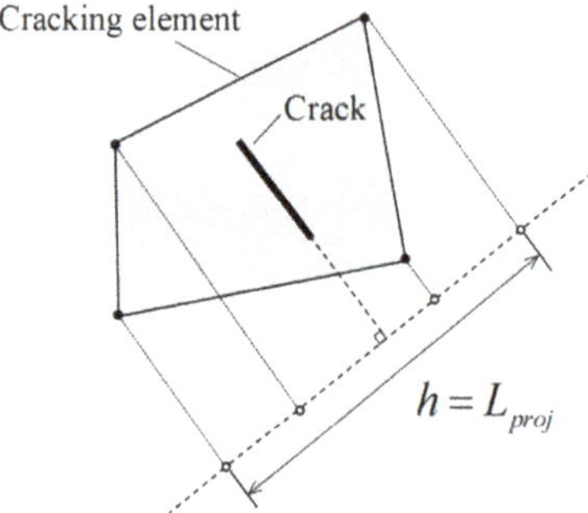

Figure 16. Calculation method of crack band width.

4.3. Simulation Results of Direct Tension Test

4.3.1. The Stress-Deformation Curve

The test and simulation results of the tensile stress-deformation softening curve of concrete are shown in Figure 17. The test results include data from monotonic tensile tests and the tensile-softening envelopes obtained from cyclic loading tests. The deformation in the test results was obtained by averaging the elongation measured by all the extensometers mounted on the surface of the specimens. Accordingly, the deformation of the simulation results was obtained by calculating the average longitudinal relative displacement of the area where the extensometers were distributed. The simulated stress-deformation curve reproduces the nonlinear characteristics during the stress rise phase and the softening characteristics after the peak stress. A good agreement is observed between the simulated curve and the test curves.

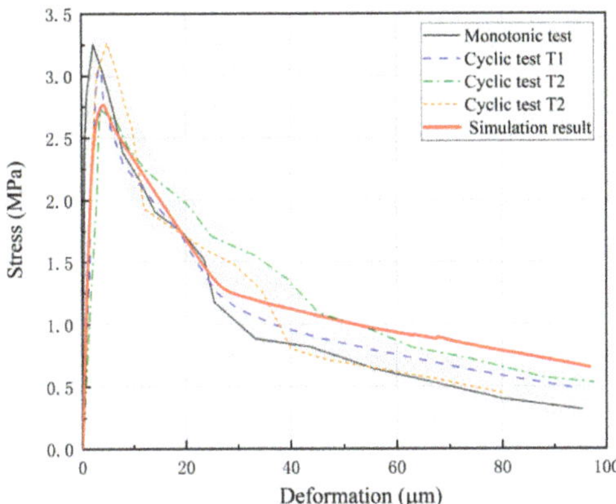

Figure 17. Comparison of the stress-deformation softening curves.

4.3.2. Evolution of Deformation Distribution

Figure 18 shows the test and simulation results of the evolution of the deformation distribution in the monotonic tensile test. The deformation distribution curves along the cross section at the groove in Figure 18b correspond to the characteristic points marked on the tensile stress-deformation curves in Figure 18a. Eight test curves and eleven simulation curves were plotted to compare the evolution characteristics of the deformation distribution during the tensile test.

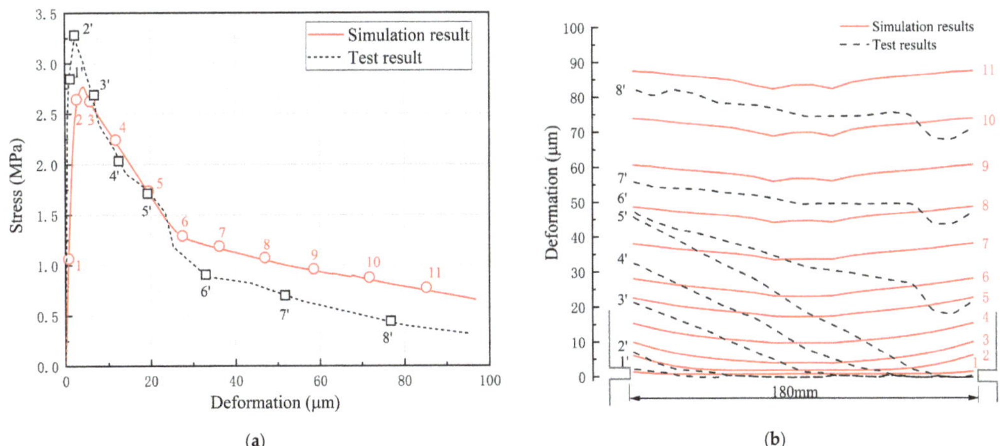

Figure 18. Comparison of deformation distribution curves: (**a**) the characteristic points and (**b**) the tensile stress distribution curves.

In the simulation results, it can be seen that the deformation distribution curves are always symmetrical. In the elastic stage of initial loading (point 1), the deformation is evenly distributed along the section of the groove. As the tensile stress increases, the deformation near the groove increases rapidly, and the deformation distribution curve evolves into a "U" shape before the stress reaches the peak (point 2). The large deformation near the groove indicates the initiation of the crack. When the stress-deformation curve enters the softening stage, the deformation increases rapidly and gradually evolves to an approximatively uniform distribution (points 3–11). The evolution characteristics of the simulated deformation distribution curves show that the crack originated from the groove at both ends and extended to the middle simultaneously.

In the test results, an asymmetric distribution of deformation is observed, which differs significantly from the simulated results. From initial loading to the first stage of softening, the deformation near the groove on one side increases rapidly, but the deformation near the groove on the other side is almost zero (points $1'$ to $5'$). It is not until the stress deformation curve enters the second stage of softening that the deformation gradually evolves to an approximatively uniform distribution (points $6'$ to $8'$). The evolution of the test deformation distribution curves shows that the crack initiated from one side of the groove and gradually expanded to the other side. This result may be due to the unevenness of the tensile force applied to the specimen during the test.

4.3.3. Simulation Results of Stress and Strain Distribution

The simulation results of the contour plots of stress and strain in the y-direction corresponding to state points 1, 3, and 5 in Figure 18a are shown in Figures 19–21.

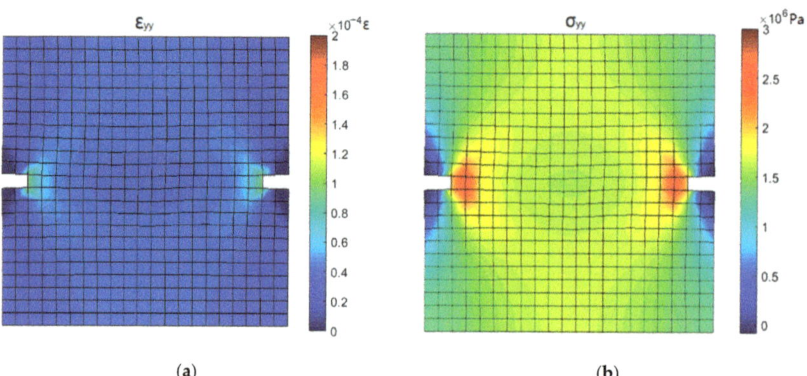

(a)   (b)

Figure 19. Stress and strain contours corresponding to point 1: (**a**) y-directional strain and (**b**) y-directional stress.

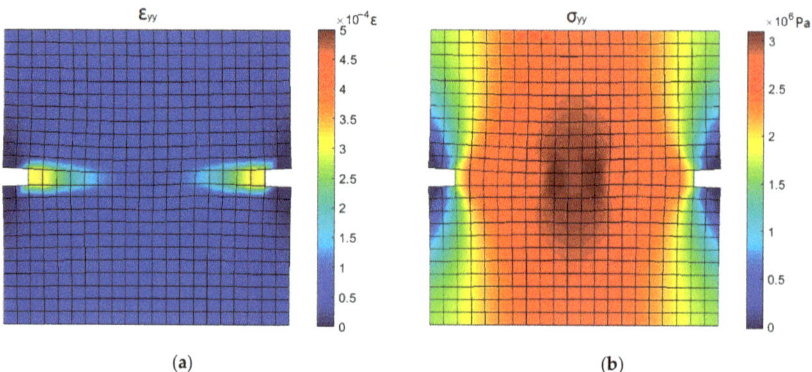

(a)   (b)

Figure 20. Stress and strain contours corresponding to point 3: (**a**) y-directional strain and (**b**) y-directional stress.

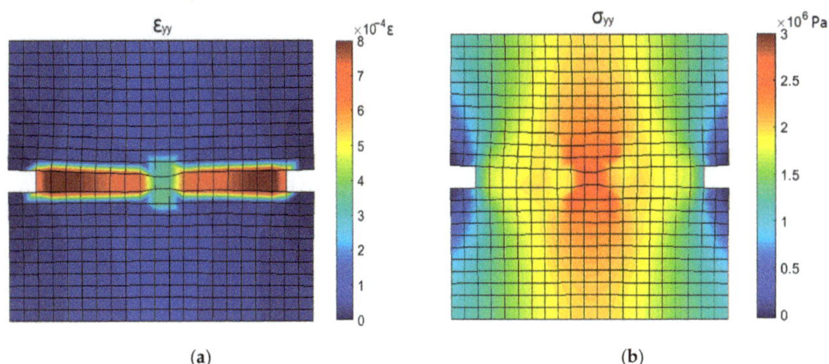

Figure 21. Stress and strain contours corresponding to point 5: (**a**) *y*-directional strain and (**b**) *y*-directional stress.

Figure 19 shows the stress and strain contours corresponding to the linear elastic stage of the stress-deformation curve (point 1). A significant concentration of stress and strain is observed near the groove. At the top and bottom ends of the specimen, the stress in the middle is obviously higher than that on both sides due to the uneven resistance caused by the groove.

Figure 20 shows the stress and strain contours when the stress crosses the peak and begins to decrease (point 3). The strain at the grooves increases rapidly and the high strain zone is extending from the groove to the middle of the specimen. The stress near the groove decreases due to the occurrence of a crack and the position of the maximum stress shifts to the middle of the specimen. High stress is spread throughout the specimen, and the load carried by the specimen reaches the maximum value.

Figure 21 shows the stress and strain contours corresponding to the softening stage (point 5). The strain at the groove section increases rapidly and a distinct "crack band" is observed almost across the groove section. As the crack extends toward the middle of the specimen, the stress near the groove decreases, and an "X"-shaped high-stress zone is observed in the middle of the specimen.

4.4. Simulation Results of Cyclic Tension–Compression Test

4.4.1. The Stress-Deformation Curves

Figure 22 shows comparisons between the test and simulation results of cyclic tests. Figure 22a shows the test and simulation results of the stress-deformation curves under cyclic loading condition T1. It can be seen that the unloading and reloading curves do not coincide, and they enclose a spindle-shaped closed loop, which is commonly known as the "hysteresis loop". The presence of the hysteresis loop indicates that there is additional energy dissipated during the unloading–reloading process. According to the mechanism of the hysteretic behavior of concrete, the dissipated energy is considered to be related to crack propagation and friction behavior of crack surfaces. The simulation results of the hysteresis loop during the first softening stage agree very well with the test results. During the second softening stage, deviations between simulated hysteresis loops and test results are observed, which are caused by the offset of the simulated softening envelope. Although the unloading and reloading paths in the model are specified as linear segments, the simulated stress-deformation curve exhibits nonlinear characteristics similar to the test results.

Figure 22b shows the test and simulation results of the stress-deformation curves under cyclic loading condition T2. In this cyclic test, the loading direction changes during the unloading and reloading process, and the stress state alternates between tension and compression. The hysteresis loops under the loading condition T2 are more significant compared to the loading condition T1, indicating that more energy is dissipated during the unloading and reloading process. As can be seen, the four hysteresis loops predicted by the model are in good agreement with the test results.

Figure 22c shows the test and simulation results of the stress-deformation curves under cyclic loading condition T3. In this condition, the compression stiffness recovers to the initial value when the compressive stress reaches the crack-closing stress. The stress-deformation curve presents complete unloading and reloading paths with a large amount of energy dissipated. It can be seen that the simulation results of the hysteresis loop are greatly influenced by the simulation accuracy of the stress-deformation envelope. Because the unrecoverable tensile deformation is not considered in the model, the simulated curve in the compression zone deviates from the test results.

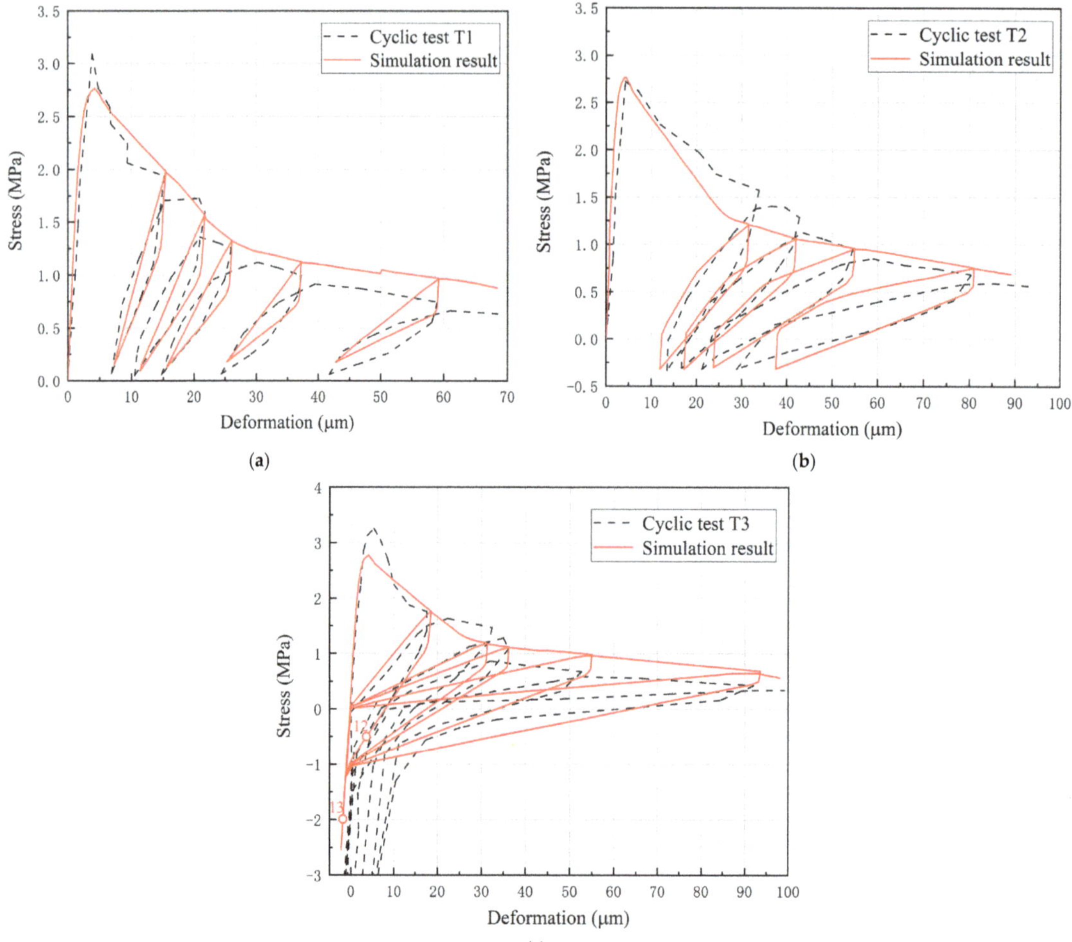

Figure 22. Comparison between test and simulation results of (**a**) cyclic test T1, (**b**) cyclic test T2, and (**c**) cyclic test T3.

4.4.2. Stress and Strain Contours of Crack-Closure Process

In cyclic test T3, with the increase in compressive stress, the open crack surfaces gradually closed. The stress and strain contours corresponding to the state points 12 and 13 during the process of crack closure in Figure 22c are shown in Figures 23 and 24.

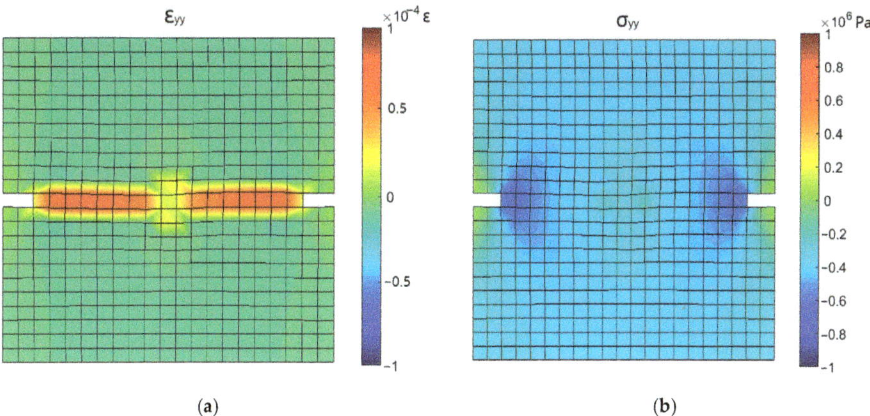

(a)　　　　　　　　　　　　　　　　　　　(b)

Figure 23. Stress and strain contours corresponding to point 12: (**a**) y-directional strain and (**b**) y-directional stress.

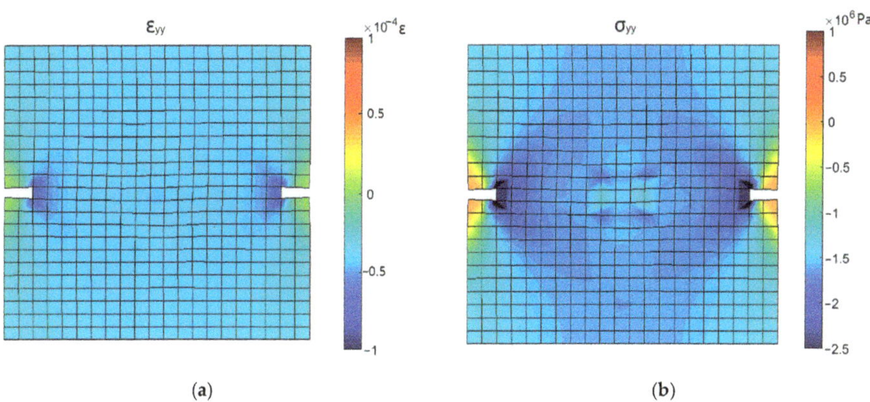

(a)　　　　　　　　　　　　　　　　　　　(b)

Figure 24. Stress and strain contours corresponding to point 13: (**a**) y-directional strain and (**b**) y-directional stress.

During the initial stage of reversed compression (point 12), the specimen is in compression and the concentration of compressive stress is observed near the groove. Since the crack is not completely closed, the cross-section at the groove still retains a large tensile strain, while the tensile strain of other parts has been released. When the compressive stress increases to 2 MPa (point 13), the compressive strain is observed in the cross-section at the groove, indicating the complete closure of the crack surface. The distribution of compressive strain in the specimen is the same as that of compressive stress, which means that the sample entered the state of elastic compression.

4.4.3. Evolution of Stiffness and Dissipated Energy

The test and simulated results of the reloading stiffness ratio are compared in Figure 25a. The reloading stiffness ratio is defined as the ratio of the reloading stiffness E to the initial loading stiffness E_0 in the cyclic load test, as illustrated in Figure 25a. The reloading stiffness ratios in the cyclic load tests T1, T2, and T3 were calculated and summarized in Figure 25a. It can be seen that the reloading stiffness ratio decreases rapidly with the increase in deformation. When the deformation reaches 20 μm, the reloading stiffness declines to 15% of the initial loading stiffness. When the deformation reaches 50 μm, the reloading stiffness declines to less than 5% of the initial loading stiffness. The simulated reloading stiffness ratio presents good agreement with the test results.

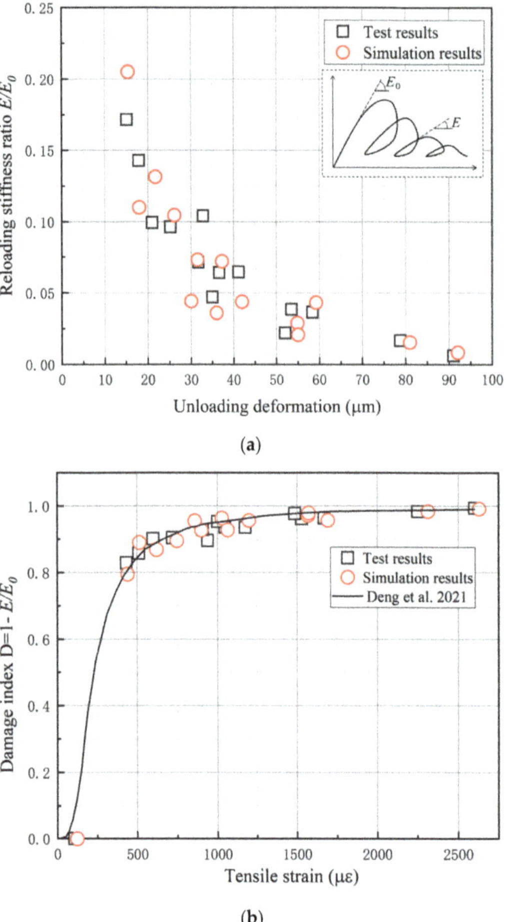

Figure 25. The test and simulation results of (**a**) reloading stiffness ratio and (**b**) damage index [42].

According to the continuum damage mechanics theory, the damage of material can be characterized by reloading stiffness ratio. Figure 25b shows the test and simulation results of the damage corresponding to different unloading strains in the cyclic tests. The unloading tensile strain in the figure was obtained by dividing the unloading deformation by the extensometer length of 35 mm. In addition, the damage at peak strain is assumed to be zero. As can be seen from the figure, the damage increases rapidly after entering the softening stage, reaching 0.8 at 500 $\mu\varepsilon$. After that, the growth rate gradually decreases and gradually tends to 1. As a comparison, the damage evolution curve based on the analytic envelope expression in the literature [42] was also drawn in the figure. It can be seen that the damage simulation results at different unloading strains are very consistent with the test data and the damage evolution curve. This indicates that the hysteretic model has a good ability to predict the damage evolution feature under cyclic tension.

Further comparison is performed to validate the proposed model. The predicted dissipated energy during each cycle of the tests T1, T2, and T3 is compared with the test results in Table 1. The magnitude of the dissipated energy of each cycle was obtained by calculating the area of the corresponding hysteresis loop. For the cyclic tests T2 and T3, the hysteretic dissipated energy increases with the increase in the cyclic ordinal number (i.e., unloading deformation). Good agreement is observed between the dissipated energy results of the test and simulation. In the case of cyclic tests T1 and T3, the relative errors of the total cumulative dissipated energy are only 2.37% and 7.05%, respectively.

Table 1. Comparisons of test and simulated dissipated energy results.

Cycle Number	Cyclic Test T1			Cyclic Test T2			Cyclic Test T3		
	Test (10^{-3} N/mm)	Simulation (10^{-3} N/mm)	Error (%)	Test (10^{-3} N/mm)	Simulation (10^{-3} N/mm)	Error (%)	Test (10^{-3} N/mm)	Simulation (10^{-3} N/mm)	Error (%)
1	2.45	2.44	−0.16	7.37	9.37	−21.4	15.68	15.23	−2.94
2	3.31	2.96	−11.6	7.74	11.32	−31.6	25.11	22.54	−11.4
3	2.12	2.45	13.3	8.75	13.43	−34.8	27.1	26.27	−3.18
4	2.28	2.24	−1.66	12.39	17.33	−28.5	32.77	40.5	19.1
5	2.87	2.62	−9.27	-	-	-	51.05	58.69	13.0
Total	13.03	12.72	−2.37	36.24	51.44	−29.5	151.71	163.22	7.05

5. Conclusions

Based on the smeared crack theory, this paper presents a hysteretic model to describe the nonlinear constitutive relationship of concrete under cyclic tension and tension–compression reversals. The numerical predictions of the proposed model are compared with several test results, and the following conclusions are obtained.

(1) By modifying the cracking modulus, the opening–closing behavior of the crack surface that produces the hysteretic phenomenon of concrete under cyclic tension–compression is directly modeled in the framework of the smeared crack theory.

(2) The proposed model is able to reproduce the hysteretic curves of concrete under complex tensile cyclic load conditions, as well as the degradation of reloading stiffness, energy dissipation, and stiffness recovery due to the crack closure.

(3) The model can simulate the initiation and propagation of concrete cracks under uniaxial cyclic tensile load, as well as the opening and closing behavior of the crack surface during the unloading–reloading process.

(4) The model adopts linear unloading–reloading paths and has only two parameters, which make the model easy to use and suitable for introduction into finite element programs.

(5) The model was verified by comparing the results with several cyclic tests. In addition, the proposed model can be applied to the nonlinear analysis of real concrete structures.

Author Contributions: Conceptualization, P.Z.; Methodology, P.Z.; Validation, S.W.; Investigation, S.W. and L.H.; Writing—original draft, P.Z.; Visualization, L.H.; Funding acquisition, P.Z. All authors have read and agreed to the published version of the manuscript.

Funding: This research was funded by the National Natural Science Foundation of China (52108249).

Data Availability Statement: Some or all data, models, or code that support the findings of this study are available from the corresponding author upon reasonable request.

Acknowledgments: The financial supports provided by the National Natural Science Foundation of China (52108249) are gratefully acknowledged.

Conflicts of Interest: The authors declare that they have no known competing financial interests or personal relationships that could have appeared to influence the work reported in this paper.

References

1. Sharma, R.; Ren, W.; Mcdonald, S.A.; Yang, Z.J. Micro-mechanisms of concrete failure under cyclic compression: X-ray tomographic in-situ observations. In Proceedings of the 9th International Conference on Fracture Mechanics of Concrete and Concrete Structures (FraMCoS-9), Berkeley, CA, USA, 29 May 2016.
2. Li, B.; Xu, L.; Chi, Y.; Huang, B.; Li, C. Experimental investigation on the stress-strain behavior of steel fiber reinforced concrete subjected to uniaxial cyclic compression. *Constr. Build. Mater.* **2017**, *140*, 109–118. [CrossRef]
3. Isojeh, B.; El-Zeghayar, M.; Vecchio, F.J. Concrete damage under fatigue loading in uniaxial compression. *ACI Mater. J.* **2017**, *114*, 225–235. [CrossRef]
4. Isojeh, B.; El-Zeghayar, M.; Vecchio, F.J. Simplified constitutive model for fatigue behavior of concrete in compression. *J. Mater. Civ. Eng.* **2017**, *29*, 04017028. [CrossRef]
5. Sima, J.F.; Roca, P.; Molins, C. Cyclic constitutive model for concrete. *Eng. Struct.* **2008**, *30*, 695–706. [CrossRef]
6. De Maio, U.; Greco, F.; Leonetti, L.; Blasi, P.N.; Pranno, A. A cohesive fracture model for predicting crack spacing and crack width in reinforced concrete structures. *Eng. Fail. Anal.* **2022**, *139*, 106452. [CrossRef]
7. Foster, S.J.; Marti, P. Cracked membrane model: Finite element implementation. *J. Struct. Eng.* **2003**, *129*, 1155–1163. [CrossRef]
8. Grassl, P.; Jirásek, M. Damage-plastic model for concrete failure. *Int. J. Solids Struct.* **2006**, *43*, 7166–7196. [CrossRef]
9. Zhang, J.; Li, J.; Ju, J.W. 3D elastoplastic damage model for concrete based on novel decomposition of stress. *Int. J. Solids Struct.* **2016**, *94*, 125–137. [CrossRef]
10. Zhang, L.; Zhao, L.H.; Liu, Z.; Jia, M. An elastic-plastic damage constitutive model of concrete under cyclic loading and its numerical implementation. *Eng. Mech.* **2023**, *40*, 152–161. [CrossRef]
11. Yankelevsky, D.Z.; Reinhardt, H.W. Uniaxial behavior of concrete in cyclic tension. *J. Struct. Eng.* **1989**, *115*, 166–182. [CrossRef]
12. Chang, G.A.; Mander, J.B. *Seismic Energy Based Fatigue Damage Analysis of Bridge Columns: Part 1—Evaluation of Seismic Capacity*; NCEER Technical Report NCEER-94-6; University at Buffalo: Buffalo, NY, USA, 1994.
13. Aslani, F.; Jowkarmeimandi, R. Stress–strain model for concrete under cyclic loading. *Mag. Concr. Res.* **2012**, *64*, 673–685. [CrossRef]
14. Chen, X.D.; Bu, J.B.; Xu, L.Y. Effect of strain rate on post-peak cyclic behavior of concrete in direct tension. *Constr. Build. Mater.* **2016**, *124*, 746–754. [CrossRef]
15. McCall, K.R.; Guyer, R.A. Equation of state and wave propagation in hysteretic nonlinear elastic materials. *J. Geophys. Res. Solid Earth* **1994**, *99*, 23887–23897. [CrossRef]
16. Chen, X.D.; Huang, Y.B.; Chen, C.; Lu, J.; Fan, X.Q. Experimental study and analytical modeling on hysteresis behavior of plain concrete in uniaxial cyclic tension. *Int. J. Fatigue* **2017**, *96*, 261–269. [CrossRef]
17. Chen, X.D.; Xu, L.Y.; Bu, J.W. Dynamic tensile test of fly ash concrete under alternating tensile–compressive loading. *Mater. Struct.* **2018**, *51*, 7667–7674. [CrossRef]
18. Long, Y.C.; He, Y.M. An anisotropic damage model for concrete structures under cyclic loading-uniaxial modeling. *J. Phys. Conf. Ser.* **2017**, *842*, 012040. [CrossRef]
19. Long, Y.C.; Yu, C.T. Numerical Simulation of the Damage Behavior of a Concrete Beam with an Anisotropic Damage Model. *Strength Mater.* **2018**, *50*, 735–742. [CrossRef]
20. Liu, Z.; Zhang, L.; Zhao, L.; Wu, Z.; Guo, B. A Damage Model of Concrete including Hysteretic Effect under Cyclic Loading. *Materials* **2022**, *15*, 5062. [CrossRef] [PubMed]
21. Tudjono, S.; Lie, H.A.; As'ad, S. Reinforced concrete finite element modeling based on the discrete crack approach. *Civ. Eng. Dimens.* **2016**, *18*, 72–77. [CrossRef]
22. Areias, P.; Rabczuk, T.; de Sá, J.C. A novel two-stage discrete crack method based on the screened Poisson equation and local mesh refinement. *Comput. Mech.* **2016**, *58*, 1003–1018. [CrossRef]
23. Fujiwara, Y.; Takeuchi, N.; Shiomi, T.; Kambayashi, A. Discrete crack analysis for concrete structures using the hybrid-type penalty method. *Comput. Concr.* **2015**, *16*, 587–604. [CrossRef]
24. Shi, Z.; Nakano, M.; Nakamura, Y.; Liu, C. Discrete crack analysis of concrete gravity dams based on the known inertia force field of linear response analysis. *Eng. Fract. Mech.* **2014**, *115*, 122–136. [CrossRef]
25. Thybo, A.E.A.; Michel, A.; Stang, H. Smeared crack modelling approach for corrosion-induced concrete damage. *Mater. Struct.* **2017**, *50*, 146. [CrossRef]
26. Edalat-Behbahani, A.; Barros, J.A.; Ventura-Gouveia, A. Three dimensional plastic-damage multidirectional fixed smeared crack approach for modelling concrete structures. *Int. J. Solids Struct.* **2017**, *115*, 104–125. [CrossRef]
27. Edalat-Behbahani, A.; Barros, J.; Ventura-Gouveia, A. Application of plastic-damage multidirectional fixed smeared crack model in analysis of RC structures. *Eng. Struct.* **2016**, *125*, 374–391. [CrossRef]
28. Amir, P. Seismic improvement of gravity dams using isolation layer in contact area of dam–reservoir in smeared crack approach. *Iran. J. Sci. Technol. Civ. Eng.* **2018**, *43*, 137–155. [CrossRef]
29. Rots, J.G.; Nauta, P.G.; Kuster, G.; Blaauwendraad, J. Smeared crack approach and fracture localization in concrete. *Heron* **1985**, *30*, 1–48.
30. Bažant, Z.P.; Oh, B.H. Crack band theory for fracture of concrete. *Mater. Struct.* **1983**, *16*, 155–177. [CrossRef]
31. Wang, G.; Pekau, O.A.; Zhang, C.; Wang, S. Seismic fracture analysis of concrete gravity dams based on nonlinear fracture mechanics. *Eng. Fract. Mech.* **2000**, *65*, 67–87. [CrossRef]

32. Jirásek, M.; Bauer, M. Numerical aspects of the crack band approach. *Comput. Struct.* **2012**, *110-111*, 60–78. [CrossRef]
33. Xu, W.; Waas, A.M. Modeling damage growth using the crack band model; effect of different strain measures. *Eng. Fract. Mech.* **2016**, *152*, 126–138. [CrossRef]
34. Reinhardt, H.W.; Cornelissen, H. Post-peak cyclic behaviour of concrete in uniaxial tensile and alternating tensile and compressive loading. *Cem. Concr. Res.* **1984**, *14*, 263–270. [CrossRef]
35. Reinhardt, H.W.; Cornelissen, H.; Hordijk, D.A. Tensile tests and failure analysis of concrete. *J. Struct. Eng.* **1986**, *112*, 2462–2477. [CrossRef]
36. Mazars, J.; Berthaud, Y.; Ramtani, S. The unilateral behaviour of damaged concrete. *Eng. Fract. Mech.* **1990**, *35*, 629–635. [CrossRef]
37. Zheng, F.G.; Wu, Z.R.; Gu, C.S.; Bao, T.F.; Hu, J. A plastic damage model for concrete structure cracks with two damage variables. *Sci. China Technol. Sci.* **2012**, *55*, 2971–2980. [CrossRef]
38. Nouailletas, O.; La Borderie, C.; Perlot, C.; Rivard, P.; Ballivy, G. Experimental Study of Crack Closure on Heterogeneous Quasi-Brittle Material. *J. Eng. Mech.* **2015**, *141*, 1–11. [CrossRef]
39. Zhang, P.; Ren, Q.; Lei, D. Hysteretic model for concrete under cyclic tension and tension-compression reversals. *Eng. Struct.* **2018**, *163*, 388–395. [CrossRef]
40. Reinhardt, H.W. Fracture Mechanics of an Elastic Softening Material Like Concrete. *Heron* **1984**, *29*, 1–42.
41. Červenka, J.; Červenka, V.; Laserna, S. On crack band model in finite element analysis of concrete fracture in engineering practice. *Eng. Fract. Mech.* **2018**, *197*, 27–47. [CrossRef]
42. Deng, F.Q.; Chi, Y.; Xu, L.H.; Huang, L.; Hu, X. Constitutive behavior of hybrid fiber reinforced concrete subject to uniaxial cyclic tension: Experimental study and analytical modeling. *Constr. Build. Mater.* **2021**, *295*, 123650. [CrossRef]

materials

MDPI

Article

A Modified Bond-Associated Non-Ordinary State-Based Peridynamic Model for Impact Problems of Quasi-Brittle Materials

Jing Zhang [1,2], Yaxun Liu [1,2], Xin Lai [1,2,*], Lisheng Liu [1,2,*], Hai Mei [1,2] and Xiang Liu [1,2]

[1] Hubei Key Laboratory of Theory and Application of Advanced Materials Mechanics, Wuhan University of Technology, Wuhan 430070, China
[2] Department of Engineering Structure and Mechanics, Wuhan University of Technology, Wuhan 430070, China
* Correspondence: laixin@whut.edu.cn (X.L.); liulish@whut.edu.cn (L.L.)

Abstract: In this work, we have developed a novel bond-associated non-ordinary state-based peridynamic (BA-NOSB PD) model for the numerical modeling and prediction of the impact response and fracture damage of quasi-brittle materials. First, the improved Johnson-Holmquist (JH2) constitutive relationship is implemented in the framework of BA-NOSB PD theory to describe the nonlinear material response, which also helps to eliminate the zero-energy mode. Afterwards, the volumetric strain in the equation of state is redefined by the introduction of the bond-associated deformation gradient, which can effectively improve the stability and accuracy of the material model. Then, a new general bond-breaking criterion is proposed in the BA-NOSB PD model, which is capable of covering various failure modes of quasi-brittle materials, including the tensile-shear failure that is not commonly considered in the literature. Subsequently, a practical bond-breaking strategy and its computational implementation are presented and discussed by means of energy convergence. Finally, the proposed model is verified by two benchmark numerical examples and demonstrated by the numerical simulation of edge-on impact and normal impact experiments on ceramics. The comparison between our results and references shows good capability and stability for impact problems of quasi-brittle materials. Numerical oscillations and unphysical deformation modes are effectively eliminated, showing strong robustness and bright prospects for relevant applications.

Keywords: peridynamics; impact response of ceramics; zero-energy mode; kinetic energy; bond-breaking criterion

Citation: Zhang, J.; Liu, Y.; Lai, X.; Liu, L.; Mei, H.; Liu, X. A Modified Bond-Associated Non-Ordinary State-Based Peridynamic Model for Impact Problems of Quasi-Brittle Materials. *Materials* **2023**, *16*, 4050. https://doi.org/10.3390/ma16114050

Academic Editors: Andres Sotelo and A. Javier Sanchez-Herencia

Received: 2 April 2023
Revised: 13 May 2023
Accepted: 26 May 2023
Published: 29 May 2023

1. Introduction

In recent decades, ceramic materials have attracted great attention from researchers and engineers and have been extensively adopted for applications in the fields of aerospace, military equipment, biomedicine, and industrial manufacturing, due to its excellent performance, such as high hardness, high specific strength, high specific modulus, high temperature resistance and high abrasion resistance. In the initial stage, ceramics are applied in the field of armor protection with great success due to their excellent anti-penetration properties [1], which has great prospects. In military and protection applications, ceramics are often subjected to extreme loading conditions [2], such as shock waves from explosions or projectile penetration. Therefore, extensive research and testing have been devoted to the understanding of the fundamental nature and mechanism of the impact response and anti-penetration behavior of ceramic materials. Various approaches and numerical simulations have been carried out to study the damage and fracture of ceramics [3–5], including but not limited to the finite element method (FEM), cohesive zone model (CZM), and extended finite element method (XFEM). However, due to the mesh dependence, the damage or fracture of materials leads to grid distortion, and the calculation cannot continue in mesh-based approaches. On the other hand, the cohesive zone model only allows cracks to propagate along element boundaries, which may not be able to accurately describe the crack path. In XFEM, although cracks can propagate along any direction, the introduction of additional degrees of freedom brings considerable computational complexity. Damage or fracture will result in discontinuity of the displacement field, which creates inherent limitations for the above numerical simulation methods based on continuum mechanics. To date, it is still challenging for numerical approaches to give a precise prediction of the complex fracture and damage patterns on ceramics.

To naturally simulate the discontinuous behavior of materials such as damage and cracking, Silling [6] proposed the peridynamic (PD) theory with nonlocal interactions as the core. This theory is also called bond-based PD (BB PD). In the PD theory, a body is described by a set of discretized material points with mass and other physical properties. The description of macroscopic material properties is formed by nonlocal bonding and interaction between material points. The damage of materials is part of PD theory, which is described by breaking bonds and truncating interactions between material points. PD is a theory of nonlocal continuum mechanics, which uses integral equations instead of integral–differential equations and does not require the displacement field to satisfy the requirement of continuous derivability. Therefore, the PD theory is well suited to modeling discontinuity problems without the difficulties encountered by classical local theory. Further, Silling [7] successfully proved the ability of PD theory to model complex fracture problems using the Kalthoff–Winkler experiment. In addition, Silling and Askari [8] developed the PD prototype microelastic brittle (PMB) material model and its numerical implementation technology, and derived the critical stretch formulation. The BB PD theory proposed earlier does not include stress and strain measurements, encountered the restriction of fixed Poisson's ratio, and cannot describe the shear deformation of materials, which had certain limitations. Hence, Silling et al. [9] proposed the state-based PD (SB PD) theory, which extends the PD theory and the material responses it can describe. The theory can be divided into the ordinary state-based PD (OSB PD) and the non-ordinary state-based PD (NOSB PD), depending on the form of interactions.

The PD theory has significant advantages in modeling discontinuity problems, and is widely used in simulating material damage, cracks, and fractures. Some promising applications of the PD theory in damage prediction and fracture modeling are outlined below. Ha and Bobaru [10,11] reproduced the crack growth and fracture behavior of glass materials in dynamic tensile experiments. Zhu and Ni [12] proposed the bond rotation effect in BB PD, which removed the limitation of Poisson's ratio. Chu et al. [13] presented a rate-dependent BB PD model to simulate the anti-penetration behavior of ceramics. Liu et al. [14] proposed a comprehensive tensile-shear and compressive-shear failure criterion to evaluate the impact damage and fracture characteristics of ceramics. Erdogan and Oterkus [15] proposed an OSB PD model based on the Mises yield criterion. Zhang and Qiao [16] established a new bond-breaking criterion based on the critical relative rotation angle. Foster et al. [17] implemented an elastic viscoplastic constitutive relationship in NOSB PD and reproduced the Taylor impact test of 6061-T6 aluminum. O'Grady and Foster [18] developed a Euler–Bernoulli beam model. Lai et al. [19,20] implemented the Drucker–Prager and JH2 constitutive models in NOSB PD to model dynamic brittle fracture of quasi-brittle materials. Wu et al. [21] introduced the Holmquist–Johnson–Cook (HJC) constitutive theory to analyze the dynamic mechanical behavior of concrete during impact. Wang et al. [22] studied the thermo-viscoplastic responses of metals under impact loads. Zhu and Zhao [23] simulated single-hole rock blasting with the JH2 damage and tensile failure models. Yang et al. [24] modeled the impact spalling of concrete based on the BA-NOSB PD model for quasi-brittle materials. Li et al. [25] presented a PD model using the Mindlin–Reissner shell theory and simulated the brittle fracture of thin shells.

Although the capability of the NOSB PD model incorporating constitutive models of classical continuum mechanics to simulate complex material responses has been demonstrated, there are still urgent problems to be addressed. The zero-energy mode is one of the major challenges in the NOSB PD theory, which leads to numerical oscillations, affects calculation accuracy, and even produces incorrect calculation results. To date, a number of control schemes and stabilization strategies have been proposed. For example, a stable supplemental force state [26–29] has been added to the original force vector state to control the zero-energy mode. Yaghoobi and Chorzepa [30] suppressed numerical oscillations by introducing a higher-order version of the conventional deformation gradient. The stress-point method [31] and the higher-order stress-point method [32] were also developed to remove numerical oscillations. In particular, zero-energy mode oscillations are believed to be caused by non-unique mapping relationships from the deformation state to the force vector state, due to the conventional deformation gradient. Therefore, Chen [33] proposed the bond-associated deformation gradient to suppress or eliminate numerical oscillations. The BA-NOSB PD model has also been proved to meet the requirements for stability and convergence [34]. Gu et al. [35] further studied and discussed the possible causes for numerical oscillations in the model. On the other hand, the treatment of broken bonds is also a key problem that deserves attention and needs to be addressed. For the SB PD theory, the force vector state at a material point depends on the deformation behavior of all bonds in its family over time. Therefore, the treatment of broken bonds has a crucial effect on the calculation of deformation states for unbroken bonds. This further affects the stability of the material model, and needs to be taken into account. Silling [27] and Li et al. [36] introduced damage to the influence function so that a broken bond was no longer included in subsequent calculations. In order to make better use of NOSB PD to simulate complex material responses, the material model stability and the treatment of broken bonds will be part of our work.

Following the above introduction, in Section 2, we first briefly review the original PD models for quasi-brittle materials. After that, a modified BA-NOSB PD model is established, and its constitutive update scheme is presented. Then, the other four zero-energy mode control schemes are introduced. Subsequently, the numerical discretization and implementation used in this paper are discussed. In Section 3, the modified model is verified using two benchmark examples, and numerical simulations of edge impact and normal impact experiments on ceramics are carried out. Finally, the conclusions of this work are summarized in Section 4.

2. Methodology

2.1. Brief Review of Peridynamic Models for Quasi-Brittle Materials

To date, a variety of theoretical models [13,14,19–21,23,24,37–39] have been developed for damage prediction and fracture modeling of quasi-brittle materials. In this section, two PD models for quasi-brittle materials [20,24] are briefly reviewed. First, the original and bond-associated NOSB PD theory and their basic formulas are presented in Sections 2.1.1 and 2.1.2, respectively. Then, the JH2 material model used for the constitutive update of the above two PD theories is reviewed in Section 2.1.3.

2.1.1. Original Non-Ordinary State-Based Peridynamics

In the PD theory, a body is modeled by a finite number of discretized material points. Further, the material point $\mathbf{X}$ interacts only with the material point $\mathbf{X}'$ within a limited range, and the limited range of influence is called horizon δ. The material point $\mathbf{X}'$ is called the neighbor of its center point $\mathbf{X}$, and all the neighbors of the material point $\mathbf{X}$ form the family $H_\mathbf{X}$, i.e., $H_\mathbf{X} = H(\mathbf{X}, \delta) = \{\mathbf{X}' \in \mathcal{B} : \{\|\mathbf{X}' - \mathbf{X}\| \leq \delta\}\}$. In the reference configuration, the reference position state is defined as

$$\underline{\mathbf{X}}\langle \boldsymbol{\xi} \rangle = \mathbf{X}' - \mathbf{X} = \boldsymbol{\xi}. \tag{1}$$

In the current configuration, the relative displacement state and the deformation state are defined as

$$\underline{\mathbf{U}}\langle \boldsymbol{\xi} \rangle = \mathbf{u}' - \mathbf{u} = \boldsymbol{\eta}, \tag{2}$$

$$\underline{\mathbf{Y}}\langle\xi\rangle = \mathbf{y}' - \mathbf{y} = \xi + \eta,\tag{3}$$

where $\mathbf{y}$ and $\mathbf{y}'$ are the spatial positions of the two material points, η is a relative displacement between them, and ξ is called a bond.

The NOSB PD theory [9] is a generalization of BB PD, which provides generality in modeling material responses, such as volumetric strain or shear deformation. In NOSB PD, the equation of motion for the material point $\mathbf{X}$ at the instant of t can be expressed as

$$\rho_0(\mathbf{X})\ddot{\mathbf{u}}(\mathbf{X},t) = \int_{H_\mathbf{X}}\left[\underline{\mathbf{T}}(\mathbf{X},t)\langle\xi\rangle - \underline{\mathbf{T}}(\mathbf{X}',t)\langle-\xi\rangle\right]\mathrm{d}V_{\mathbf{X}'} + \mathbf{b}(\mathbf{X},t),\tag{4}$$

in which $\rho_0(\mathbf{X})$ represents the mass density of materials, $\ddot{\mathbf{u}}(\mathbf{X},t)$ represents the acceleration, $\mathbf{b}(\mathbf{X},t)$ represents the external body force density, and $\underline{\mathbf{T}}(\mathbf{X},t)$ represents the force vector state. In particular, the force state $\underline{\mathbf{T}}(\mathbf{X},t)$ is a function of the deformation state $\underline{\mathbf{Y}}$.

In NOSB PD, the material response of a material point will depend on the deformation behavior of all bonds within its horizon. The mapping relationship between the deformation state and the force vector state is the constitutive model of materials in the PD framework. The force density vector at the material point $\mathbf{X}$ can be written as

$$\mathbf{t}(\mathbf{X},t) = \underline{\mathbf{T}}(\mathbf{X},t)\langle\xi\rangle = \underline{\omega}\langle\|\xi\|\rangle\mathbf{PK}^{-1}\underline{\mathbf{X}}\langle\xi\rangle,\tag{5}$$

in which $\underline{\omega}\langle\|\xi\|\rangle$ is the influence function on a bond.

The shape tensor $\mathbf{K}$ is defined as

$$\mathbf{K} = \int_{H_\mathbf{X}}\underline{\omega}\langle\|\xi\|\rangle\underline{\mathbf{X}}\langle\xi\rangle \otimes \underline{\mathbf{X}}\langle\xi\rangle\mathrm{d}V_{\mathbf{X}'}\tag{6}$$

The first Piola–Kirchhoff stress tensor can be calculated by classical constitutive models, as

$$\mathbf{P} = \mathcal{J}\sigma\mathbf{F}^{-\mathrm{T}},\tag{7}$$

in which $\mathcal{J} = \det(\mathbf{F})$, T is the matrix transposition operator, and σ is the Cauchy stress. The approximate nonlocal deformation gradient tensor $\mathbf{F}$ is defined as

$$\mathbf{F} = \left(\int_{H_\mathbf{X}}\underline{\omega}\langle\|\xi\|\rangle\underline{\mathbf{Y}}\langle\xi\rangle \otimes \underline{\mathbf{X}}\langle\xi\rangle\mathrm{d}V_{\mathbf{X}'}\right)\mathbf{K}^{-1}.\tag{8}$$

2.1.2. Bond-Associated Non-Ordinary State-Based Peridynamics

The zero-energy mode can lead to instability of the material model, which is manifested in numerical oscillations of solutions or completely erroneous calculation results. To eliminate the numerical oscillations caused by the zero-energy mode, Chen [33] proposed the BA-NOSB PD theory and proved the convergence and stability of the PD model [34].

In the original NOSB PD theory, the point-associated deformation gradient results in a non-unique mapping relationship from the deformation vector state to the force vector state. The definition of "bond-associated horizon" is presented to meet the requirement of the unique mapping relationship. The bond-associated horizon is defined as $H_\xi = H_\mathbf{X} \cap H_{\mathbf{X}'}$, as shown in Figure 1. In BA-NOSB PD, an individual bond corresponds to a unique deformation gradient tensor. Thus, the bond-associated deformation gradient tensor with respect to the bond ξ at the material point $\mathbf{X}$ is defined as

$$\mathbf{F}_\mathrm{b} = \left(\int_{H_\xi}\underline{\omega}\langle\|\xi\|\rangle\underline{\mathbf{Y}}\langle\xi\rangle \otimes \underline{\mathbf{X}}\langle\xi\rangle\mathrm{d}V_{\mathbf{X}'}\right)\mathbf{K}_\mathrm{b}^{-1},\tag{9}$$

where the subscript b indicates that the physical quantity is bond-associated. The bond-associated deformation gradients ensure the desired unique mapping relationship, and the bond-associated shape tensor can be defined as

$$\mathbf{K}_b = \int_{H_{\xi}} \underline{\omega}\langle\|\xi\|\rangle\underline{\mathbf{X}}\langle\xi\rangle \otimes \underline{\mathbf{X}}\langle\xi\rangle dV_{\mathbf{X}'}. \tag{10}$$

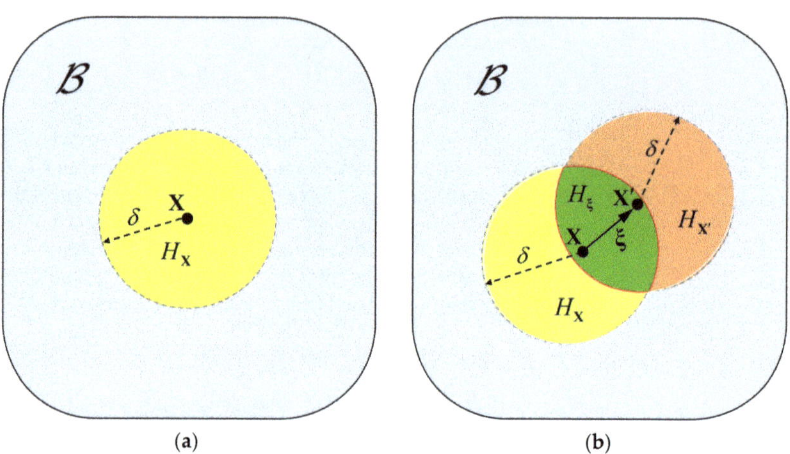

Figure 1. Definitions of different horizons. (**a**) Point-associated horizons; (**b**) bond-associated horizons.

The nonlocal deformation gradient tensor $\mathbf{F}$ is redefined based on Equation (9) by weighted averaging as

$$\mathbf{F} = \frac{\sum\limits_{n=1}^{N_b} \underline{\omega}\langle\|\xi^n\|\rangle\mathbf{F}_b^n}{\sum\limits_{n=1}^{N_b} \underline{\omega}\langle\|\xi^n\|\rangle}. \tag{11}$$

where N_b represents the number of bond-associated neighbors at the material point $\mathbf{X}$.

The bond-associated force density vector is recalculated as

$$\mathbf{t}_b(\mathbf{X}, t) = \underline{\mathbf{T}}_b(\mathbf{X}, t)\langle\xi\rangle = \frac{\int_{H_{\xi}} 1 dV_{\mathbf{X}'}}{\int_{H_{\mathbf{X}}} 1 dV_{\mathbf{X}'}}\underline{\omega}\langle\|\xi\|\rangle\mathbf{P}_b\mathbf{K}_b^{-1}\underline{\mathbf{X}}\langle\xi\rangle, \tag{12}$$

The bond-associated first Piola–Kirchhoff stress tensor is defined as

$$\mathbf{P}_b = \mathcal{J}_b\sigma_b\mathbf{F}_b^{-T}, \tag{13}$$

in which $\mathcal{J}_b = \det(\mathbf{F}_b)$, and σ_b is the bond-associated Cauchy stress tensor.

According to Gu's work [35], the bond-associated deformation gradient not only satisfies a unique mapping relation from the deformation state to the force vector state, but also satisfies the kinematic constraints of the bonds, as follows:

$$\begin{cases} \xi + \eta = \mathbf{F}_b(\mathbf{X}, \xi)\xi \\ \xi + \eta = \mathbf{F}_b(\mathbf{X}', -\xi)\xi \end{cases}. \tag{14}$$

Moreover, both the bond-associated shape tensors and the bond-associated deformation gradient tensors satisfy the symmetry, i.e., $\mathbf{K}_b(\mathbf{X}, \xi) = \mathbf{K}_b(\mathbf{X}', -\xi)$ and $\mathbf{F}_b(\mathbf{X}, \xi) = \mathbf{F}_b(\mathbf{X}', -\xi)$. The bond-associated first Piola–Kirchhoff stress tensors satisfies symmetry as well:

$$\mathbf{P}_b(\mathbf{X}, \xi) = \mathbf{P}_b(\mathbf{X}', -\xi). \tag{15}$$

As shown in Figure 2, similar to the BB PD model with rotation effect [12], the two bond-associated force density vectors in a bond are also equal in magnitude and opposite in direction, but not parallel to the bond, which can be written as

$$\underline{\mathbf{T}_b}(\mathbf{X},t)\langle \boldsymbol{\xi} \rangle = -\underline{\mathbf{T}_b}(\mathbf{X}',t)\langle -\boldsymbol{\xi} \rangle. \tag{16}$$

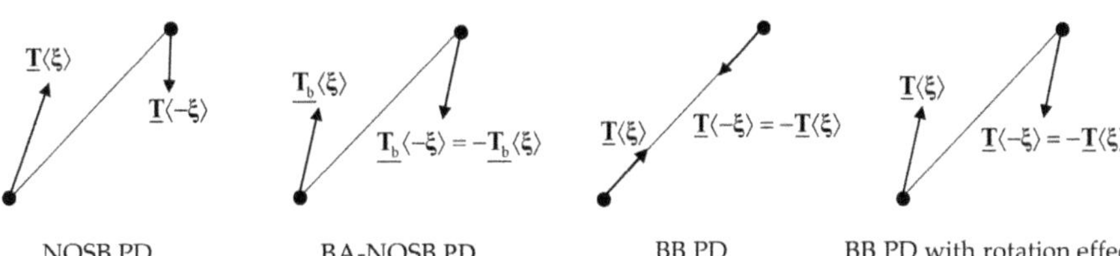

Figure 2. Force vector state for PD models.

Finally, in BA-NOSB PD, the equation of motion for the material point $\mathbf{X}$ at time t can be expressed as

$$\rho_0(\mathbf{X})\ddot{\mathbf{u}}(\mathbf{X},t) = \int_{H_\mathbf{X}} \left[\underline{\mathbf{T}_b}(\mathbf{X},t)\langle \boldsymbol{\xi} \rangle - \underline{\mathbf{T}_b}(\mathbf{X}',t)\langle -\boldsymbol{\xi} \rangle\right]\mathrm{d}V_{\mathbf{X}'} + \mathbf{b}(\mathbf{X},t). \tag{17}$$

2.1.3. JH2 Constitutive Model

The quasi-brittle materials also have important engineering applications, such as armored ceramics, bulletproof glass, concrete dams, and rock materials. The mechanical behavior and material response of these materials in different applications is an important basis for evaluating and measuring their safety in use and working life. To date, some constitutive models [40–44] describing quasi-brittle materials have been developed and used. The original Johnson–Holmquist (JH1) model [45] does not allow gradual softening, may damage self-healing, and is too sensitive to material model parameters. The JH2 model [46] is an improved version of the JH1 model, which is a widely used constitutive model for brittle or quasi-brittle materials, including ceramics, concrete, and rocks.

In the JH2 constitutive model, the equivalent stress is normalized as

$$\sigma^* = \sigma_i^* - D_{\mathrm{JH2}}(\sigma_i^* - \sigma_f^*), \tag{18}$$

in which σ_i^* is the normalized intact equivalent stress of materials, σ_f^* is the normalized fracture equivalent stress of materials, and D_{JH2} is dimensionless damage ($0 \leq D_{\mathrm{JH2}} \leq 1.0$).

The true equivalent stress can be normalized as

$$\sigma^* = \sigma/\sigma_{\mathrm{HEL}}, \tag{19}$$

in which σ represents the true equivalent stress, σ_{HEL} represents the equivalent stress at HEL, and HEL represents the Hugoniot elastic limit.

The normalized intact equivalent strength of materials is expressed as

$$\sigma_i^* = A(P^* + T^*)^{\mathrm{N}}\left(1 + C \cdot \ln \dot{\varepsilon}^*\right). \tag{20}$$

The normalized fracture equivalent strength of materials is expressed as

$$\sigma_f^* = B(P^*)^{\mathrm{M}}\left(1 + C \cdot \ln \dot{\varepsilon}^*\right). \tag{21}$$

Also, the fracture equivalent strength of the material should satisfy $\sigma_f^* \leq$ SFMAX. Moreover, A, B, C, M, N, and SFMAX are material parameters. $\dot{\varepsilon}^* = \dot{\varepsilon}/\dot{\varepsilon}_0$ is the dimensionless strain rate, in which $\dot{\varepsilon}$ is the true strain rate and $\dot{\varepsilon}_0 = 1.0\,\mathrm{s}^{-1}$.

The normalized hydrostatic pressure is expressed as

$$P^* = P/P_{\text{HEL}}, \tag{22}$$

where P is the true pressure, and P_{HEL} is the pressure at HEL.

The maximum tensile hydrostatic pressure is normalized to

$$T^* = T/P_{\text{HEL}}, \tag{23}$$

where T is the maximum hydrostatic tension that can be sustained.

Similar to the JH1 model [45], the damage in the JH2 model is accumulated as

$$D_{\text{JH2}} = \sum \Delta \varepsilon^P / \varepsilon_f^P, \tag{24}$$

in which $\Delta \varepsilon^P$ is the plastic strain accumulated under the pressure P. The fracture plastic strain can be expressed as

$$\varepsilon_f^P = D_1 (P^* + T^*)^{D_2}, \tag{25}$$

in which D_1 and D_2 are material parameters. In addition, the material can no longer withstand any plastic strain when $P^* = -T^*$, but ε_f^P can increase with the increase in P^*.

Before materials fracture ($D_{\text{JH2}} = 0$), the energy loss can be negligible. At this moment, the hydrostatic pressure is

$$P = \begin{cases} K_1 \mu + K_2 \mu^2 + K_3 \mu^3, & \text{If } \mu \geq 0 \\ K_1 \mu, & \text{otherwise} \end{cases}. \tag{26}$$

After fractured ($D_{\text{JH2}} > 0$), damage begins to accumulate. As the volumetric strain increases, volume bulking can occur. Due to energy loss, incremental pressure ΔP is added to the pressure P as

$$P = K_1 \mu + K_2 \mu^2 + K_3 \mu^3 + \Delta P, \tag{27}$$

in which K_1, K_2, and K_3 are material parameters, and μ is the volumetric strain. The material is in tension when $\mu \geq 0$, otherwise it is in compression.

The elastic internal energy can be written as

$$U = \sigma^2 / 6G, \tag{28}$$

where σ is the equivalent stress, and G is the shear modulus.

The incremental energy loss can be calculated as

$$\Delta U = U_{D_{\text{JH2}}(t-\Delta t)} - U_{D_{\text{JH2}}(t)}, \tag{29}$$

where $U_{D_{\text{JH2}}(t-\Delta t)}$ and $U_{D_{\text{JH2}}(t)}$ are the elastic internal energies calculated from Equation (28) for the corresponding time step.

The updated incremental pressure ΔP is calculated as

$$\Delta P_t = -K_1 \mu_t + \sqrt{(K_1 \mu_t + \Delta P_{t-\Delta t})^2 + 2\beta K_1 \Delta U}, \tag{30}$$

in which β is the material parameter, and ΔP is the incremental pressure.

2.2. Modified Bond-Associated Non-Ordinary State-Based Peridynamic Model

In this section, the original BA-NOSB PD model based on the JH2 constitutive relationship is modified to more accurately model the impact problems of quasi-brittle materials. First, the concept of "bond-associated volumetric strain" and its new calculation scheme are proposed via the introduction of the bond-associated deformation gradient. Then, a general tensile-shear coupling bond-breaking criterion is proposed in the framework of the BA-NOSB PD theory. Finally, a new treatment strategy for broken bonds is also given.

2.2.1. Bond-Associated Volumetric Strain

It is worth noting that all physical quantities are bond-associated within the BA-NOSB PD theory. Some improvements are proposed in this section to implement a fully bond-associated JH2 constitutive model.

The deformation of an object under the action of an external force can generally be classified as volume change and shape change. In the theory of plasticity, it is generally considered that the volume change is caused by spherical stresses, while the shape change is caused by deviatoric stresses. Therefore, the stress state at a point can be decomposed into a spherical stress state and a deviatoric stress state. In addition, the stress state at the point is determined by the stress tensor $\boldsymbol{\sigma}$, which can be expressed as

$$\boldsymbol{\sigma} = P\mathbf{I} + \mathbf{S}, \tag{31}$$

in which $P\mathbf{I}$ indicates the spherical stress tensor, P is the pressure, $\mathbf{I}$ is a second-order identity tensor, and $\mathbf{S}$ indicates the deviatoric stress tensor.

In the framework of BA-NOSB PD theory, the above equation can be rewritten as

$$\boldsymbol{\sigma}_b = P_b\mathbf{I} + \mathbf{S}_b, \tag{32}$$

where $\mathbf{S}_b$ represents the bond-associated deviatoric stress tensor, which is given in Section 2.3. According to Equation (26), the bond-associated hydrostatic pressure P_b can be calculated as

$$P_b = \begin{cases} K_1\mu_b + K_2\mu_b^2 + K_3\mu_b^3, & \text{If } \mu_b \geq 0 \\ K_1\mu_b, & \text{otherwise} \end{cases}. \tag{33}$$

As seen in Equation (33), the bond-associated hydrostatic pressure P_b depends on the bond-associated volumetric strain μ_b. In the original JH2 constitutive model [45], the volumetric strain μ in the equation of state is defined as

$$\mu = \rho/\rho_0 - 1, \tag{34}$$

where ρ_0 is the initial mass density, and ρ is the current mass density.

The volumetric strain μ in Equation (34) depends only on the mass density of a point, and can also be expressed as

$$\mu = \mu_p = dV/dv - 1, \tag{35}$$

in which the subscript b indicates that the physical quantity is point-associated, dV is the reference volume element, and dv is the current volume element.

Referring to classical continuum mechanics, and according to the third invariant of the bond-associated deformation gradient tensor $\mathbf{F}_b$ in Section 2.1.2, $\mathcal{J}_b = \det(\mathbf{F}_b)$, the transformation relationship between volume elements under different configurations can be expressed as

$$dV = \mathcal{J}_b dv. \tag{36}$$

The equation above holds for any shaped volume element since it can be approximated by an infinite number of parallel hexahedra. Substituting Equation (36) into Equation (35), the bond-associated volumetric strain μ_b can be defined as

$$\mu = \mu_b = 1/\mathcal{J}_b - 1. \tag{37}$$

Obviously, μ_p is a point-associated nonlocal physical quantity, while μ_b is a bond-associated nonlocal physical quantity. Without losing generality, in addition to the JH2 model [45], the defined bond-associated volumetric strain μ_b can also be used in other constitutive models [46–48] implemented in the framework of the BA-NOSB PD theory. Further details of the constitutive update using the JH2 model will be given in Section 2.3.

2.2.2. Bond-Breaking Criterion

If the external force exceeds the fracture strength of materials, local damage, cracks or fractures will occur. The PD can naturally model discontinuous behavior such as damage or cracks, and damage is also incorporated into its constitutive model. In the PD theory, the bonds are carriers that transmit non-local interactions, while damage is defined by breaking bonds. In addition, broken bonds will no longer transmit PD interaction. Accurately assessing and describing the discontinuity behavior of materials usually requires a robust bond-breaking criterion.

In most engineering problems, materials usually suffer shear failure, not just tensile and compressive failure. According to Equation (16), the bond-associated force density vector can describe not only tensile-compressive deformation, but also shear deformation. However, the critical stretch criterion [8] proposed earlier cannot evaluate material damage caused by shear deformation. Therefore, the assessment of shear damage needs to be refined in BA-NOSB PD. Based on the comprehensive failure criterion proposed by Liu et al. [14], a new general tensile-shear coupling bond-breaking criterion is proposed and implemented in the framework of BA-NOSB PD theory. As shown in Figure 3, under the condition of small deformation, the relative rotation angle vector [12] is approximated as

$$\gamma = \frac{\eta - s\zeta \mathbf{n}}{\zeta} \tag{38}$$

where $\mathbf{n} = (\xi + \eta)/\|\xi + \eta\|$ is an identity direction vector of the bond ξ in the current configuration, and $\zeta = \|\xi\|$.

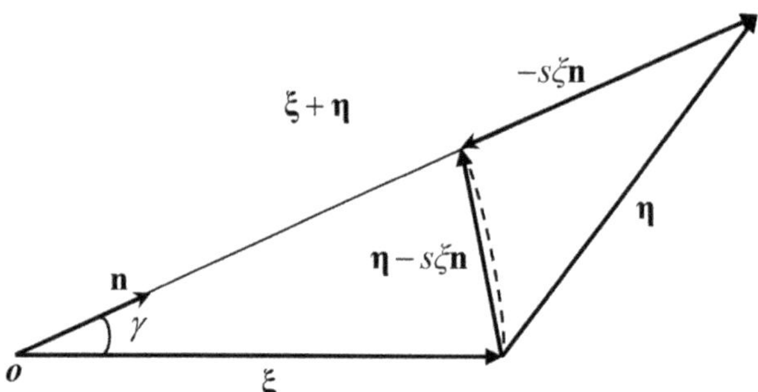

Figure 3. Schematic diagram of the relative rotation angle [14].

The bond stretch in PD is defined as

$$s(\xi, \eta) = \frac{\|\xi + \eta\| - \|\xi\|}{\|\xi\|}. \tag{39}$$

Similarly, the bond-breaking threshold used to assess the tensile-shear failure of materials in this work is defined by the dimensionless quantity λ, which can be expressed as

$$\lambda = \sqrt{\alpha_0 \left(\frac{s}{s_0}\right)^2 + \beta_0 \left(\frac{\gamma}{\gamma_0}\right)^2}, \tag{40}$$

in which s_0 is the critical stretch, γ_0 is the critical relative rotation angle, and $\gamma = \|\gamma\|$ is the relative rotation angle. The difference between the anti-tensile capability and anti-shear capability of materials is considered in this work by defining the anti-tensile weight α_0 and anti-shear weight β_0 respectively, as follows:

$$\begin{cases} \alpha_0 = \frac{s_0}{s_0 + \gamma_0} \\ \beta_0 = \frac{\gamma_0}{s_0 + \gamma_0} \end{cases}. \tag{41}$$

In particular, the tensile-shear coupling bond-breaking criterion degrades to the critical stretch criterion when the anti-tensile weight and anti-shear weight are taken as $\alpha_0 = 1$ and $\beta_0 = 0$, respectively.

Further, the critical stretch s_0 and the critical relative rotation angle γ_0 are defined by the fracture energy of materials. In the PD theory, the energy release rate G_0 [8] and the shear fracture energy G_s [14] are respectively expressed as

$$
\begin{cases}
G_0 = \frac{\pi c s_0^2 \delta^5}{10} \\
G_S = \frac{\pi c \gamma_0^2 \delta^5}{10}
\end{cases}.
\tag{42}
$$

For the three-dimensional case, the critical stretch is calculated as

$$
s_0 = \sqrt{5G_0/(9K\delta)},
\tag{43}
$$

where K represents the bulk modulus.

Similar to the derivation of the critical stretch s_0 by Silling and Askari [8], substituting the bond constant $c = 18K/\pi\delta^4$ of the PMB model into Equation (42), the critical relative rotation angle γ_0 is expressed as

$$
\gamma_0 = \sqrt{5G_s/(9K\delta)}.
\tag{44}
$$

In the PD theory, the local damage of the material point $\mathbf{X}$ at the instant of t can be defined as

$$
\varphi(\mathbf{X}, t) = 1 - \frac{\int_{H_\mathbf{X}} \mu(\xi, t)\, dV_{\mathbf{X}'}}{\int_{H_\mathbf{X}} dV_{\mathbf{X}'}},
\tag{45}
$$

where the scalar function $\mu(\xi, t)$ is

$$
\mu(\xi, t) = \begin{cases}
1, & \lambda < 1.0 \\
0, & \lambda \geq 1.0
\end{cases}
\tag{46}
$$

2.2.3. Treatment Strategy of Broken Bonds

Unlike BB PD theory, the force density vectors at a material point depend on the deformation behavior of all bonds in SB PD. Therefore, the description of broken bonds in the state-based theory is very important, since it will affect the calculation update of the force vector state for unbroken bonds. The following two treatment strategies are given for a broken bond, which will be compared and discussed in the subsequent Section 3.2. If the bond ξ breaks, in each subsequent time-integral cycle, the relative position vector between two material points in the current configuration is considered as

$$
\underline{\mathbf{Y}}_B\langle\xi\rangle = \xi,
\tag{47}
$$

$$
\underline{\mathbf{Y}}_B\langle\xi\rangle = \xi + \eta_B,
\tag{48}
$$

in which the subscript B indicates that the bond has been broken, and η_B is the relative displacement vector between two material points when the bond ξ breaks.

2.3. Constitutive Update Scheme

In this section, the constitutive update scheme using the JH2 model is given in the framework of the BA-NOSB PD theory. First, based on the bond-associated deformation gradient tensor $\mathbf{F}_b$, the bond-associated spatial velocity gradient tensor $\mathbf{L}_b$ can be calculated as

$$
\mathbf{L}_b = \dot{\mathbf{F}}_b \mathbf{F}_b^{-1},
\tag{49}
$$

where $\dot{\mathbf{F}}_b$ is the material derivative of the deformation gradient $\mathbf{F}_b$, which can be calculated as

$$\dot{\mathbf{F}}_b = \int_{H_\xi} \underline{\omega}\langle\|\xi\|\rangle\underline{\dot{\mathbf{Y}}}\langle\xi\rangle \otimes \underline{\mathbf{X}}\langle\xi\rangle dV_{\mathbf{X}'}\mathbf{K}_b^{-1}. \tag{50}$$

The spatial velocity gradient $\mathbf{L}_b$ is decomposed into a bond-associated deformation rate tensor $\mathbf{D}_b$ and a bond-associated rotation rate tensor $\mathbf{W}_b$:

$$\mathbf{D}_b = \tfrac{1}{2}\left(\mathbf{L}_b + \mathbf{L}_b^{\mathrm{T}}\right) \quad \text{and} \quad \mathbf{W}_b = \tfrac{1}{2}\left(\mathbf{L}_b - \mathbf{L}_b^{\mathrm{T}}\right). \tag{51}$$

To ensure that the resulting constitutive relationship satisfies the principle of coordinate invariance, the algorithm proposed by Rubinstein and Atluri [49] is used to update the bond-associated Cauchy stress σ_b. Before this, the bond-associated unrotated deformation rate of the tensor $\mathbf{d}_b$ can be expressed as

$$\mathbf{d}_b = \left(\mathbf{R}_b^t\right)^{\mathrm{T}}\mathbf{D}_b\mathbf{R}_b^t, \tag{52}$$

where $\mathbf{R}_b^t$ is the bond-associated orthogonal rotation tensor that describes the rigid-body rotation, which can be calculated as

$$\mathbf{R}_b^t = \left[\mathbf{I} + \frac{\sin(\Delta t\Omega)}{\Omega}\Omega - \frac{1 - \cos(\Delta t\Omega)}{\Omega^2}\Omega^2\right]\mathbf{R}_b^{t-\Delta t}, \tag{53}$$

in which the rigid-body rotation rate tensor Ω can be calculated as

$$\begin{cases} z_i = e_{ikj}(D_b)_{jm}(V_b)_{mk} \\ w_i = -\tfrac{1}{2}e_{ijk}(W_b)_{jk} \\ \varpi = \mathbf{w} + [\mathrm{tr}(\mathbf{V}_b)\mathbf{I} - \mathbf{V}_b]^{-1}\mathbf{z}' \\ \Omega_{ij} = e_{ikj}\varpi_k \end{cases} \tag{54}$$

where e_{ikj} is the permutation symbol, and the left stretch tensor is

$$\mathbf{V}_b^t = \mathbf{V}_b^{t-\Delta t} + \Delta t\dot{\mathbf{V}}_b^{\Delta t}, \tag{55}$$

in which $\dot{\mathbf{V}}_b^{\Delta t} = \mathbf{L}_b\mathbf{V}_b^t - \mathbf{V}_b^t\Omega$ is the left stretch tensor rate.

As of now, we have calculated the bond-associated unrotated deformation rate of the tensor $\mathbf{d}_b$. In the JH2 constitutive relationship, the bond-associated unrotated Cauchy stress τ_b is calculated using the bond-associated unrotated deformation rate of the tensor $\mathbf{d}_b$. At each time step, we will compute and store the rotated Cauchy stress $\sigma_b^{t-\Delta t}$ at the previous step through the stored unrotated Cauchy stress $\tau_b^{t-\Delta t}$, as

$$\tau_b^{t-\Delta t} = \left(\mathbf{R}_b^{t-\Delta t}\right)^{\mathrm{T}}\sigma_b^{t-\Delta t}\mathbf{R}_b^{t-\Delta t}, \tag{56}$$

where $\mathbf{V}_b = \mathbf{I}$ and $\mathbf{R}_b = \mathbf{I}$ when $t = 0$.

We first assume that the material deformation is in the elastic stage, then the total elastic strain increment $\Delta\mathbf{e}$ and the partial elastic strain increment $\Delta\mathbf{e}^{\mathrm{dev}}$ are respectively expressed as

$$\Delta\mathbf{e} = \mathbf{d}_b\Delta t, \quad \Delta\mathbf{e}^{\mathrm{dev}} = \Delta\mathbf{e} - \tfrac{1}{3}\mathbf{I}. \tag{57}$$

At the current time t, the bond-associated unrotated trial Cauchy stress is

$$\tau_b^{t,\mathrm{trial}} = \tau_b^{t-\Delta t} + \lambda \cdot \frac{1}{3}\mathrm{tr}(\Delta\mathbf{e})\mathbf{I} + 2G\mathbf{e}^{\mathrm{dev}} \tag{58}$$

where λ and G are the Lamé constants of solids. Based on Von–Mises plasticity theory, the bond-associated trial deviatoric stress tensor can be calculated as

$$\mathbf{S_b} = \boldsymbol{\tau}_b^{t,\text{trial}} - \frac{1}{3}\text{tr}\left(\boldsymbol{\tau}_b^{t,\text{trial}}\right)\mathbf{I} \tag{59}$$

The bond-associated equivalent Von–Mises yield stress is

$$S_b^{\text{VM}} = \sqrt{\frac{3}{2}(S_b)_{ij}(S_b)_{ij}} \tag{60}$$

The normalized strain rate is $\dot{\varepsilon}^* = d_b^{\text{VM}}/\dot{\varepsilon}_0$, in which $d_b^{\text{VM}} = \sqrt{\frac{2}{3}(d_b)_{ij}(d_b)_{ij}}$ is the total equivalent strain rate. Furthermore, the bond-associated hydrostatic pressure is calculated as

$$P_b^t = \begin{cases} K_1\mu_b^t + K_2\left(\mu_b^t\right)^2 + K_3\left(\mu_b^t\right)^3 + \Delta P_b^{t-\Delta t}, & \text{If } \mu_b^t \geq 0 \\ K_1\left(\mu_b^t\right) + \Delta P_b^{t-\Delta t}, & \text{otherwise} \end{cases} \tag{61}$$

Based on Equation (60), the bond-associated yield function is expressed as

$$f(\sigma_b, \dot{\varepsilon}_b) = S_b^{\text{VM}} - \sigma_b^* \cdot \sigma_{\text{HEL}}. \tag{62}$$

If $f(\sigma_b, \dot{\varepsilon}_b) < 0$, the material is in the elastic stage, and the bond-associated unrotated Cauchy stress is equal to the bond-associated unrotated trail Cauchy stress. Otherwise, the material yields, and we need to recalculate the bond-associated true elastic strain increment, plastic strain increment, and unrotated Cauchy stress.

For the three-dimensional stress state, the bond-associated plastic strain rate vector is calculated according to Prandtl–Reuss flow law as

$$\dot{\varepsilon}_b^p = \dot{\lambda}\frac{\partial f(\sigma_b, \dot{\varepsilon}_b)}{\partial S_b} = \dot{\lambda}a, \tag{63}$$

in which $\dot{\lambda}$ is the bond-associated plastic strain rate multiplier, and a is an orthogonal vector. When the plastic deformation occurs, the stress rate or stress increment is perpendicular to the yield surface, as follows:

$$f(\sigma_b, \dot{\varepsilon}_b) = a^{\text{T}}\dot{\sigma}_b. \tag{64}$$

When the material yields, the total strain rate is decomposed into the elastic component and the plastic component. The relationship between stress and stress rate can be expressed as

$$\dot{\sigma}_b = \mathbf{C}\left(\dot{\varepsilon}_b - \dot{\varepsilon}_b^p\right) = \mathbf{C}\left(\dot{\varepsilon}_b - \dot{\lambda}a\right), \tag{65}$$

where $\mathbf{C}$ is a stiffness matrix of the material. The bond-associated plastic strain rate multiplier $\dot{\lambda}$ is

$$\dot{\lambda} = \frac{a^{\text{T}}\mathbf{C}\dot{\varepsilon}_b}{a^{\text{T}}\mathbf{C}a}. \tag{66}$$

By considering Equations (64)–(66), the bond-associated equivalent plastic strain increment is

$$\Delta\varepsilon_b^p = \frac{a^{\text{T}}\mathbf{C}\dot{\varepsilon}_b\Delta\text{T}}{\sqrt{2}a^{\text{T}}\mathbf{C}a}\sqrt{\left(S_b^{xx} - S_b^{yy}\right)^2 + \left(S_b^{xx} - S_b^{zz}\right)^2 + \left(S_b^{yy} - S_b^{zz}\right)^2 + 6\left[\left(S_b^{xy}\right)^2 + \left(S_b^{yz}\right)^2 + \left(S_b^{zx}\right)^2\right]}. \tag{67}$$

If the bond-associated equivalent plastic strain increment is not currently zero, the bond-associated fracture damage factor accumulates as

$$D_b^t = D_b^{t-\Delta t} + \frac{\Delta \varepsilon_b^P}{\varepsilon_{b,f}^P} \tag{68}$$

The incremental energy loss is

$$\Delta U = U^{t-\Delta t} - \frac{\left(\sigma_{\text{HEL}} \cdot \sigma_b^*\right)^2}{6G}. \tag{69}$$

In the current configuration, the bond-associated final hydrostatic pressure is recalculated as

$$P_b^t = \frac{1}{3}\text{tr}\left(\boldsymbol{\tau}_b^{t,\text{trail}}\right) + \Delta P_b^t. \tag{70}$$

Finally, the bond-associated unrotated Cauchy stress tensor and rotated Cauchy stress at the current time step can be updated to

$$\boldsymbol{\tau}_b^t = P_b^t \mathbf{I} + \mathbf{S}_b, \tag{71}$$

$$\boldsymbol{\sigma}_b^t = \mathbf{R}_b^t \boldsymbol{\tau}_b^t \left(\mathbf{R}_b^t\right)^{\text{T}}. \tag{72}$$

Thereafter, we can calculate the stress tensor $\mathbf{P}_b$ in Equation (13), and then update the force state $\underline{\mathbf{T}}_b$ at the current time step in Equation (12).

2.4. Other Zero-Energy Mode Control Schemes

In addition to the bond-associated deformation gradient scheme proposed in BA-NOSB PD, different zero-energy mode control schemes have been proposed to suppress or eliminate numerical oscillations. In this section, we briefly introduce four zero-energy mode control schemes based on the supplemental force state. In the NOSB model based on the supplemental force state, the total force state can be written as

$$\underline{\mathbf{T}}^{\text{total}}(\mathbf{X},t)\langle \boldsymbol{\xi} \rangle = \underline{\mathbf{T}}(\mathbf{X},t)\langle \boldsymbol{\xi} \rangle + \underline{\mathbf{T}}^{\text{s}}(\mathbf{X},t)\langle \boldsymbol{\xi} \rangle, \tag{73}$$

in which the original force state $\underline{\mathbf{T}}$ is defined in Equation (5), and $\underline{\mathbf{T}}^{\text{s}}$ is the corresponding supplemental force state.

- Control Scheme I:

In Littlewood's work [26], a penalty scheme is used to control the zero-energy mode. This is similar to the method of controlling deformable hourglass modes in finite element analysis by adding the hourglass force to the original force state. The added hourglass force is expressed as

$$\underline{\mathbf{T}}^{\text{s}}(\mathbf{X},t)\langle \boldsymbol{\xi} \rangle = -c_{\text{hg}} c \frac{h_{\text{proj}}}{\|\boldsymbol{\xi}\|} \frac{\boldsymbol{\xi}+\boldsymbol{\eta}}{\|\boldsymbol{\xi}+\boldsymbol{\eta}\|} \Delta V_{\mathbf{X}} \Delta V_{\mathbf{X}'}, \tag{74}$$

in which c_{hg} is a user parameter controlling the level of the hourglass force, c is the bond constant of the PMB model, and $c = 18K/\pi\delta^4$.

The projection of the hourglass vector $\mathbf{h}$ on the relative position vector $\boldsymbol{\xi}+\boldsymbol{\eta}$ under the current configuration can be expressed as

$$h_{\text{proj}} = \mathbf{h} \cdot (\boldsymbol{\xi}+\boldsymbol{\eta}), \tag{75}$$

where the hourglass vector $\mathbf{h}$ is

$$\mathbf{h} = \mathbf{y}'^* - \mathbf{y}', \tag{76}$$

in which the position of the neighbor $\mathbf{X}'$ in the current configuration is calculated by the approximative deformation gradient $\mathbf{F}$ as:

$$\mathbf{y}'^{*} = \mathbf{y} + \mathbf{F}\boldsymbol{\xi}. \tag{77}$$

- Control Scheme II:

In addition, Silling [27] derived the stability conditions of the NOSB PD model and proposed a new supplemental force state that can be written as

$$\underline{\mathbf{T}}^{\mathrm{s}}(\mathbf{X},t)\langle\boldsymbol{\xi}\rangle = \underline{\omega}\langle\|\boldsymbol{\xi}\|\rangle \frac{Gc_1}{\omega_0}\underline{\mathbf{z}}\langle\boldsymbol{\xi}\rangle, \tag{78}$$

where G is a user parameter controlling the level of the supplemental force state, and $c_1 = c/\delta$ is the micro-modulus in BB PD. The sum of the influence function values of all bonds at the material point $\mathbf{X}$ is

$$\omega_0 = \int_{H_\mathbf{X}} \underline{\omega}\langle\|\boldsymbol{\xi}\|\rangle \mathrm{d}V_{\mathbf{X}'}. \tag{79}$$

Furthermore, the non-uniform deformation state can be expressed as

$$\underline{\mathbf{z}}\langle\boldsymbol{\xi}\rangle = \underline{\mathbf{Y}}\langle\boldsymbol{\xi}\rangle - \mathbf{F}\boldsymbol{\xi}. \tag{80}$$

- Control Scheme III:

Li et al. [28] proposed a novel supplemental force state characterizing the non-uniform deformation state based on the linearized BB PD theory, which can be written as

$$\underline{\mathbf{T}}^{\mathrm{s}}(\mathbf{X},t)\langle\boldsymbol{\xi}\rangle = \underline{\omega}\langle\|\boldsymbol{\xi}\|\rangle\underline{\mathbf{C}}\langle\boldsymbol{\xi}\rangle\underline{\mathbf{z}}\langle\boldsymbol{\xi}\rangle, \tag{81}$$

where $\underline{\mathbf{C}}\langle\boldsymbol{\xi}\rangle$ is the elasticity coefficient tensor, defined as

$$\underline{\mathbf{C}}\langle\boldsymbol{\xi}\rangle = c\boldsymbol{\xi} \otimes \boldsymbol{\xi}/\|\boldsymbol{\xi}\|^3. \tag{82}$$

Compared with the control schemes I and II, this scheme does not require the user parameter, and avoids the complicated parameter adjustment process.

- Control Scheme IV:

Wan et al. [29] proposed a new supplemental force state directly within the NOSB PD model, which can be expressed as

$$\underline{\mathbf{T}}^{\mathrm{s}}(\mathbf{X},t)\langle\boldsymbol{\xi}\rangle = \underline{\omega}\langle\|\boldsymbol{\xi}\|\rangle\mathbf{C}_{\mathrm{e}}\mathbf{K}^{-1}\underline{\mathbf{z}}\langle\boldsymbol{\xi}\rangle, \tag{83}$$

where $\mathbf{C}_{\mathrm{e}}$ is the elastic modulus tensor. Similar to the Control Scheme III, this supplemental force state also does not contain the user parameter.

2.5. Numerical Discretization and Implementation

2.5.1. Numerical Discretization

The solution to the PD governing equation of motion usually requires discrete spatial domain and time domain for numerical integration. In the PD theory, the geometric model of the problem domain is presented by discretized material points. Accordingly, in the BA-NOSB PD model, the discrete form of the equation of motion for the material point $\mathbf{X}_i$ at the instant of t can be written as

$$\rho_0(\mathbf{X}_i)\ddot{\mathbf{u}}(\mathbf{X}_i,t) = \sum_{j}^{N}\left[\underline{\mathbf{T}}_{\mathrm{b}}(\mathbf{X}_i,t)\langle\boldsymbol{\xi}_{ij}\rangle - \underline{\mathbf{T}}_{\mathrm{b}}(\mathbf{X}_j,t)\langle\boldsymbol{\xi}_{ji}\rangle\right]V_{\mathbf{X}_j} + \mathbf{b}(\mathbf{X}_i,t), \tag{84}$$

where N is the number of neighbors for the material point $\mathbf{X}_i$, and $\boldsymbol{\xi}_{ij} = \mathbf{X}_j - \mathbf{X}_i$.

The discrete forms of the bond-associated shape tensor and deformation gradient tensor about the bond $\boldsymbol{\xi}_{ij}$ at the material point $\mathbf{X}_i$ is

$$\mathbf{K}_b(\mathbf{X}_i, \boldsymbol{\xi}_{ij}) = \sum_{k=1}^{N_b} \omega \langle \|\boldsymbol{\xi}_{ik}\| \rangle \boldsymbol{\xi}_{ik} \otimes \boldsymbol{\xi}_{ik} \Delta V_{\mathbf{X}_k}, \tag{85}$$

$$\mathbf{F}_b(\mathbf{X}_i, \boldsymbol{\xi}_{ij}) = \left[\sum_{k=1}^{N_b} \omega \langle \|\boldsymbol{\xi}_{ik}\| \rangle (\boldsymbol{\xi}_{ik} + \boldsymbol{\eta}_{ik}) \otimes \boldsymbol{\xi}_{ik} \Delta V_{\mathbf{X}_k} \right] \mathbf{K}_b^{-1}(\mathbf{X}_i, \boldsymbol{\xi}_{ij}), \tag{86}$$

where N_b is the number of bond-associated neighbors with respect to the bond $\boldsymbol{\xi}_{ij}$ at the material point $\mathbf{X}_i$.

Finally, we use the explicit Velocity–Verlet integration algorithm to implement the time-integral scheme, which can be expressed as

$$\dot{\mathbf{u}}_{t+\Delta t} = \dot{\mathbf{u}}_t + \frac{1}{2}(\ddot{\mathbf{u}}_t + \ddot{\mathbf{u}}_{t+\Delta t})\Delta t, \tag{87}$$

$$\mathbf{u}_{t+\Delta t} = \mathbf{u}_t + \dot{\mathbf{u}}_t \Delta t + \frac{1}{2}\ddot{\mathbf{u}}_t (\Delta t)^2, \tag{88}$$

where Δt is the time step size of the time-integral cycle.

2.5.2. Artificial Viscosity

In numerical simulations of the impact or penetration problems, jump oscillations often occur in the impact domain, which may lead to unphysical deformation modes. Artificial viscosity [50,51] is widely used in penetration or impact simulations to improve the stability of numerical algorithms and to prevent unphysical overlapping or interpenetration between particles. In this work, the artificial viscosity implemented in PD theory by Lai et al. [20] is used for impact problems, which can be expressed as

$$\prod_{ij} = \begin{cases} \dfrac{-\alpha_{\prod}\bar{c}_{ij}\phi_{ij} + \beta_{\prod}\phi_{ij}^2}{\bar{\rho}_{ij}}, & \mathbf{v}_{ij} \cdot \mathbf{x}_{ij} < 0 \\ 0, & \mathbf{v}_{ij} \cdot \mathbf{x}_{ij} \geq 0 \end{cases}, \tag{89}$$

where

$$\phi_{ij} = \frac{\delta_{ij}\mathbf{v}_{ij} \cdot \mathbf{x}_{ij}}{\|\mathbf{x}_{ij}\|^2 + \varphi^2}, \tag{90}$$

$$\bar{c}_{ij} = \frac{1}{2}(c_i + c_j), \tag{91}$$

$$\bar{\rho}_{ij} = \frac{1}{2}(\rho_i + \rho_j), \tag{92}$$

$$\delta_{ij} = \frac{1}{2}(\delta_i + \delta_j), \tag{93}$$

$$\mathbf{v}_{ij} = \mathbf{v}_i - \mathbf{v}_j \text{ and } \mathbf{x}_{ij} = \mathbf{x}_i - \mathbf{x}_j. \tag{94}$$

In the above Equations, $\alpha_{\prod}$ and $\beta_{\prod}$ are constants with values of about 1.0. The factor $\varphi = 0.1\delta_{ij}$ is taken into account to avoid numerical divergence. The velocity vector of the particle is $\mathbf{v}$. The sound speed, mass density, and horizon of the particle are c, ρ, and δ, respectively.

The artificial viscosity state introduced can be written as

$$\underline{\prod}(\mathbf{X}_i, t)\langle \boldsymbol{\xi}_{ij} \rangle = \omega \langle \|\boldsymbol{\xi}_{ij}\| \rangle \nabla_i \prod_{ij}\langle \boldsymbol{\xi}_{ij} \rangle, \tag{95}$$

in which ∇ is the vector differential operator. When $\mathbf{v}_{ij} \cdot \mathbf{x}_{ij} < 0$,

$$\nabla_i \prod_{ij} = \frac{-\alpha_\Pi \bar{c}_{ij} + 2\phi_{ij}\beta_\Pi}{\bar{\rho}_{ij}}\frac{\partial \phi_{ij}}{\partial x_i}, \tag{96}$$

where

$$\frac{\partial \phi_{ij}}{\partial x_i} = \delta_{ij}\left(-\frac{\partial \mathbf{v}_i}{\partial x_i}\underline{\mathbf{Y}}\langle \boldsymbol{\xi}_{ij}\rangle - \dot{\underline{\mathbf{Y}}}\langle \boldsymbol{\xi}_{ij}\rangle\right)\frac{1}{\|\underline{\mathbf{Y}}\langle \boldsymbol{\xi}_{ij}\rangle\|^2 + \varphi^2}$$
$$+ \delta_{ij}\left(\dot{\underline{\mathbf{Y}}}\langle \boldsymbol{\xi}_{ij}\rangle \cdot \underline{\mathbf{Y}}\langle \boldsymbol{\xi}_{ij}\rangle\right)\frac{2\underline{\mathbf{Y}}\langle \boldsymbol{\xi}_{ij}\rangle}{\left(\|\underline{\mathbf{Y}}\langle \boldsymbol{\xi}_{ij}\rangle\|^2 + \varphi^2\right)^2}, \tag{97}$$

where

$$\frac{\partial \mathbf{v}_i}{\partial x_i} = \left(\sum_{j=1}^{N} \underline{\omega}\langle \|\boldsymbol{\xi}_{ij}\|\rangle \dot{\underline{\mathbf{Y}}}\langle \boldsymbol{\xi}_{ij}\rangle \otimes \boldsymbol{\xi}_{ij}\Delta V_{\mathbf{X}_j}\right)\mathbf{K}_{\mathbf{X}_i}^{-1}\mathbf{F}_{\mathbf{X}_i}^{-1} \tag{98}$$

$$\dot{\underline{\mathbf{Y}}}\langle \boldsymbol{\xi}_{ij}\rangle = \mathbf{v}(\mathbf{X}_j, t) - \mathbf{v}(\mathbf{X}_i, t) \tag{99}$$

It is worth noting that the introduced artificial viscosity is not physical viscosity, but only numerical correction. This correction prevents the unphysical deformation behavior of particles in dynamic fracture problems such as impact or penetration.

2.5.3. Contact Algorithm

The impact or penetration simulation is a critical problem often encountered in fields such as industrial manufacturing, military equipment, and scientific research. Such problems often involve contact and collision analysis between two or more objects. Among the interface contact algorithms [52–54] that have been developed and used, the penalty method is the most commonly used [55]. In this work, the penalty method implemented in PD theory by Lai et al. [20] will be used to model impact problems. Figure 4 shows the normal impact problem for two objects $\mathcal{A}$ and $\mathcal{B}$. The material points $\mathbf{X}_i$ and $\mathbf{X}_j$ are on the surfaces of the objects $\mathcal{A}$ and $\mathcal{B}$, respectively. At the initial moment, the material point $\mathbf{X}_j$ remains stationary, and the velocity of the material point $\mathbf{X}_i$ is $\mathbf{v}_i$. The relative position vector at the current time step is $\mathbf{g}_n = \mathbf{x}_j - \mathbf{x}_i$, and $l = \|\mathbf{g}_n\|$.

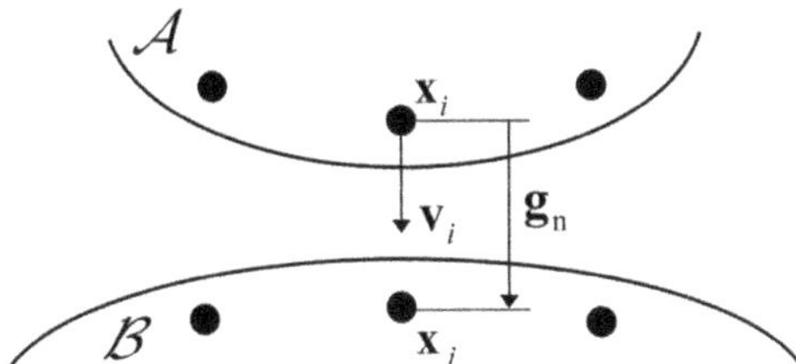

Figure 4. Schematic diagram of the contact surfaces under normal impact [20].

When $l < l_c$, contacts occur between the material points of the master and slave, and a contact bond $\boldsymbol{\xi}_{ij} = \mathbf{X}_j - \mathbf{X}_i$ is formed. At the current moment t, the contact force of the material points $\mathbf{X}_j$ on $\mathbf{X}_i$ can be expressed as

$$f_c(\mathbf{X}_i, t)\langle \boldsymbol{\xi}_{ij}\rangle = \begin{cases} -\mathbf{g}_n/l \cdot C_\xi V_{\mathbf{X}_i}\ln(l/l_c), & \text{if } l < l_c \\ 0, & \text{otherwise} \end{cases} \tag{100}$$

where l_c is a standard value and is usually chosen to be slightly smaller than the grid size of the discretization. The contact stiffness C_ξ is defined on the contact bond, and can be calculated as

$$C_\xi = \min(\rho_i, \rho_j) \times 2 \times 10^n, \tag{101}$$

where ρ_i and ρ_j are the mass densities, and n is the penalty factor. It is worth noting that contact forces can only be repulsive forces.

Therefore, the total contact force at the material point $\mathbf{X}_i$ is expressed as

$$\mathbf{F}_c(\mathbf{X}_i, t) = \sum_{i=1}^{M} \mathbf{f}_c(\mathbf{X}_i, t)\langle \xi_{ij} \rangle, \quad \xi_{ij} < \delta_c, \tag{102}$$

where M is the number of material points forming contact bonds, and δ_c is a contact horizon.

3. Numerical Examples

In this section, four numerical examples are established to verify and demonstrate the modified BA-NOSB PD model. In the first part, two numerical examples are used to numerically model the uniaxial tension experiments of elastic bars under different loading forms to verify the effectiveness and robustness of the modified model. In the second part, two numerical experiments of the edge-on impact and normal impact of ceramics are carried out to demonstrate the ability of the modified model for impact problems of quasi-brittle materials. Good agreement is shown by comparing PD simulation results with references.

3.1. Uniaxial Tension of an Elastic Bar under Stress Loading

In this section, the numerical modeling of the uniaxial tension test of a three-dimensional elastic bar under stress loading is carried out, and the computational model is seen in Figure 5a. The configuration of the model is $80 \times 10 \times 10$ mm^3. A fixed constraint is applied at one end of the bar, and a dynamic tensile stress $\sigma_x = 5.0$ kPa is applied to the other end. The specimen material used in this work is an Al$_2$O$_3$ ceramic with elastic modulus E $= 220$ GPa. The parameters of the JH2 constitutive model for the Al$_2$O$_3$ ceramic are listed in Table 1. As shown in Figure 5b, the loading end consists of one layer of red material points, and the fixed end consists of m layers of light green material points. In the PD simulation, the boundary conditions are as follows: the dynamic tensile stress is applied to red material points, and the fixed constraint is applied to light green material points, as seen in Figure 5b. The ceramic bar is discretized into cubic particles with the grid size Δx, where $\Delta x = 1.0 \times 10^{-3}$ m. The horizon size of the discretization model is $\delta = m \cdot \Delta x$, and the time step size is $\Delta t = 1.0 \times 10^{-8}$ s. The model is discretized into 9801 material points.

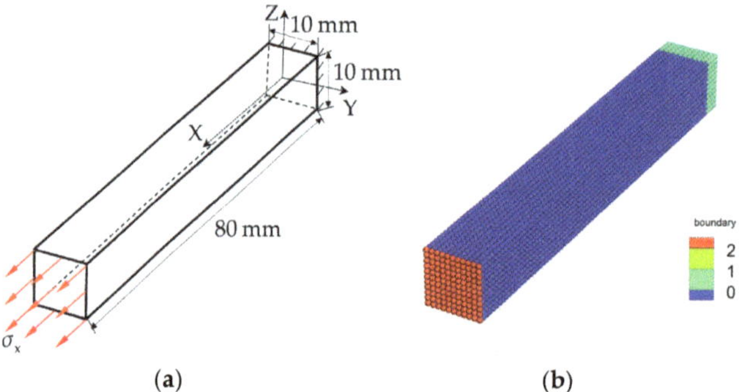

Figure 5. Schematic diagram of uniaxial tension for an elastic bar under stress loading. (a) Computational model. (b) Discretization.

Table 1. JH2 constitutive model parameters of the Al$_2$O$_3$ ceramic [56].

Parameter	Value	Parameter	Value
Density ρ (kg/m^3)	3700	T (GPa)	0.2
Shear modulus G (GPa)	90.16	P$_{HEL}$ (GPa)	1.46
Poisson's ratio ν	0.22	σ_{HEL} (GPa)	2.0
A	0.93	D$_1$	0.005
B	0.31	D$_2$	1.0
C	0.0	K$_1$ (GPa)	130.95
M	0.6	K$_2$ (GPa)	0.0
N	0.6	K$_3$ (GPa)	0.0
$\dot{\varepsilon}_0$	1.0	β	1.0

This section will verify the effectiveness of controlling zero-energy modes of the modified BA-NOSB PD model and compare the numerical stability of different control schemes. The axial displacement simulation results from different control schemes are compared with those of the original NOSB PD model. In comparison and analysis, simulation results at $t = 100$ μs are selected. In this example, the grid size Δx of the discretization model remains fixed, and five cases with horizon factors m of 1.5, 2.0, 3.0, 4.0, and 5.0 are selected for comparison and analysis of the results.

Figure 6 shows the axial displacement contours of the 3D ceramic bar calculated by the original NOSB PD model, four other zero-energy mode control schemes, and the modified BA-NOSB PD model when $m = 1.5$. The displacement solution obtained from the original NOSB PD model in Figure 6a shows very violent oscillations. The result based on the Control Scheme I in Figure 6b is similar to that of the original NOSB PD model, and there are also severe oscillations. As shown in Figure 6d,e, the displacement solutions obtained by the Control Schemes III and IV are obviously incorrect, and there are obvious oscillations near the fixed end. The displacement solutions of the Control Scheme II and the modified BA-NOSB PD model show discontinuity near the fixed end, and the results of the Control Scheme II are smoother and more continuous than that of the modified BA-NOSB PD model, as shown in Figure 6c,f.

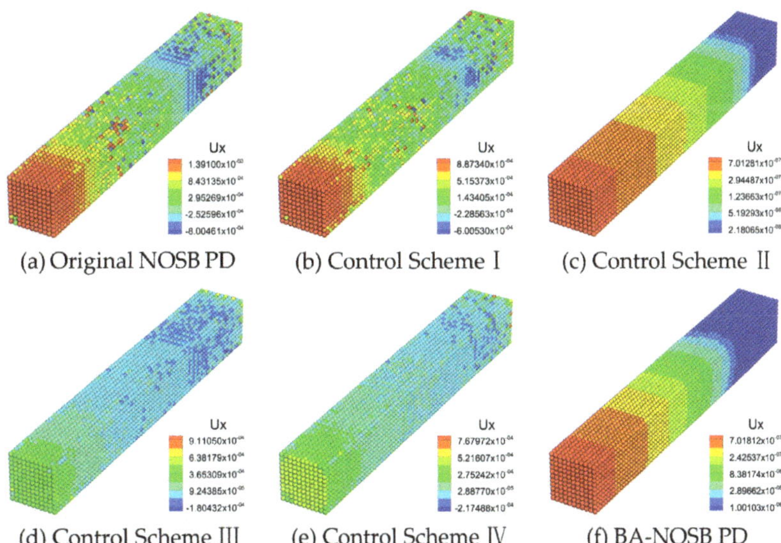

(a) Original NOSB PD   (b) Control Scheme I   (c) Control Scheme II

(d) Control Scheme III   (e) Control Scheme IV   (f) BA-NOSB PD

Figure 6. Comparison of axial displacement contours with $m = 1.5$.

Figure 7 shows the axial displacement contours calculated by the different PD models when $m = 2.0$. The displacement solution obtained from the original NOSB PD model in Figure 7a shows slight oscillations, which are more significant near the fixed end, and the displacement contour in the middle of the bar is not smooth and continuous. The results of the Control Schemes I, III, and IV are similar to that of the original NOSB PD model, all with slight oscillations, as seen in Figure 7b,d,e. In addition, the Control Scheme II and the modified BA-NOSB PD model both obtain smooth displacement solutions, but the results of the Control Scheme II have slight oscillations near the fixed end, as seen in Figure 7c,f.

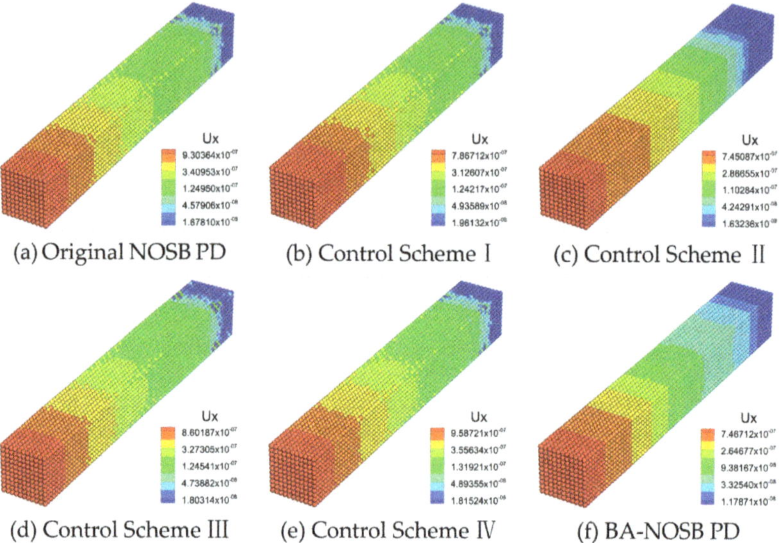

(a) Original NOSB PD   (b) Control Scheme I   (c) Control Scheme II

(d) Control Scheme III   (e) Control Scheme IV   (f) BA-NOSB PD

Figure 7. Comparison of axial displacement contours with $m = 2.0$.

Figure 8 shows the axial displacement contours calculated by the different PD models when $m = 3.0$. The displacement solution obtained from the original NOSB PD model in Figure 8a shows obvious oscillations near the fixed end and slight oscillations near the loading end. In addition, the displacement contour in the middle of the bar is not smooth and continuous. The displacement solutions obtained by the Control Schemes I, III, and IV are similar. These results have obvious oscillations near the fixed end, and the displacement contours in the middle of the bar are not smooth and continuous, as shown in Figure 8b,d,e. The result obtained by the Control Scheme II in Figure 8c has slight oscillations near the fixed end. In addition, the modified BA-NOSB PD model provides a smooth displacement solution that can effectively eliminate numerical oscillations, as shown in Figure 8f.

Figure 9 shows the axial displacement contours calculated by the different PD models when $m = 4.0$. There are slight oscillations in the displacement solution obtained from the original NOSB PD model in Figure 9a, and the displacement contour in the middle of the bar is not smooth and continuous. The results of the Control Schemes I, III, and IV are similar. Moreover, these results have slight oscillations near the fixed end, and the displacement contours in the middle of the bar are not smooth and continuous, as shown in Figure 9b,d,e. The result obtained by the Control Scheme II in Figure 9c shows slight oscillations near the fixed end. The modified BA-NOSB PD model provides smooth and continuous displacement contours, as shown in Figure 9f.

Figure 10 shows the axial displacement contours calculated by the different PD models when $m = 5.0$. The displacement solution obtained from the original NOSB PD model in Figure 10a shows obvious oscillations near both the fixed end and the loading end. The results of the Control Schemes I, III, and IV all show obvious oscillations at the fixed end and in the middle of the bar, as shown in Figure 10b,d,e. In addition, the result of the Control Scheme II has slight oscillations near the fixed end, and the displacement contour in the middle of the bar is not smooth and continuous, as shown in Figure 10c. The displacement solution obtained from the modified BA-NOSB PD model in Figure 10f has no obvious oscillations, but it does show discontinuous features near the fixed end.

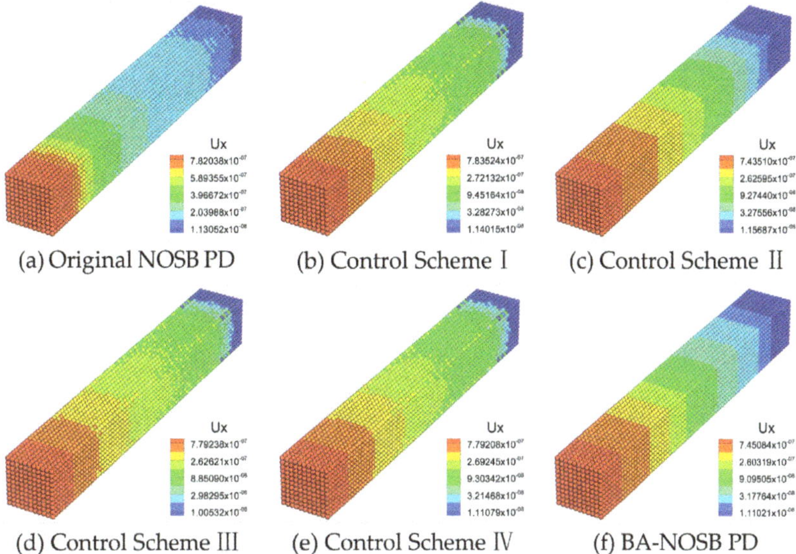

Figure 8. Comparison of axial displacement contours with $m = 3.0$.

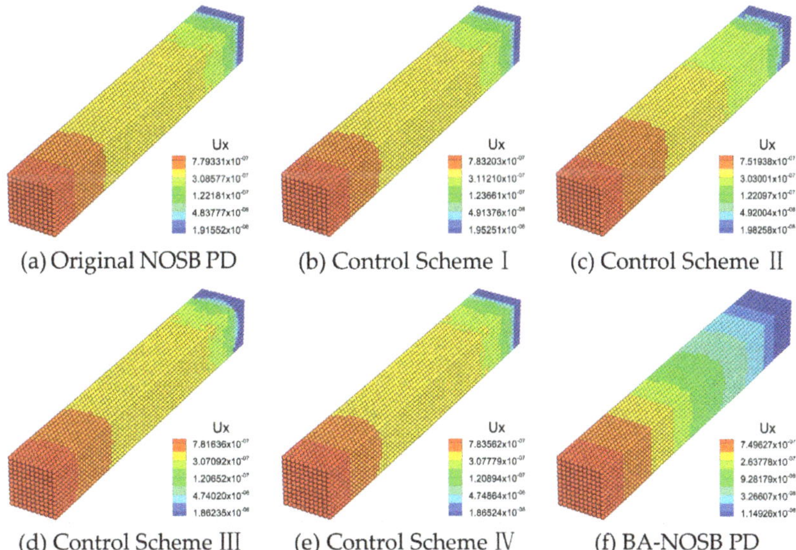

Figure 9. Comparison of axial displacement contours with $m = 4.0$.

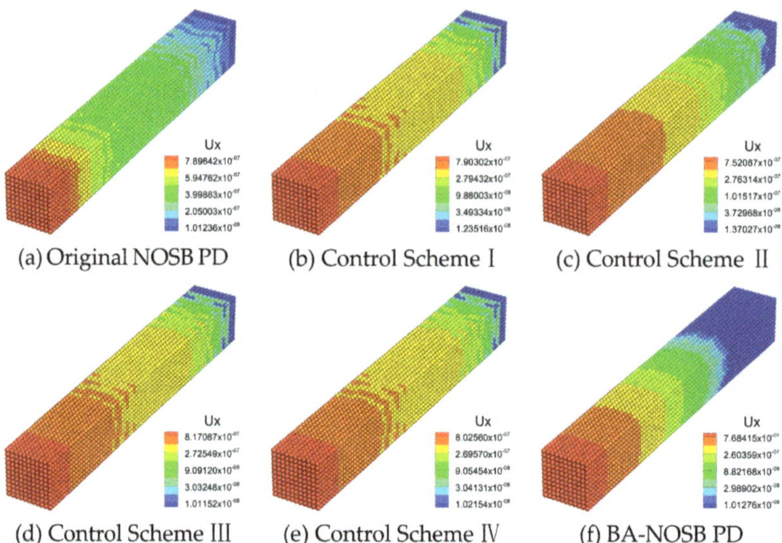

(a) Original NOSB PD (b) Control Scheme I (c) Control Scheme II

(d) Control Scheme III (e) Control Scheme IV (f) BA-NOSB PD

Figure 10. Comparison of axial displacement contours with $m = 5.0$.

This section simulates a uniaxial tension experiment of a 3D ceramic bar under stress loading, and compares the effectiveness of different zero-energy mode control schemes to suppress or eliminate numerical oscillations. The following conclusions can be drawn from the above comparison and analysis. First, the simulation results of the original NOSB PD model show different degrees of oscillations for different values of m. Similarly, the results of the Control Schemes I, III, and IV all show different degrees of oscillations for different values of m, which fail effectively to suppress or eliminate numerical oscillations. In addition, the Control Scheme II effectively eliminates oscillations at smaller values of m ($1.5 \leq m \leq 3.0$), while slight oscillations exist at larger values of m ($4.0 \leq m \leq 5.0$). Finally, the modified BA-NOSB PD model can effectively eliminate oscillations for different values of m, showing slight discontinuous characteristics in the case of $m = 1.5$ or $m = 5.0$.

On the other hand, by analyzing the above simulation results with numerical oscillations, two conclusions can be drawn. First, the oscillations of the result are more obvious when a large or small horizon factor m ($m \leq 2.0$ or $m \geq 4.0$) is selected, while the oscillations of the result are lighter when a moderate horizon factor m ($m = 3.0$) is selected. Secondly, the oscillations of the simulation results mainly exist in the region near the boundaries, such as the fixed end or the loading end.

In summary, the modified BA-NOSB PD model can effectively eliminate the numerical oscillations and unphysical deformation modes. Compared to four other zero-energy mode control schemes, the modified model has a more robust zero-energy mode control capability.

3.2. Uniaxial Tension of an Elastic Bar under Velocity Loading

In order to more accurately implement the JH2 constitutive model in the framework of BA-NOSB PD theory, the bond-associated volumetric strain μ_b is defined in Section 2.2.1. In addition, for the SB PD theory, a new treatment strategy for broken bonds is proposed in Section 2.2.3. In this section, for the volumetric strain in the equation of state and the broken bonds, the following three solutions are presented to modify the original BA-NOSB PD model, as shown in Table 2.

Table 2. Solutions to modify the original BA-NOSB PD model.

Solution A	Solution B	Solution C
$\mu = \mu_p = \rho/\rho_0 - 1$	$\mu = \mu_b = 1/\mathcal{J}_b - 1$	$\mu = \mu_b = 1/\mathcal{J}_b - 1$
$\underline{\mathbf{Y}}_B\langle\xi\rangle = \xi$	$\underline{\mathbf{Y}}_B\langle\xi\rangle = \xi$	$\underline{\mathbf{Y}}_B\langle\xi\rangle = \xi + \eta_B$

To evaluate the above three solutions, the uniaxial tension experiment of the 3D elastic bar under velocity loading will be numerically modeled in this section, and the computational model is shown in Figure 11a. The configuration of the model is $80 \times 10 \times 10$ mm^3. At the initial moment, an instantaneous tensile velocity load of the same magnitude is exerted on both ends of the bar, and the magnitude of the velocity load is $v_X = 0.1$ m/s. The sample material is the Al$_2$O$_3$ ceramic with the same material parameters as in Section 3.1. The JH2 constitutive model parameters of Al$_2$O$_3$ ceramic are shown in Table 1. In the PD simulation, the tensile velocity loads are applied to red and light green material points, respectively, as seen in Figure 11b. The grid size of the discretization is $\Delta x = 1.0 \times 10^{-3}$ m, and the horizon factor is chosen as $m = 3.0$. The time step size is $\Delta t = 1.0 \times 10^{-8}$ s. The model is discretized into 9801 material points.

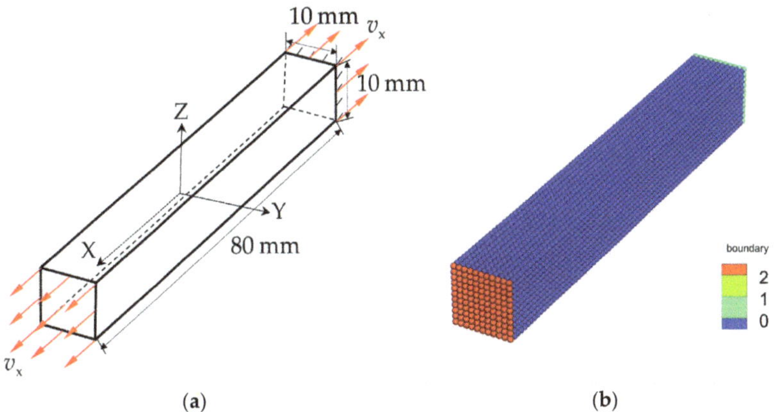

Figure 11. Schematic diagram of uniaxial tension of an elastic bar under velocity loading. (**a**) Computational model. (**b**) Discretization.

From the perspective of kinetic energy convergence, the change in the system kinetic energy over time will be compared to discuss and evaluate the above three solutions. At the initial moment, both ends of the bar are subjected to a tensile velocity load, and the initial system kinetic energy is 1.85×10^{-6} J. This initial kinetic energy is also the total energy of the whole system. Figure 12 shows the change in the total kinetic energy of the PD simulation system based on the three solutions over time. For Solution A, the ceramic bar begins to show damage at $t = 65$ μs, and the system kinetic energy increases sharply to 9.902×10^{-4} J from the damage occurrence to complete brittle fracture. For Solution B, the bar begins to show damage at $t = 82.5$ μs, and the system kinetic energy increases significantly to 2.701×10^{-4} J from the damage occurrence to complete brittle fracture. The final system kinetic energy for Solutions A and B is much higher than the initial value, and the system kinetic energy will increase significantly after the damage occurs, as shown in Figure 12a. Completely different from Solutions A and B, the system kinetic energy of Solution C does not change significantly from the damage appearance to brittle fracture, as shown in Figure 12b,c. Further, as shown in Figure 12c, the system kinetic energy of Solution B increases rapidly after the damage occurs, while the system kinetic energy of Solution C remains stable. Compared to the initial value, it can be seen that the system kinetic energy of Solution C is always lower than the initial value and does not increase significantly or far exceed the initial value, as shown in Figure 12d.

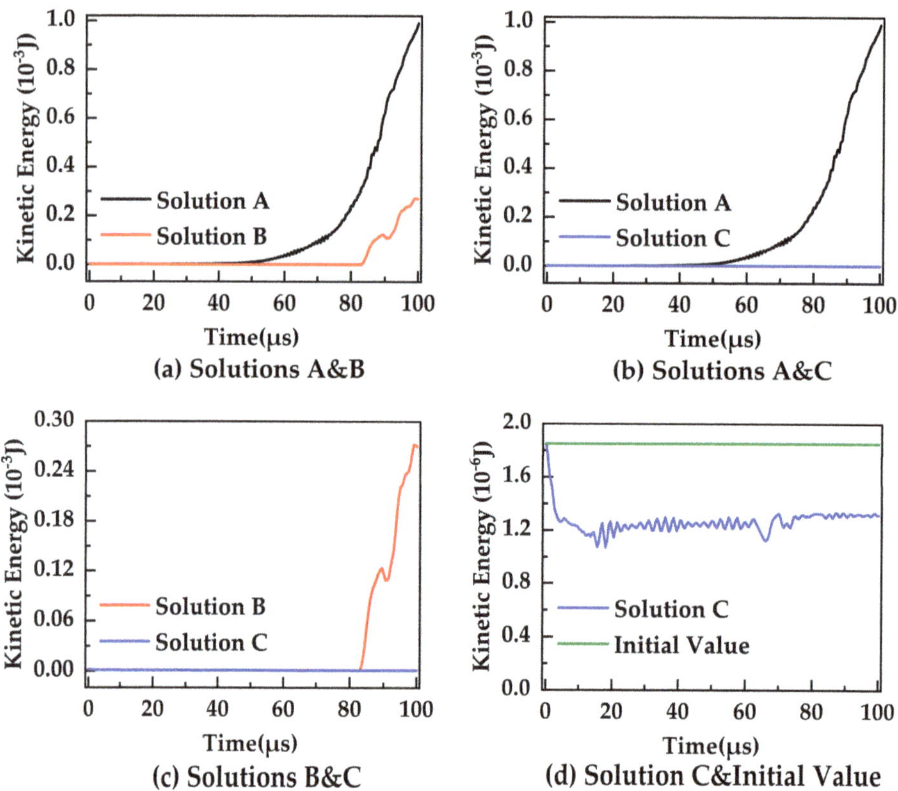

Figure 12. Convergence curves for the system kinetic energy with different solutions.

By comparing and analyzing the convergence curves of the system kinetic energy with the three solutions, the results are summarized as follows. First, the system kinetic energy of Solutions A and B both increase dramatically after the damage occurs, and the final values are much larger than the initial value. Second, the system kinetic energy of Solution C changes steadily over time and is always below the initial value. The results show that only Solution C satisfies the requirement of kinetic energy convergence, while both Solutions A and B do not. In addition, the system kinetic energy of Solutions A and B in the final stage is much greater than the total energy, which means that neither meets the requirement of energy conservation. From the perspective of kinetic energy convergence, Solution C provides better accuracy and robustness for the BA-NOSB PD model and is chosen as the final modified solution.

3.3. Edge-On Impact Simulation of Ceramics

In this section, the numerical experiments on the edge-on impact of ceramic plates will be carried out. The contact algorithm in Section 2.5.3 will be used to describe the interaction between the cylindrical projectile and the target plate. Further, the artificial viscosity introduced in Section 2.5.2 will be used to prevent unphysical oscillations or interpenetration in the impact zone. Figure 13 shows the computational model of a ceramic plate under edge-on impact. The configuration of the target plate is $70 \times 70 \times 10$ mm^3. In addition, the projectile is a cylinder with a diameter of 16 mm and a height of 21 mm. The target plate is the Al_2O_3 ceramic, and its material parameters and JH2 constitutive model parameters are the same as those in Section 3.1. For simplicity, the projectile is regarded as a rigid body with density $\rho = 8060 \text{ kg/m}^3$. With the same set-up as in the reference, free boundary conditions are applied to the ceramic target plate, and the impact velocity of the projectile is $v = 85 \text{ m/s}$. In the PD simulation, the grid size of the discretization is $\Delta x = 1.0 \times 10^{-3}$ m, and the horizon factor is chosen as $m = 3.0$. The time step size used in the edge-on impact is $\Delta t = 2.0 \times 10^{-9}$ s. The computational model is discretized into 55,510 material points.

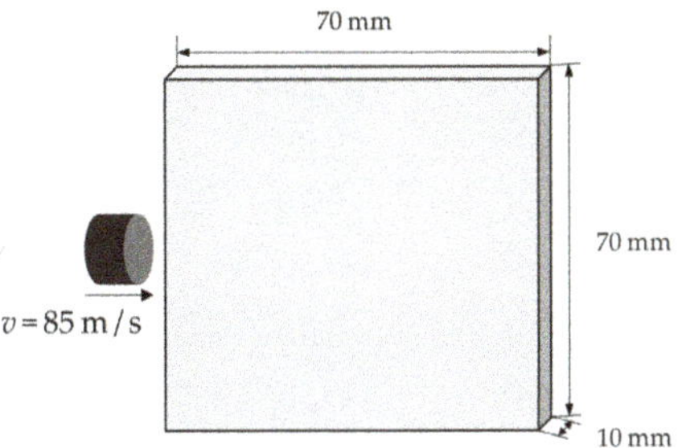

Figure 13. Computational model of ceramics under edge-on impact.

Figure 14 shows the fracture patterns of the Al_2O_3 ceramic plate under edge-on impact in the experiment and numerical simulation. Comparing the PD simulation result with the fracture image provided by Strassburger [57], good agreement can be found. Specifically, both the delta-like fracture pattern and the radial crack propagation paths in Figure 14a can be observed from the simulation result in Figure 14b. In addition, the simulation result shows the characteristics of the fragmentation of the target edge and the semi-elliptical damage zone, which is very similar to the fracture behavior observed in the experiment. The results demonstrate that the modified BA-NOSB PD model can accurately reproduce the fracture pattern and the crack propagation paths of the Al_2O_3 ceramic plate under the edge-on impact experiment.

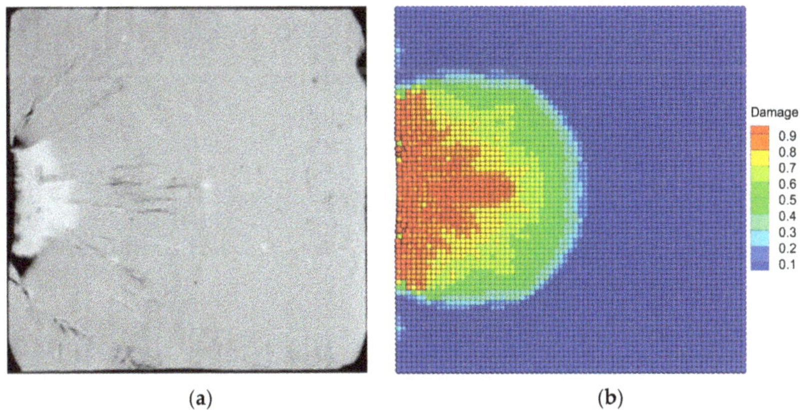

(**a**)   (**b**)

Figure 14. Comparison of fracture patterns for the Al_2O_3 ceramic plate under edge-on impact when $t = 9.7 \text{ μs}$ (**a**) Experimental [57]; (**b**) PD simulation.

The damage evolution and crack propagation of the Al_2O_3 ceramic plate during edge-on impact are shown in Figure 15. After the cylindrical projectile hit the ceramic target plate, it began to penetrate the ceramic plate. At $t = 2.0 \text{ μs}$, the left edge of the target plate began to display gear-shaped damage, as shown in Figure 15a. At $t = 4.0 \text{ μs}$, the damage of the target plate further increases and extends along the impact direction, as shown in Figure 15b. With the continuous action of the projectile, five radial cracks form on the edge of the target plate, as shown at the instant of $t = 6.0$ μs in Figure 15c. At $t = 8.0$ μs, a triangular crack extension zone and a semicircular damage zone are formed on the target plate, as shown in Figure 15d. Subsequently, at $t = 8.8$ μs, a corrugated damage zone and radial cracks are formed on the edge of the target plate, as shown in Figure 15e. Finally, with the weakening of the projectile action, the damage tended to be stable, and a fracture pattern similar to an alluvial delta is formed on the front surface of the ceramic plate, as shown at the instant of $t = 9.7$ μs in Figure 15f.

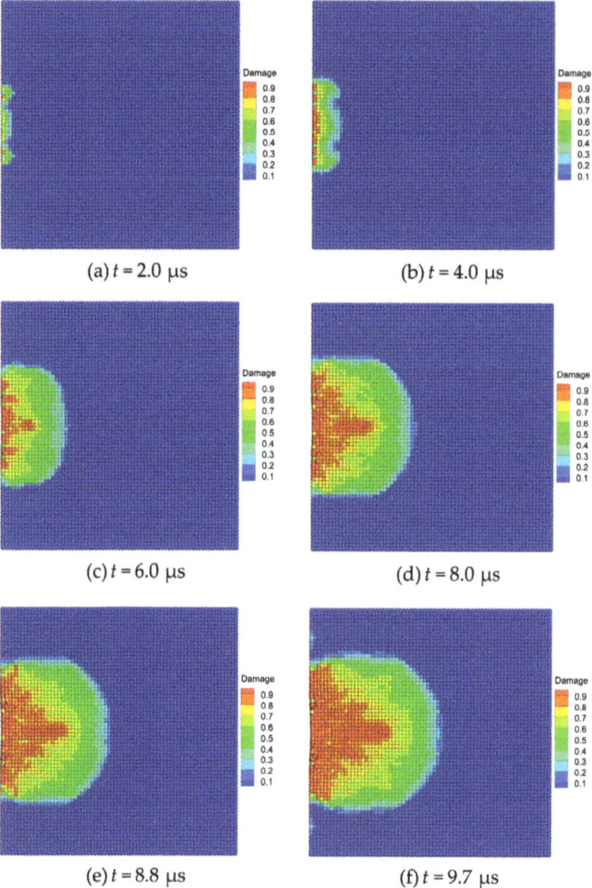

Figure 15. Damage evolution patterns of the Al_2O_3 ceramic plate under edge-on impact.

3.4. Normal Impact Simulation of Ceramics

In this section, the normal impact experiment on a ceramic plate will be numerically modeled. The contact algorithm in Section 2.5.3 will be used to describe the interaction of the spherical projectile with the target plate. The artificial viscosity in Section 2.5.2 is incorporated to avoid unphysical oscillations or interpenetration in the impact area. The computational model of the normal impact simulation for the ceramic plate is shown in Figure 16. The configuration of the ceramic plate is $70 \times 70 \times 9$ mm^3. Moreover, the projectile is a sphere with a diameter of 6.4 mm. The material of the target plate is the Al_2O_3 ceramic with the same material parameters as in Section 3.1. The JH2 constitutive model parameters of the Al$_2$O$_3$ ceramic are shown in Table 1. For simplicity, the projectile material is set to be rigid, and its density is $\rho = 7800$ kg/m^3. With the same set-up as in the reference, the ceramic plate is subjected to free boundary conditions, and the impact velocity of the projectile is $v = 200$ m/s. In the PD simulation, the grid size of the discretization is $\Delta x = 1.0 \times 10^{-3}$ m, and the horizon factor is chosen as $m = 3.0$. The time step size in normal impact is $\Delta t = 2.0 \times 10^{-9}$ s. The model is discretized into 47,256 material points.

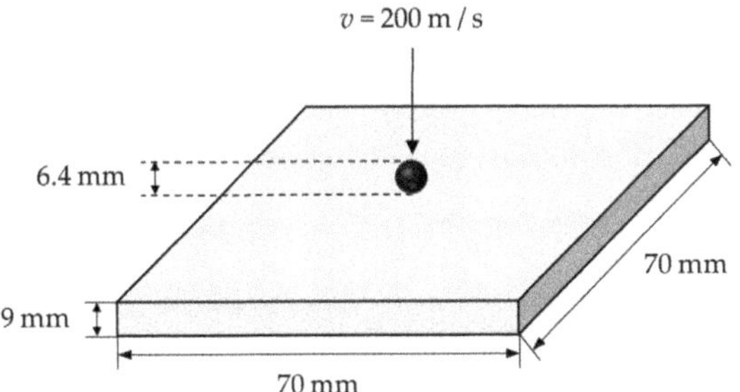

Figure 16. Computational model of ceramics under normal impact.

Figure 17 shows the fracture patterns of the upper surface of the Al$_2$O$_3$ ceramic plate under normal impact obtained from the experiment and PD simulation. Comparing the experimental result [58] and the simulation result, good agreement can be found on the fracture patterns. The PD simulation result in Figure 17b accurately reproduces the eight radial cracks along the $0°$, $45°$, $90°$, $135°$, $180°$, $225°$, $270°$, and $315°$ directions of the ceramic plate in the experiment. In addition, the PD simulation result can accurately reproduce the characteristics of fragmentation and collapse in the center of the plate caused by projectile penetration. These fracture characteristics agree well with the experimental result in Figure 17b. The results show that the modified BA-NOSB PD model can accurately capture the crack propagation paths of the Al$_2$O$_3$ ceramic plate under normal impact, demonstrating the ability to model impact problems and the reliability of damage prediction. In particular, the established model does not take into account the asymmetric factors that may exist in the experiment such as the projectile deflection and the microscopic defects of the sample. Therefore, the PD simulation result does not show the asymmetry of the crack path in the experiment.

The damage evolution and crack growth of the Al$_2$O$_3$ ceramic plate during normal impact are shown in Figure 18. With the impact of the steel ball on the ceramic plate, the compression stress waves generated in the center of the target constantly spread to the boundary surfaces. The compression stress waves are reflected as the tensile stress waves once they have propagated to a free surface. When $t = 2.0$ μs, the upper surface of the target forms a square damage zone. Two circumferential cracks appear in the central damage zone, and cracks initiate outward at the boundary of the damage zone, as shown in Figure 18a. With the continuous propagation, reflection, and overlapping of stress waves, the square damage zone on the target surface expands further, and a collapse is observed in the central area of the target. In addition, four radial cracks along the directions of $45°$, $135°$, $225°$, and $315°$ begin to form, as shown at the instant of $t = 4.0$ μs in Figure 18b. When $t = 6.0$ μs, an obvious circular crack forms in the center of the target, and four radial cracks along the directions of $0°$, $90°$, $180°$, and $270°$ initiate and grow outward, as seen in Figure 18c. Under the continuous penetration of the steel projectile, the crack tips along the directions of $0°$, $90°$, $180°$, and $270°$ grow outward, and then obvious radial cracks form, as shown at the instant of $t = 8.0$ μs in Figure 18d. When $t = 8.8$ μs, four radial cracks form on the upper surface of the target along the directions of $0°$, $90°$, $180°$, and $270°$, as shown in Figure 18e. Finally, the damage tends to be stable at $t = 9.5$ μs, while the center of the target produces fracture behaviors such as fragmentation and collapse, and the top surface produces eight radial cracks along the 0°, 45°, 90°, 135°, 180°, 225°, 270°, and 315° directions, as shown in Figure 18f.

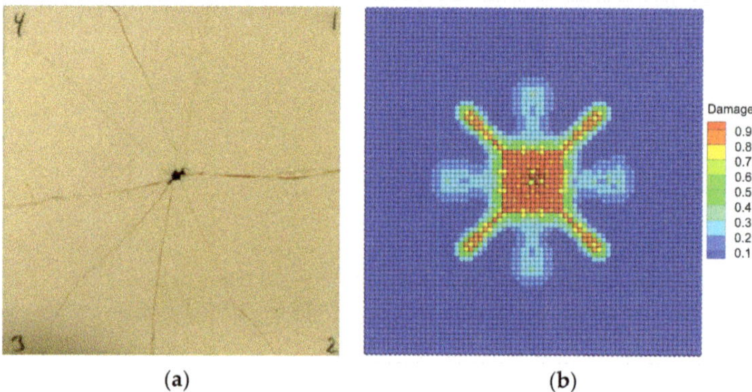

Figure 17. Comparison of fracture patterns for the Al$_2$O$_3$ ceramic plate under normal impact. (**a**) Experimental [58]; (**b**) PD simulation.

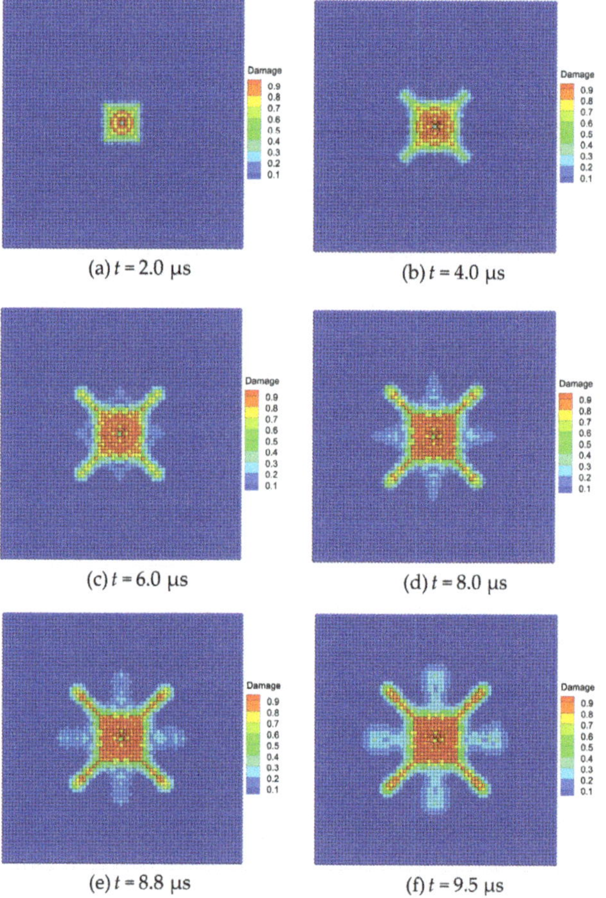

Figure 18. Damage evolution patterns of the Al$_2$O$_3$ ceramic plate under normal impact.

4. Conclusions

In this work, a novel bond-associated non-ordinary state-based peridynamic model for quasi-brittle materials is proposed. The JH2 constitutive model is implemented in the framework of the bond-associated non-ordinary state-based peridynamic theory in the first place, to describe the mechanical behavior of quasi-brittle materials. A new calculation scheme of the volumetric strain in the constitutive model is proposed with the introduction of the bond-associated deformation gradient to improve stability and accuracy in describing the bulking effect. A new treatment strategy for broken bonds is proposed as well. In addition, considering the effect of tensile-shear deformation on material damage, a universal tensile-shear coupling bond-breaking criterion is proposed and implemented in the bond-associated non-ordinary state-based peridynamics.

The effectiveness and robustness of the modified bond-associated non-ordinary state-based peridynamic model are verified to some extent by two benchmark examples. The edge-on impact and normal impact experiments on ceramics are numerically modeled and simulated. The results reveal that the proposed model can effectively control zero-energy mode and eliminate numerical oscillations. Compared with the other four control schemes listed in this work, the modified bond-associated model has a more stable zero-energy mode control capability and convenience. In addition, from the energy conservation analysis of the system, one can find better accuracy and robustness from the modified bond-associated non-ordinary state-based peridynamic model. Good agreement can be found with the experimental data in the two demonstration benchmark tests, which implies that the modified bond-associated non-ordinary state-based peridynamic model can accurately reproduce the fracture patterns and crack growth paths of ceramics in the experiment.

Author Contributions: Conceptualization, J.Z., Y.L., X.L. (Xin Lai) and L.L.; Methodology, J.Z., Y.L., X.L. (Xin Lai) and L.L.; Software, J.Z.; Validation, J.Z.; Formal analysis, J.Z., H.M. and X.L. (Xiang Liu); Investigation, J.Z., Y.L. and H.M.; Resources, X.L. (Xin Lai), L.L. and H.M.; Writing—original draft, J.Z.; Writing—review & editing, X.L. (Xin Lai); Visualization, H.M. and X.L. (Xiang Liu); Supervision, X.L. (Xin Lai), L.L., H.M. and X.L. (Xiang Liu); Project administration, X.L. (Xin Lai) and L.L.; Funding acquisition, X.L. (Xin Lai) and L.L. All authors have read and agreed to the published version of the manuscript.

Funding: This research was funded by the National Natural Science Foundation of China (Nos. 11802214, 11972267, and 51932006) and the Fundamental Research Funds for the Central Universities (WUT: 223114010).

Institutional Review Board Statement: Not applicable.

Informed Consent Statement: Not applicable.

Data Availability Statement: The data used to support the findings of this study are available upon request from the corresponding author.

Conflicts of Interest: The authors declare no conflict of interest.

References

1. Walley, S.M. Historical review of high strain rate and shock properties of ceramics relevant to their application in armour. *Adv. Appl. Ceram.* **2010**, *109*, 446–466. [CrossRef]
2. Chen, W.W.; Rajendran, A.M.; Song, B.; Nie, X. Dynamic fracture of ceramics in armor applications. *J. Am. Ceram. Soc.* **2007**, *90*, 1005–1018. [CrossRef]
3. Song, J.H.; Wang, H.; Belytschko, T. A comparative study on finite element methods for dynamic fracture. *Comput. Mech.* **2008**, *42*, 239–250. [CrossRef]
4. Pezzotta, M.; Zhang, Z.L.; Jensen, M.; Grande, T.; Einarsrud, M.A. Cohesive zone modeling of grain boundary microcracking induced by thermal anisotropy in titanium diboride ceramics. *Comput. Mater. Sci.* **2008**, *43*, 440–449. [CrossRef]
5. Krishnan, K.; Sockalingam, S.; Bansal, S.; Rajan, S.D. Numerical simulation of ceramic composite armor subjected to ballistic impact. *Compos. B Eng.* **2010**, *41*, 583–593. [CrossRef]
6. Silling, S.A. Reformulation of elasticity theory for discontinuities and long-range forces. *J. Mech. Phys. Solids.* **2000**, *48*, 175–209. [CrossRef]

7. Silling, S.A. Dynamic Fracture Modeling with a Meshfree Peridynamic Code. In *Computational Fluid and Solid Mechanics*; Elsevier Science Ltd.: Amsterdam, The Netherlands, 2003; pp. 641–644.

8. Silling, S.A.; Askari, E. A meshfree method based on the Peridynamic model of solid mechanics. *Comput. Struct.* **2005**, *83*, 1526–1535. [CrossRef]

9. Silling, S.A.; Epton, M.; Weckner, O.; Xu, J.; Askari, E. Peridynamic states and constitutive modeling. *J. Elast.* **2007**, *88*, 151–184. [CrossRef]

10. Ha, Y.D.; Bobaru, F. Studies of dynamic crack propagation and crack branching with Peridynamics. *Int. J. Fract.* **2010**, *162*, 229–244. [CrossRef]

11. Ha, Y.D.; Bobaru, F. Characteristics of dynamic brittle fracture captured with Peridynamics. *Eng. Fract. Mech.* **2011**, *78*, 1156–1168. [CrossRef]

12. Zhu, Q.Z.; Ni, T. Peridynamic formulations enriched with bond rotation effects. *Int. J. Eng. Sci.* **2017**, *121*, 118–129. [CrossRef]

13. Chu, B.F.; Liu, Q.W.; Liu, L.S.; Lai, X.; Mei, H. A rate-dependent Peridynamic model for the dynamic behavior of ceramic materials. *CMES-Comp. Model. Eng. Sci.* **2020**, *124*, 151–178. [CrossRef]

14. Liu, Y.X.; Liu, L.S.; Mei, H.; Liu, Q.W.; Lai, X. A modified rate-dependent Peridynamic model with rotation effect for dynamic mechanical behavior of ceramic materials. *Comput. Meth. Appl. Mech. Eng.* **2022**, *388*, 114246. [CrossRef]

15. Kudryavtsev, O.A.; Sapozhnikov, S.B. Numerical simulations of ceramic target subjected to ballistic impact using combined DEM/FEM approach. *Int. J. Mech. Sci.* **2016**, *114*, 60–70. [CrossRef]

16. Zhang, Y.; Qiao, P.Z. A new bond failure criterion for ordinary state-based Peridynamic mode II fracture analysis. *Int. J. Fract.* **2019**, *215*, 105–128. [CrossRef]

17. Foster, J.T.; Silling, S.A.; Chen, W.W. Viscoplasticity using Peridynamics. *Int. J. Numer. Methods Eng.* **2010**, *81*, 1242–1258. [CrossRef]

18. O'Grady, J.; Foster, J. Peridynamic beams: A non-ordinary, state-based model. *Int. J. Solids Struct.* **2014**, *51*, 3177–3183. [CrossRef]

19. Lai, X.; Ren, B.; Fan, H.F.; Li, S.F.; Wu, C.T.; Regueiro, R.A.; Liu, L.S. Peridynamics simulations of geomaterial fragmentation by impulse loads. *Int. J. Numer. Anal. Methods Geomech.* **2015**, *39*, 1304–1330. [CrossRef]

20. Lai, X.; Liu, L.S.; Li, S.F.; Zeleke, M.; Liu, Q.W.; Wang, Z. A non-ordinary state-based Peridynamics modeling of fractures in quasi-brittle materials. *Int. J. Impact Eng.* **2018**, *111*, 130–146. [CrossRef]

21. Wu, L.W.; Huang, D.; Xu, Y.P.; Wang, L. A non-ordinary state-based Peridynamic formulation for failure of concrete subjected to impacting loads. *ES-Comp. Model. Eng. Sci.* **2019**, *118*, 561–581. [CrossRef]

22. Wang, H.; Xu, Y.P.; Huang, D. A non-ordinary state-based Peridynamic formulation for thermo-visco-plastic deformation and impact fracture. *Int. J. Mech. Sci.* **2019**, *159*, 336–344. [CrossRef]

23. Zhu, F.; Zhao, J.D. Peridynamic modelling of blasting induced rock fractures. *J. Mech. Phys. Solids.* **2021**, *153*, 104469. [CrossRef]

24. Yang, S.Y.; Gu, X.; Zhang, Q.; Xia, X.Z. Bond-associated non-ordinary state-based Peridynamic model for multiple spalling simulation of concrete. *Acta Mech. Sin.* **2021**, *37*, 1104–1135. [CrossRef]

25. Li, S.; Lai, X.; Liu, L.S. Peridynamic Modeling of Brittle Fracture in Mindlin-Reissner Shell Theory. *CMES-Comp. Model. Eng. Sci.* **2022**, *131*, 715–746. [CrossRef]

26. Littlewood, D.J. A Nonlocal Approach to Modeling Crack Nucleation in AA 7075-T651. In Proceedings of the ASME International Mechanical Engineering Congress and Exposition, Sandia National Laboratory, Albuquerque, NM, USA, 10–18 November 2011.

27. Silling, S.A. Stability of Peridynamic correspondence material models and their particle discretizations. *Comput. Meth. Appl. Mech. Eng.* **2017**, *322*, 42–57. [CrossRef]

28. Li, P.; Hao, Z.M.; Zhen, W.Q. A stabilized non-ordinary state-based Peridynamic model. *Comput. Meth. Appl. Mech. Eng.* **2018**, *339*, 262–280. [CrossRef]

29. Wan, J.; Chen, Z.; Chu, X.H.; Liu, H. Improved method for zero-energy mode suppression in Peridynamic correspondence model. *Acta Mech. Sin.* **2019**, *35*, 1021–1032. [CrossRef]

30. Yaghoobi, A.; Chorzepa, M.G. Higher-order approximation to suppress the zero-energy mode in non-ordinary state-based Peridynamics. *Comput. Struct.* **2017**, *188*, 63–79. [CrossRef]

31. Luo, J.; Sundararaghavan, V. Stress-point method for stabilizing zero-energy modes in non-ordinary state-based Peridynamics. *Int. J. Solids Struct.* **2018**, *150*, 197–207. [CrossRef]

32. Cui, H.; Li, C.G.; Zheng, H. A higher-order stress point method for non-ordinary state-based Peridynamics. *Eng. Anal. Bound. Elem.* **2020**, *117*, 104–118. [CrossRef]

33. Chen, H.L. Bond-associated deformation gradients for peridynamic correspondence model. *Mech. Res. Commun.* **2018**, *90*, 34–41. [CrossRef]

34. Chen, H.L.; Spencer, B.W. Peridynamic bond-associated correspondence model: Stability and convergence properties. *Int. J. Numer. Methods Eng.* **2019**, *117*, 713–727. [CrossRef]

35. Gu, X.; Zhang, Q.; Madenci, E.; Xia, X.Z. Possible causes of numerical oscillations in non-ordinary state-based Peridynamics and a bond-associated higher-order stabilized model. *Comput. Meth. Appl. Mech. Eng.* **2019**, *357*, 112592. [CrossRef]

36. Li, P.; Hao, Z.M.; Yu, S.R.; Zhen, W.Q. Implicit implementation of the stabilized non-ordinary state-based Peridynamic model. *Int. J. Numer. Methods Eng.* **2020**, *121*, 571–587. [CrossRef]

37. May, S.; Vignollet, J.; De Borst, R. A numerical assessment of phase-field models for brittle and cohesive fracture: Γ-convergence and stress oscillations. *Eur. J. Mech. A-Solids* **2015**, *52*, 72–84. [CrossRef]

38. Wu, J.Y. A unified phase-field theory for the mechanics of damage and quasi-brittle failure. *J. Mech. Phys. Solids* **2017**, *103*, 72–99. [CrossRef]

39. Narayan, S.; Anand, L. A gradient-damage theory for fracture of quasi-brittle materials. *J. Mech. Phys. Solids* **2019**, *129*, 119–146. [CrossRef]

40. Costin, L.S. A microcrack model for the deformation and failure of brittle rock. *J. Geophys. Res.* **1983**, *88*, 9485–9492. [CrossRef]

41. Addessio, F.L.; Johnson, J.N. A constitutive model for the dynamic response of brittle materials. *J. Appl. Phys.* **1990**, *67*, 3275–3286. [CrossRef]

42. Rajendran, A.M.; Grove, D.J. Modeling the shock response of silicon carbide, boron carbide and titanium diboride. *Int. J. Impact Eng.* **1996**, *18*, 611–631. [CrossRef]

43. Zuo, Q.H.; Addessio, F.L.; Dienes, J.K.; Lewis, M.W. A rate-dependent damage model for brittle materials based on the dominant crack. *Int. J. Solids Struct.* **2006**, *43*, 3350–3380. [CrossRef]

44. Zuo, Q.H.; Disilvestro, D.; Richter, J.D. A crack-mechanics based model for damage and plasticity of brittle materials under dynamic loading. *Int. J. Solids Struct.* **2010**, *47*, 2790–2798. [CrossRef]

45. Johnson, G.R.; Holmquist, T.J. *A Computational Constitutive Model for Brittle Materials Subjected to Large Strains, High Strain Rates and High Pressures. Shock Wave and High-Strain-Rate Phenomena in Materials*; Marcel Dekker Inc.: New York, NY, USA, 1992; pp. 1075–1081.

46. Johnson, G.R.; Holmquist, T.J. An Improved Computational Constitutive Model for Brittle Materials. In *AIP Conference Proceedings*; American Institute of Physics: College Park, MD, USA, 1994.

47. Johnson, G.R.; Cook, W.H. A Constitutive Model and Data for Materials Subjected to Large Strains, High Strain Rates, and High Temperatures. In Proceedings of the Seventh International Symposium on Ballistics, The Hague, The Netherlands, 19–21 April 1983.

48. Johnson, G.R.; Holmquist, T.J.; Beissel, S.R. Response of aluminum nitride (including a phase change) to large strains, high strain rates, and high pressures. *J. Appl. Phys.* **2003**, *94*, 1639–1646. [CrossRef]

49. Rubinstein, R.; Atluri, S.N. Objectivity of incremental constitutive relations over finite time steps in computational finite deformation analyses. *Comput. Meth. Appl. Mech. Eng.* **1983**, *36*, 277–290. [CrossRef]

50. Monaghan, J.J.; Gingold, R.A. Shock simulation by the particle method SPH. *J. Comput. Phys.* **1983**, *52*, 374–389. [CrossRef]

51. Li, S. Smoothed particle hydrodynamics-a meshfree method, by G.R. Liu and M.B. Liu. *Comput. Mech.* **2004**, *33*, 491. [CrossRef]

52. Belytschko, T.; Lin, J.I. A three-dimensional impact-penetration algorithm with erosion. *Int. J. Impact Eng.* **1987**, *5*, 111–127. [CrossRef]

53. Heinstein, M.W.; Attaway, S.W.; Swegle, J.W.; Mello, F.J. *A General-Purpose Contact Detection Algorithm for Nonlinear Structural Analysis Codes*; Sandia National Labs: Albuquerque, NM, USA, 1993.

54. Whirley, R.G.; Engelmann, B.E. Automatic contact algorithm in DYNA3D for crashworthiness and impact problems. *Nucl. Eng. Des.* **1994**, *150*, 225–233. [CrossRef]

55. Bourago, N.G.; Kukudzhanov, V.N. A review of contact algorithms. *Mech. Solids* **2005**, *40*, 35–71.

56. Cronin, D.S.; Bui, K.; Kaufmann, C.; McIntosh, G.; Berstad, T.; Cronin, D. Implementation and Validation of the Johnson-Holmquist Ceramic Material Model in LS-Dyna. In Proceedings of the 4th European LS-DYNA User's Conference, Ulm, Germany, 22–23 May 2003.

57. Strassburger, E. Visualization of impact damage in ceramics using the edge-on impact technique. *Int. J. Appl. Ceram. Technol.* **2004**, *1*, 235–242. [CrossRef]

58. Simons, E.C.; Weerheijm, J.; Toussaint, G.; Sluys, L.J. An experimental and numerical investigation of sphere impact on alumina ceramic. *Int. J. Impact Eng.* **2020**, *145*, 103670. [CrossRef]

Article

Numerical Simulation of Failure Behavior of Reinforced Concrete Shear Walls by a Micropolar Peridynamic Model

Feng Shen [1],* , Zihan Chen [1], Jia Zheng [2] and Qing Zhang [3]

[1] School of Civil Engineering, Suzhou University of Science and Technology, Suzhou 215011, China; chenzihan1998@126.com
[2] Suzhou Jcon Greenbuild Fabricated Co., Ltd., Suzhou 215004, China; zhengjia1996912@163.com
[3] Department of Engineering Mechanics, Hohai University, Nanjing 211100, China; lxzhangqing@hhu.edu.cn
* Correspondence: shenfeng1023@163.com

Abstract: A reinforced concrete shear wall is an important building structure. Once damage occurs, it not only causes great losses to various properties but also seriously endangers people's lives. It is difficult to achieve an accurate description of the damage process using the traditional numerical calculation method, which is based on the continuous medium theory. Its bottleneck lies in the crack-induced discontinuity, whereas the adopted numerical analysis method has the continuity requirement. The peridynamic theory can solve discontinuity problems and analyze material damage processes during crack expansion. In this paper, the quasi-static failure and impact failure of shear walls are simulated by improved micropolar peridynamics, which provides the whole process of microdefect growth, damage accumulation, crack initiation, and propagation. The peridynamic predictions are in good match with the current experiment observations, filling the gap of shear wall failure behavior in existing research.

Keywords: peridynamics; improved micropolar model; RC shear walls; numerical simulation; impact failure

Citation: Shen, F.; Chen, Z.; Zheng, J.; Zhang, Q. Numerical Simulation of Failure Behavior of Reinforced Concrete Shear Walls by a Micropolar Peridynamic Model. *Materials* **2023**, *16*, 3199. https://doi.org/10.3390/ma16083199

Academic Editor: René de Borst

Received: 10 March 2023
Revised: 26 March 2023
Accepted: 8 April 2023
Published: 18 April 2023

1. Introduction

Reinforced concrete (RC) shear walls are widely used in concrete structures for their low cost and high stiffness advantages. However, RC shear walls have deficiencies such as poor ductility, early cracking, and apparent shear damage, which limit their use to some extent. As the study of the damage behavior of RC shear walls mainly focused on two aspects of seismic performance study and plane impact resistance of shear walls, Jiang et al. [1] analyzed the law of influence of axial compression ratio on the seismic capacity and damage development of an RC shear wall. Su et al. [2] proposed a two-parameter seismic damage model based on a combination of stiffness degradation and deformation. Lopes et al. [3,4] proposed a new method for calculating the shear bearing capacity by conducting repeated loading tests on low RC shear walls. Paulay et al. [5] proposed that a proper arrangement of diagonal reinforcement in making low RC shear walls obtain a better energy dissipation effect and achieve bending damage. Lefas et al. [6] proposed that the shear capacity of RC shear walls is related to the triaxial compressive stress conditions generated in the bottom section of the compressive zone. Sittipunt et al. [7] considered the effect of loading mechanisms on RC shear wall experiments. They proposed a finite element model that can reflect the performance of RC shear wall members with different reinforcement under reciprocal loading and their possible damage patterns. Vallenas et al. [8] proposed an analytical model for high RC shear walls under monotonic loading because shear deformation occupies a relatively large proportion of high RC shear walls. Terzioglu et al. [9] studied the effect of shear, bending, and slip deformation on the lateral deformation of RC shear walls. Salonikios et al. [10,11] introduced a theoretical formula for the ultimate strength of RC shear walls and proposed a method for calculating the plastic hinge length of RC shear wall members with small aspect ratios. Looi et al. [12] found that axial loading significantly affects crack distribution, failure modes, and deformation properties and proposed two improved empirical prediction models. Houshmand-Sarvestani et al. [13] found that a steel plate with added damping and stiffness (ADAS) dampers can improve the ductility of the steel shear walls in addition to dissipating energy. Yi et al. [14] designed a pendulum tester with adjustable hammer weight and speed to perform out-of-plane impact tests on RC shear walls. Tan et al. [15] discussed the cyclic shear performance of the buckling-restrained cold-formed steel plate shear wall (SPSW) under different buckling constraints. Zhang et al. [16] predicted the seismic performance of shear walls through mechanical learning (ML) methods, and these have a high accuracy in predicting shear wall failure patterns, and predicting the shear strength and flexural strength of reinforced concrete walls more accurately and effectively than existing design formulas.

In order to explore the damage mechanism of shear walls, it is necessary to describe their damage process and damage pattern under load accurately. The finite element method (FEM) is the main tool for the numerical simulation of concrete structures. Simple modeling and fast operation. However, the finite element method is based on the theory of continuous media. Hence the calculation of cracks in the structure will deviate from reality. Extended finite element (X-FEM) is a major improvement in the finite element method [17]. The calculation grid used is not related to the internal composition of the structure. There is no need to perform a grid weight division many times. Therefore, crack expansion simulation can be performed. However, it is still challenging to accurately and efficiently model complex cracking problems, such as three-dimensional composite crack growth problems with random tracks or complex loads. Discrete elements (DEM) solve the problem of discontinuity by dividing the structure into many elements and defining the bonding constitutive between the elements [18]. However, its data structure is complex, and how to distinguish adjacent elements, and the contact determination of the element will make the calculation very complicated. It is difficult to achieve an accurate description of the damage process using the traditional numerical calculation method, which is based on the continuous medium theory. Its bottleneck lies in the crack-induced discontinuity, whereas the adopted numerical analysis method has the continuity requirement. This difficulty is expected to be solved with the emergence of the peridynamic theory. The peridynamic theory is a set of theories no longer based on the traditional continuity assumption and local contact action. The peridynamic theory uses spatial integration and time-differential equations to describe the interaction of masses and solves the crack sprouting and extension problem by equations of motion in integral form, which fundamentally avoids the singularity caused by discontinuity problems and eliminates the need to introduce additional fracture criteria. By introducing the concept of damage to the material and judging the cumulative number of bond fractures to describe the formation and extension of cracks, it enables cracks to emerge, extend and branch naturally in the object. Thus, it has obvious advantages in the study of structural damage problems, and many excellent research results have emerged.

In 2000, Silling et al. [19] proposed the peridynamics theory and algorithms, demonstrating their unique advantages in solving damage problems in solid materials and structures and providing a promising theoretical framework for developing robust numerical models. Gerstle et al. [20,21] have improved based on the PD theory based on "bond". Based on the radial movement of material points, the relative rotation of the material point pairs is considered, and the micropolar peridynamic model is proposed. Zhang et al. [22] proposed a new bond-slip model and demonstrated its powerful ability to simulate the damage behavior of reinforced concrete structures. Yaghoobi et al. [23,24] introduced the semi-discrete method into the micropolar PD model to study the effect of reinforcing fibers on the analysis of cracks in cementitious materials. Shi et al. [25] proposed homogeneous and heterogeneous bonding for the same and different materials of adjacent material points. Zheng et al. [26] improved the micropolar peridynamics model by considering concrete materials' low tensile strength and high compressive strength and weakened the interaction between rebar and concrete in reinforced concrete structures. Zhang et al. [27] and Zheng et al. [28] coupled PD with an atomistic (AM) model and classical continuum mechanics (CCM) model, respectively, which guaranteed accuracy and enhanced computational efficiency. Shen et al. [29] established a new hybrid model of peridynamics (PD) and finite element method (FEM) based on implicit schemes to improve the computational efficiency. This paper focuses on the failure behavior of reinforced concrete walls under lateral loading and impact loading through a reasonable modeling approach and improved principal structure, providing an effective method for solving the shear wall force damage problem in engineering practice.

The remaining parts of this article are as follows: Section 2 summarizes the basic theory of peridynamics, improved micropolar peridynamic model, and modeling technique. Section 3 introduces the numerical implementation of the explicit dynamic algorithm. The proposed method provides two examples to verify its reliability, including the damage of the reinforced concrete shear wall under the static action in Section 4. It also includes data compared with experiments. For the conclusion, see Section 5.

2. Basic Theory of Peridynamics

2.1. Basic Ideas

The peridynamic theory was introduced by Silling [19] in 2000, which can be used at different scales, from micro to macro, does not assume continuous or small deformation behavior, and has no requirement for the concepts of stress and strain. The peridynamic equation of motion is an integral formulation rather than a partial differential equation, which faces many difficulties at discontinuities. Internal forces are expressed through nonlocal interactions between pairs of material points, and damage can be regarded as a part of the constitutive model. The peridynamic model allows discontinuities and cracks to emerge naturally.

As shown in Figure 1, the object in the spatial domain R is uniformly discretized into material points with material information, and at a certain moment t, each material point x_i does not interact only with its neighboring material points, but there is an intrinsic force of equal magnitude and opposite direction force between all other material points in its near-field range H, which can be obtained as

$$\rho\frac{\partial^2 u(x_i,t)}{\partial t^2} = \int_H f\big(u(x_j,t) - u(x_i,t), x_j - x_i\big)dV_{x_j} + b(x_i,t) \tag{1}$$

where ρ is the material point material density, u is the displacement of the material point, f is the interaction force between the two material points, $\partial^2 u(x_i,t)/\partial t^2$ is the acceleration, and b is the prescribed external body force density of the material point x_i.

$$\xi = x_j - x_i, \eta = u(x_j,t) - u(x_i,t) \tag{2}$$

where ξ is the length of the bond, and η is the relative displacement of the bond; the expression is as follows:

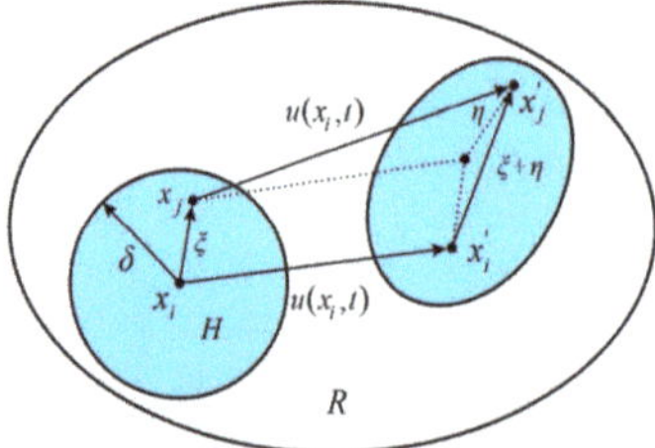

Figure 1. Interaction between material points.

The interaction force f can be written as

$$f(\xi,\eta) = \frac{(\xi+\eta)}{\parallel \xi+\eta \parallel}c(\xi)s\mu(\xi,t) \tag{3}$$

where $c(\xi)$ is a micromodulus function of the bond, and the relative stretch s of the bond can be expressed as

$$s = \frac{\parallel \xi+\eta \parallel - \parallel \xi \parallel}{\parallel \xi \parallel} \tag{4}$$

Represents the historically related scalar value function between the connection status of the bond between the material points, μ is given as follows:

$$\mu(\xi,t) = \begin{cases} 1, & s(\xi,t') < s_0(t) \text{ for all } 0 \le t' \le t \\ 0, & \text{otherwise} \end{cases} \tag{5}$$

where s_0 is a critical stretch of the material for failure. Failure occurs when the stretch between two particles exceeds the critical stretch, and these two particles stop interacting with each other. Even though S is referred to as a particle property, bond breaking must be symmetric for all pairs of particles that share a bond. The elastic modulus E, tensile strength f_t, and compressive strength f_c can be used to express the critical stretch of concrete material [26] as follows:

$$\begin{cases} s_t = 0.8f_t/E, & s(\xi) > 0 \\ s_c = -f_c/E, & s(\xi) < 0 \end{cases} \tag{6}$$

When the stretch s between material points exceeds the critical stretch s_t, it means that the bond between them is broken.

The broken condition of the bonds to the rest material points in the horizon determines the damage to the material point. This method is closer to the determination of real damage. The inexplicit nature of local damage at a particle arising from the introduction of failure in the constitutive model, an unambiguous notion of local damage at a point is described as a scalar-valued function $\varphi(x,t)$:

$$\varphi(x,t) = 1 - \frac{\int_H \mu(x,\xi,t)dH}{\int_H dH} \tag{7}$$

where H is the neighborhood of the particle x, which is assumed to be a spherical region centered at x with the radius of the material horizon, and is analogous to the cutoff radius used in molecular dynamics.

The function $\varphi(x,t)$ reflects the degree of damage, $\varphi \in [0,1]$. $\varphi = 0$ indicates that the material point is undamaged whereas $\varphi = 1$ means the material point has been completely damaged.

2.2. Micropolar Peridynamics

The microscopic elastic-brittle model (PMB) based on the "bond" theory only considers the radial movement between the material points and thus has a Poisson's ratio limitation. To solve this problem, the radial movement and rotation of the material points are considered simultaneously, and the equation of motion is expressed as

$$\begin{cases} \rho\frac{\partial^2 u(x_i,t)}{\partial t^2} = \int_H f\left(u(x_j,t) - u(x_i,t), x_j - x_i, \theta_i, \theta_j\right)dV_{x_j} + b(x_i,t) \\ I\frac{\partial^2 \theta(x_i,t)}{\partial t^2} = \int_H m\left(u(x_j,t) - u(x_i,t), x_j - x_i, \theta_i, \theta_j\right)dV_{x_j} + M(x_i,t) \end{cases} \tag{8}$$

where I denotes the rotational moment of inertia, θ_i, θ_j denotes the rotational displacement of the material point, m is the moment of the material points, and M is the external moment.

The force function between the material point pairs f is related to the displacement u between the material points by the expression

$$f = ku \tag{9}$$

Equation (9) can also be represented in the form of a matrix

$$\begin{pmatrix} f_{xi} \\ f_{yi} \\ m_{zi} \\ f_{xj} \\ f_{yj} \\ m_{zj} \end{pmatrix} = \begin{pmatrix} EA/\zeta & 0 & 0 & -EA/\zeta & 0 & 0 \\ 0 & 12EI/\zeta^3 & 6EI/\zeta^2 & 0 & -12EI/\zeta^3 & 6EI/\zeta^2 \\ 0 & 6EI/\zeta^2 & 4EI/\zeta & 0 & -6EI/\zeta^2 & 2EI/\zeta \\ -EA/\zeta & 0 & 0 & EA/\zeta & 0 & 0 \\ 0 & -12EI/\zeta^3 & -6EI/\zeta^2 & 0 & 12EI/\zeta^3 & -6EI/\zeta^2 \\ 0 & 6EI/\zeta^2 & 2EI/\zeta & 0 & -6EI/\zeta^2 & 4EI/\zeta \end{pmatrix} \begin{pmatrix} u_i \\ v_i \\ \theta_{zi} \\ u_j \\ v_j \\ \theta_{zj} \end{pmatrix} \tag{10}$$

where $EA = c$, $EI = d$, c is the tensile micromodulus, and d is the bending micromodulus of the bond. I is the rotational inertia, and E is the elastic modulus.

Considering the material point pair between the remote force action, after correction to obtain the tensile and bending micromodulus coefficients in the case of plane stress problem are as follows [30]:

$$\begin{cases} c = \frac{105E}{4\pi\delta^3(1-v)}\left(1-\left(\frac{\zeta}{\delta}\right)^2\right)^2 \\ d = \frac{5E(1-3v)}{16\pi\delta(1-v^2)}\left(1-\left(\frac{\zeta}{\delta}\right)^2\right)^2 \end{cases} \tag{11}$$

The micropolar peridynamics break through the limitation of a certain Poisson's ratio. The Poisson's ratio for the plane stress problem is changed from 0.33 to a range of 0–0.33, whereas the Poisson's ratio for the three-dimensional problem is changed from 0.25 to a range of 0–0.25.

2.3. Rebar Constitutive Model

This article assumes that the rebars in the reinforced concrete structure cannot break due to damage. The rebars use a damage model similar to the ideal elastoplastic model. Their tensile and compressive properties are assumed to be the same. The force between the material points is linearly elastic when the stretch between the material points of the rebars is between the critical stretch and the critical compression rate. The force between the material points is constant when the stretch between the material points exceeds the critical stretch. The horizontal line in the model signifies that the bar has reached yield, as shown in Figure 2.

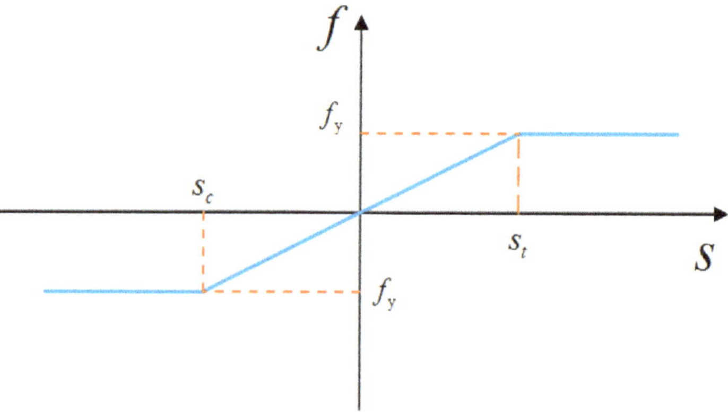

Figure 2. Reinforcement constitutive model.

2.4. Concrete Constitutive Model

To describe the damage behavior of concrete, the micropolar force vector-bond stretch relationship is proposed [23], which improves the concrete instanton of the micropolar peridynamic model by introducing a correction factor for bond damage with the following expression: $F = (1 - \lambda)KU$, λ is a damage factor between 0 and 1. $\lambda = 0$ means no damage, whereas $\lambda = 1$ means complete damage. s_{ft} controls the slope of the softening branch. Considering that the compressive strength of concrete is much higher than the tensile strength, concrete will experience fatigue, and its bearing capacity will be reduced when suffering loads in the opposite direction. This paper reduces s_{uc}, s_c, s_t, and s_{ut} when the direction of the bond force changes, to achieve the role of simulating concrete fatigue. The specific operation is as follows: when the bond force positive and negative sign changes, s_{uc}, s_{0c}, s_{0t}, and s_{ut} will multiply a weakening parameter α, thus reflecting concrete fatigue, concrete is more easily destroyed. As shown in Equation (12) and the bond force-stretch function is shown in Figure 3.

$$
f = \begin{cases}
1 & (s < \alpha s_{uc}) \\
0 & (\alpha s_{uc} \leq s \leq \alpha s_{0t}) \\
1 - \dfrac{\alpha s_{0t}}{s} e^{-\frac{s - \alpha s_{0t}}{\alpha s_{ft} - \alpha s_{0t}}} & (\alpha s_{0t} \leq s < \alpha s_{ut}) \\
1 & (\alpha s_{ut} \leq s)
\end{cases}
\tag{12}
$$

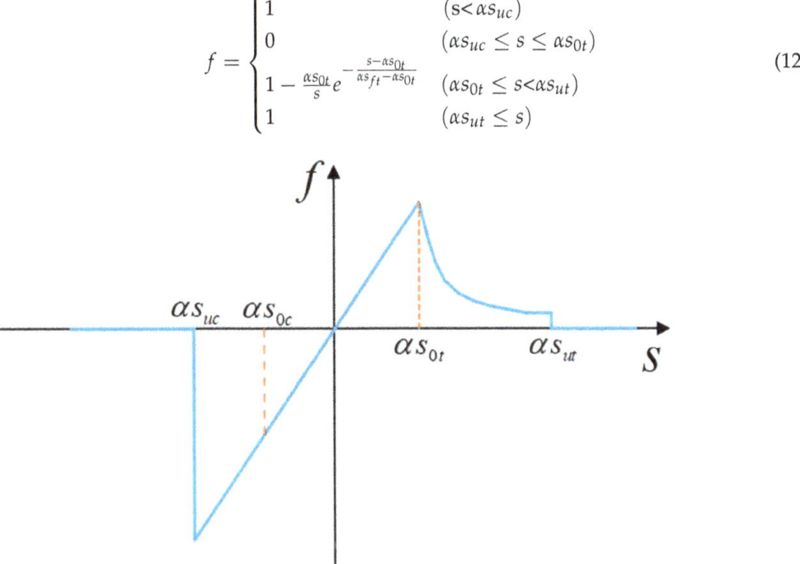

Figure 3. Adhesion-tension relationship for concrete material model.

In the present constitute model, the whole damage of the connection can be expressed in two ways. The bond breaks completely when the bond extension s is less than αs_{uc} or greater than αs_{ut}, then the concrete is damaged. When the bond extension s is between αs_{uc} and αs_{0t}, the intrinsic force and stretch vary linearly. When the bond extension s is between αs_{0t} and αs_{ut} the intrinsic force decreases exponentially.

2.5. Interface Processing

How to describe the interface between steel and concrete is a great challenge in peridynamics. As the strength of the bond at the interface is lower in concrete, this paper adopts a weakened connection property. In reinforced concrete structures, the two materials with different physical properties mainly rely on the bond stress between them to work together. Sliding damage to the reinforced concrete bonding interface is the reason for the performance to failure. Bond failure is a cross-scale process in which a large amount of fine-scale damage develops and induces macroscopic failure [31]. To solve the problem, it should be distinguished from bonds between homogeneous materials. A little plastic deformation begins to occur in concrete, microcracks begin to sprout, and strains grow rapidly faster when the tensile stress σ reaches 40–60% of the peak stress f_y. Therefore, the interaction between rebar and concrete in reinforced concrete structures is weakened and expressed as follows [32]:

$$\begin{cases} c = \min\left\{ c_{x_i}, 0.4\left(c_{x_i} + c_{x_j} \right), c_{x_j} \right\} \\ d = \min\left\{ d_{x_i}, 0.4\left(d_{x_i} + d_{x_j} \right), d_{x_j} \right\} \end{cases} \tag{13}$$

where c_{x_i} is the tensile micromodulus, d_{x_i} is the bending micromodulus of material point x_i, c_{x_j} is the tensile micromodulus, and d_{x_j} is the bending micromodulus of material point x_j.

3. Numerical Implementation

3.1. Artificial Damping

According to the basic idea of peridynamics, the object is uniformly discretized into material points, and the material point spacing is taken as Δx. In solving the quasi-static problem, the dynamic relaxation algorithm is borrowed, and artificial damping is added to the equations of motion, and the equations converge rapidly, then the equations are as follows:

$$\begin{cases} \rho \ddot{u}(x,t) + C\dot{u}(x,t) = \sum\limits_{j=1}^{p} f(\eta,\xi,\theta_i,\theta_j)V_j + b(x,t) \\ I\ddot{\theta}(x,t) + C\dot{\theta}(x,t) = \sum\limits_{j=1}^{p} m(\eta,\xi,\theta_i,\theta_j)V_j + M(x,t) \end{cases} \tag{14}$$

where p is the number of other material points in the horizon, and V_j is the volume of material point j. C is the artificial damping, whose size only affects the speed of computational convergence.

3.2. Impact Contact Algorithm

As shown in Figure 4, the impacted concrete beam is stationary when the falling hammer first starts to contact the concrete beam. The combined force at each material point is zero. When the falling hammer penetrates the concrete beam, the material point occupied by the falling hammer is squeezed away to the surface of the falling hammer [33].

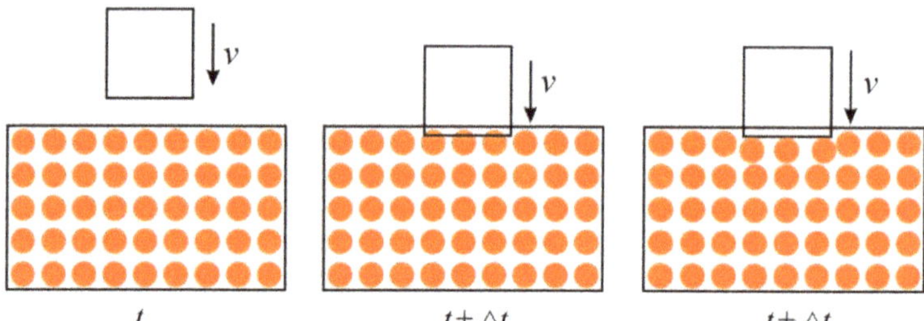

Figure 4. Rigid projectile impact contact model.

The coordinate of a material point at time t is x_k, $u^t_{(k)}$ is the displacement, and $v^t_{(k)}$ is the velocity. The falling hammer is in contact with the concrete beam at moment t, and the material point overlaps with the falling hammer at moment $t + \Delta t$. In the process of impact, the falling hammer continuously moved the material points occupied by the concrete beam from its track to the closest position on the surface of the falling hammer, and the new displacement after the displacement is calculated as $u^{t+\Delta t}_{(k)}$, and the new velocity is $\overline{v}^{t+\Delta t}_{(k)}$:

$$\overline{v}^{t+\Delta t}_{(k)} = \frac{\overline{u}^{t+\Delta t}_{(k)} - u^t_{(k)}}{\Delta t} \tag{15}$$

The combined force on the falling hammer is

$$F^{t+\Delta t} = \sum_{k=1} -\rho_k \frac{\overline{v}_{(k)}^{t+\Delta t} - v_{(k)}^t}{\Delta t} V_{(k)} \lambda_{(k)}^{t+\Delta t} \tag{16}$$

where $\lambda_{(k)}^{t+\Delta t} = 1$ means that the falling hammer and the material point overlap, $\lambda_{(k)}^{t+\Delta t} = 0$ means that the two do not overlap situation.

To prevent the occurrence of mutual penetration deformation between material points when simulating concrete structures, a short-range repulsive force term needs to be added between material points, and the short-range repulsive force equation is as follows:

$$f_s(y',y) = \min\left\{0, \frac{135E}{\pi\delta^4}(\|y'-y\|) - d\right\} \frac{y'-y}{\|y'-y\|} \tag{17}$$

The expression of the short-range force action distance between material points is written as:

$$d = \min\left\{0.9\|x'-x\|, 1.35|\Delta x|\right\} \tag{18}$$

4. Numerical Examples

4.1. Quasi-Static Damage of RC Shear Walls

The RC shear wall is shown in Figure 5. Concrete properties are stated as follows. The concrete strength grade is C30, Young's modulus $E = 21.4$ GPa, Poisson's ratio $v = 0.2$, the mass density $\rho = 2400$ kg/m^3, the tensile strength of 2.94 N/mm^2, and the compressive strength of 29.1 N/mm^2. Steel properties are also described as follows. The rebar grade is HRB400, Young's modulus $E = 211$ GPa, Poisson's ratio $v = 0.3$, and the mass density $\rho = 7800$ kg/m^3. The model contains a total of 140,751 cell points with a time step of 10^{-7} s. The loading zone is 350 mm $\times$ 50 mm in the upper part, and 350 mm $\times$ 350 mm in the lower part for observing displacement and damage. The axial pressure ratio is controlled by applying vertical force in the upper half zone, and for comparison with the experimental data, the axial pressure ratio is controlled to be 0.1, and then lateral force is applied to observe the damage pattern of the RC shear walls.

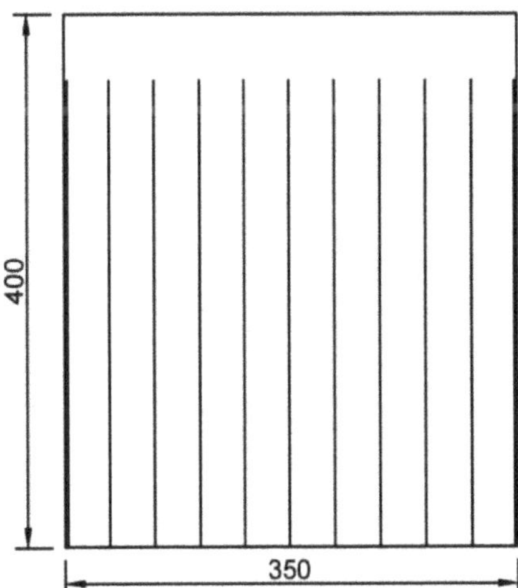

Figure 5. Geometrical dimensions of the RC shear walls.

Figure 6a–d are the schematic diagrams of the crack expansion during the damage process of the RC shear walls. At first, there is no damage to the member. Then cracks appear in the wall from the bottom with increased applied load. Cracks appear at the bond between bars and concrete and develop rapidly. As the load continues to increase, cracks in the corner of the wall start to develop upward, and significant damage appears on the compressed side. Finally, cracks connecting the diagonal of the shear wall appeared, meanwhile significant damage appeared in the compressed area, and the RC shear walls were damaged.

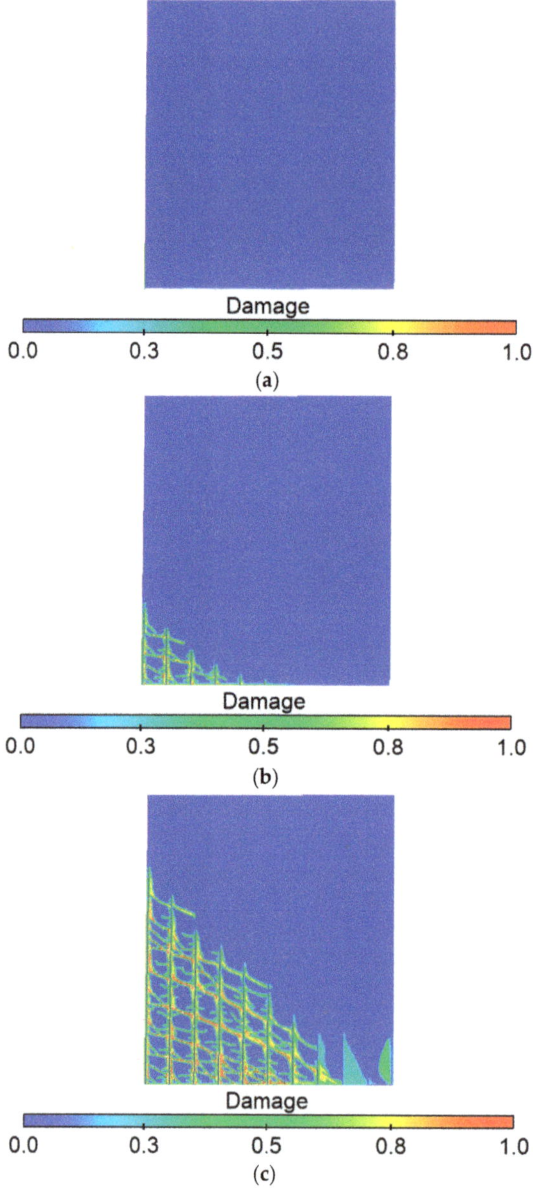

Figure 6. *Cont.*

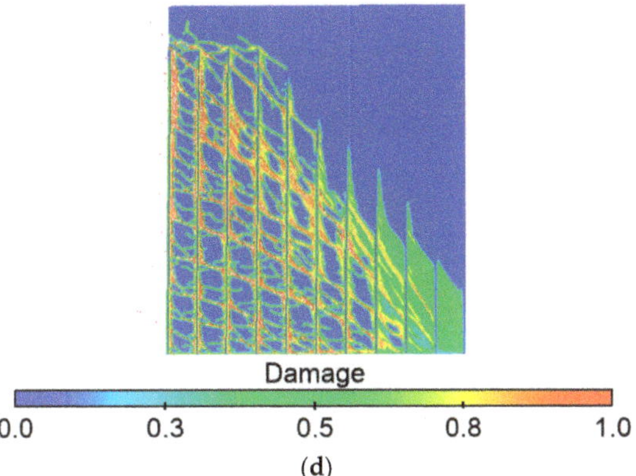

Figure 6. RC shear walls unilateral force damage process. (**a**) Initial stage; (**b**) Crack initiation; (**c**) Crack propagation and (**d**) Structure failure.

As shown in Figure 7, the rack development and final damage results of the peridynamic are similar to the experimental results [12]. The numerical result of the load-displacement curve is consistent with the experiment as shown in Figure 8.

Figure 7. Damage modes of RC shear walls [12]. Copyright 2017, Engineering Structures.

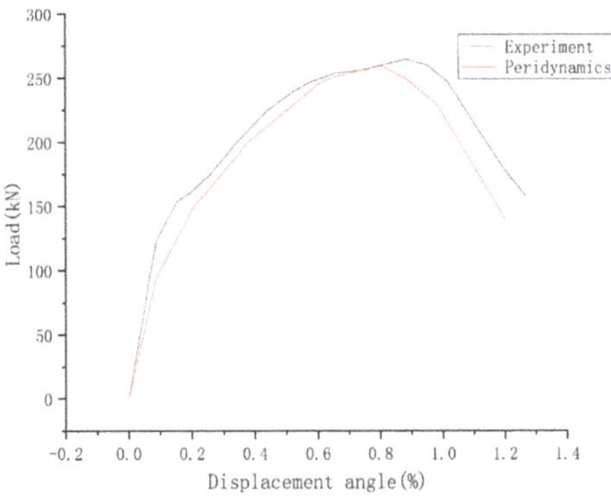

Figure 8. Load-displacement curve [12].

4.2. Failure Behavior of the RC Shear Walls under Impact Loading

As shown in Figure 9, the RC shear wall's dimensions are 1.1 m × 2.1 m × 0.16 m. Material properties for concrete are Young's modulus E = 30 GPa, Poisson's ratio v = 0.2, and density ρ = 2400 kg/m^3. The compressive strength of concrete is 30.7 N/mm^2, which is assumed to be a homogeneous, isotropic, elastic, and brittle material. The properties of the rebars: Young's modulus E = 200 GPa, Poisson's ratio v = 0.3, yield strength f_y = 442 MPa and mass density ρ = 7800 kg/m^3. The model contains a total of 398,157 cell points with a time step of 10^{-7} s.

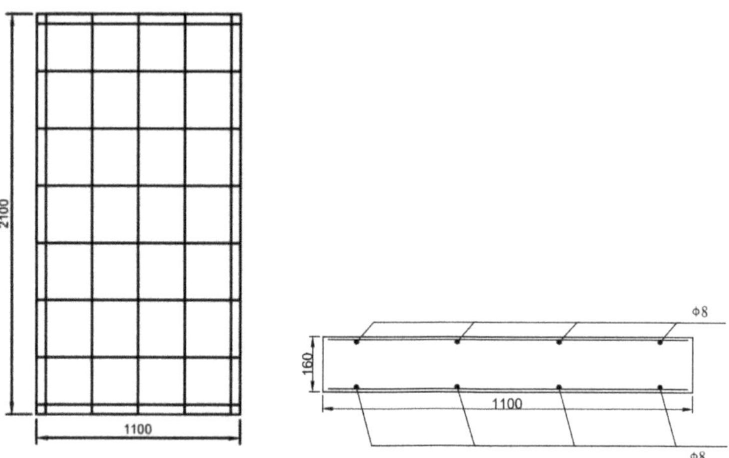

Figure 9. RC shear walls dimensions and reinforcement distribution.

A 640 kg drop hammer falls freely from 0.75 m. The impact side of the shear wall is front and the other side is back. The falling hammer is only along the impact direction movement to ensure that the falling hammer in the impact process does not occur in other directions of displacement, to avoid causing the loss of impact energy. The constraint is achieved by fixing the displacement of the two short sides of the RC shear walls.

Figure 10 is a schematic diagram of the crack expansion process on the impacted side. The failure process is shown in Figure 10a–e. During the initial stage of simulation, no microcracks were observed. When the falling hammer contacted the wall, the cracks arose in the center and propagated in all directions. As the damage increased, transverse cracks appeared, and damage grew at the fixed position. As the simulation proceeded, cracks covered the surface. Then the forked cracks on the wall surface deepened, and the damage increased rapidly. Figure 11 shows the crack expansion process on the back side. Compared with the front side, the back-side cracks appear later, and there are fewer cracks and less damage.

It is consistent with the crack development pattern of the test results in Figure 12. The impact force-time course curve is shown in Figure 13. The peak test impact force is at 562 kN, and the peridynamic theory simulation result is at 612 kN, with an error of 8.9%. The time displacement curve at the midpoint of the RC shear walls is shown in Figure 14. The maximum displacement of the experimental midpoint is 33 mm, and the peridynamic theory simulation results in 33.8 mm, with an error of 2.4%.

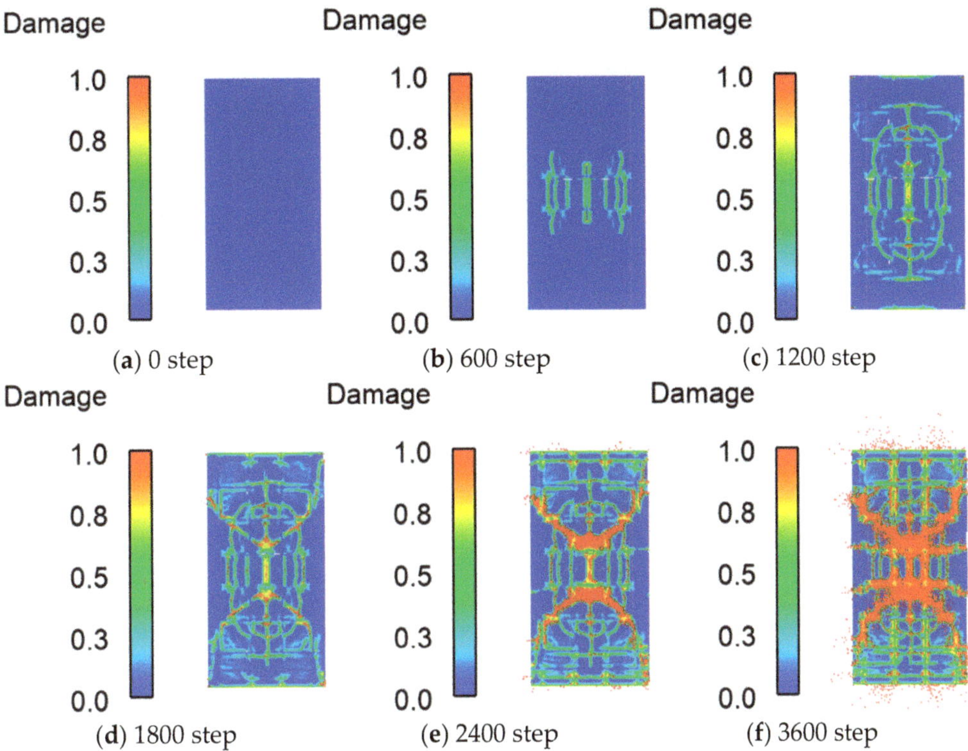

Figure 10. Crack development process at the front of the RC shear wall after falling hammer contact with the RC shear wall.

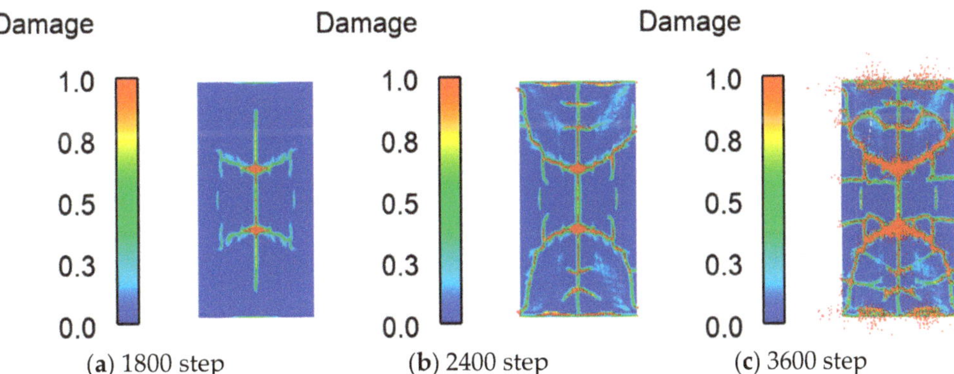

Figure 11. Crack development process at the back of the RC shear wall after falling hammer contact with the RC shear wall.

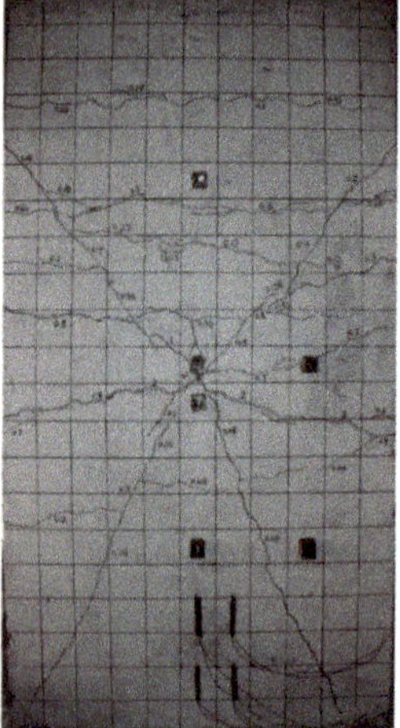

Figure 12. Cracks at the back of the RC shear wall [14]. Copyright 2019, Vibration and Shock.

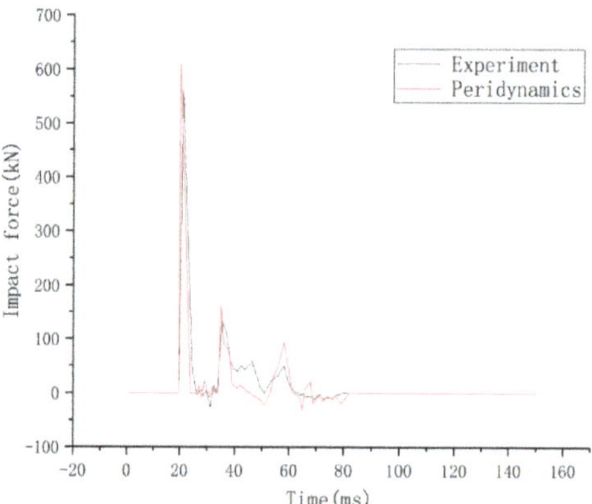

Figure 13. Time histories of impact force [14].

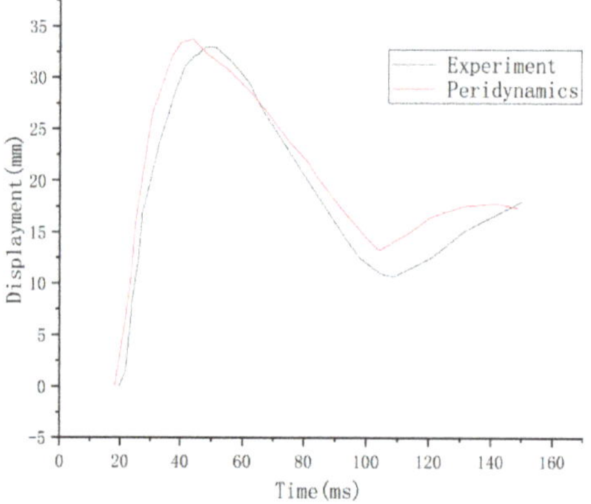

Figure 14. Midpoint displacement of the wall [14].

5. Conclusions

To accurately describe the failure behavior and damage evolution of reinforced concrete shear walls, this paper improves the micropolar peridynamic model. Considering that the compressive strength of concrete is much higher than the tensile strength and the effect of concrete fatigue on concrete strength, the role of the "bond" between reinforcement and concrete is weakened. Based on the improved micropolar peridynamic, the failure behavior of the quasi-static reinforced concrete shear wall is simulated. The cracks appeared from the bond between the rebars and the concrete in the shear zone and developed rapidly. Finally, cracks connecting the diagonal of the shear wall appeared; meanwhile, significant damage appeared in the compressed area, and the RC shear walls were damaged. Since the reinforcement and concrete are idealized native structures, there is a specific difference in the simulated results and an inevitable error in the load-displacement angle curve.

In addition, through the contact algorithm method, a whole failure process of RC shear walls was acquired, which from cracks appearing from the center of the shear wall, the cracks develop around, transverse cracks appear, and finally, the forked cracks deepen, and lead to the failure of the shear wall. The appearance and development of front and back cracks are consistent with the experiment, the impact displacement curve is well-fitted, and the maximum impact force error is within a small range.

The results show that the peridynamic method has a better ability to simulate the cumulative accumulation and gradual destruction process of RC shear walls, which provides a compelling new method to solve the RC shear walls damage problem in engineering. The improved micropolar peridynamics model is accurate and effective to calculate the failure behavior of RC shear walls. However, the peridynamics has the problems of low computational efficiency and difficult results processing, which need to be further studied.

Author Contributions: Conceptualization, F.S. and Z.C.; methodology, F.S. and J.Z.; software, Z.C. and J.Z.; validation, Z.C.; formal analysis, Z.C.; data curation, Z.C.; writing—original draft preparation, F.S. and Z.C.; writing—review and editing, J.Z. and Q.Z.; visualization, Z.C.; supervision, F.S. and Q.Z.; funding acquisition, F.S. and Q.Z. All authors have read and agreed to the published version of the manuscript.

Funding: This research was funded by the National Natural Science Foundation of China (51709194, U1934206, 11932006, 12172121); the Qinglan Project of Jiangsu University, and the Priority Academic Program Development of Jiangsu Higher Education Institutions.

Institutional Review Board Statement: Not applicable.

Informed Consent Statement: Not applicable.

Data Availability Statement: Not applicable.

References

1. Jiang, H.J.; Ying, Y.; Wang, B.; Zhang, Y.X. Experiment on seismic damage behavior of RC shear walls. *Build. Struct.* **2012**, *42*, 113–117. (In Chinese)
2. Cao, R.J. Seismic Damage Analysis of Reinforced Concrete Shear Walls. Master's Thesis, Xi'an University of Science and Technology, Xi'an, China, 2016. (In Chinese).
3. Lopes, M.S. Experimental shear-dominated response of RC walls: Part I: Objectives, methodology and results. *Eng. Struct.* **2001**, *23*, 229–239. [CrossRef]
4. Lopes, M.S. Experimental shear-dominated response of RC walls. Part II: Discussion of results and design implications. *Eng. Struct.* **2001**, *23*, 564–574. [CrossRef]
5. Paulay, T.; Priestley, M.J.N.; Synge, A.J. Ductility in earthquake resisting squat shearwalls. *J. Proc.* **1982**, *79*, 257–269.
6. Lefas, I.D.; Kotsovos, M.D.; Ambraseys, N.N. Behavior of reinforced concrete structural walls: Strength, deformation characteristics, and failure mechanism. *Struct. J.* **1990**, *87*, 23–31.
7. Sittipunt, C.; Wood, S.L. Influence of web reinforcement on the cyclic response of structural walls. *Struct. J.* **1995**, *92*, 745–756.
8. Vallenas, J.M.; Bertero, V.V.; Popov, E.P. Hysteric behavior of reinforced concrete structural walls. *Recon. Tech. Rep. N* **1979**, *80*, 27533.

9. Terzioglu, T.; Orakcal, K.; Massone, L.M. Cyclic lateral load behavior of squat reinforced concrete walls. *Eng. Struct.* **2018**, *160*, 147–160. [CrossRef]
10. Salonikios, T.N. Shear strength and deformation patterns of R/C walls with aspect ratio 1.0 and 1.5 designed to Eurocode 8 (EC8). *Eng. Struct.* **2002**, *24*, 39–49. [CrossRef]
11. Salonikios, T.N. Analytical prediction of the inelastic response of RC walls with low aspect ratio. *J. Struct. Eng.* **2007**, *133*, 844–854. [CrossRef]
12. Looi, D.T.W.; Su, R.K.L.; Cheng, B.; Tsang, H.H. Effects of axial load on seismic performance of reinforced concrete walls with short shear span. *Eng. Struct.* **2017**, *151*, 312–326. [CrossRef]
13. Houshmand-Sarvestani, A.; Totonchi, A.; Shahmohammadi, M.A. Numerical assessment of the effects of ADAS yielding metallic dampers on the structural behavior of steel shear walls (SSWs). *Mech. Based Des. Struc.* **2023**, *51*, 1626–1644. [CrossRef]
14. Yi, J.W.; Shi, X.D. Numerical simulation analysis for RC shear walls under impact load. *Vibr. Shock* **2019**, *38*, 102–110. (In Chinese)
15. Tan, J.K.; Su, M.N.; Wang, Y.H. Experimental study on cyclic shear performance of steel plate shear wall with different buckling restraints. *Structures.* **2022**, *35*, 469–482. [CrossRef]
16. Zhang, H.; Cheng, X.; Li, Y. Prediction of failure modes, strength, and deformation capacity of RC shear walls through machine learning. *J. Build. Eng.* **2022**, *50*, 104145. [CrossRef]
17. Grégoire, D.; Maigre, H.; Rethore, J.; Combescure, A. Dynamic crack propagation under mixed-mode loading–comparison between experiments and X-FEM simulations. *Int. J. Solid Struct.* **2007**, *44*, 6517–6534. [CrossRef]
18. Meng, Q.; Xue, H.; Song, H.; Zhuang, X.; Rabczuk, T. Rigid-Block DEM Modeling of Mesoscale Fracture Behavior of Concrete with Random Aggregates. *J. Eng. Mech.* **2023**, *149*, 2. [CrossRef]
19. Silling, S.A. Reformulation of elasticity theory for discontinuities and long-range forces. *J. Mech. Phys. Solids* **2000**, *48*, 175–209. [CrossRef]
20. Gerstle, W.; Sau, N.; Silling, S. Peridynamic modeling of concrete structures. *Nuclear Eng. Design* **2007**, *237*, 1250–1258. [CrossRef]
21. Gerstle, W.; Sau, N.; Aguilera, E. Micropolar peridynamic constitutive model for concrete 2007. In Proceedings of the 19th International Conference on Structural Mechanics in Reactor Technology, Toronto, Japan, 7 August 2007.
22. Zhang, N.; Gu, Q.; Huang, S.; Xue, X.; Li, S. A practical bond-based peridynamic modeling of reinforced concrete structures. *Eng. Struct.* **2021**, *244*, 112748. [CrossRef]
23. Yaghoobi, A.; Chorzepa, M.G. Fracture analysis of fiber reinforced concrete structures in the micropolar peridynamic analysis framework. *Eng. Fract. Mech.* **2017**, *169*, 238–250. [CrossRef]
24. Yaghoobi, A.; Chorzepa, M.G. Meshless modeling framework for fiber reinforced concrete structures. *Comput. Struct.* **2015**, *161*, 43–54. [CrossRef]
25. Shi, H.S.; Qian, S.R.; Xu, T. Study on reinforced concrete structure failure based on peridynamic theories. *Guizhou Sci.* **2016**, *34*, 64–68. (In Chinese)
26. Zheng, J.; Shen, F.; Gu, X.; Zhang, Q. Simulating failure behavior of reinforced concrete T-beam under impact loading by using peridynamics. *Int. J. Impact Eng.* **2022**, *165*, 104231. [CrossRef]
27. Zhang, J.; Han, F.; Yang, Z.; Cui, J. Coupling of an atomistic model and bond-based peridynamic model using an extended Arlequin framework. *Comput. Methods Appl. Mech. Eng.* **2023**, *403*, 115663. [CrossRef]
28. Zheng, G.; Li, L.; Han, F.; Xia, Y.; Shen, G.; Hu, P. Coupled peridynamic model for geometrically nonlinear deformation and fracture analysis of slender beam structures. *Int. J. Numer. Methods Eng.* **2022**, *123*, 3658–3680. [CrossRef]
29. Shen, F.; Yu, Y.; Zhang, Q.; Gu, X. Hybrid model of peridynamics and finite element method for static elastic deformation and brittle fracture analysis. *Eng. Anal. Bound. Elements.* **2020**, *113*, 17–25. [CrossRef]
30. Zhu, H.Y.; Liu, L.S.; Liu, Q.W.; Lai, X. Application of an Improved Bond-based Peridynamic Method to Analysis of Flow around a Cylinder. *Saf. Environ. Eng.* **2018**, *5*, 174–180. (In Chinese)
31. Bai, Y.L.; Wang, H.Y.; Xia, M.F.; Ke, J.F. Statistical mesomechanics of solid, linking coupled multiple space and time scales. *Appl. Mech. Rev.* **2005**, *58*, 372–388. [CrossRef]
32. Guo, Z.H.; Shi, X.D. *Reinforced Concrete Theory and Analysis*; Tsinghua University Press: Beijing, China, 2003. (In Chinese)
33. Madenci, E.; Oterkus, E. *Peridynamic Theory and its Applications*; Springer: New York, NY, USA, 2013.

Article

Traction-Associated Peridynamic Motion Equation and Its Verification in the Plane Stress and Fracture Problems

Ming Yu, Zeyuan Zhou and Zaixing Huang *

State Key Laboratory of Mechanics and Control of Mechanical Structures, Nanjing University of Aeronautics and Astronautics, 29 Yudao Street, Nanjing 210016, China
* Correspondence: huangzx@nuaa.edu.cn

Abstract: How to prescribe traction on boundary surface is still an open question in peridynamics. This problem is investigated in this paper. Through introducing the induced body force defined by boundary traction, the Silling's peridynamic motion equation is extended to a new formulation called the traction-associated peridynamic motion equation, which is verified to be compatible with the conservation laws of linear momentum and angular momentum. The energy conservation equation derived from the traction-associated peridynamic motion equation has the same form as that in the original peridynamics advanced by Silling. Therefore, the constitutive models of the original peridynamics can be directly applied to the traction-associated peridynamic motion equation. Some benchmark examples in the plane stress problems are calculated. The numerical solutions agree well with the classical elasticity solutions, and the volume correction and the surface correction are no longer needed in the numerical algorithm. These results show that the traction-associated peridynamic motion equation not only retains all advantages of the original peridynamics, but also can conveniently deal with the complex traction boundary conditions.

Keywords: peridynamics; traction-associated peridynamic motion equation; traction boundary condition; bond-based constitutive model

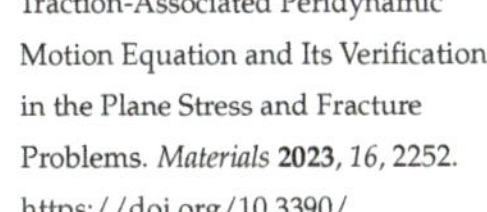

Citation: Yu, M.; Zhou, Z.; Huang, Z. Traction-Associated Peridynamic Motion Equation and Its Verification in the Plane Stress and Fracture Problems. *Materials* **2023**, *16*, 2252. https://doi.org/10.3390/ma16062252

Academic Editor: Angelo Marcello Tarantino

Received: 31 January 2023
Revised: 28 February 2023
Accepted: 5 March 2023
Published: 10 March 2023

1. Introduction

Peridynamics (PD) is a nonlocal continuum theory of mechanics developed in the recent two decades [1–5]. Its core consists in that a weighted integral of relative displacement over spatial domain is used instead of the gradient of displacement (strain) in the governing equations of deformation. Therefore, peridynamics can be used to conveniently and effectively analyze deformation companied with evolution of discontinuities. Peridynamics was firstly proposed by Silling [1] and then further improved by Silling and his collaborators [2]. Since then, it has been applied to investigate various problems associated with wave, damage, fracture, and impact breakage [3–9]. However, the traction boundary condition is incompatible with peridynamics because the governing equation of original peridynamics appears in the form of an integro-differential equation and does not involve the gradient with respect to spatial position. Therefore, boundary conditions cannot be imposed as naturally in the original peridynamics as in Classical Continuum Mechanics (CCM). A question therefore arises: how to incorporate suitable traction boundary condition?

Recalling the ways of the traction boundary condition is imposed in peridynamics, we can roughly divide them into three types. The first method [10] is to convert the traction on the boundary surface into the body force in an inner boundary layer according to the static force equivalence. The body force is usually supposed to uniformly distribute in the boundary layer, and the thickness of the inner boundary layer is taken as the spacing between material points. The second type of method is still based on the original peridynamics motion equation, but some modifications or operations will be performed to achieve the purpose of directly imposing traction boundary conditions. For example, a method only applicable to the non-ordinary correspondence models in peridynamics is proposed in [11], a weak form of peridynamic governing equations is proposed in [12,13], and a way based on Taylor expansion strategy to impose traction boundary conditions in ordinary state-based peridynamics is proposed in [14,15]. As the third method, Huang [16] suggested modifying the peridynamic motion equation to introduce the surface traction.

The traction boundary condition based on the first method is very popular in theoretical analysis and numerical calculation. Using this traction boundary condition, the well-posedness of the linear peridynamics with a given nonlocal kernel has been proved in mathematics [17,18]. Many numerical computation and analysis for practical problems can be found in [4,5,8]. However, it is unnatural and unhandy to convert the traction on the boundary surface to body forces in the inner boundary layer. Specially, when the material particles are close to the boundary surface or interface, the constitutive parameters need to be corrected [4,10,19,20]. Consequently, the traction boundary condition based on the first method is impractical for the sophisticated loading cases and geometrical surface.

If the non-ordinary correspondence models are adopted, the traction boundary condition can be specified directly. However, the non-ordinary correspondence models are limited due to the zero energy mode [21,22], and there are still some problems to be solved. The weak form of peridynamic governing equations has some changes comparing with the original peridynamic governing equation. Although the method of imposing traction boundary conditions based on Taylor expansion strategy can be successfully used in bond-based peridynamics, it still needs to set fictitious nodes. As for the model established by the third method, it is new and promising. Zhou [23] simplified the boundary transfer functions and proposed a method of imposing traction boundary conditions suitable for two-dimensional problems for bond-based peridynamics. However, since the three scalar-typical boundary transfer functions involved in this idea are difficult to determine, it has not been widely applied yet. Overall, how to characterize the traction boundary condition in peridynamics is still an open question and thus requires further investigation.

Meshfree discretization has been widely used to solve PD problems [24–26]. However, the standard discretization scheme is still very expensive for calculating the large-scale problems such as 3-dimensional problems. A new meshfree scheme [27] is expected to solve these difficulties.

In PD theory, there is no complete non-local neighborhood of a material point near a material boundary, resulting in the so-called skin or surface effects. Generally, the surface effect needs to be corrected to obtain the correct physical results. Various surface effect correction methods were studied in [10]. In addition, there are also many attempts to eliminate the surface correction [28–33].

The outline of the paper is as follows. In Section 2, through introducing the induced body force defined by boundary traction, we propose the traction-associated peridynamic motion equation and show that it is compatible with the conservation laws of linear momentum and angular momentum. In Section 3, the energy conservation law is derived from the traction-associated peridynamic motion equation. Two kinds of the bond-based constitutive models are discussed. The concrete form of the induced body force is determined. The prototype microelastic (PMB) constitutive model with the local damage is briefly introduced. The numerical algorithm is discussed in Section 4. In Section 5, the traction-associated peridynamic motion equation is used to calculate three benchmark examples in the plane stress problems. A fracture problem is simulated to verify the effectiveness of traction-associated peridynamic motion equation in the failure analysis. Finally, we close this paper with summary and comment.

2. The Induced Body Force and Extension of Peridynamic Motion Equation

Let $\mathbf{p}(\mathbf{x}''', t)$ is a traction exerted at the point $\mathbf{x}'''$ on the boundary surface $\partial\Omega_p \subset \partial\Omega$ of a peridynamic media Ω. Due to the nonlocality of peridynamic media, $\mathbf{p}(\mathbf{x}''', t)$ will permeate the interior of Ω and induce a body force $\mathbf{b_i}(\mathbf{x}, t)$ acting at the material particles. We call $\mathbf{b_i}(\mathbf{x}, t)$ the induced body force, which is represented as

$$\mathbf{b_i}(\mathbf{x}, t) = \int_{\partial\Omega_p} \left[\mathbf{G}\left(\mathbf{x}, \mathbf{x}''', t\right) \cdot \mathbf{p}\left(\mathbf{x}''', t\right) \right] dA_{\mathbf{x}'''}. \tag{1}$$

In Equation (1), $\mathbf{G}\left(\mathbf{x}, \mathbf{x}''', t\right)$ is a second-order tensor field called the transfer function of boundary traction and with the dimension of $1/\mathrm{m}^3$, which reads

$$\mathbf{G}\left(\mathbf{x}, \mathbf{x}''', t\right) = \mathbf{P}\left(\mathbf{x}, \mathbf{x}''', t\right) \cdot \mathbf{S}\left(\mathbf{x}''', t\right), \tag{2}$$

where

$$\begin{cases} \mathbf{P}\left(\mathbf{x}, \mathbf{x}''', t\right) = \alpha\left(\mathbf{x}''', \mathbf{x}\right) \dfrac{\left(\mathbf{y}''' - \mathbf{y}\right) \otimes \left(\mathbf{y}''' - \mathbf{y}\right)}{\left(\mathbf{y}''' - \mathbf{y}\right) \cdot \left(\mathbf{y}''' - \mathbf{y}\right)} \\ \mathbf{S}\left(\mathbf{x}''', t\right) = \left[\int_{\Omega} \mathbf{P}\left(\mathbf{x}, \mathbf{x}''', t\right) dV_{\mathbf{x}} \right]^{-1} \end{cases}. \tag{3}$$

In Equation (3), $\mathbf{y}''' = \mathbf{x}''' + \mathbf{u}'''(\mathbf{x}''', t)$ is the position vector of $\mathbf{x}'''$ in the deformed configuration and $\mathbf{u}''' = \mathbf{u}'''(\mathbf{x}''', t)$ is a displacement field on $\partial\Omega_p$. $\alpha(\mathbf{x}''', \mathbf{x})$ is a weight function determining the characteristic of $\mathbf{G}(\mathbf{x}, \mathbf{x}''', t)$ and will be discussed in the following. By Equations (2) and (3), it is easy to verify

$$\int_{\Omega} \mathbf{G}\left(\mathbf{x}, \mathbf{x}''', t\right) dV_{\mathbf{x}} = \mathbf{I}, \tag{4}$$

where $\mathbf{I}$ is the second order unit tensor. Equation (4) is a sufficient and necessary condition to ensure the compatibility of peridynamic motion equation with total equilibrium of linear momentum and angular momentum. Next, let us discuss this argument.

By Equation (4), the integrals of Equation (1) and $\mathbf{y}(\mathbf{x}, t) \times$ Equation (1) over Ω lead to

$$\int_{\Omega} \mathbf{b_i}(\mathbf{x}, t) dV_{\mathbf{x}} = \int_{\partial\Omega_p} \mathbf{p}\left(\mathbf{x}''', t\right) dA_{\mathbf{x}'''} \tag{5}$$

$$\int_{\Omega} \mathbf{y}(\mathbf{x}) \times \mathbf{b_i}(\mathbf{x}, t) dV_{\mathbf{x}} = \int_{\partial\Omega_p} \mathbf{y}'''\left(\mathbf{x}'''\right) \times \mathbf{p}\left(\mathbf{x}''', t\right) dA_{\mathbf{x}'''}. \tag{6}$$

After the induced body force $\mathbf{b_i}(\mathbf{x}, t)$ is introduced, the motion equation of peridynamic media subjected simultaneously to boundary traction and external body force can be written as

$$\rho(\mathbf{x})\ddot{\mathbf{u}}(\mathbf{x}, t) = \mathbf{b_i}(\mathbf{x}, t) + \int_{H_{\mathbf{x}}} \left\{ \underline{\mathbf{T}}[\mathbf{x}, t]\langle\boldsymbol{\xi}\rangle - \underline{\mathbf{T}}\left[\mathbf{x}', t\right]\langle-\boldsymbol{\xi}\rangle \right\} dV_{\mathbf{x}'} + \mathbf{b}(\mathbf{x}, t), \tag{7}$$

where $\boldsymbol{\xi} = \mathbf{x}' - \mathbf{x}$, $H_{\mathbf{x}}$ is a spherical neighborhood of $\mathbf{x}$ with radius δ, $\underline{\mathbf{T}}$ the force vector state field [2], $\mathbf{b}(\mathbf{x}, t)$ the external body force, and $\rho(\mathbf{x})$ the mass density. We call Equation (7) the traction-associated peridynamic motion equation. Clearly, Equation (7) is an extension of the Silling's peridynamic motion equation. Let $\mathbf{x}''' \in \partial\Omega_p$ and $\mathbf{x} \in \Omega$. If we take $\mathbf{G}(\mathbf{x}, \mathbf{x}''', t) = \mathrm{H}(|\mathbf{x}''' - \mathbf{x}|)\mathbf{I}/V_b$ where $\mathrm{H}(|\mathbf{x}''' - \mathbf{x}|)$ is the dimensionless square wave function and V_b the volume of the boundary layer of $\partial\Omega_p$ with the thickness of Δ, then Equation (1) will degenerate into a common formula when ones deal with the boundary traction in peridynamics.

By Equations (5) and (6), the integrals of Equation (7) and $\mathbf{y}(\mathbf{x}, t) \times$ Equation (7) over Ω yield

$$\int_{\Omega} \rho(\mathbf{x})\ddot{\mathbf{u}}(\mathbf{x}, t) dV_{\mathbf{x}} = \int_{\partial\Omega_p} \mathbf{p}(\mathbf{x}, t) dA_{\mathbf{x}} + \int_{\Omega} \mathbf{b}(\mathbf{x}, t) dV_{\mathbf{x}} \tag{8}$$

$$\int_{\Omega} \rho(\mathbf{x})\overline{\mathbf{y}(\mathbf{x}) \times \mathbf{v}(\mathbf{x}, t)} dV_{\mathbf{x}} = \int_{\partial\Omega_p} \mathbf{y}(\mathbf{x}) \times \mathbf{p}(\mathbf{x}, t) dA_{\mathbf{x}} + \int_{\Omega} \mathbf{y}(\mathbf{x}) \times \mathbf{b}(\mathbf{x}, t) dV_{\mathbf{x}}, \tag{9}$$

which describe total equilibrium of linear momentum and angular momentum, respectively. Therefore, Equation (7) is consistent with the conservation laws of linear momentum and angular momentum. In addition, it is easy to see that Equation (7) is form-invariant under the Galileo transformation.

3. Peridynamic Constitutive Model

3.1. Balance Equation of Energy

Let $\mathbf{v} = \mathbf{v}(\mathbf{x})$ is the velocity field within material, and ε is the internal energy density. Only elastic deformation is concerned, in peridynamics, total energy conservation can be represented as

$$\frac{D}{Dt}\int_\Omega \left(\frac{1}{2}\rho\mathbf{v}^2 + \rho\varepsilon\right)dV = \int_\Omega \mathbf{b_i}\cdot\mathbf{v}dV_\mathbf{x} + \int_\Omega \mathbf{b}\cdot\mathbf{v}dV. \tag{10}$$

Equation (10) can be further written as

$$\int_\Omega \rho\mathbf{a}\cdot\mathbf{v}dV + \int_\Omega \rho\dot{\varepsilon}dV = \int_\Omega \mathbf{b_i}\cdot\mathbf{v}dV_\mathbf{x} + \int_\Omega \mathbf{b}\cdot\mathbf{v}dV, \tag{11}$$

where $\mathbf{a}$ is acceleration field. In terms of Equation (7), Equation (11) reduces to

$$\int_\Omega \rho\dot{\varepsilon}dV = \int_\Omega \left\{\int_{H_\mathbf{x}} \left\{\underline{\mathbf{T}}\left[\mathbf{x}',t\right]\langle-\boldsymbol{\xi}\rangle - \underline{\mathbf{T}}[\mathbf{x},t]\langle\boldsymbol{\xi}\rangle\right\}dV_{\mathbf{x}'}\right\}\cdot\mathbf{v}dV_\mathbf{x}. \tag{12}$$

Since $H_\mathbf{x} \subset \Omega$ is a compact supported set of $\underline{\mathbf{T}}[\mathbf{x}', t]\langle-\boldsymbol{\xi}\rangle$ and $\underline{\mathbf{T}}[\mathbf{x}, t]\langle\boldsymbol{\xi}\rangle$, Equation (12) can be written as

$$\int_\Omega \rho\dot{\varepsilon}dV = \int_\Omega \left\{\int_\Omega \left\{\underline{\mathbf{T}}\left[\mathbf{x}',t\right]\langle-\boldsymbol{\xi}\rangle - \underline{\mathbf{T}}[\mathbf{x},t]\langle\boldsymbol{\xi}\rangle\right\}dV_{\mathbf{x}'}\right\}\cdot\mathbf{v}dV_\mathbf{x}. \tag{13}$$

Exchanging $\mathbf{x}'$ and $\mathbf{x}$, and then using definition of the compact supported set, we have

$$\int_\Omega \rho\dot{\varepsilon}dV = \int_\Omega \left\{\int_{H_\mathbf{x}} \underline{\mathbf{T}}[\mathbf{x},t]\langle\boldsymbol{\xi}\rangle\left[\mathbf{v}\left(\mathbf{x}'\right) - \mathbf{v}(\mathbf{x})\right]dV_{\mathbf{x}'}\right\}dV_\mathbf{x}. \tag{14}$$

By the localized hypothesis [34,35], the balance equation of energy is given as follows

$$\rho\dot{\varepsilon} = \int_{H_\mathbf{x}} \underline{\mathbf{T}}[\mathbf{x},t]\langle\boldsymbol{\xi}\rangle\left[\mathbf{v}\left(\mathbf{x}'\right) - \mathbf{v}(\mathbf{x})\right]dV_{\mathbf{x}'}. \tag{15}$$

Clearly, Equation (15) has the same form as that in the original peridynamics advanced by Silling and his collaborators [1,2]. Equation (15) is a basis to determine the peridynamic constitutive models of hyperelastic material. Therefore, the hyperelastic constitutive models in the original peridynamics can be inherited without modification by the traction-associated peridynamics.

3.2. Bond-Based Constitutive Models

Bond-based (BB) constitutive models are simplified versions of state-based (SB) constitutive models. For brevity, only the BB constitutive models are considered, not concerned the SB constitutive models. The BB constitutive models have been systematically established by Silling [1], and their forms are not unique. The microelastic models [5] are used to describe the elastic deformation of isotropic materials. Two commonly used microelastic models are listed below.

1. General microelastic models:

$$\mathbf{f}\left(\mathbf{u}' - \mathbf{u}, \boldsymbol{\xi}\right) = \begin{cases} c\left(\left|\mathbf{y}' - \mathbf{y}\right| - |\boldsymbol{\xi}|\right)\frac{\mathbf{y}'-\mathbf{y}}{|\mathbf{y}'-\mathbf{y}|} & |\boldsymbol{\xi}| \leq \delta \\ 0 & \text{otherwise} \end{cases}. \tag{16}$$

The linear form of Equation (16) is

$$\mathbf{f}\left(\mathbf{u}' - \mathbf{u}, \boldsymbol{\xi}\right) = \begin{cases} c\frac{\boldsymbol{\xi}\otimes\boldsymbol{\xi}}{|\boldsymbol{\xi}|^2} \cdot \left(\mathbf{u}' - \mathbf{u}\right) & |\boldsymbol{\xi}| \leq \delta \\ 0 & \text{otherwise} \end{cases}, \tag{17}$$

where $\mathbf{f}(\mathbf{u}'-\mathbf{u}, \boldsymbol{\xi})$ is the force density vector [1] with the dimension of N/m^6. The relation between $\mathbf{f}(\mathbf{u}'-\mathbf{u}, \boldsymbol{\xi})$ and the force vector state $\underline{\mathbf{T}}$ in Equation (7) is represented as $\underline{\mathbf{T}}[\mathbf{x}, t] <\boldsymbol{\xi}> = -\underline{\mathbf{T}}\,[\mathbf{x}', t] <-\boldsymbol{\xi}> == \mathbf{f}(\mathbf{u}'-\mathbf{u}, \boldsymbol{\xi})/2$. The parameter c is called the spring constant or the bond-constant, which reads [5]

$$c = \begin{cases} \frac{15E}{\pi\delta^5} & 3 - \text{dimension} \\ \frac{12E}{\pi h\delta^4} & 2 - \text{dimension plane stress} \\ \frac{64E}{5\pi h\delta^4} & 2 - \text{dimension plane strain} \\ \frac{3E}{h_1\delta^3} & 1 - \text{dimension} \end{cases}, \tag{18}$$

where E is the Young's modulus, h the thickness of plate and h_1 the cross-sectional area of rod.

2. The prototype microelastic (PM) model:

PM model is another special form of the microelastic models

$$\mathbf{f}\left(\mathbf{u}' - \mathbf{u}, \boldsymbol{\xi}\right) = \begin{cases} c\frac{|\mathbf{y}'-\mathbf{y}|-|\boldsymbol{\xi}|}{|\boldsymbol{\xi}|}\frac{\mathbf{y}'-\mathbf{y}}{|\mathbf{y}'-\mathbf{y}|} & |\boldsymbol{\xi}| \leq \delta \\ 0 & \text{otherwise} \end{cases}. \tag{19}$$

The linear form of Equation (19) is

$$\mathbf{f}\left(\mathbf{u}' - \mathbf{u}, \boldsymbol{\xi}\right) = \begin{cases} c\frac{\boldsymbol{\xi}\otimes\boldsymbol{\xi}}{|\boldsymbol{\xi}|^3} \cdot \left(\mathbf{u}' - \mathbf{u}\right) & |\boldsymbol{\xi}| \leq \delta \\ 0 & \text{otherwise} \end{cases}. \tag{20}$$

The parameter c in Equations (19) and (20) is still the bond-constant and it takes the value below [4,5]

$$c = \begin{cases} \frac{12E}{\pi\delta^4} & 3 - \text{dimension} \\ \frac{9E}{\pi h\delta^3} & 2 - \text{dimension plane stress} \\ \frac{48E}{5\pi h\delta^3} & 2 - \text{dimension plane strain} \\ \frac{2E}{h_1\delta^2} & 1 - \text{dimension} \end{cases}. \tag{21}$$

It should be noted that in the BB models, the Poisson's ratio of 3D and 2D plane strain problems are fixed at $1/4$, while that of 2D plane stress problem are fixed at $1/3$.

3.3. Transfer Function of Boundary Traction

According to Equations (2) and (3), the transfer function $\mathbf{G}(\mathbf{x}, \mathbf{x}''', t)$ of the boundary traction characterizes the change of intensity when the traction is transferred from the boundary surface to the interior of the body, which is determined by the weight function $\alpha(\mathbf{x}''', \mathbf{x})$. In physics, $\alpha(\mathbf{x}''', \mathbf{x})$ should attenuates to zero as it moves away from the boundary surface. Therefore, $\alpha(\mathbf{x}''', \mathbf{x})$ can be defined a function with compact support, i.e.,

$$\alpha\left(\mathbf{x}''', \mathbf{x}\right) = \begin{cases} q\left(|\mathbf{x}''' - \mathbf{x}|\right) \neq 0 & |\mathbf{x}''' - \mathbf{x}| \leq \iota \\ 0 & \text{otherwise} \end{cases}, \tag{22}$$

where $\mathbf{x}''' \in \partial\Omega_p$ while $\mathbf{x} \in \Omega$. ι is a scale parameter. For simplicity, we take $\iota = \delta$. As thus, Equation (22) means that the traction on the boundary surface is dispersed in the boundary layer with the thickness of δ.

Consider the quasi-static uniaxial tension of a rod subjected to tensile force p at two ends. Through the inverse method [36] and the undetermined coefficient method, we find that when the general microelastic constitutive Equations (16) and (17) are adopted, if $\alpha(x''', x)$ takes the form below

$$\alpha(x''', x) = \begin{cases} k\left(\delta^2 - |x''' - x|^2\right) & |x''' - x| \leq \delta \\ 0 & \text{otherwise} \end{cases} \tag{23}$$

then the classical elasticity solution of the uniaxial tension can be acquired.

Similarly, when the PM constitutive Equations (19) and (20) are used, if $\alpha(x''', x)$ is written as

$$\alpha(x''', x) = \begin{cases} k(\delta - |x''' - x|) & |x''' - x| \leq \delta \\ 0 & \text{otherwise} \end{cases} \tag{24}$$

the same result is also given. It should be noted that k in Equations (23) and (24) is any constant and is not zero. For simplicity, we can take $k = 1$.

Equations (23) and (24) can be extended to 2-dimensional and 3-dimensional form below

$$\alpha\left(\mathbf{x}''', \mathbf{x}\right) = \begin{cases} \delta^2\left|\mathbf{x}''' - \mathbf{x}\right|^2 & \left|\mathbf{x}''' - \mathbf{x}\right| \leq \delta \\ 0 & \text{otherwise} \end{cases} \tag{25}$$

$$\alpha\left(\mathbf{x}''', \mathbf{x}\right) = \begin{cases} \delta - \left|\mathbf{x}''' - \mathbf{x}\right| & \left|\mathbf{x}''' - \mathbf{x}\right| \leq \delta \\ 0 & \text{otherwise} \end{cases} \tag{26}$$

Thus, the concrete form of the induced body force is determined due to Equation (25) or (26), In the following, they will be directly used to analyze benchmark examples of 2D plane stress problems.

3.4. Prototype Microelastic Brittle Damage Model

In PD theory, local damage at a point is defined as the weighted ratio of the number of eliminated interactions to the total number of initial interactions of the material point with its family members, that is [4,19]

$$\varphi(\mathbf{x}, t) = 1 - \frac{\int_{H_x} \mu(\boldsymbol{\xi}, t) dV_{\mathbf{x}'}}{\int_{H_x} dV_{\mathbf{x}'}}. \tag{27}$$

It should be noted that the local damage φ ranges from 0 to 1. When $\varphi = 1$, all the interactions initially associated with the point have been eliminated, while $\varphi = 0$ means that all interactions exist. The measurement of the local damage value is an indicator of the possible formation of cracks within a body. In Equation (27), μ is a history-dependent scalar-valued function, which reads

$$\mu(\boldsymbol{\xi}, t) = \begin{cases} 1 & s(\boldsymbol{\xi}, t') < S_c \text{ for all } 0 \leq t' \leq t \\ 0 & \text{otherwise} \end{cases}, \tag{28}$$

where S_c is the critical stretch of bond failure, while s is bond stretch defined by

$$s = \frac{\left|\mathbf{y}' - \mathbf{y}\right| - |\boldsymbol{\xi}|}{|\boldsymbol{\xi}|}. \tag{29}$$

Therefore, a simple way to introduce failure into the constitutive model to allow bonds (springs) to break when they are stretched beyond a predefined limit. After bond failure, there is no tensile force sustainable in the bond, and once a bond fails, it is failed forever (there is no provision for "healing" of a failed bond). As thus, the PMB model to characterize brittle damage [4,5,19] can be written as

$$\mathbf{f}\left(\mathbf{u}' - \mathbf{u}, \boldsymbol{\xi}, t\right) = \begin{cases} cs\mu(\boldsymbol{\xi}, t)\frac{\mathbf{y}'-\mathbf{y}}{|\mathbf{y}'-\mathbf{y}|} & |\boldsymbol{\xi}| \leq \delta \\ 0 & \text{otherwise} \end{cases}. \tag{30}$$

4. Numerical Algorithm

4.1. Spatial Discretization

Meshfree spatial discrete method [19] is used to discretize continuum into a range of arbitrary shaped subdomains, in which, the collocation points (nodes) are placed. With one-point Gauss quadrature strategy [37], the spatial discrete form of Equation (7) can be written as

$$\rho(\mathbf{x}_i)\ddot{\mathbf{u}}(\mathbf{x}_i, t)$$
$$= \sum_{\partial\Omega_p} \mathbf{G}(\mathbf{x}_i, \mathbf{x}_k, t) \cdot \overline{\mathbf{p}}(\mathbf{x}_k, t) A_{\mathbf{x}_k} + \sum_{H_{\mathbf{x}_i}} \left\{ \underline{\mathbf{T}}[\mathbf{x}_i, t]\langle \mathbf{x}_j - \mathbf{x}_i \rangle - \underline{\mathbf{T}}[\mathbf{x}_j, t]\langle \mathbf{x}_i - \mathbf{x}_j \rangle \right\} V_{\mathbf{x}_j} + \mathbf{b}(\mathbf{x}_i, t). \tag{31}$$

The same spatial discrete strategy can be applied to acquire the integral value of $\mathbf{G}(\mathbf{x}, \mathbf{x}''')$. It is worth noting that the collocation points must also be set on the traction boundary surface due to the introduction of the boundary integration.

4.2. Time Integration

The adaptive dynamic relaxation (ADR) method [38] is used to solve Equation (31). After introducing new fictitious inertia and damping terms, Equation (31) is rewritten as

$$\lambda_i \ddot{\mathbf{u}}_i^n + c^n \lambda_i \dot{\mathbf{u}}_i^n = \mathbf{F}_i^n, \tag{32}$$

where n is n-th time (iterative) step, λ_i the fictitious density of the i-th node, and c^n the damping coefficient. The vector $\mathbf{F}_i^n$ in Equation (32) is the summation of internal and external forces, which reads

$$\mathbf{F}_i^n$$
$$= \sum_{\partial\Omega_p} \mathbf{G}(\mathbf{x}_i, \mathbf{x}_k, t) \cdot \overline{\mathbf{p}}(\mathbf{x}_k, t) A_{\mathbf{x}_k} + \sum_{H_{\mathbf{x}_i}} \left\{ \underline{\mathbf{T}}[\mathbf{x}_i, t]\langle \mathbf{x}_j - \mathbf{x}_i \rangle - \underline{\mathbf{T}}[\mathbf{x}_j, t]\langle \mathbf{x}_i - \mathbf{x}_j \rangle \right\} V_{\mathbf{x}_j} + \mathbf{b}(\mathbf{x}_i, t). \tag{33}$$

By the explicit central-difference integration, the iterative scheme of displacements and velocities can be written as

$$\begin{cases} \dot{\mathbf{u}}_i^{n+\frac{1}{2}} = \dfrac{(2-c^n\Delta t)\dot{\mathbf{u}}_i^{n-\frac{1}{2}} + \frac{2\Delta t \mathbf{F}_i^n}{\lambda_i}}{2+c^n\Delta t} \\ \mathbf{u}_i^{n+1} = \mathbf{u}_i^n + \Delta t \dot{\mathbf{u}}_i^{n+\frac{1}{2}} \end{cases}. \tag{34}$$

As Equation (34) has an unknown velocity field at $t^{-1/2}$, the iterative process can be not started. However, if we assume $\mathbf{u}_i^0 \neq 0$ and $\mathbf{v}_i^0 = 0$, the iteration can be completed by using

$$\dot{\mathbf{u}}_i^{\frac{1}{2}} = \frac{\Delta t \mathbf{F}_i^0}{2\lambda_i}. \tag{35}$$

In this algorithm, since the fictitious density λ_i, the damping coefficient c^n and the time step Δt are not actual physical quantities, their values can be chosen so as to make the convergence of numerical solution as fast as possible. Therefore, we take $\Delta t = 1$, while the calculation of λ_i and c^n can refer to [4,38–40].

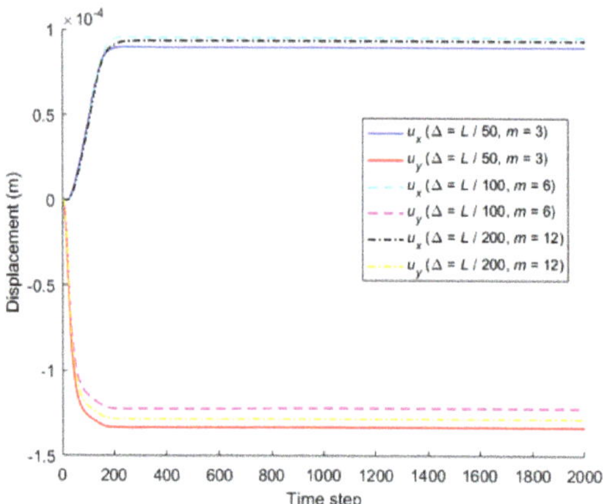

Figure 5. *m*-convergence of displacement with time step at different nodes.

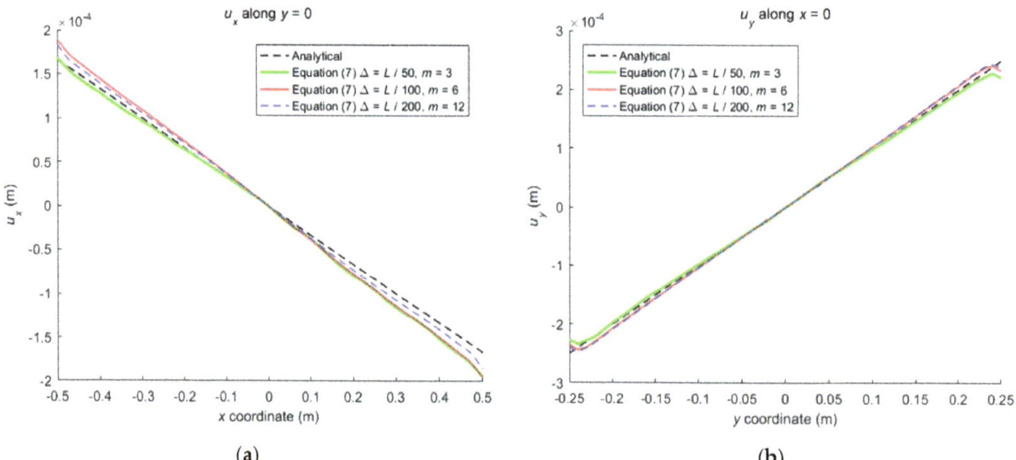

(**a**)　　　　　　　　　　　　(**b**)

Figure 6. The displacements along central lines of the rectangular plate for three different combinations: (**a**) u_x along $y = 0$; (**b**) u_y along $x = 0$.

Figure 5 that the calculation converges very quickly with time step, no matter which combination mode is adopted. From Figure 6, we see that the numerical results of the three different combinations agree with each other, and the errors between them and classical solutions are all within 5%. Therefore, the numerical algorithm is of *m*-convergence [41–44].

Comparison between Equation (7) and the original PD is carried out through numerical calculation. Figures 3 and 7 show that the displacements calculated by Equation (7) and the original PD converges very quickly with time step, and the convergence modes are similar. As can be seen from Figures 8–10, there exist good matches between predictions of traction-associated peridynamic motion equation and analytical solutions as well as original PD predictions.

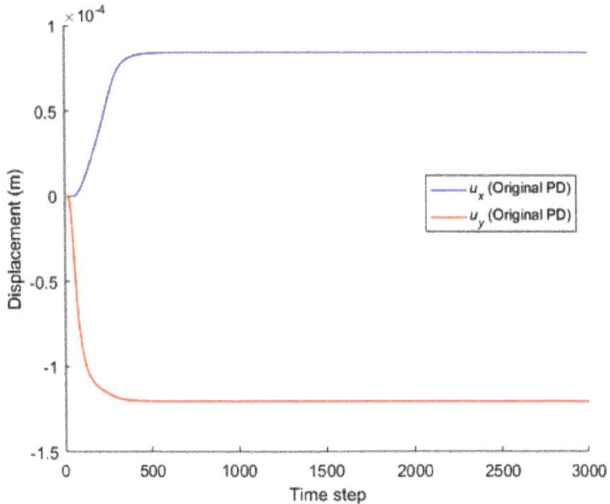

Figure 7. Convergence of original PD with time step.

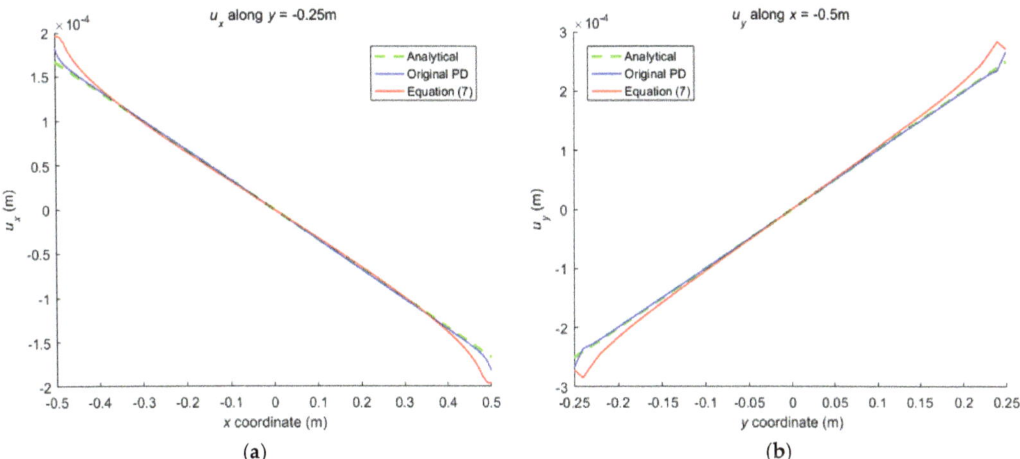

(**a**)

(**b**)

Figure 8. The displacement along the bottom side and left side of the rectangular plate subjected to tension: (**a**) u_x along $y = -0.25$ m; (**b**) u_y along $x = -0.5$ m.

In order to balance the computational accuracy and efficiency, in the following, the horizon size is specified as $\delta = 3.015\Delta$ and grid size $\Delta = L/100$. The total time step is set to 3000. The material parameters of plates in all examples are the same as those of the plate in Figure 1, and the same discretization as Figure 2 is adopted.

5.2. Example 2: A Rectangular Plate Subjected to Bending

As shown in Figure 11, a rectangular plate with the same size as the plate in Figure 1 is subjected to an anti-symmetrically distributed loads with a maximal value $q = 200$ Mpa. This is a pure bending problem with the stress boundary condition.

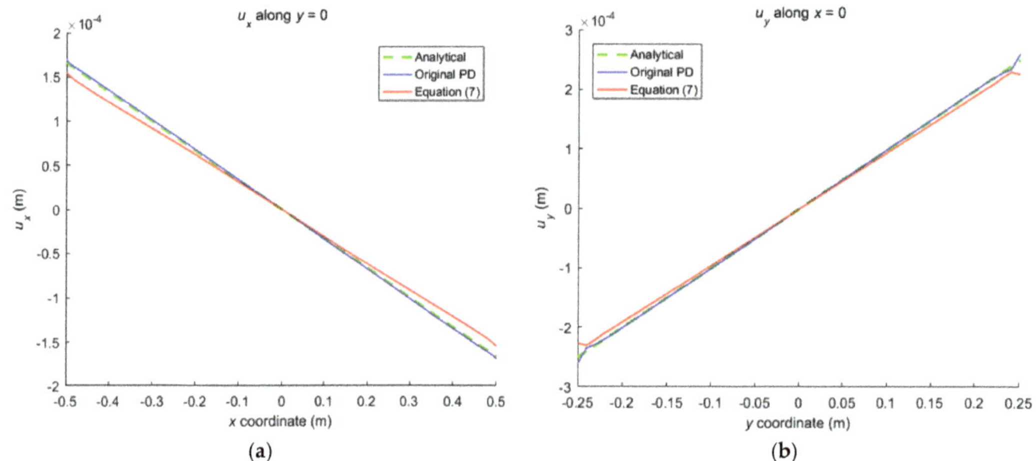

Figure 9. The displacement along central lines of the rectangular plate subjected to tension: (**a**) u_x along $y = 0$; (**b**) u_y along $x = 0$.

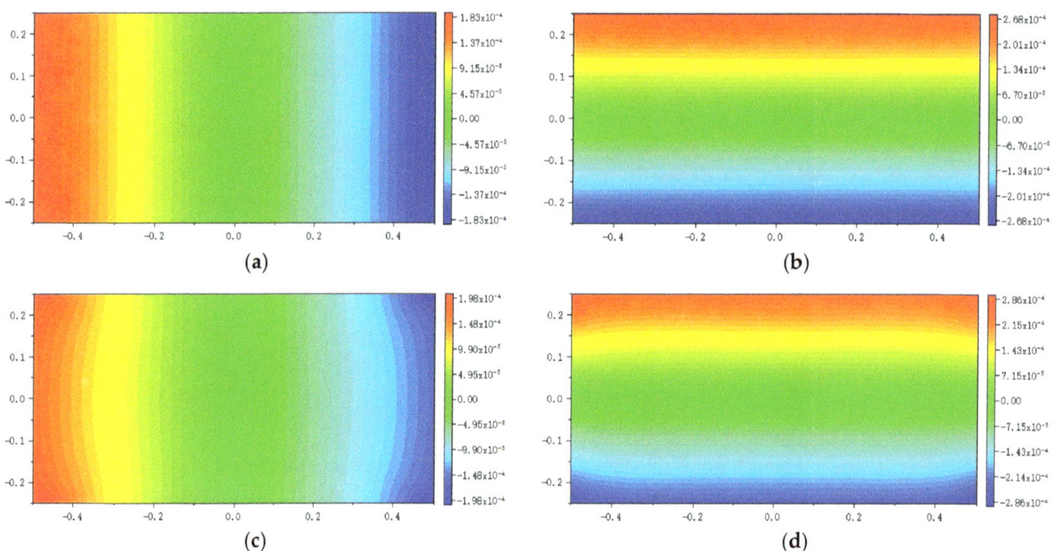

Figure 10. The distribution of displacement in the rectangular plate subjected to tension: (**a**) u_x calculated by original PD; (**b**) u_y calculated by original PD; (**c**) u_x calculated by Equation (7); (**d**) u_y calculated by Equation (7).

The convergence analysis is shown in Figure 12. It can be seen that the displacements calculated by the original PD and Equation (7) converge with time step in the same way, and completely converge at the 1000th time step.

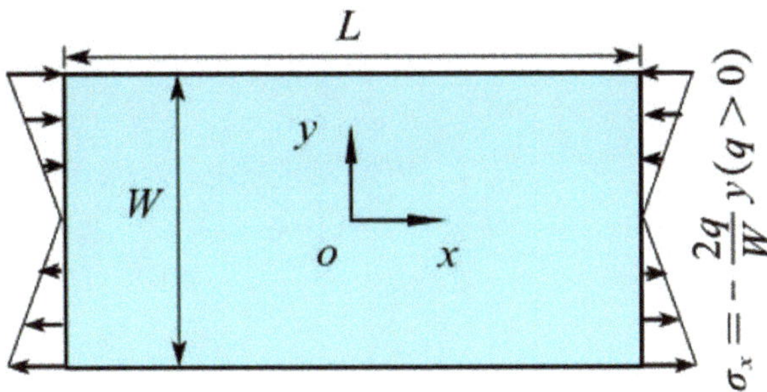

Figure 11. The plate subjected to bending.

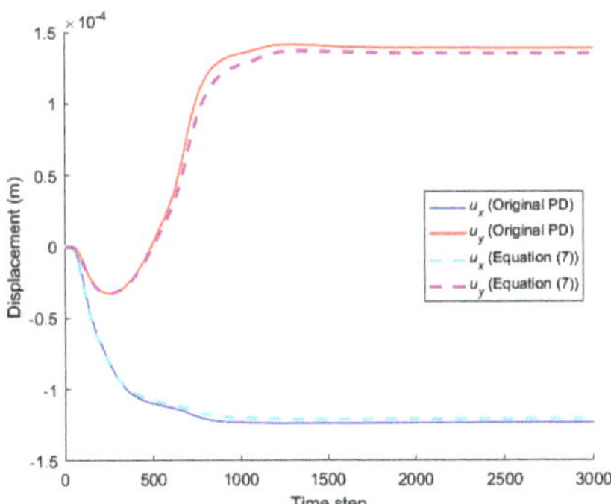

Figure 12. Convergence of displacement at $x = -0.25$ m and $y = -0.12$ m with time step.

The numerical results are illustrated in Figures 13–17, From which, it can be seen that the displacement distribution predicted by Equation (7) agrees with that by original PD, and the two have better matching with analytical solutions.

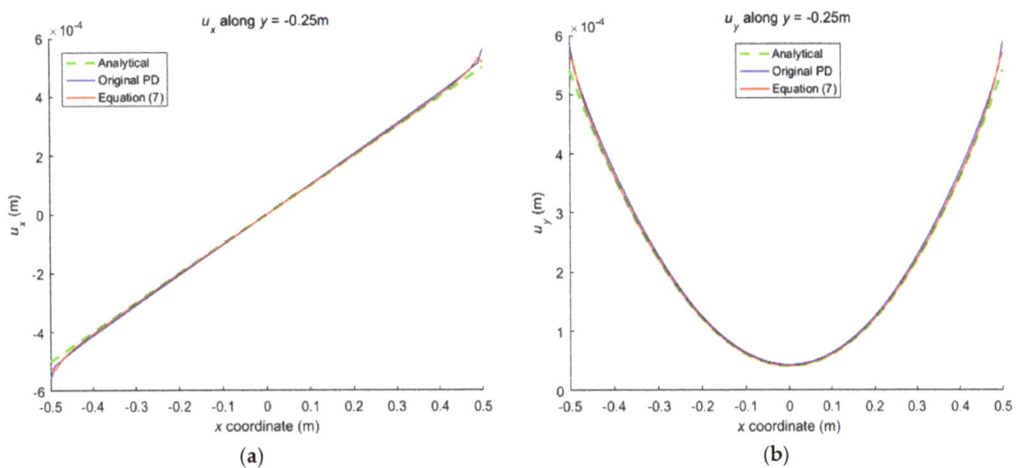

Figure 13. The displacement along the bottom side when the plate bending: (**a**) u_x along $y = -0.25$ m; (**b**) u_y along $y = -0.25$ m.

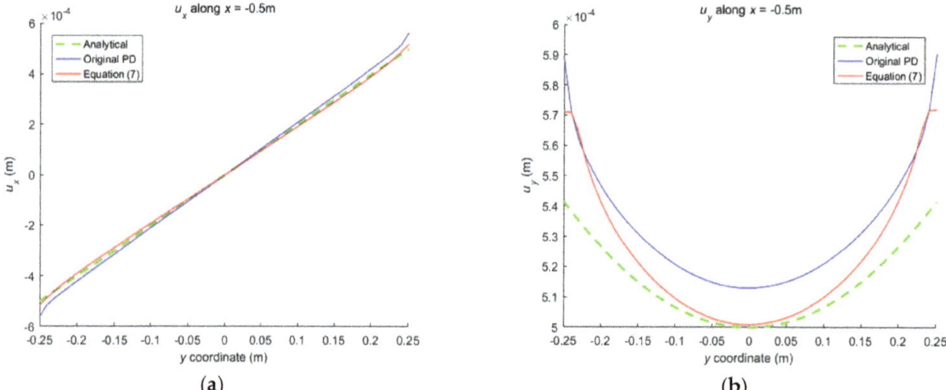

Figure 14. The displacement along the left side when the plate bending: (**a**) u_x along $x = -0.5$ m; (**b**) u_y along $x = -0.5$ m.

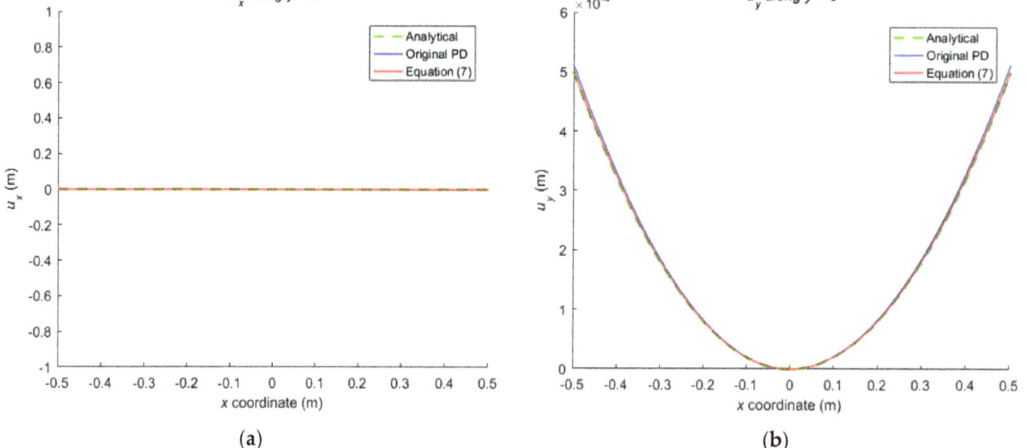

Figure 15. The displacement along the horizontal line $y = 0$ when the plate bending: (**a**) u_x along $y = 0$; (**b**) u_y along $y = 0$.

5.3. Example 3: A Square Plate with A Circular Hole Subjected to Tension by Two Opposite Edges

As shown in Figure 18, a squared plate with central circular hole is subjected to uniform tension $q = 200$ MPa. The side length of the plate $L = 0.5$ m and the radius of the circular hole $r = 0.05$ m.

The convergence analysis is shown in Figure 19. It can be seen that the displacements calculated by the original PD and Equation (7) converge with time step in the same way, and completely converge at the 500th time step.

The results in Figure 20 show that the displacements given by FEA, PD, and Equation (7) are close to each other in distribution. The relative error between them is less than 6.7%.

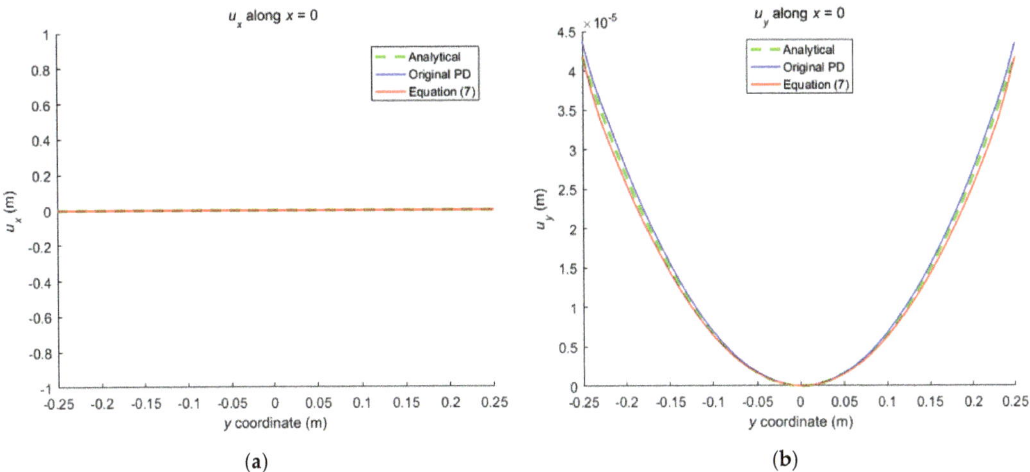

(**a**)   (**b**)

Figure 16. The displacement along the vertical line $x = 0$ when the plate bending: (**a**) u_x along $x = 0$; (**b**) u_y along $x = 0$.

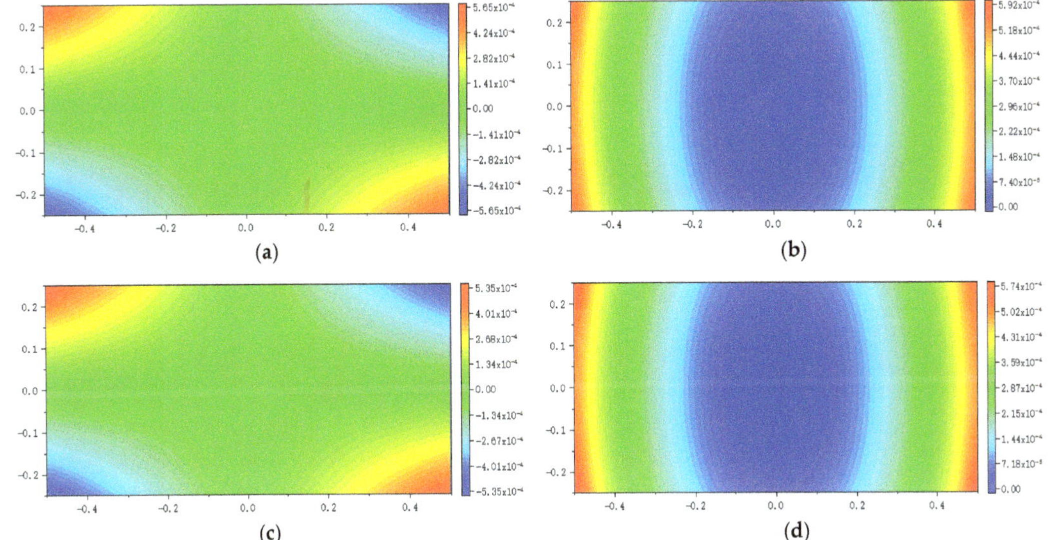

(**a**)   (**b**)

(**c**)   (**d**)

Figure 17. The distribution of displacement when the plate bending: (**a**) u_x calculated by original PD; (**b**) u_y calculated by original PD; (**c**) u_x calculated by Equation (7); (**d**) u_y calculated by Equation (7).

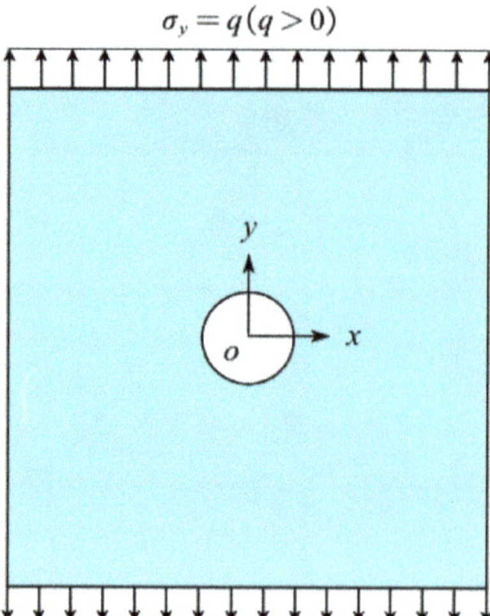

Figure 18. Squared plate with central circular hole subjected to uniform tension.

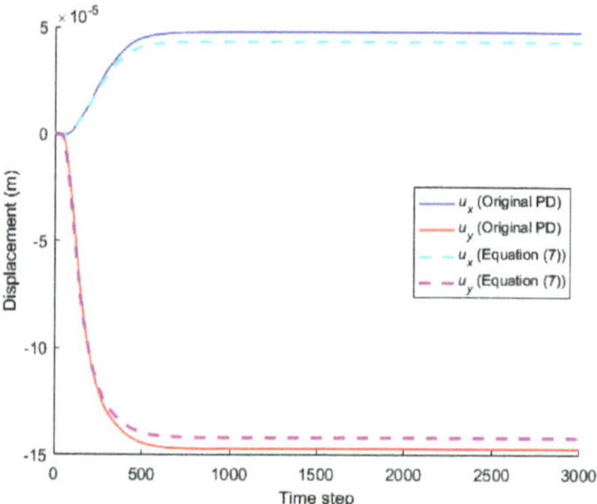

Figure 19. Convergence of displacement at $x = -0.125$ m and $y = -0.125$ m with time step.

5.4. Example 4: Failure of A Square Plate with A Circular Hole under Quasi-Static Loading

We continue to investigate the fracture of the plate with central circular hole under tension. As shown in Figure 21, in order to avoid nodes occurring at the propagation path of cracks, all nodes (red and black dots) have been anew collocated. The grid size is taken as $\Delta = L/100$. The nodes on the boundary surface (red dots) are only involved in the integration of the boundary traction and correspond to a volume of 0. In calculation, the critical stretch S_c of bond failure takes 0.0058.

The propagation of crack is characterized by the value of the damage φ. When the load q arrives at 380 MPa, the plate breaks due to cracking. Figure 22 shows the damage φ calculated by the original PD. At the 500th time step, the crack initiates from two sides of the hole. As the time step increases to 700, the crack propagates and the damage φ reaches 0.479. As the time step continues to increase, the plate breaks at the 900th time step.

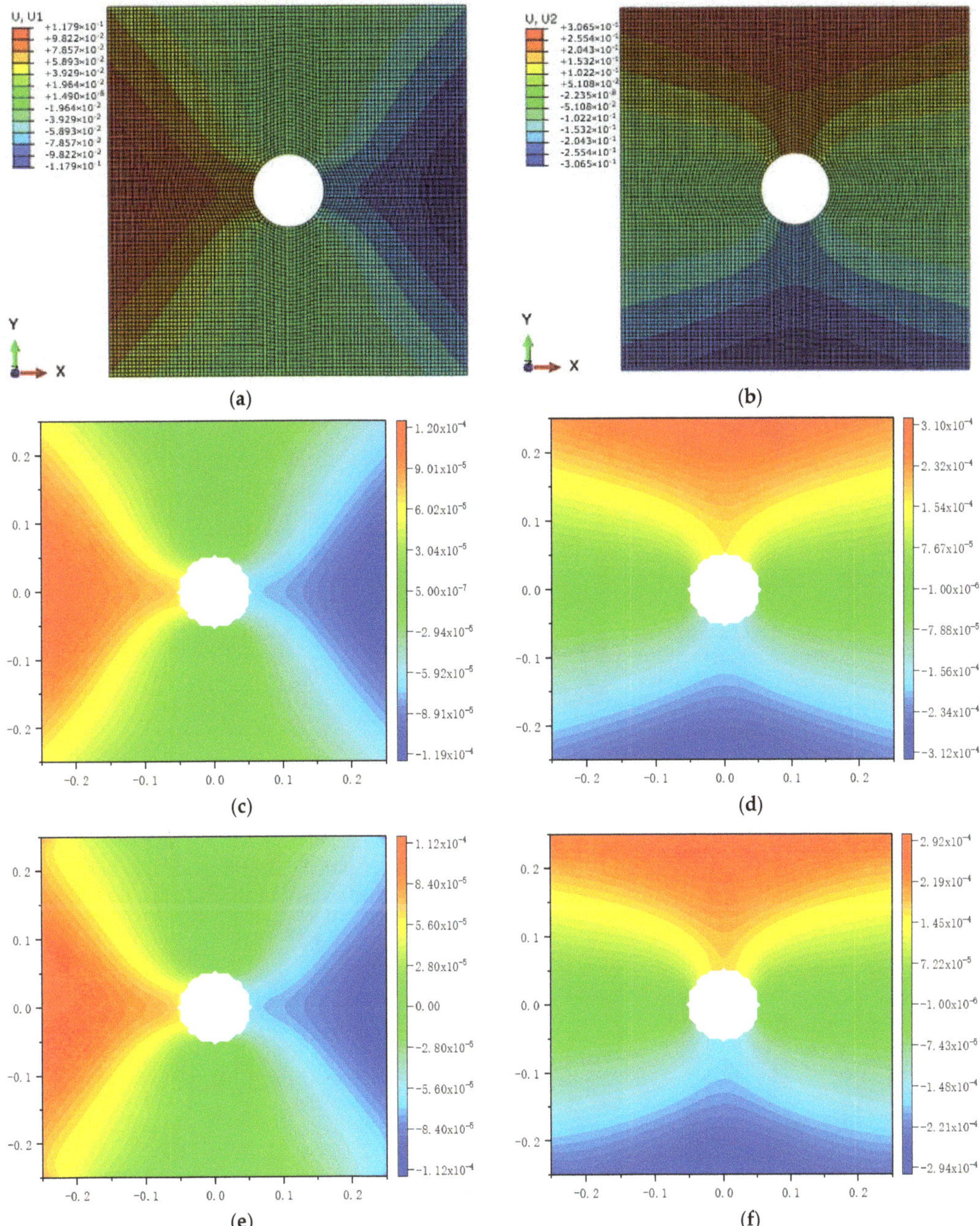

Figure 20. The distribution of displacement in the plate with central circular hole: (**a**) u_x calculated by ABAQUS; (**b**) u_y calculated by ABAQUS; (**c**) u_x calculated by original PD; (**d**) u_y calculated by original PD; (**e**) u_x calculated by Equation (7); (**f**) u_y calculated by Equation (7).

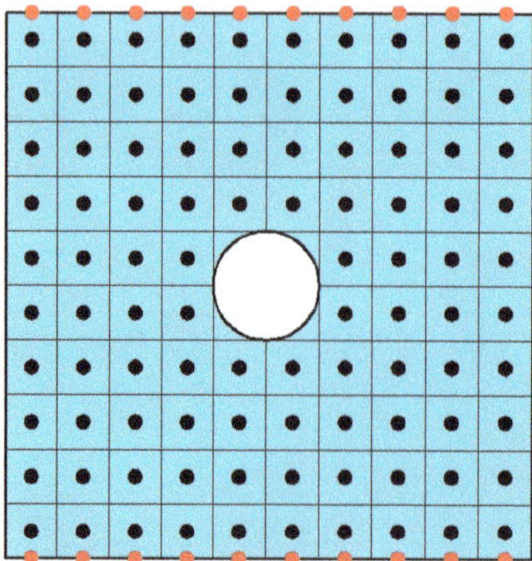

Figure 21. The discretization of the square plate.

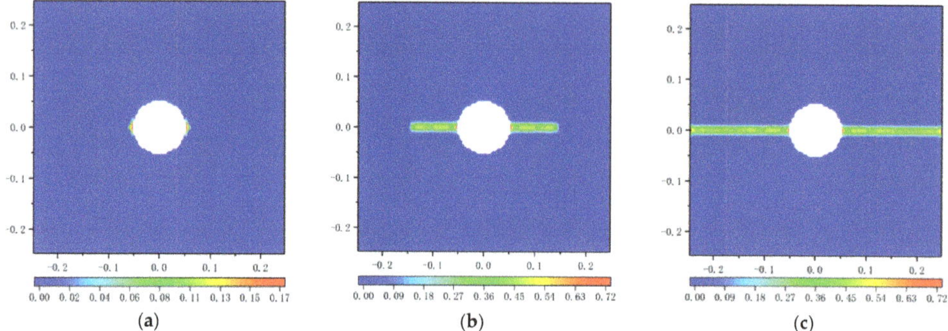

Figure 22. Damage plots for the square plate with a circular hole at the end of different time steps based on original PD: (**a**) 500 time steps; (**b**) 700 time steps; (**c**) 900 time steps.

The damage φ calculated by Equation (7) is illustrated in Figure 23. The results show that the crack initiates at the 700th time step, and at the 900th time step, the damage φ of the crack propagation reaches 0.42. When the time step arrives at 1125, the plate fails due to cracking.

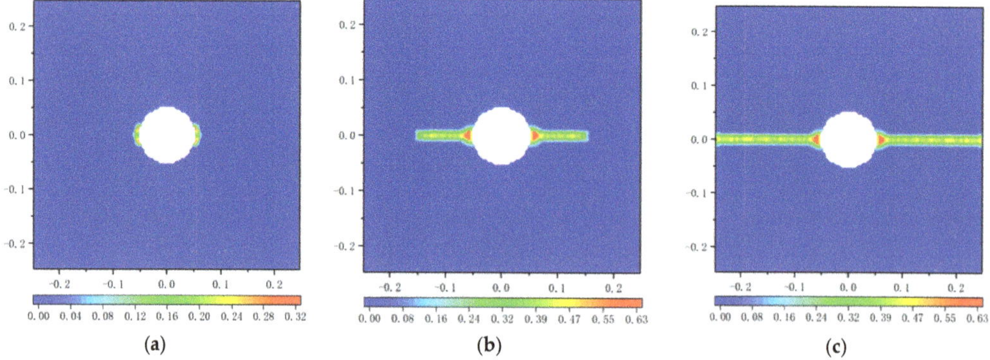

Figure 23. Damage plots for the square plate with a circular hole at the end of different time steps based on Equation (7): (**a**) 700 time steps; (**b**) 900 time steps; (**c**) 1125 time steps.

6. Conclusions

Through introducing the induced body force defined by boundary traction, the Silling's peridynamic motion equation is extended to the traction-associated peridynamic motion equation. From investigation on this equation, the conclusions are summarized as follows.

- The traction-associated peridynamic motion equation is consistent with the conservation laws of linear and angular momentum, and it is form-invariant under the Galileo transformation.
- The constitutive models in the original peridynamics can be inherited without modification by the traction-associated peridynamics. The concrete form of the induced body force is determined by matching with the constitutive models.
- Numerical calculations for the typical plane stress problems are in good agreement with the classical elasticity solutions, and the volume correction and the surface correction are no longer needed in the numerical algorithm.

Author Contributions: Conceptualization, Z.H.; Methodology, Z.H.; Software, M.Y. and Z.Z.; Validation, Z.Z.; Formal analysis, M.Y. and Z.H.; Investigation, M.Y. and Z.H.; Resources, Z.Z.; Data curation, M.Y.; Writing—original draft, M.Y.; Writing—review & editing, Z.H.; Visualization, Z.Z.; Supervision, Z.H.; Project administration, Z.H.; Funding acquisition, Z.H. All authors have read and agreed to the published version of the manuscript.

Funding: This research was funded by the National Natural Science Foundation of China, Grant Nos. 12072145 and 11672129.

Institutional Review Board Statement: Not applicable.

Informed Consent Statement: Not applicable.

Data Availability Statement: Not applicable.

Conflicts of Interest: The authors declare no conflict of interest.

References

1. Silling, S.A. Reformulation of elasticity theory for discontinuities and long-range forces. *J. Mech. Phys. Solids* **2000**, *48*, 175–209. [CrossRef]
2. Silling, S.A.; Epton, M.; Weckner, O.; Xu, J.; Askari, E. Peridynamic states and constitutive modeling. *J. Elast.* **2007**, *88*, 151–184. [CrossRef]
3. Silling, S.A.; Lehoucq, R.B. Peridynamic theory of solid mechanics. *Adv. Appl. Mech.* **2010**, *44*, 73–168.
4. Madenci, E.; Oterkus, E. *Peridynamic Theory and Its Applications*; Springer: New York, NY, USA, 2014.
5. Bobaru, F.; Foster, J.T.; Geubelle, P.H.; Silling, S.A. *Handbook of Peridynamic Modeling*; CRC Press: Boca Raton, USA, 2016.
6. Javili, A.; Morasata, R.; Oterkus, E.; Oterkus, S. Peridynamics review. *Math. Mech. Solids* **2019**, *24*, 3714–3739. [CrossRef]
7. Ladanyi, G.; Gonda, V. Review of peridynamics: Theory, applications and future perspectives. *Stroj. Vestn. J. Mech. Eng.* **2021**, *67*, 666–681. [CrossRef]
8. Zhou, X.P.; Wang, Y.T. State-of-the-art review on the progressive failure characteristics of geomaterials in peridynamic theory. *J. Eng. Mech.* **2021**, *147*, 03120001. [CrossRef]
9. Han, D.; Zhang, Y.; Wang, Q.; Lu, W.; Jia, B. The review of the bond-based peridynamics modeling. *J. Micromech. Mol. Phys.* **2019**, *4*, 1830001. [CrossRef]
10. Le, Q.V.; Bobaru, F. Surface corrections for peridynamic models in elasticity and fracture. *Comput. Mech.* **2018**, *61*, 499–518. [CrossRef]
11. Wu, C.T.; Ren, B. A stabilized non-ordinary state-based peridynamics for the nonlocal ductile material failure analysis in metal machining process. *Comput. Methods Appl. Mech. Eng.* **2015**, *291*, 197–215. [CrossRef]
12. Madenci, E.; Dorduncu, M.; Barut, A.; Nam, P. Weak form of peridynamics for nonlocal essential and natural boundary conditions. *Comput. Methods Appl. Mech. Eng.* **2018**, *337*, 598–631. [CrossRef]
13. Madenci, E.; Dorduncu, M.; Nam, P.; Gu, X. Weak form of bond-associated non-ordinary state-based peridynamics free of zero energy modes with uniform or non-uniform discretization. *Eng. Fract. Mech.* **2019**, *218*, 106613. [CrossRef]
14. Scabbia, F.; Zaccariotto, M.; Galvanetto, U. A novel and effective way to impose boundary conditions and to mitigate the surface effect in state-based Peridynamics. *Int. J. Numer. Methods Eng.* **2021**, *122*, 5773–5811. [CrossRef]
15. Scabbia, F.; Zaccariotto, M.; Galvanetto, U. A new method based on Taylor expansion and nearest-node strategy to impose Dirichlet and Neumann boundary conditions in ordinary state-based Peridynamics. *Comput. Mech.* **2022**, *70*, 1–27. [CrossRef]
16. Huang, Z.X. Revisiting the peridynamic motion equation due to characterization of boundary conditions. *Acta Mech. Sin.* **2019**, *35*, 972–980. [CrossRef]

17. Aksoylu, B.; Mengesha, T. Results on nonlocal boundary value problems. *Numer. Funct. Anal. Optim.* **2010**, *31*, 1301–1317. [CrossRef]

18. Zhou, K.; Du, Q. Mathematical and numerical analysis of linear peridynamic models with nonlocal boundary conditions. *SIAM J. Numer. Anal.* **2010**, *48*, 1759–1780. [CrossRef]

19. Silling, S.A.; Askari, E. A meshfree method based on the Peridynamic model of solid mechanics. *Comput. Struct.* **2005**, *83*, 1526–1535. [CrossRef]

20. Liu, W.Y.; Hong, J.W. Discretized peridynamics for linear elastic solids. *Comput. Mech.* **2012**, *50*, 579–590. [CrossRef]

21. Tupek, M.R.; Radovitzky, R. An extended constitutive correspondence formulation of peridynamics based on nonlinear bond-strain measures. *J. Mech. Phys. Solids* **2014**, *65*, 82–92. [CrossRef]

22. Behzadinasab, M.; Foster, J.T. On the stability of the generalized, finite deformation correspondence model of peridynamics. *Int. J. Solids Struct.* **2019**, *182*, 64–76. [CrossRef]

23. Zhou, Z.; Yu, M.; Wang, X.; Huang, Z. Peridynamic analysis of 2-dimensional deformation and fracture based on an improved technique of exerting traction on boundary surface. *Arch. Mech.* **2022**, *74*, 441–461.

24. Silling, S.A. Dynamic fracture modeling with a meshfree peridynamic code. In *Computational Fluid and Solid Mechanics*; Elsevier Science Ltd.: Amsterdam, The Netherlands, 2003; pp. 641–644.

25. Parks, M.L.; Seleson, P.; Plimpton, S.J.; Silling, S.A. *Peridynamics with LAMMPS: A User Guide v0.3 Beta*; Sandia Report 2011–8253; Sandia National Laboratories: Albuquerque, NM, USA, 2011.

26. Ha, Y.D. An extended ghost interlayer model in peridynamic theory for high-velocity impact fracture of laminated glass structures. *Comput. Math. Appl.* **2020**, *80*, 744–761. [CrossRef]

27. Shojaei, A.; Hermann, A.; Cyron, C.J.; Seleson, P.; Silling, S.A. A hybrid meshfree discretization to improve the numerical performance of peridynamic models. *Comput. Methods Appl. Mech. Eng.* **2022**, *391*, 114544. [CrossRef]

28. Shojaei, A.; Hermann, A.; Seleson, P.; Silling, S.A.; Rabczuk, T.; Cyron, C.J. Peridynamic elastic waves in two-dimensional unbounded domains: Construction of nonlocal Dirichlet-type absorbing boundary conditions. *Comput. Methods Appl. Mech. Eng.* **2023**, *407*, 115948. [CrossRef]

29. Shojaei, A.; Hermann, A.; Seleson, P.; Cyron, C.J. Dirichlet absorbing boundary conditions for classical and peridynamic diffusion-type models. *Comput. Mech.* **2020**, *66*, 773–793. [CrossRef]

30. Madenci, E.; Barut, A.; Phan, N. Bond-based peridynamics with stretch and rotation kinematics for opening and shearing modes of fracture. *J. Peridyn. Nonlocal Model.* **2021**, *3*, 211–254. [CrossRef]

31. Zhang, Y.; Madenci, E. A coupled peridynamic and finite element approach in ANSYS framework for fatigue life prediction based on the kinetic theory of fracture. *J. Peridyn. Nonlocal Model.* **2022**, *4*, 51–87. [CrossRef]

32. Trask, N.; You, H.; Yu, Y.; Parks, M.L. An asymptotically compatible meshfree quadrature rule for nonlocal problems with applications to peridynamics. *Comput. Methods Appl. Mech. Eng.* **2018**, *343*, 151–165. [CrossRef]

33. Yu, Y.; You, H.; Trask, N. An asymptotically compatible treatment of traction loading in linearly elastic peridynamic fracture. *Comput. Meth. Appl. Mech. Eng.* **2021**, *377*, 113691. [CrossRef]

34. Reddy, J.N. *An Introduction to Continuum Mechanics*, 2nd ed.; Cambridge University Press: Cambridge, UK, 2013.

35. Holzapfel, G.A. *Nonlinear Solid Mechanics: A Continuum Approach for Engineering*; Ringgold Inc.: Portland, OR, USA, 2000.

36. Nishawala, V.V.; Ostoja-Starzewski, M. Peristatic solutions for finite one- and two-dimensional systems. *Math. Mech. Solids* **2017**, *22*, 1639–1653. [CrossRef]

37. Chen, Z.G.; Bobaru, F. Selecting the kernel in a peridynamic formulation: A study for transient heat diffusion. *Comput. Phys. Commun.* **2015**, *197*, 51–60. [CrossRef]

38. Kilic, B.; Madenci, E. An adaptive dynamic relaxation method for quasi-static simulations using the peridynamic theory. *Theor. Appl. Fract. Mech.* **2010**, *53*, 194–204. [CrossRef]

39. Kilic, B. Peridynamic Theory for Progressive Failure Prediction in Homogeneous and Heterogeneous Materials. Doctor Thesis, The University of Arizona, Tucson, AZ, USA, 2008.

40. Ni, T.; Zhu, Q.Z.; Zhao, L.Y.; Li, P.F. Peridynamic simulation of fracture in quasi brittle solids using irregular finite element mesh. *Eng. Fract. Mech.* **2018**, *188*, 320–343. [CrossRef]

41. Bobaru, F.; Yang, M.J.; Alves, L.F.; Silling, S.A.; Askari, E.; Xu, J.F. Convergence, adaptive refinement, and scaling in 1D peridynamics. *Int. J. Numer. Methods Eng.* **2009**, *77*, 852–877. [CrossRef]

42. Ha, Y.D.; Bobaru, F. Studies of dynamic crack propagation and crack branching with peridynamic. *Int. J. Fract.* **2010**, *162*, 229–244. [CrossRef]

43. Bobaru, F.; Zhang, G.F. Why do cracks branch? A peridynamic investigation of dynamic brittle fracture. *Int. J. Fract.* **2015**, *196*, 59–98. [CrossRef]

44. Liu, M.H.; Wang, Q.; Lu, W. Peridynamic simulation of brittle-ice crushed by a vertical structure. *Int. J. Nav. Archit. Ocean Eng.* **2017**, *9*, 209–218. [CrossRef]

materials

Article

Thermal Regulation Performance of Shape-Stabilized-Phase-Change-Material-Based Prefabricated Wall for Green Grain Storage

Changnv Zeng, Chaoxin Hu and Wanwan Li *

School of Civil Engineering, Henan University of Technology, Zhengzhou 450001, China
* Correspondence: liww2008@163.com

Abstract: In order to meet the great demand for green grain storage and low carbon emissions, paraffin, high-density polyethylene (HDPE), and expanded graphite (EG) were used to produce shape-stabilized phase change material (SSPCM) plates, which were then used to reconstruct building walls for existing granaries. A new type of SSPCM plate was then prefabricated with different thermal conductivities and a high latent heat. This plate could be directly adhered to the existing granary walls. In order to evaluate the thermal regulation performance of these phase change granary walls, experiments and numerical methods were established, specifically for the summer condition. The thermal behavior of the SSPCM granary wall was compared with that of the common concrete granary wall to obtain the optimal parameters. It was concluded that increasing the thickness of the SSPCM layer can reduce the temperature rise of the wall. However, the maximum latent heat utilization rate and energy storage effects were obtained when the SSPCM thickness was at an intermediate level of 30 mm. The thermal conductivity of the SSPCM had a controversial effect on the thermal resistance and latent heat utilization behaviors of the SSPCM. Considering the temperature level and energy saving rate, a 30 mm thick SSPCM plate with a thermal conductivity of 0.2 W/m·K provided a superior performance. When compared to the common wall, the optimized energy-saving rate was greatly enhanced by 35.83% for the SSPCM granary wall with a thickness of 30 mm and a thermal conductivity of 0.2 W/m·K.

Keywords: prefabricated SSPCM granary wall; green grain storage; heat transfer characteristics; thermal conductivity

Citation: Zeng, C.; Hu, C.; Li, W. Thermal Regulation Performance of Shape-Stabilized-Phase-Change-Material-Based Prefabricated Wall for Green Grain Storage. *Materials* **2023**, *16*, 964. https://doi.org/10.3390/ma16030964

Academic Editor: Antonio Caggiano

Received: 15 December 2022
Revised: 9 January 2023
Accepted: 18 January 2023
Published: 20 January 2023

1. Introduction

The green granary uses green, environmental-protection technology to ensure the quality and safety of grain storage, importantly to avoid the pollution of the grains and the environment [1]. The granary temperature is affected by the outdoor ambient air temperature and solar radiation. For most old granaries, the poor thermal insulation performance of the envelope enclosure easily leads to the penetration of substantial ambient heat into granary space. This will cause the grain temperature to exceed 40 °C [2]. A further consequence is that grain mildew and insect pests are induced. In order to provide a safe temperature for stored grain, most old granaries use an air-conditioning system to maintain a low temperature condition. Alternatively, old granaries use pesticides to avoid insect pests. These will cause a high electric energy consumption or chemical pollution to the grain and the environment. Regardless, high-quality grain cannot be provided, especially during long-term grain storage in China. Therefore, it is of great importance to transform these old granaries into green granaries to ensure energy saving and the environmental protection of grain storage.

The addition of thermal insulation boards to improve the thermal resistance of the building envelope has been commonly mostly used in the renovation of old granaries. Derradji et al. [3] used an expanded polystyrene insulation board to enhance the internal insulation of the building wall to control the indoor temperature. Different kinds of insulation materials, such as extruded polystyrene, mineral wool, polyurethane, etc., were applied to the granary roof [4]. The addition of thermal insulation layers increased the thermal resistance of the granary envelope and effectively reduced the granary temperature, which was influenced by the environment temperature. However, the existing renovations, performed by adding different thermal insulation materials to the old granaries, are high-energy consumption processes. For example, it is easy for the insulation to become damaged, fall off, and reduce performance when laid on the wall. Most importantly, the existing methods cannot regulate the self-induced grain temperature, especially in the large-capacity granaries. Thus, extra air-conditioning is usually used to regulate the grain temperature during long-term storage; this also leads to high energy consumption. Therefore, it is of great importance to develop new granary-renovation technologies for low energy consumption.

Phase change materials (PCMs), called energy-saving and self-temperature-regulated materials, can be integrated into the building envelope. PCMs can be used for different climates or seasons in different countries to manage the thermal environment and save energy with low energy and decreased carbon emissions [5]. They can store and release a large amount of latent heat during their phase change process, which makes them maintain a relatively constant temperature when temperature ranges among the phase change temperature of PCMs. Thus, reconstructed granary walls utilizing PCMs will enable granaries to have a high energy-storage capacity, effectively improving the walls' temperature-controlling and heat-insulating performance [6]. The PCM granary wall may provide a promising green technology for the wall itself.

Many relevant theories and practices regarding PCMs are used in civil buildings to maintain a comfortable indoor environment [7,8]. For example, PCM particles were mixed into cement mortar and concrete to enhance their heat-storage capacity for regulating the indoor temperature of buildings [9–11]. However, the latent heat of the phase change mortar and concrete were lower than that of the PCM, which restricted its regulation ability on reducing the indoor temperature [12].

Although relevant theories of PCMs in civil buildings can be applied to the granary wall, granaries need to be controlled to a lower temperature. For example, the grain temperature needs to be below 25 °C to decrease insect harm to the stored grain and to maintain a stable temperature for long-term storage. Primarily, the interior temperature is specified to be no more than 25 °C for the quasi-low temperature granary [13], while the interior temperature should be for comfortable civil buildings 28 °C [14]. This causes different demands of the phase change temperature and heat-storage capacity.

More importantly, the grain storage industry focuses on the energy-saving reconstruction of existing granary buildings rather than launching new construction, which is quite different from civil buildings [15]. This leads to the conclusion that prefabricated wall construction technology should be used in the granary. However, the existing phase change walls in civil buildings are usually constructed by mixing phase change particles with common building materials, such as mortar or concrete, which is inapplicable to prefabricated walls. Given that, in previous studies, the PCM plate could be applied to the light walls to increase their thermal inertia and adjust the indoor temperature of buildings [16,17], we propose a prefabricated, shape-stabilized PCM (SSPCM) plate with a high latent heat integrated into the granary wall.

Moreover, a common problem is that the high cost of PCM for a low phase change temperature was unsuitable for large-scale applications in civil buildings and granaries.

To deal with the above problems, a prefabricated SSPCM plate with a lower phase change temperature and a lower cost was prepared for direct use in the granary wall. Paraffin is one of the most popular PCMs because of its high thermal stability, high latent heat capacity, consistent phase-transition temperature, low price, availability without flammability, toxic distortion, and crystallization problems [18]. However, paraffin leaks easily during the melting process, and its thermal conductivity is relatively low: approximately

0.2 W·m^{-1} K^{-1} [19]. To deal with the leakage problem of paraffin, an HDPE with a similar straight-molecular-chain structure to paraffin was used as a supporting material to wrap the paraffin, which can greatly reduce the melting leakage of paraffin [20]. Additionally, the high tenacity of HDPE may enhance the mechanical strength and structural stability of the SSPCM.

In order to enhance the thermal conductivity, EG was added into the mixture of paraffin and HDPE. It may improve the phase change rate and the utilization of the whole phase change latent capacity [21].

Therefore, a new type of prefabricated SSPCM plate with paraffin, HDPE, and EG was prepared for the granary wall in this study. Based on the prefabricated SSPCM plate, the experimental and physical models were established to investigate the thermal performance of the SSPCM granary wall. Its heat transfer behavior was then tested and simulated to select the optimal parameters of the SSPCMs. Furthermore, taking the common granary wall as a reference, the energy-saving effect of the SSPCM granary wall was evaluated.

2. Materials and Method

2.1. Experimental Section

2.1.1. Preparation of SSPCM Plate

In this paper, the SSPCM was prepared with paraffin, high-density polyethylene (HDPE), and expanded graphite (EG), shown in in Figure 1. Considering that mixing paraffin has a much lower cost than the existing low-temperature pure paraffin, the solid–liquid paraffin mixture was selected for the phase change matrix, as is shown in Figure 1a. The mixture was prepared using solid and liquid paraffin with a mass ratio of 7:3. The phase change temperature and latent heat of solid paraffin were 52 °C and 252.86 J/g, respectively, while the pour point of liquid paraffin was −10 °C [22]. The phase change temperature and the latent heat of mixed paraffin were then 28.7 °C and 181.6 J/g, respectively. Due to its high compatibility with the paraffin mixture, HDPE was used as the supporting material. It had a a melting point of 160 °C, as is shown in Figure 1b. The EG was selected as a thermal-conductivity enhancer for the prepared SSPCM [23], as is shown in Figure 1c. The EG possessed a specific surface area of 32 m^2/g, a micro thickness of 350 nm, and a particle size of 38 μm.

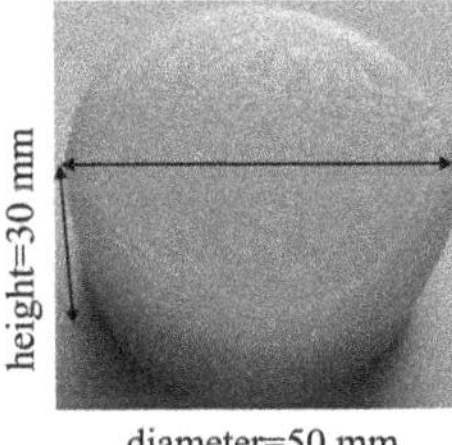
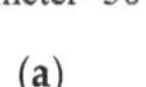

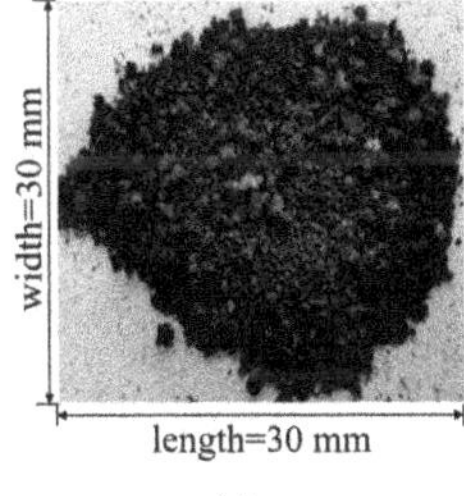

Figure 1. The raw materials of SSPCM: (**a**) solid–liquid mixed paraffin, (**b**) HDPE particle, (**c**) expanded graphite powder.

The SSPCM was prepared by the melt blending method, as in our previous study [24]. The mass ratio of paraffin to HDPE was 8:2, and the mass of EG accounted for 2% of the total mass of paraffin and HDPE. First, paraffin was heated to 60 °C in an oil bath basin until it was melted completely. Second, the HDPE was mixed with the molten paraffin in an oil bath basin. The temperature of the oil was set to 160 °C with continuous stirring of 30 rpm for 30 min. The EG powder was then added into the oil bath basin and stirred at 30 rpm for 30 min. Finally, the mixture was poured into a rectangular mold and cooled to room temperature, forming the 800 × 800 × 30 mm SSPCM plate shown in Figure 2.

Figure 2. The prefabricated SSPCM plate.

2.1.2. The Properties of SSPCM

The microstructure of the SSPCM was observed using a scanning electron microscope (SEM). The $10 \times 10 \times 3$ mm SSPCM sample was dried and pre-processed by the metal-spraying process to enhance its conductivity [24]. Figure 3 displays the SEM of the EG and the SSPCM. As is shown in the Figure 3a, the EG presented a worm-like structure; the molten HDPE and paraffin can be absorbed well by the pores of the EG. In Figure 3b, the brighter part is the HDPE, while the darker parts are paraffin and EG [25]. The paraffin and EG were uniformly wrapped by the net structure of the HDPE, which proved that HDPE presented great compatibility with paraffin and EG [25].

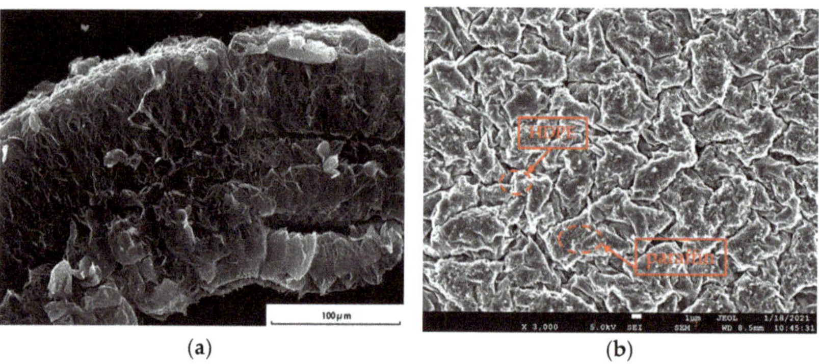

Figure 3. The SEM photographs of (**a**) EG and (**b**) the SSPCM.

A DSC-100 differential scanning calorimeter was used to measure the latent heat and phase change temperature of the SSPCM. The 10 mg SSPCM was placed into an aluminum crucible to be heated from room temperature to 150 °C at a heating rate of 10 °C/min. The development of heat flux with temperature was recorded automatically. As is shown in Figure 4, the latent heat of the SSPCM is 150.7 kJ/kg. The onset melting temperature and phase change temperature were 16.6 °C and 28.5 °C, respectively.

The thermal conductivities of the samples were measured using a DZDR-S Thermal Constants Analyzer. Two pieces of $100 \times 100 \times 30$ mm SSPCM samples were set on the test plate, and the test probe was placed between the two SSPCM samples. The thermal conductivity of SSPCM was tested three times, and the values in the permissible error range of 3% were averaged as the final measurement value. The thermal conductivity of the SSPCM is 0.8 W/m·K, which is about four times higher than that of pure, solid paraffin.

The thermal stability of the SSPCM was measured using a high and low temperature-alternating damp heat test chamber with an accelerated thermal cycling analysis. An amount of 20 g of SSPCM was heated and cooled from room temperature (20 °C) to 80 °C during the 1000 cycles. The mass-loss ratio of PCM was only 0.9%, indicating the high thermal stability property of this SSPCM.

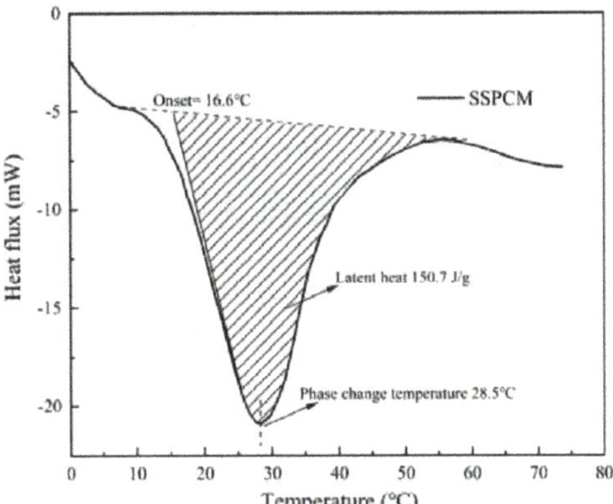

Figure 4. The DSC curve of the SSPCM.

2.1.3. The Establishment of the Thermal Experimental Platform

An experimental platform representing the SSPCM granary wall was built, as is shown in Figure 5, which consisted of an SSPCM granary wall, a temperature-acquisition instrument (Agilent34970A), heat source equipment, and environment simulation boards. The SSPCM granary wall was constructed with a concrete layer of 800 × 800 × 210 mm and a prefabricated SSPCM plate of 800 × 800 × 30 mm. The four sides of the concrete layer were wrapped by 100 mm of insulated cotton. This ensured that the concrete layer was closed to an adiabatic boundary along its height and width. The heat source was mainly transferred from the thickness of the concrete layer. Moreover, the environment simulation boards were used to positively simulate the external temperature environment. During the experimental process, the heat source equipment was utilized to heat the surrounding air for two days. The external temperature environment was mainly regulated by adjusting the power of the heating source equipment and the distance from the SSPCM granary wall. The indoor and outdoor temperature data were effectively collected by the temperature-acquisition instrument in real time.

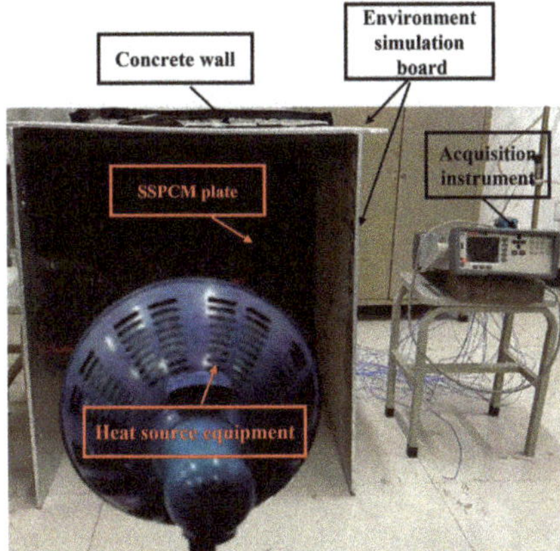

Figure 5. The experimental platform of the SSPCM granary wall.

2.2. Numerical Section

2.2.1. The Physical Model of the Granary Walls

As is shown in Figure 6, the common granary wall consists of a 240 mm concrete layer. The SSPCM granary wall was formed by 30 mm of the SSPCM plate and 210 mm of the concrete layer, corresponding to its widths of L1 and L2, respectively. According to the DSC test, the phase change temperature of the SSPCM was 28.5 °C. The other parameters of SSPCM are listed in Table 1.

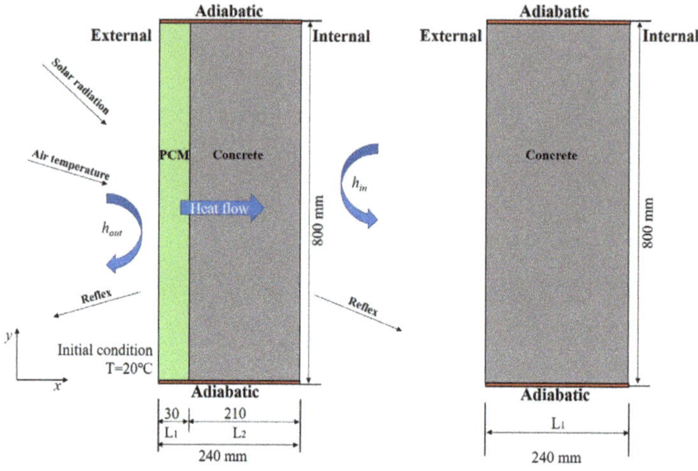

Figure 6. The heat transfer physical model of the granary walls.: (**left**) the SSPCM granary wall and (**right**) the common granary wall.

Table 1. Material properties of concrete and SSPCM [26,27].

Material	Density (kg/m³)	Specific Heat Capacity (J/kg·K)	Phase Change Temperature (°C)	Latent Heat (J/g)	Thermal Conductivity (W/m·K)	PCM Thickness (mm)
Concrete	2400	1030	N/A	N/A	1.74	N/A
SSPCM	790	1500	28.5	150	0.8 0.2, 0.4, 0.6, 0.8, 1.2, 1.6	10, 20, 30, 40 30

The ANSYS Fluent software was used to simulate the heat-transfer process of the SSPCM granary wall and the common granary wall. The internal and external boundaries of the granary wall were assumed to have convective heat transfer. Additionally, the top and bottom boundaries were set as adiabatic [28]. The initial temperature of the granary wall was 20 °C. According to the national standard of grain storage in China, the granary local temperature should be lower than 25 °C [13].

In order to preferably perform the heat transfer in the SSPCM granary wall, a structural, quadrilateral mesh with 4800 element numbers was employed to discretize the physical model of the SSPCM granary wall. The PISO (Pressure Implicit Split Operator) algorithm was used for coupling pressure and velocity. In the spatial discretization section, a least-squares cell-based evaluation was used for the gradient, and the second order was used for the pressure. The second order upwind was used for the momentum and energy equation. Furthermore, the first order implicit function was used for the transient formulation.

Considering the thermal effect of solar radiation, a comprehensive solar–air temperature was adopted, shown in Figure 7, which was selected from a typical week in the summer in Zhengzhou city. The solar–air temperature ($T_{sol\text{-}air}$) consists of solar radiation and ambient temperature (T_{air}), as is shown in Equations (1) and (2).

$$T_{sol\text{-}air} = T_{air} + \frac{\alpha I_t}{h_{out}} \tag{1}$$

$$I_t = I_D + I_d + I_{Earth} \tag{2}$$

where α is the solar radiation absorption coefficient of the outdoor wall surface, 0.48; h_{out} is the heat transfer coefficient of the outdoor wall surface, 18.3 W/m^2·K [29]; I_t is the total intensity of solar radiation; and W/m^2 consists of the direct solar irradiance, I_D, the diffuse horizontal irradiation, I_d, and the ground-reflected irradiation I_{Earth}, W/m^2 [29].

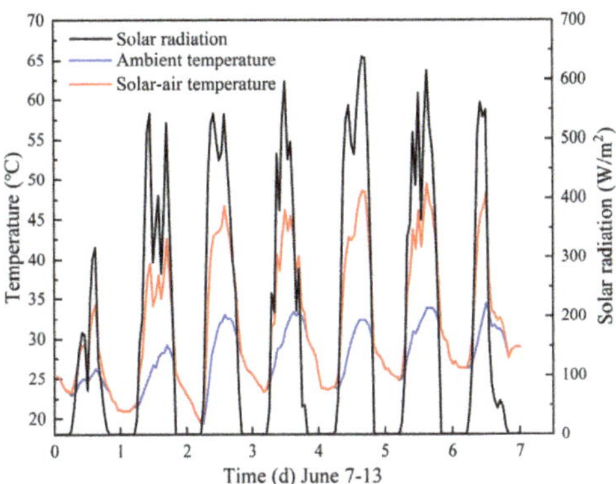

Figure 7. The solar–air temperature of Zhengzhou city [30].

2.2.2. Mathematical Model

As was mentioned above, the SSPCM granary wall consisted of the SSPCM layer and the concrete layer. The bottom and top wall surfaces were under adiabatic conditions and the heat was applied in a uniform distribution along the y-direction. Here, a one-dimensional heat-transfer mathematical model along the x-direction was established for both the concrete and SSPCM layer; it is similar to those found in the literature [31–33]. The apparent heat capacity method was used to describe the latent heat transfer of the SSPCM.

For simplicity, the physical model of the SSPCM granary wall was established under the following assumptions:

(a) All layers of the wall system were assumed to be homogenous and isotropic;
(b) The heat transfer through the wall was one-dimensional;
(c) The thermal expansion of the materials was not considered;
(d) The thermophysical properties of the materials were constant, except for the change in the material properties of the SSPCM during the phase change interval;
(e) For multi-layered wall systems, the contact resistance between different layers was negligible.

The heat-transfer process nonlinearly changed with the time and ambient temperature. The heat-conduction equations of concrete and SSPCM can be expressed by Equations (3) and (4), respectively.

$$\rho_1 c_1 \frac{\partial T}{\partial t} = \frac{\partial}{\partial x}\left(k_1 \frac{\partial T}{\partial x}\right) \tag{3}$$

$$\rho_{SSPCM} c_{SSPCM} \frac{\partial T}{\partial t} = \frac{\partial}{\partial x}\left(k_{SSPCM} \frac{\partial T}{\partial x}\right) \tag{4}$$

where ρ_1 and ρ_{SSPCM} represent the densities of concrete and SSPCM, respectively, kg/m^3; c_1 and c_{SSPCM} represent the specific heat capacity of concrete and SSPCM, respectively,

J/kg·K; and k_1 and k_{SSPCM} represent the thermal conductivity of concrete and SSPCM, respectively, W/m·K. Temperature is represented by T, °C; and t is time, s.

Equations (5) and (6) are used to account for the change in material properties during the phase-transition process of SSPCM. The specific heat capacity is presented by Equation (7) for different phase-change states.

$$\rho_{SSPCM} = \theta_s \rho_s + \theta_l \rho_l \tag{5}$$

$$k_{SSPCM} = \theta_s k_s + \theta_l k_l \tag{6}$$

$$c_{SSPCM}(T) = \begin{cases} c_{SSPCM}, T < T_m - \Delta T, \, T \geq T_m + \Delta T \\ \delta(T)\Delta H + c_{SSPCM}, T_m - \Delta T \leq T < T_m + \Delta T \end{cases} \tag{7}$$

$$\delta(T) = \frac{\exp(-(T - T_m)^2)/(\Delta T)^2)}{\Delta T \sqrt{\pi}} \tag{8}$$

where θ is the volume proportion of the SSPCM; θ_s and θ_l are the volume proportions of the solid phase and liquid phase in the SSPCM, respectively; ρ_s and ρ_l are the densities of solid SSPCM and liquid SSPCM, respectively, kg/m^3; k_s and k_l are the thermal conductivity of solid SSPCM and liquid SSPCM, respectively, W/m·K; $c_{SSPCM}(T)$ is the equivalent specific heat capacity of SSPCM, which depends on temperature; T_m is the phase-change temperature of SSPCM; $2\Delta T$ is the range of phase-change temperature; and ΔH is the latent heat of SSPCM. T_m, ΔT, and ΔH can be obtained from the DSC test. The Gaussian function of the equivalent specific heat is represented by $\delta(T)$, which is expressed by Equation (8).

The boundary conditions of the external and internal wall surfaces can be described by Equations (9) and (10), respectively.

$$-k_{SSPCM}\frac{\partial T}{\partial x} = h_{out}(T_{solar\text{-}air} - T) \tag{9}$$

$$-k_1\frac{\partial T}{\partial x} = h_{in}(T_{in} - T) \tag{10}$$

where h_{in} is the heat transfer coefficient of the internal surface, 8.7 W/m^2·K [30]; and T_{in} is the internal air temperature, °C.

Here, the adiabatic boundaries of the bottom and top wall surfaces can be formulated in Equations (11) and (12).

$$\frac{\partial T}{\partial y} = 0, \, y = 0 \tag{11}$$

$$\frac{\partial T}{\partial y} = 0, \, y = 800 \tag{12}$$

where y is the height direction of the SSPCM granary wall, in mm.

The initial condition is described in Equations (13) and (14).

$$T = T_0, 0 \leq x \leq 240 \text{ mm}, t = 0 \tag{13}$$

$$T_{in} = T_0, t = 0 \tag{14}$$

where T_0 is the initial temperature, set at 20 °C.

2.2.3. Thermal Performance Evaluation Indexes of the SSPCM Granary Wall

In order to quantitatively evaluate the heat-transfer performance and the temperature-regulation ability of the granary wall, some evaluation indexes [34–36] were used, including the temperatures of the internal wall surface and the SSPCM layer, heat flux, and energy saving.

(1) The temperature of the internal wall surface and the SSPCM layer

The external heat energy was gradually transferred to the internal wall surface and the SSPCM layer. The greater the decrease in temperature in the internal wall surface, the better the energy storage efficiency of the SSPCM was. The phase change process can be observed from the temperature development of the SSPCM layer.

(2)   Heat flux on the internal surface of the wall

This is also known as the heat flux (ϕ) on the inner surface of the wall, referring to the heat energy passing through the inner surface per unit of time. The smaller the heat flux, the less heat transfers to the inner surface.

$$\phi = -kA\frac{dT}{dx} \tag{15}$$

(3)   Energy saving rate

The cooling load is the total amount of reduced heat through the inner surface of the wall within one week, 168 h. The energy saving value is expressed by the ratio of the difference in cooling load between the common granary wall and the SSPCM granary wall to the common granary wall [12]. The greater the ratio, the greater the energy saving was.

$$Q = \int_0^{168} \phi dt \tag{16}$$

$$E_{saving} = \frac{Q_0 - Q_{PCM}}{Q_0} \tag{17}$$

where Q is the cooling load, $W \cdot h/m^2$; Q_0 is the cooling load of the common granary wall, $W \cdot h/m^2$; Q_{PCM} is the cooling load of the SSPCM granary wall, $W \cdot h/m^2$; and E_{saving} is the energy saving rate, %.

2.3. The Experiment Result and Model Validation

To validate the reasonability of the physical model, the experimental results were compared with the simulation results. As is shown in Figure 8, the calculated results of the internal surface temperatures on the SSPCM granary wall were close to the experimental values. The internal surface temperature varied with the variation of the external ambient temperature. For a comparison at 30 h, the indoor surface temperature of the simulation was, at maximum, 1.64 °C higher than the temperature measured in the experiment. The maximum relative error was approximately 7.39%. The relative errors were mainly caused by the permissible error of ±0.8 °C in the accuracy of the temperature-acquisition instrument and the neglected contact thermal resistance between the SSPCM plate and the concrete wall.

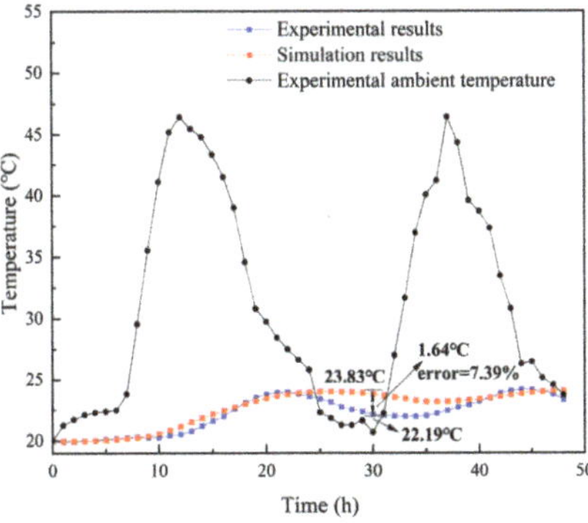

Figure 8. The comparison between the experiment and calculated results.

3. Results and Discussion

In order to optimize the design parameters of the granary, the heat-transfer performance of SSPCM granary walls was investigated by comparing it to that of the common granary walls. Figure 9 shows the results of the temperature contour between the common and SSPCM granary walls. Here, the SSPCM layer was 30 mm thick with a latent heat of 150 kJ/kg, a phase-transition temperature of 28.5 °C, and a thermal conductivity of 0.8 W/m·K.

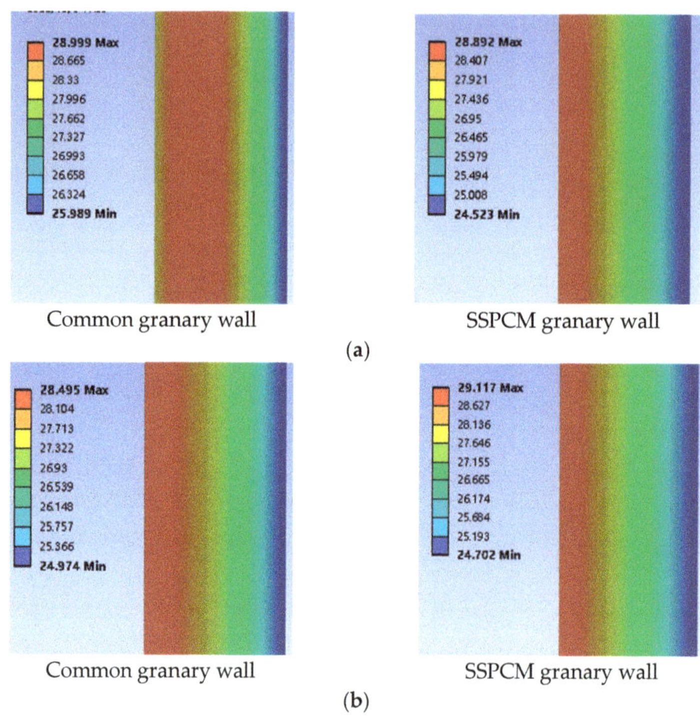

Figure 9. The temperature distribution of the common and SSPCM granary walls at (**a**) T = 72 h (third day) and (**b**) T = 168 h (seventh day).

As can be seen from Figure 9a, on the third day, the red margin in the SSPCM granary wall was smaller than the red margin in the common granary wall, showing a lower temperature and a narrower distribution range in the red region. This is because the latent heat of the SSPCM can effectively absorb the external heat energy to decrease the amount of heat transfer into the indoor space. Further, the thermal conductivity of the SSPCM was lower than that of concrete; therefore, the heat resistance and heat inertia of the SSPCM granary wall was higher than that of the common granary wall. Thus, less heat was conducted to the SSPCM granary wall when compared to the common granary wall. The SSPCM granary wall demonstrated a superior temperature-regulation performance.

In Figure 9b, at the end of the 7th day, the red high-temperature region and the internal surface temperature of the common granary wall were still higher than that of the SSPCM granary wall, although the distribution range gradually became smaller. Moreover, the maximum external surface temperature of the common granary was lower than that of the SSPCM granary wall, due to the fact that concrete can quickly release heat energy. Summarily, the SSPCM plays a key role in self-regulating the temperature of the SSPCM granary wall to greatly decrease the granary energy consumption. Thus, it may provide green grain storage without extra energy consumption, which meets the requirement of green grain storage.

3.1. The Effect of the SSPCM Thickness on the Heat Transfer

It is necessary to optimize the thickness of the phase change material for maximum use and economic efficiency. To compare the effect of the SSPCM thickness on the heat-transfer performance of granary walls, four SSPCM granary-wall models were simulated in this paper. These were named SSPCM-0, SSPCM-10, SSPCM-20, SSPCM-30, and SSPCM-40. SSPCM-0 represented the common granary wall without the SSPCM plate. SSPCM-10, SSPCM-20, SSPCM-30, and SSPCM-40 represented the SSPCM granary wall with 10 mm, 20 mm, 30 mm, and 40 mm SSPCM layers, respectively.

Figure 10 displays the temperature variations of the internal wall surface with different SSPCM layer thicknesses. The temperature of the internal wall surface decreased with the increasing thickness of the SSPCM layer. On the sixth day, the maximum temperature of the internal wall surface was 27.33 °C for SSPCM-0 without any SSPCM. Additionally, the temperatures of SSPCM-30 and SSPCM-40 were 26.3 °C and 26 °C, which were 1.03 °C and 1.33 °C lower than that of SSPCM-0, respectively. The SSPCM layer can effectively decrease the internal surface temperature of the SSPCM granary wall.

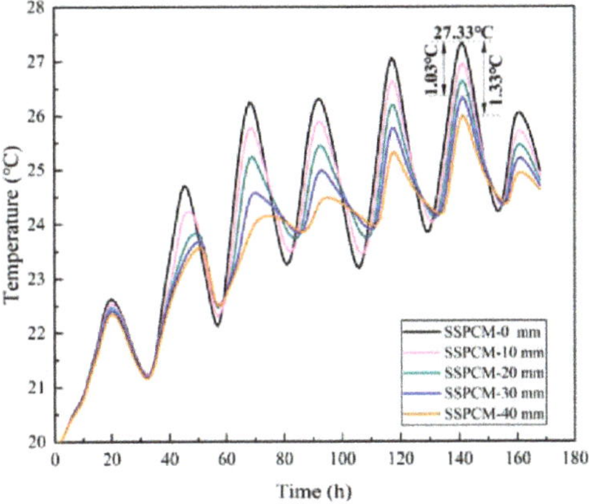

Figure 10. The internal surface temperature of SSPCM granary wall with different SSPCM layer thicknesses.

The peak temperature in the SSPCM layer also showed a similar trend to that of the internal wall surface, as is shown in Figure 11. On the sixth day, the SSPCM layers had completed their phase transition with a total utilization of their latent heat. The peak temperatures of the SSPCM layer surfaces were 37.84 °C, 36.47 °C, 35.13 °C, and 33.82 °C, corresponding to SSPCM-10, SSPCM-20, SSPCM-30, and SSPCM-40, respectively. In comparison to the SSPCM-10, the internal surface temperature of SSPCM-30 and SSPCM-40 were reduced by 2.71 °C and 4.02 °C, respectively. With the increasing increment of SSPCM thickness, a higher external heat was absorbed in the SSPCM layer.

Figure 12 displayed the liquid fraction of four kinds of SSPCM layers. The liquid fraction of the SSPCM layer gradually increased with the rise of external temperature. During the 32 h to 50 h period, the maximum liquid fraction of SSPCM-30 and SSPCM-40 was 82.73% and 64.74%, respectively. Both values then completely recovered to the solid state. However, during the 120 h to 140 h period, all kinds of SSPCM layers were quickly and completely changed from a solid state to a liquid state. Furthermore, none could entirely recover to a solid state due to the continuously high ambient temperature. Thus, based on the reduction in the internal surface peak temperature of the SSPCM granary wall and the SSPCM layer and the liquid fraction of the SSPCM layer, the SSPCM plate with a 30 mm thickness is optimal.

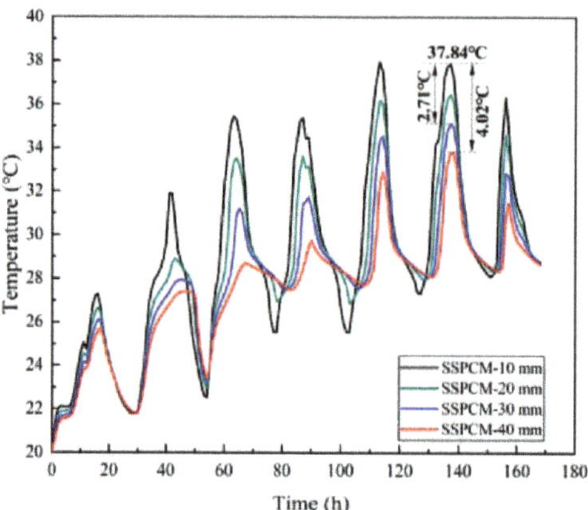

Figure 11. The internal surface temperature of the SSPCM layer with different SSPCM layer thicknesses.

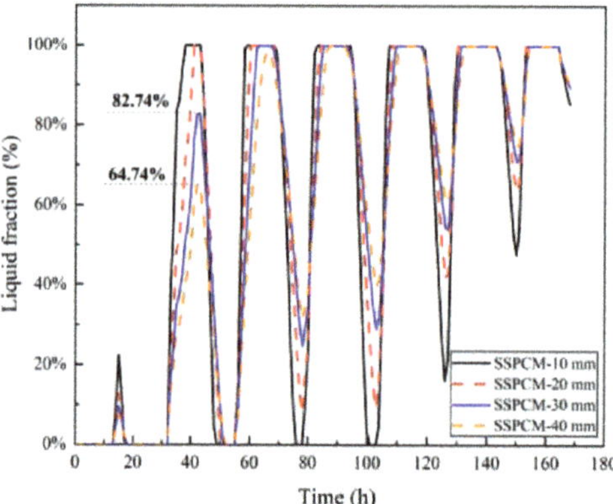

Figure 12. The liquid fraction of the SSPCM layer with different SSPCM thicknesses.

3.2. The Effects of the Prefabricated SSPCM Plate Location on the Heat Transfer

Figure 13 displays the temperature variations of the internal wall surface when the SSPCM layer was at different locations on the granary wall. Here, the SSPCM thickness is 30 mm, and the thermal conductivity of the SSPCM is 0.8 W/m·K. The SSPCM plate was installed on the outer and inner sides, respectively.

As can be seen from Figure 13, the SSPCM layer location indeed affected the heat transfer of the wall. During the 65 h to 72 h period, the peak internal surface temperature of the common granary wall was 26.24 °C. When compared with the common granary wall, the maximum temperatures of the SSPCM granary wall surface with inner and outer locations were 25.88 °C and 24.46 °C, corresponding to the temperature differences of 0.44 °C and 1.78 °C, respectively. However, in the two cases of the SSPCM wall, the temperature was still lower than that of the common granary wall. During the 140 h to 150 h period, the maximum internal surface temperatures were 27.33 °C, 26.29 °C, and 26.35 °C, corresponding to the common granary wall, SSPCM granary wall with the SSPCM layer installed on the outer location, and inner location, respectively. The temperatures of the SSPCM granary walls were reduced by 1.04 °C and 0.98 °C when compared to the common granary wall. It was indicated that the SSPCM installed on the external side of the granary wall could provide a lower temperature for the granary wall in Zhengzhou city. Furthermore, the internal wall surface was near the grain storage requirement of 25 °C, which could save more energy in achieving the requirement [13].

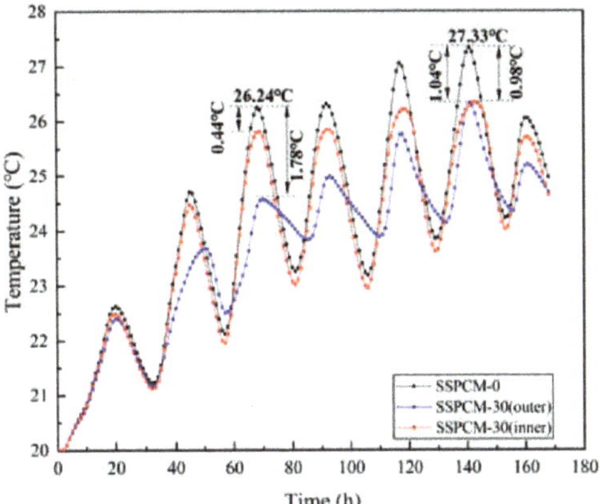

Figure 13. The internal surface temperature of SSPCM granary wall with different position.

Figure 14 represents the internal surface temperature distribution of the SSPCM layer at different SSPCM positions. When installed on the outer side, the SSPCM layer had a higher internal surface temperature than the SSPCM layer located on the internal side. During the 135 h to 145 h period, the maximum temperature difference was 8.79 °C between the external and internal SSPCM layer locations. The temperature difference between the SSPCM layer locations was mainly caused by the thermal insulation effect of the concrete layer. When the SSPCM layer was installed on the external side of the granary wall, the solar radiation and ambient temperature directly affected the SSPCM layer. The heat energy was primarily absorbed by the SSPCM latent heat and was then gradually transferred to the internal surface. However, when the SSPCM layer was installed on the internal side of the granary wall, the solar–air temperature was greatly decreased by the concrete layer. In this case, its temperature variation was slower than that of the SSPCM layer installed on the external side.

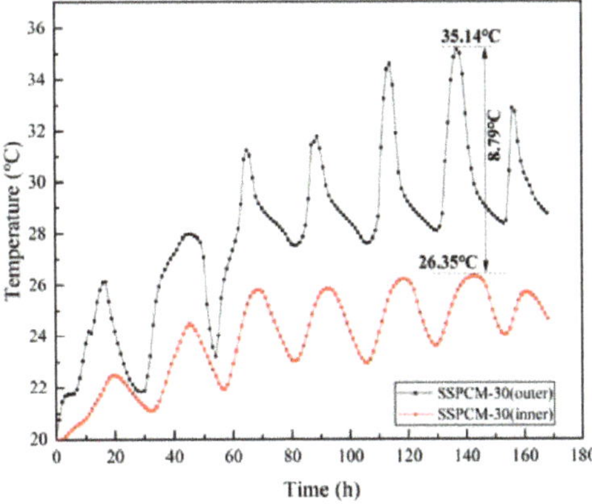

Figure 14. The internal surface temperature of the SSPCM layer at different SSPCM locations.

In summary, when the SSPCM layer is applied to the external granary wall, the internal surface temperature is lower than that of the inner wall. In Zhengzhou city, an SSPCM plate located on an external wall may provide better temperature regulation for a PCM granary wall.

3.3. The Effects of the Thermal Conductivity

Thermal conductivity has a controversial effect on the temperature regulation of the walls. It is well known that decreasing thermal conductivity will prevent more heat flux from being conducted into the indoor space [37]. However, with respect to the SSPCM plate with a large volume, enhancing the thermal conductivity can improve the phase change rate and the utilization of the whole phase change latent capacity (improving the liquid ratio), whereby the SSPCM plate will restrain more heat from entering indoors [33]. Therefore, the comprehensive effects of thermal conductivity need to be considered emphatically.

The internal surface temperature of SSPCM at different thermal conductivity conditions is displayed in Figure 15. In these cases, the SSPCM thickness was 30 mm and the SSPCM layer was located on the outer wall. The thermal conductivity of SSPCM was set as 0.2, 0.4, 0.6, 0.8, 1.2, and 1.6 W/m·K, respectively.

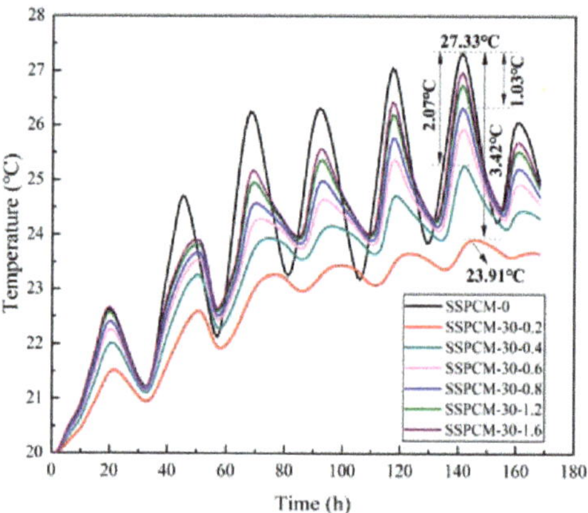

Figure 15. The internal surface temperatures of the SSPCM granary wall with different thermal conductivities.

As can be seen from Figure 15, the internal surface temperature decreased with the decreasing of the thermal conductivity of SSPCM. Under the same ambient conditions, the lower the thermal conductivity, the lower the internal surface temperature of the wall was. When the thermal conductivity was 0.2 W/m·K, the internal surface temperature of the wall was the minimum 23.91 °C, which was 3.42 °C lower than when the thermal conductivity was 1.6 W/m·K. This was followed by 2.07 °C and 1.03 °C differences when the thermal conductivities of SSPCM layers were 0.4 W/m·K and 0.8 W/m·K, respectively. This phenomenon can be also proven using the temperature revolution of the SSPCM plate at different thermal conductivities, as is shown in Figure 16. The maximum peak temperature difference of the SSPCM layer was 8.27 °C between the thermal conductivities of 0.2 W/m·K and 1.6 W/m·K. This indicates that lowering the thermal conductivity did indeed decrease the heat conduction flux into the walls.

However, lowering the thermal conductivity decreased the liquid fraction and the utilization of the phase-change latent capacity. Figure 17 shows the liquid fractions of the SSPCM layers at different thermal conductivities. During the 55 h to 100 h period, the SSPCM layers completely changed to a liquid state when the thermal conductivity of the SSPCM layer was higher than 0.4 W/m·K. For example, the peak liquid fraction of SSPCM layers was 78.75% and 82.75% at 72 h and 96 h, respectively, for a thermal conductivity of the SSPCM plates of 0.2 W/m·K. However, for the case of a thermal conductivity of 0.4 W/m·K, the peak liquid fractions of the SSPCM layers were 94.94% and 99.50% at the same time. This indicates that enhancing the thermal conductivity did indeed increase the utilization of the phase-change latent capacity.

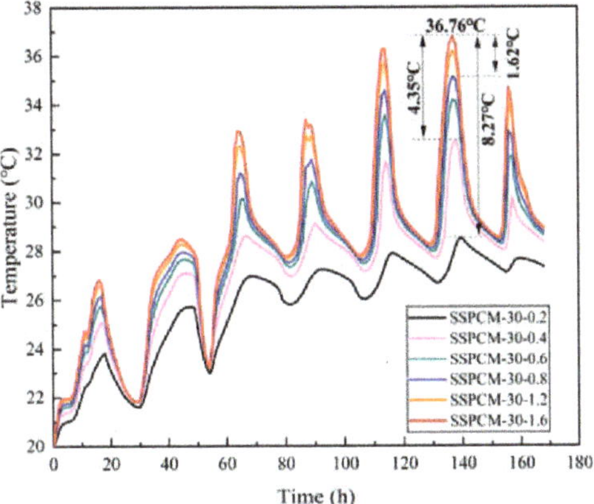

Figure 16. The internal surface temperature of the SSPCM layer at different thermal conductivities.

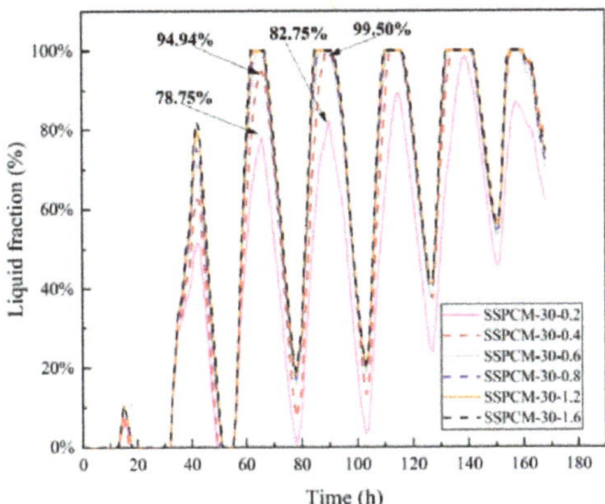

Figure 17. The liquid fraction of the SSPCM layer with different thermal conductivities.

Together, compared with the fact that enhancing thermal conductivity can increase the utilization rate of the latent capacity, the heat resistance caused by low thermal conduction plays a major role in controlling the temperature rise of the internal wall. This is due to the fact that the SSPCM was incorporated in the form of flat plate whose surface direction uniformly received the ambient heat. While the thickness magnitude (30 mm) was quite small compared to the direction of the plane (800 mm), the phase change resistance along the thickness direction was relatively low, leading to little demand on the thermal conductivity enhancement.

Furthermore, the effect of the thermal conductivity of the SSPCM on the energy-saving effect is investigated in this section. As is detailed in Table 2, the energy-saving rate with the effect of SSPCM being incorporated into the granary wall can be obtained by Equation (14). The cooling load (Q) of the common granary wall and the SSPCM granary wall at different thermal conductivities was 5996.43, 3848.16, 4729.11, 5139.63, 5375.09, 5634.28, and 5774.07 W·h/m^2, respectively. When compared to the common granary wall, the energy-saving rates of the SSPCM granary wall were 35.83%, 20.13%, 14.29%, 10.36%, 6.04%, and 3.71%. The energy-saving effect was gradually improved with a decrease in the thermal conductivity of the SSPCM layers. When the SSPCM layer was installed on the external side of the SSPCM granary wall, the SSPCM layer not only used the latent heat to store the heat energy, but also resisted the heat-energy transition to the indoor environment. The lower thermal conductivity can resist more heat energy transfer, which mainly caused the downtrend of the energy-saving effect with the increase of thermal conductivity in the SSPCM layer.

Table 2. Cooling load and energy saving rate for the PCM granary wall.

Parameters	Common Granary Wall	PCM Granary Wall Thermal Conductivity of PCM (W/m·K)					
		0.2	0.4	0.6	0.8	1.2	1.6
Q (W·h·m^{-2})	5996.43	3848.16	4729.11	5139.63	5375.09	5634.28	5774.07
ΔQ (W·h·m^{-2})	/	2148.27	1267.32	856.8	621.34	326.15	222.36
E_{saving} (%)	/	35.83	21.13	14.29	10.36	6.04	3.71

Summarily, in these cases, the optimal parameters of the granary wall are: a 30 mm thick SSPCM plate installed on the outer location of the granary wall. A smaller thermal conductivity is recommended. In this paper, with all the optimal parameters used in the granary walls, the requirements for quasi-low-temperature grain storage can be satisfied.

4. Conclusions

In this paper, the prefabricated SSPCM plate was used to construct a SSPCM granary wall. When compared with the common granary wall, the heat transfer characteristics of the SSPCM granary wall were simulated to optimize its thermal regulation performance for green grain storage. The main conclusions are as follows:

(1) The SSPCM granary wall was prefabricated with the self-temperature-regulated behavior of the PCMs. The optimal parameters of the SSPCM may meet the great demand for green grain storage for large-capacity granaries in China;

(2) Increasing the thickness of the SSPCM layer can reduce the temperature and heat flux of the internal wall surface to achieve the energy-saving effect. In Zhengzhou city, the SSPCM layer located on the outer wall surface can enhance the temperature-regulation effect of the SSPCM granary wall;

(3) With the increasing thickness of the SSPCM, the internal surface temperature decreased. However, in comparison to the maximum thickness of 40 mm, both the maximum latent heat utilization rate and the energy-storage effect are obtained when the SSPCM layer thickness is 30 mm;

(4) Thermal conductivity has a controversial effect on the temperature regulation of the walls. A smaller thermal conductivity dominates in the thermal regulation of walls by restraining the heat flux conducted to the indoors. The maximum energy-saving rate was up to 35.83% for the outer SSPCM layer with a thickness of 30 mm and a thermal conductivity of 0.2 W/m·K. The optimal utilization of the SSPCM layer was present when the thermal conductivity was 0.4 W/m·K.

Author Contributions: Conceptualization, C.Z. and W.L.; methodology, C.Z. and W.L.; software, C.H.; validation, C.Z. and C.H.; formal analysis, C.Z., W.L. and C.H.; investigation, C.Z. and C.H.; resources, C.Z. and W.L.; data curation, W.L.; writing—original draft preparation, C.Z. and C.H.; writing—review and editing, C.Z.; visualization, C.H.; supervision, C.Z.; project administration, W.L.; funding acquisition, C.Z. All authors have read and agreed to the published version of the manuscript.

Funding: This research was funded by the Henan Province Key Specialized Research and Development Breakthrough Plan (212102110027), Henan Province Joint Fund Project of Science and Technology (222103810075), Henan Province Key Specialized Research and Development Breakthrough Plan (222102520030), Henan Key Laboratory Open Fund Key Project (24400009), and the Innovative Funds Plan of Henan University of Technology (2022ZKCJ07).

Institutional Review Board Statement: Not applicable.

Informed Consent Statement: Informed consent was obtained from all subjects involved in the study. Written informed consent has been obtained from the patient.

Data Availability Statement: Not applicable.

Conflicts of Interest: The author declares no potential conflict of interest regarding the publication of this article.

References

1. Ziegler, V.; Paraginski, R.T.; Ferreira, C.D. Grain storage systems and effects of moisture, temperature and time on grain quality-A review. *J. Stored Prod. Res.* **2021**, *91*, 101770. [CrossRef]
2. Zhang, X.X.; Zhang, H.; Wang, Z.Q.; Chen, X.; Chen, Y. Research on temperature field of grain piles in underground grain silos lined with plastic. *J. Food Process Eng.* **2022**, *45*, e13971. [CrossRef]
3. Derradji, L.; Imessad, K.; Amara, M.; Boudali, E.F. A study on residential energy requirement and the effect of the glazing on the optimum insulation thickness. *Appl. Therm. Eng.* **2017**, *112*, 975–985. [CrossRef]
4. Wang, H.T.; Hang, Y.G.; Liu, Y. Integrated economic and environmental assessment-based optimization design method of building roof thermal insulation. *Buildings* **2022**, *12*, 916. [CrossRef]
5. Rinawa, M.L.; Pitchandi, P.; Vigneshkumar, N.; Sharma, R.; Singh, M.K.; Subbiah, R.; Kumar, P.M. Experimental analysis of the metal roofed industrial building using nano-silica disbanded crude wax (NDCW). *Mater. Today Proc.* **2022**, *62*, 1746–1751. [CrossRef]
6. Qudama, A.Y.; Márta, S. Influential aspects on melting and solidification of PCM energy storage containers in building envelope applications. *Int. J. Green Energy.* **2021**, *18*, 966–986.
7. Austin, M.C.; Araúz, J.; Mora, D. Numerical assessment of different phase change materials as a passive strategy to reduce energy consumption in buildings under tropical climates. *Buildings* **2022**, *12*, 906. [CrossRef]
8. Boobalakrishnan, P.; Kumar, P.M.; Balaji, G.; Jenaris, D.S.; Kaarthik, S.; Babu, M.J.P.; Karthhik, K. Thermal management of metal roof building using phase change material (PCM). *Mater. Today: Proc.* **2022**, *47*, 5052–5058. [CrossRef]
9. Guo, J.W.; Jiang, Y.Q.; Wang, Y.; Zou, B. Thermal storage and thermal management properties of a novel ventilated mortar block integrated with phase change material for floor heating: An experimental study. *Energy Convers. Manag.* **2020**, *205*, 112288. [CrossRef]
10. Jin, W.Z.; Jiang, L.H.; Chen, L.; Yin, T.J.; Gu, Y.; Guo, Z.M.; Yan, X.C.; Ben, X.Q. Enhancement of thermal conductivity by graphene as additive in lauric/stearic acid/treated diatomite composite phase change materials for heat storage in building envelope. *Energy Build.* **2021**, *246*, 111087. [CrossRef]
11. Yuan, X.W.; Wang, B.M.; Chen, P.; Luo, T. Study on the Frost Resistance of Concrete Modified with Steel Balls Containing Phase Change Material (PCM). *Materials.* **2021**, *14*, 4497. [CrossRef] [PubMed]
12. Sharma, V.; Rai, A.C. Performance assessment of residential building envelopes enhanced with phase change materials. *Energy Build.* **2020**, *208*, 109664. [CrossRef]
13. *GB/T229890*; Technical Criterion for Grain and Oil-Seeds Storage. China Standards Press: Beijing, China, 2013.
14. *GB/55015*; General Code for Energy Efficiency and Renewable Energy Application in Buildings. China Building Industry Press: Beijing, China, 2021.
15. Nikiforov, A. Rational organizational and technological decisions on the grain storages construction or renovation sites. *Acad. J. Ser. Ind. Mach. Build. Civ. Eng.* **2018**, *1*, 290–297. [CrossRef]
16. Wang, C.X.; Deng, S.M.; Niu, J.L.; Long, E. A numerical study on optimizing the designs of applying PCMs to a disaster-relief prefabricated temporary-house (PTH) to improve its summer daytime indoor thermal environment. *Energy* **2019**, *181*, 239–249. [CrossRef]
17. Jia, J.J.; Liu, B.; Ma, L.Y.; Wang, H.; Li, D.; Wang, Y.R. Energy saving performance optimization and regional adaptability of prefabricated buildings with PCM in different climates. *Case Stud. Therm. Eng.* **2021**, *26*, 101164. [CrossRef]
18. Li, W.W.; Cheng, W.L.; Xie, B.; Liu, N.; Zhang, L.S. Thermal sensitive flexible phase change materials with high thermal conductivity for thermal energy storage. *Energy Convers. Manag.* **2017**, *149*, 1–12. [CrossRef]
19. Qu, Y.; Wang, S.; Tian, Y.; Zhou, D. Comprehensive evaluation of Paraffin-HDPE shape stabilized PCM with hybrid carbon nano-additives. *Appl. Therm. Eng.* **2019**, *163*, 114404. [CrossRef]
20. Mu, M.L.; Basheer, P.A.M.; Sha, W.; Bai, Y.; McNally, T. Shape stabilised phase change materials based on a high melt viscosity HDPE and paraffin waxes. *Appl. Energy* **2016**, *162*, 68–82. [CrossRef]
21. Frac, M.; Pichór, W.; Szołdra, P.; Szudek, W. Cement composites with expanded graphite/paraffin as storage heater. *Constr. Build. Mater.* **2021**, *275*, 122126. [CrossRef]
22. Xu, N. Study on New Phase Change Energy Storage Wall of the Granary and Its Heat Transfer Characteristics. Master's Thesis, Henan Technology and University, Zhengzhou, China, 2022.
23. Li, W.W.; Wang, F.; Cheng, W.L.; Chen, X.; Zhao, Q. Study of using enhanced heat-transfer flexible phase change material film in thermal management of compact electronic device. *Energy Convers. Manag.* **2020**, *210*, 122680. [CrossRef]

24. Betancourt-Jimenez, D.; Montoya, M.; Haddock, J.; Youngblood, J.P.; Martinez, C.J. Regulating asphalt pavement temperature using microencapsulated phase change materials (PCMs). *Constr. Build. Mater.* **2022**, *350*, 128924. [CrossRef]
25. Cheng, W.L.; Zhang, R.M.; Xie, K.; Liu, N.; Wang, J. Heat conduction enhanced shape-stabilized paraffin/HDPE composite PCMs by graphite addition: Preparation and thermal properties. *Sol. Energy Mater. Sol. Cells.* **2010**, *94*, 1636–1642. [CrossRef]
26. Qiu, L.; Ouyang, Y.X.; Feng, Y.H.; Zhang, X.X. Review on micro/nano phase change materials for solar thermal applications. *Renew. Energy* **2019**, *140*, 513–538. [CrossRef]
27. Lawag, R.A.; Ali, H.M. Phase change materials for thermal management and energy storage: A review. *J. Energy Storage* **2022**, *55*, 105602. [CrossRef]
28. Yuan, S.; Li, C.E.; Yu, H.; Tang, Y.; Xia, Z.G. Optimization of the phase-change wallboard test method: Experimental and numerical investigation. *J. Energy Storage* **2020**, *30*, 101559.
29. Yan, Q.S.; Zhao, Q.Z. *Building Thermal Process*; China Architecture & Building Press: Beijing, China, 1986; pp. 86–87.
30. Energy Plus. Available online: https://energyplus.net/weather (accessed on 5 August 2022).
31. Lei, J.; Yang, J.; Yang, E.H. Energy performance of building envelopes integrated with phase change materials for cooling load reduction in tropical Singapore. *Appl. Energy.* **2016**, *162*, 207–217. [CrossRef]
32. Li, M.L.; Cao, Q.; Pan, H.; Wang, X.Y. Effect of melting point on thermodynamics of thin PCM reinforced residential frame walls in different climate zones. *Appl. Therm. Eng.* **2021**, *188*, 116615. [CrossRef]
33. Li, Z.X.; Abdullah, A.A.A.A.; Rostamzadeh, M.; Kalbasi, R.; Shahsavar, A.; Afrand, M. Heat transfer reduction in buildings by embedding phase change material in multi-layer walls: Effects of repositioning, thermophysical properties and thickness of PCM. *Energy Convers. Manag.* **2019**, *195*, 43–56. [CrossRef]
34. Yu, J.H.; Yang, Q.C.; Ye, H.; Huang, J.C.; Liu, Y.X.; Tao, J.W. The optimum phase transition temperature for building roof with outer layer pcm in different climate regions of China. *Energy Procedia* **2019**, *158*, 3045–3051. [CrossRef]
35. Arıcı, M.; Bilgin, F.; Nižetić, S.; Karabei, H. PCM integrated to external building walls: An optimization study on maximum activation of latent heat. *Appl. Therm. Eng.* **2019**, *165*, 114560. [CrossRef]
36. Mechouet, A.; Oualim, E.M.; Mouhib, T. Effect of mechanical ventilation on the improvement of the thermal performance of PCM-incorporated double external walls: A numerical investigation under different climatic conditions in Morocco. *J. Energy Storage* **2021**, *38*, 102495. [CrossRef]
37. Liu, Z.A.; Hou, J.W.; Meng, X.; Dewancker, B.J. A numerical study on the effect of phase-change material (PCM) parameters on the thermal performance of lightweight building walls. *Case Stud. Constr. Mater.* **2021**, *15*, e00758. [CrossRef]

materials

Article

Investigation of the Fracture Characteristics of a Cement Mortar Slab under Impact Loading Based on the CDEM

Qunlei Zhang [1], Decai Wang [1,*], Jinchao Yue [2], Chun Feng [3] and Ruifu Yuan [4]

[1] School of Civil Engineering and Communication, North China University of Water Resources and Electric Power, Zhengzhou 450045, China
[2] School of Water Conservancy and Environment, Zhengzhou University, Zhengzhou 450001, China
[3] Institute of Mechanics, Chinese Academy of Sciences, Beijing 100190, China
[4] School of Energy Science and Engineering, Henan Polytechnic University, Jiaozuo 454003, China
* Correspondence: wangdecai@ncwu.edu.cn

Abstract: For brittle and quasi-brittle materials such as rock and concrete, the impact-resistance characteristics of the corresponding engineering structures are key to successful application under complex service environments. Modeling of concrete-like slab fractures under impact loading is helpful to analyze the failure mechanism of an engineering structure. In this paper, simulation models of impact tests of a cement mortar slab were developed, and a continuum–discontinuum element method (CDEM) was used for dynamic analysis. Concretely, the cracking simulations of a mortar slab when considering the hammer shape and impact velocity were conducted, and the impact process and failure results of the slab structure were analyzed. The results showed that the top fracture area of the mortar slab was significantly smaller than that of slab bottom under impact loadings of the drop hammer. The impact velocity was an important factor that affected the mortar slab's cracking. With the increase in the initial impact velocity, the effective fracture area of the slab structure increased significantly; the impact force and rupture degree of the mortar slab also showed a linear growth trend. The shapes of the impact hammerhead also had a significant effect on the crack model of the mortar slab. The effective fracture zones of slab structures were close under circular and square hammers, while the effective fracture zone was significantly larger under a rectangular hammer impact. The peak value (45.5 KN) of the impact force under a circular hammer was significantly smaller than the peak value (48.7 KN) of the impact force under the rectangular hammer. When considering the influence of the stress concentration of the impact hammerhead, the maximum impact stress of the rectangular hammer was 147.3 MPa, which was significantly greater than that of the circular hammer impact (maximum stress of 87.5 MPa). This may have meant that the slab structures were prone to a directional rupture that mainly propagated along the long axis of the rectangular hammerhead. This impact mode is therefore more suitable for rehabilitation and reconstruction projects of slab structures.

Keywords: cement mortar slab; impact cracking; CDEM; hammerhead shape; impact velocity

Citation: Zhang, Q.; Wang, D.; Yue, J.; Feng, C.; Yuan, R. Investigation of the Fracture Characteristics of a Cement Mortar Slab under Impact Loading Based on the CDEM. *Materials* **2023**, *16*, 207. https://doi.org/10.3390/ma16010207

Academic Editors: Zhanqi Cheng, Dan Huang, Lisheng Liu and Liwei Wu

Received: 7 November 2022
Revised: 16 December 2022
Accepted: 23 December 2022
Published: 26 December 2022

1. Introduction

Slab structures composed of rock and concrete materials are widely used in roads, airports, and various municipal and hydraulic projects due to their convenient construction and strong bearing capacity. However, many serious defects have appeared in the engineering structures with an increase in the service life. For example, due to the increase in heavy-duty vehicles, a series of defects such as broken slabs, dislocations, cracks, spalling, and potholes have appeared on concrete pavement. Currently, various techniques for rehabilitation, demolition, and reconstruction have been proposed [1]. Micro-cracking homogenization technology makes full use of the residual strength of the original pavement structure through a low-speed impact treatment, which can realize the green recycling and rapid construction of concrete pavement, as shown in Figure 1 [2]. Currently, a method to determine the specific impact mode to effectively fracture the concrete pavement is an important factor, and the understanding of the fracturing mechanism of concrete slab structures is key to the successful application of various technologies. Therefore, investigations of the impact fracture of slab structures will provide guidance for the renovation and reconstruction of old pavement structures.

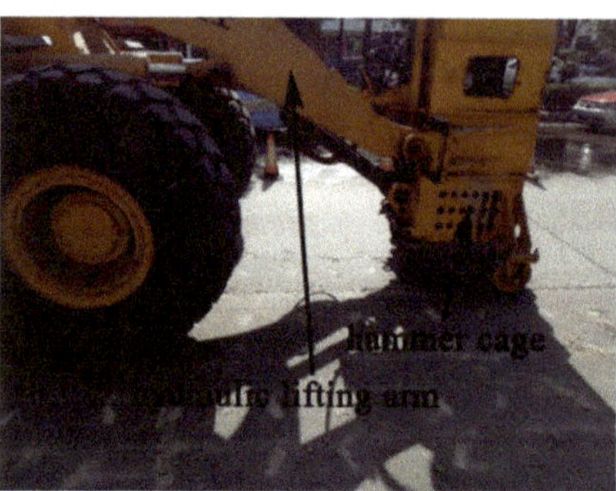

Figure 1. Micro-crack homogenization crushing of pavement structure [2].

As for the effect of the impact velocity, it has an obvious penetration effect when the velocity is greater than 500 m/s [3]. As for an impact velocity that is less than 50 m/s, elastic and plastic deformation of the collision are mainly presented [4]. For low-speed collisions, impact tests with a drop hammer are commonly used to investigate the damage effect of explosion or impact on engineering structures. A large number of experimental studies have been conducted on the impact resistance of concrete materials and structures under a low-speed collision. Chen et al. [5] investigated the high-mass, low-velocity impact behavior of concrete members including beams and slabs. Under a three-point bending configuration based on the free-fall of an instrumented impact device, Dey et al. [6] investigated the impact response of fiber-reinforced aerated concrete. Using a drop-weight impact test machine, Yoo et al. [7] conducted impact testing to estimate the impact resistance of concrete slabs strengthened with fiber-reinforced polymer sheets. Zhang et al. [8,9] investigated the flexural impact responses of steel-fiber-reinforced concrete to further analyze the effect of the hammer weight and drop height on the impact performance. The effect of support condition variation on impact behaviors of concrete slabs was investigated through impact loading tests [10]. Radnic et al. [11] investigated the impact behavior of two-way conventional reinforced concrete slabs. Sakthivel et al. [12] developed reinforced cementitious composites and studied their impact performance through a low-velocity repeated-drop-weight impact test. Yahaghi et al. [13] studied the impact resistance and crack characteristic of lightweight concrete reinforced with polypropylene fiber. Elavarasi et al. [14] investigated the response of thin concrete slabs with and without reinforcement under low velocity impacts. As for the effect of the loading rate, dry concrete had an obvious strain-rate-strengthening effect, and the compressive strength and flexural tensile strength were positively correlated with the loading rate [15]. Mei et al. found that the failure modes of concrete beams and columns under an impact load were similar and showed shear failure under high-strain-rate conditions [16]. Gu considered concrete slabs that showed flexural–tensile failure under a quasi-static load local punching shear failure under an impact load [17].

Due to the complexity of the impact-cracking process of concrete structures, successive observations of the crack-propagation process are difficult, and limited data on the impact cracking of concrete structures can be obtained in experiments. In combination with impact experiments, simulation has become an effective means to further study the fracture mechanism of concrete materials and structures due to the development of computer technology.

Nia et al. [18] conducted impact tests based on the testing procedure recommended by ACI Committee 544, and the simulation results of concrete materials were compared with experimental testing data. Gopinath et al. [19] conducted experimental and numerical investigations on concrete panels under low-velocity impact loading. Othman et al. [20–22] not only conducted an experimental investigation to understand the effect of steel reinforcement distribution on the dynamic response of concrete plates under impact load, but also established a three-dimensional finite element model to verify with the test results. Xiao et al. [23,24] conducted impact tests to study the effects of loading rates and other parameters on concrete slab performances and proposed a two-degrees-of-freedom dynamic model to predict the responses of reinforced concrete slabs subjected to low-velocity impacts. Zhao et al. [25] conducted the impact-response testing of concrete panels through drop-hammer tests and simulations. As in the above simulation studies, concrete and rock failures can be analyzed via FEM, which can accurately calculate the stress and deformation of materials and structure but cannot simulate the fracturing and crushing processes well.

Based on the continuum–discontinuum element method (CDEM), the influence of hammer elastic deformation on concrete pavement failure can be considered through the continuum element method, the cracking process of pavement structures can be well simulated via the discontinuum element method, and the interaction between the drop hammer and the pavement structure can be realized via the contact element [26]. Therefore, the CDEM is a more appropriate method to simulate the impact-cracking process of concrete-like materials and structures. However, under lower impact loading, the cracking simulation of concrete-like slabs while considering the impact-head shape and impact velocity using the CDEM has not been reported in the literature. In this study, the continuum–discontinuum element method (CDEM) was introduced, and simulation models of the drop-hammer impact test are developed. Then, the coupled calculation of the continuum element method and the discontinuum element method was used to simulate the impact-cracking process of a cement mortar slab. Further, the influences of different hammerhead shapes and impact velocities on the fracture mechanism of a mortar slab structure were analyzed.

2. Continuum–Discontinuum Element Method

The model schematic of the continuum–discontinuum element method (CDEM) is shown in Figure 2. The CDEM couples finite element calculations with discrete element calculations. It conducts the finite element calculations inside the block elements and conducts the discrete element calculations at the element boundaries [26].

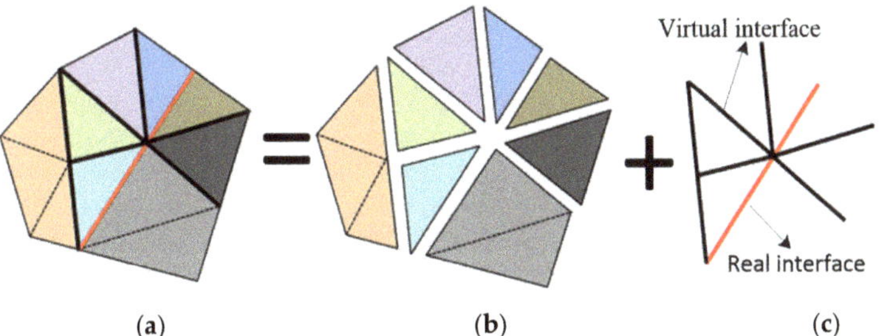

Figure 2. Model schematic of continuum–discontinuum element method: (**a**) numerical model; (**b**) block model; (**c**) interfaced model.

For every nodal mass of a block element, the equilibrium equations can be expressed in the following matrix form in an element (Equations (1) and (2)):

$$[M]\{\ddot{u}\} + [K]\{u\} + [C]\{\dot{u}\} = \{F\}_{ext} \tag{1}$$

$$\{F\}_{ext} = \{F\}_s + \{F\}_t \tag{2}$$

where $[M]$ denotes the nodal mass matrix; $[C]$ is the damping matrix; $[K]$ denotes the stiffness matrix; $\{u\}$ denotes the displacement vector; and $\{F\}_{ext}$ denotes the vector of external forces, which include the contact force $\{F\}_s$ and the external loading force $\{F\}_t$.

In this method, the acceleration is iterated by the central difference scheme, and the velocity is iterated by the unilateral difference scheme. The schemes can be written as following Equations (3)–(6):

$$\{\ddot{u}\}_i = \frac{\{u\}_{i+1} - 2\{u\}_i + \{u\}_{i-1}}{(\Delta t)^2} \tag{3}$$

$$\{\dot{u}\}_i = (\{u\}_{i+1} - \{u\}_i)/\Delta t \tag{4}$$

$$\{\dot{u}\}_{i+1} = \{\dot{u}\}_i + \{\ddot{u}\}_i \Delta t \tag{5}$$

$$\{u\}_{i+1} = \{u\}_i + \{\dot{u}\}_{i+1} \Delta t \tag{6}$$

where $\{\ddot{u}\}_i$, $\{\dot{u}\}_i$, and $\{u\}_i$ represent the acceleration, velocity, and node displacement at the i-th time step, respectively; $\{\dot{u}\}_{i+1}$ represents the node velocity at the i+1 time step; and $\{u\}_{i+1}$ represents the node displacement at the i+1 time step.

The stress and deformation of an elastic block are calculated using the following Equations (7)–(10):

$$[\Delta\varepsilon_i] = [B_i]\cdot[\Delta u] \tag{7}$$

$$[\Delta\sigma_i] = [D]\cdot[\Delta\varepsilon_i] \tag{8}$$

$$[\sigma_i] = \left[\sigma_i{}^0\right] + [\Delta\sigma_i] \tag{9}$$

$$[F_b] = \sum_{i=1}^{N} \left[B_i{}^T\right]\Delta[\sigma_i]\cdot\omega_i\cdot J_i \tag{10}$$

where i is the element Gaussian point, $[\Delta\varepsilon_i]$ is the incremental strain vector, $[B_i]$ is the strain matrix, $[\Delta u]$ is the incremental displacement vector of the node, $[\Delta\sigma_i]$ is the incremental stress vector, $[D]$ is the element elastic matrix, $[\sigma_i]$ is the total stress at the current step, $\left[\sigma_i{}^0\right]$ is the total stress at the previous step, F_b is the node internal force vector, N is the Gaussian point number, ω_i is the integral coefficient, and J_i is the Jacobian determinant value.

In the discontinuous element method, the normal and tangential forces of the contact element are calculated using Equation (11):

$$F_i(t_1) = F_i(t_0) - k_i\cdot A_c\cdot\Delta du_i, \; i = 1,2,3 \tag{11}$$

where F_1 denotes the normal force of the contact element; F_2 and F_3 denote the tangential forces of the contact element, respectively; t_0 and t_1 denote the current time step and the next time step, respectively; k_1 denotes the normal stiffness; k_2 and k_3 denote the tangential stiffnesses, respectively; A_c is the contact element area; Δdu_1 represents the relative displacements in the normal direction; and Δdu_2 and Δdu_3 represent the relative displacements in the tangential directions, respectively.

To characterize and simulate the fracture process of brittle and quasi-brittle structures composed of rock and concrete materials, the damage failure criterion of the maximum tensile stress and Mohr–Coulomb are applied to the element interface of the fracture area, which considers the tensile fracture energy and shear fracture energy of brittle and quasi-brittle materials.

The damage fracture criterion of the maximum tensile stress while considering the tensile fracture energy following Equation (12):

$$\left. \begin{array}{ll} \text{if} & -F_n(t_1) \geq \sigma_t(t_0)A_c \\ \text{then} & F_n(t_1) = -\sigma_t(t_0)A_c \\ & \sigma_t(t_1) = -(\sigma_{t0})^2 \times \Delta u_n / (2G_{ft}) + \sigma_{t0} \end{array} \right\} \tag{12}$$

where σ_{t0}, $\sigma(t_0)$, and $\sigma(t_1)$ are the tensile strength of the interface spring element in the initial time step, the current time step, and the next time step, respectively; Δu_n stands for the relative displacements in the normal directions; and G_{ft} represents the tensile fracture energy (Pa × m).

The Mohr–Coulomb damage fracture criterion while considering shear fracture energy following Equation (13):

$$\left. \begin{array}{ll} \text{if} & F_s(t_1) \geq F_n(t_1) \times \tan\phi + c(t_0)A_c \\ \text{then} & F_s(t_1) = F_n(t_1) \times \tan\phi + c(t_0)A_c \\ & c(t_1) = -(c_0)^2 \times \Delta u_s / \left(2G_{fs}\right) + c_0 \end{array} \right\} \tag{13}$$

where ϕ denotes the internal friction angle of contact element; c_0, $c(t_0)$, and $c(t_1)$ stand for the cohesion strength in the initial time step, the current time step, and the next time step, respectively; Δu_s stands for the relative displacements in the tangential directions; and G_{fs} represents the shear fracture energy (Pa × m).

3. Numerical Model, Mechanical Parameters, and Simulation Scheme

3.1. Numerical Models of Impact Test

Based on the impact test [2,27], the models of the drop hammer, bearing plate, cement mortar slab, and rigid support slab were established separately, the hammerhead shapes were circular, square, and rectangular as shown in Figure 3. The size of the mortar slab was 400 mm × 400 mm × 60 mm, the size of the impact contact area of the rectangular hammerhead was 20 mm × 54 mm, the size of the square hammerhead was 33 mm × 33 mm, and the section radius of the square hammerhead was 18.6 mm.

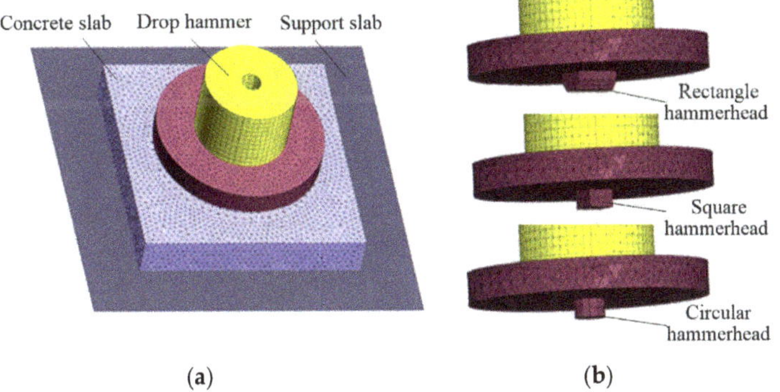

Figure 3. Numerical models of mortar slab impact test: (**a**) impact system model; (**b**) hammer model.

3.2. Mechanical Parameters in Simulation

In the simulation, the drop hammer and bearing load plate were analyzed accurately by using the continuous element method and adopting the elastic constitutive law. The calculation parameters included the density, elastic modulus, and Poisson's ratio. The discontinuous element method was used to analyze the impact cracking of the cement mortar slab. The damage fracture criteria of the maximum tensile stress and Mohr–Coulomb were applied in the element interface of the mortar slab. The calculation parameters included the normal contact stiffness, tangential contact stiffness, cohesion, internal friction angle, and tensile strength. The fracture energy parameter was 100 Pa/m. A normal contact element was used to deal with the contact between the impact hammer, mortar slab, and rigid support slab. While considering the impact rebound between the hammer and mortar slab, the strength parameters of the impact contact interface and the support interface were set to zero, and only the normal contact stiffness was required as input. For the continuous–discontinuous calculation of the mortar slab cracking, the block element parameters of the mortar slab, drop hammer, and bearing plate are shown in Table 1. The interface element parameters of the entire impact system are shown in Table 2.

Table 1. Constitutive parameters of block elements.

	Density/kg $\times$ m^{-3}	Elastic Modulus/GPa	Poisson's Ratio
Hammer	7850	200	0.25
Cement mortar	2400	28	0.2

Table 2. Constitutive parameters of contact element.

	Normal Stiffness/ Pa $\times$ m^{-1}	Tangential Stiffness/ Pa $\times$ m^{-1}	Cohesion/MPa	Internal Friction Angle/°	Tensile Strength/MPa
Concrete interface	2×10^{11}	1×10^{11}	6	54.9	5
Impact interface	2×10^{11}	0	0	54.9	0
Support interface	2×10^{10}	0	0	54.9	0

3.3. Simulation and Analysis Schemes

In the simulation of the impact cracking of the cement mortar slab, the drop hammer contained the impact kinetic energy by providing an initial impact velocity, which equivalently replaced the free-drop process of the impact hammer in the experiment. Based on the contact element and dynamic explicit algorithm of the CDEM, the simulation of impact-cracking process of the entire system could be achieved by the drop hammer with the impact kinetic energy. During the impact-cracking process of the mortar slab, the time-history curve of the impact-reaction force was obtained by accumulating the contact force between the support slab and the mortar slab. The cracking simulations under the impact loadings of the circular, square, and rectangular hammers were carried out to investigate the effect of the hammer shape on the cracking mechanism of the mortar slab. Moreover, the cracking processes of the mortar slab under the impact velocities of 3 m/s, 4 m/s, and 5 m/s were also simulated to study the effect of the impact velocities on the cracking mechanism. Further, the impact-cracking mechanisms of the mortar slab were investigated by analyzing the relationship between the impact force and the rupture degree.

4. Simulation Results and Analysis

4.1. Effect of Hammer Shapes on Cracking Mechanism

To study the cracking mechanism of a cement mortar slab under the impact loading of differently shaped hammerheads, we conducted cracking simulations using the circular, square, and rectangular hammerheads. The slab size was 400 mm $\times$ 400 mm $\times$ 60 mm, and the impact velocity of the drop hammer was 4 m/s.

4.1.1. Analysis of Impact Fracture Forms on the Cement Mortar Slab

Under the impact loadings of differently shaped hammers, the fracture mode of the mortar slab was analyzed via the crack distribution on the slab's top and bottom as shown in Figure 4.

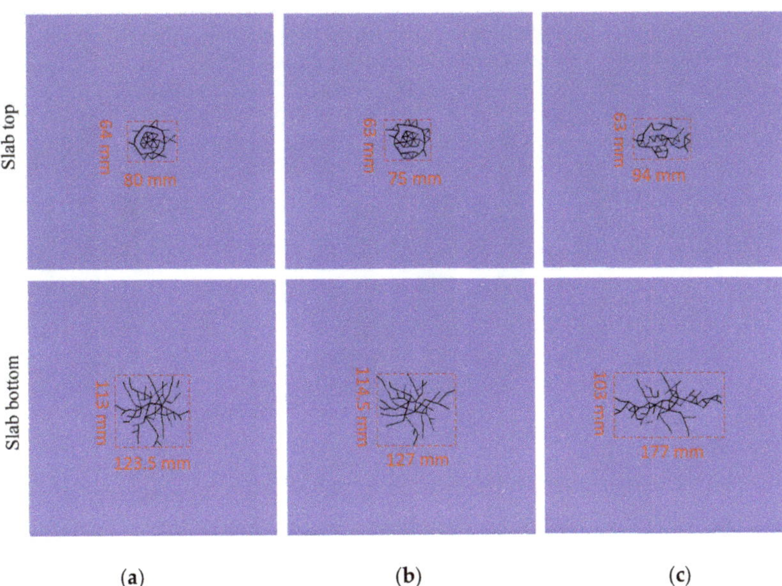

(a)   (b)   (c)

Figure 4. Crack distribution on mortar slab under differently shaped hammerheads: (**a**) circular hammer impact; (**b**) square hammer impact; (**c**) rectangular hammer impact.

As shown in Figure 4, the degree of the cement mortar slab fracture was relatively low under an impact loading of 4 m/s, and the impact fracture occurred only in the slab's middle. The slab's top was crushed in the impact contact area; however, the larger range of the slab's bottom mainly showed crack propagation due to the impact bending. The crack model on the slab top showed the circular, square, and rectangular distributions that corresponded to the shapes of the impact hammerhead. Specifically, the effective fracture area of the mortar slab's top under the impact of the circular-section hammer was 5120 mm^2, the fracture area of the slab's top under the square-section hammer was 4725 mm^2, and the fracture area of the slab's top under the rectangular hammer was 5922 mm^2. The slab's bottom produced symmetrical dispersion cracks under the circular and square hammer heads, while the slab's bottom cracks under the rectangular hammerhead mainly propagated along the longitudinal direction of the impact head. The effective fracture area of the slab's bottom under the circular-section hammer was 13,956 mm^2. The fracture area of the slab's bottom under the square-section hammer was 14,478 mm^2. The slab's bottom fracture area under the rectangular cross hammer was 18,231 mm^2.

Overall, the failure of the cement mortar slab under the impact loading of 4 m/s was relatively smaller, and the failure occurred only in the slab's middle. The effective fracture areas of the slab's top and bottom were relatively close under the circular and square hammerheads; however, the fracture area under the rectangular-section hammer was significantly greater than those of the circular and square hammers. The top fracture areas of the slab were significantly smaller than those of the slab's bottom under differently shaped hammerheads.

4.1.2. Impact-Stress Analysis of Hammerheads

When the initial impact velocity was the same, the stress concentration caused by the differently shaped hammers affected the initiation position and expansion route of the slab's top cracks, which led to the generation of different fracture forms on the concrete slab. To further study the influence of the hammerhead shapes on the cracking mechanism, the stress distribution during the impact process was obtained by monitoring the contact element force at different locations of the impact contact surface. Under the impact loadings of the differently shaped hammerheads, the effect of the stress concentration on the impact cracking was determined as shown in Figure 5.

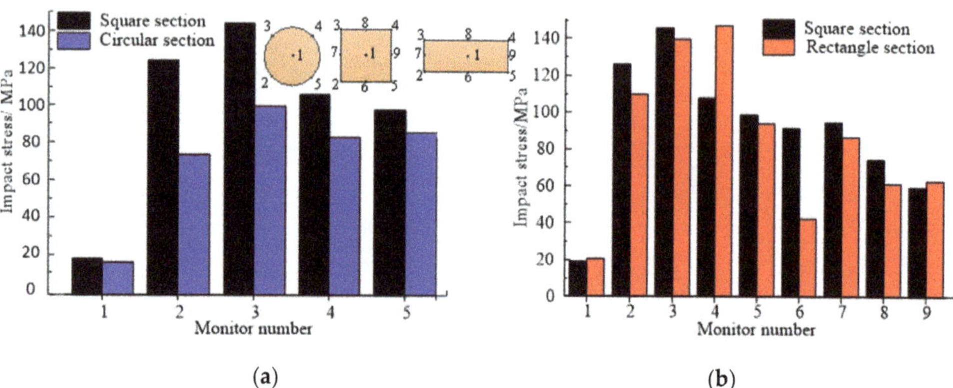

Figure 5. Stress distributions at different positions of impact section: (**a**) comparison of square and circular hammers; (**b**) comparison of square and rectangular hammers.

As shown in Figure 5, when the initial impact velocity and the impact contact area of the hammerhead were constant, the impact stress at the hammer section's center was significantly less than that of the section boundary, and the hammer stress at the midpoint of the section boundary line was significantly less than that at the impact section corner. Specifically, the center stress of the circular hammer section (15.3 MPa) was smaller than that at the section boundary (average of 78.9 MPa). The corner stress of the square hammer section (average of 119 MPa) was greater than that at the boundary line midpoint of the hammer section (average of 80 MPa), which was greater than that of the hammer section's middle (average of 19.1 MPa). The section corner stress of the rectangular hammer (average of 121 MPa) was significantly greater than those of the section edge (average of 63.3 MPa). The impact stress at the short edge midpoint of the rectangular section (average of 74.6 MPa) was significantly greater than that at the long edge's midpoint (average of 52 MPa). The minimum stress at the section midpoint of the rectangular hammer was 21 MPa. The maximum stress of the rectangular hammer (147.3 MPa) was larger than that of the square hammer (145.5 MPa), which was larger than that of the circular hammer (87.5 MPa). This showed that in the impact process of the mortar slab, the stress-concentration degree of the rectangular hammer was greater than that of the square hammer, which in turn was significantly greater than that of the circular hammer.

4.1.3. Impact Reaction and Rupture Degree Analysis

As shown in Section 4.1.1, the impact fracture area on the cement mortar slab's bottom was significantly larger than that of slab's top, and the fracture characteristics of the slab's bottom were also different under the different hammer shapes. To further study the influence of the hammer shape on the cracking mechanism, the time-history curves of the impact-reaction force between the mortar slab and the supporting plate were compared and the rupture degrees of the mortar slab were also analyzed (Figure 6).

As shown in Figure 6a, the impact process of the cement mortar slab could be divided into the fast-loading stage I, the slow-loading stage II, the unloading stage III, and the impact-termination stage IV. At the same time, it can be seen in Figure 6b that the time-history curves of the rupture degree were divided into the initial micro-cracking stage I, the rapid-cracking stage II, the slow-cracking stage III, and the cracking-termination stage IV. Under the circular, square, and rectangular hammerhead impacts at 4 m/s, the impact process lasted for about 5 ms. Conversely, in the 0–0.5 ms fast-loading stage, the impact-reaction force rapidly increased to 30 KN, but the rupture-degree curve of the mortar slab only slowly increased to 0.05%. In the slow-loading phase of 0.5–2 ms, the impact-reaction force nonlinearly increased to a peak (45 KN) at a lower rate, but the rupture-degree curves showed a rapid growth trend: the rupture degree rapidly increased from 0.05% to 0.9%. In the unloading stage of 2–5 ms, the impact reaction decreased gradually from the peak to 0; meanwhile, the rupture-degree curves of the cement mortar slab showed a slow growth trend, and the rupture degree slowly increased from 0.9% to 1.05%. In the impact-termination stage of 5–6 ms, the impact-reaction force remained 0 and the corresponding rupture degree did not increase.

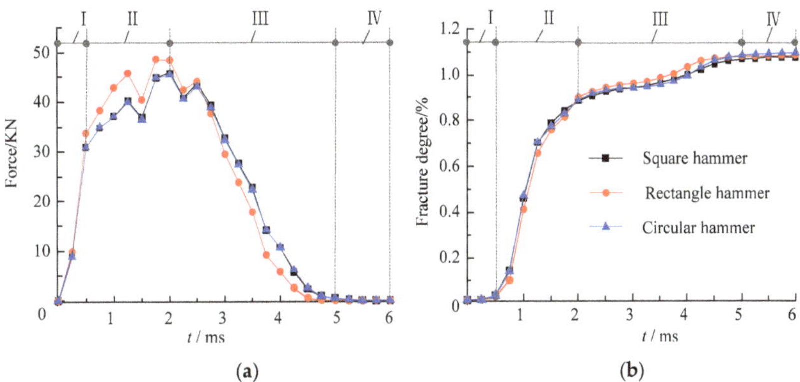

Figure 6. Impact-force and rupture-degree curves under different hammer impacts: (**a**) impact-force curves; (**b**) rupture-degree curves.

When comparing Figure 6a,b, under the 4 m/s impact of the square and circular hammers, the time-history curves of the impact force and rupture degree were basically the same. However, the impact-force and rupture-degree curves under the rectangular hammer impact were significantly different from those of the square and circular hammers. Specifically, in the rapid-loading stage of 0–0.5 ms, the curves of the impact force and rupture degree under the different hammer shapes were basically same. In the slow-loading stage of 0.5–2 ms, the magnitude and change rate of the impact force under the rectangular hammer impact were significantly greater than those under the circular and square hammers. However, at this stage, the rupture degree of the mortar slab under the rectangular hammer was slightly smaller than that under the circular and square hammers. In the unloading stage of 2–5 ms, the magnitude and change rate of the impact force under the rectangular hammer were smaller than those of the circular and square hammers, while the change rate of the impact force under the rectangular hammer impact was larger than that of the circular and square hammers. Moreover, in the unloading stage of the rectangular hammer impact, the rupture degree of the mortar slab was generally greater than that of the circular and square hammers.

To further analyze the influence of the hammer shape on the cracking mechanism of the mortar slab, the peak impact force and final rupture degree under the differently shaped hammers shown in Figure 6 were compared and analyzed as shown in Figure 7.

As shown in Figure 7, the maximum impact force (45.5 KN) of the mortar slab under the circular hammer impact was slightly less than that of the square hammer impact (45.9 KN), which in turn was significantly less than that of the rectangular hammer impact (48.7 KN). However, the ultimate rupture degree of the mortar slab under the circular hammer (1.096%) was significantly greater than that of the rectangular hammer impact (1.083%), which also was greater than that of the square hammer impact (1.075%). This indicated that the impact reaction of the support plate and the rupture degree of the mortar slab showed an opposite variation trend.

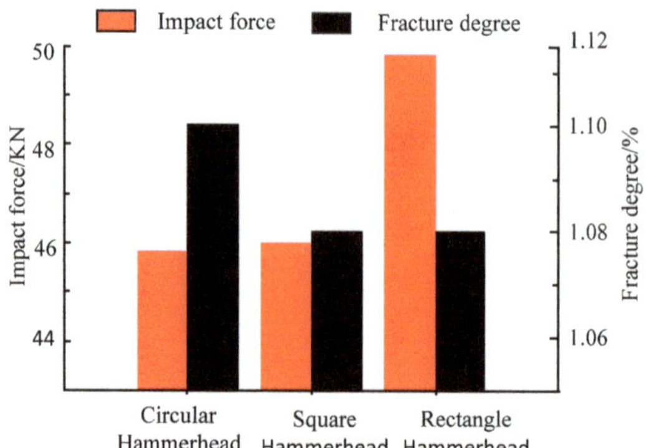

Figure 7. Comparison of impact forces and rupture degrees under different hammer impacts.

4.2. Effect of Impact Speed on Cracking Mechanism

To comprehensively investigate the cracking mechanism of the cement mortar slab, we conducted cracking simulations of the mortar slab under the rectangular hammerhead impact. The initial impact velocities of the drop hammer were 3 m/s, 4 m/s, and 5 m/s. The mortar slab's size was also 400 mm × 400 mm × 60 mm.

4.2.1. Analysis of Impact Fracture Forms on the Cement Mortar Slab

To study the influence of the impact velocity on mortar slab fracture, the slab fracture modes under the impact loadings of different velocities were analyzed via the crack distribution pattern as shown in Figure 8.

As shown in Figure 8, under the impact loadings of different initial velocities, the top of the mortar slab was crushed in the impact contact area. However, the slab's bottom produced cracks that mainly propagated along the longitudinal direction of the rectangular hammerhead. Moreover, the crushing degree of the slab's top became more severe with an increase in the impact velocity, and the propagation length of the slab's bottom crack also obviously increased. Under the impact loading with a low initial velocity (3 m/s), the main cracks on the slab's bottom propagated along the long axis of the rectangular hammerhead. Under the impact loading at the initial velocity of 4 m/s, the main crack length on the slab's bottom did not increase significantly, but multiple branch cracks on the slab's bottom were generated along the short-axis direction of the rectangular hammer. Under the impact loading of a 5 m/s initial velocity, the lengths of the main crack and branch crack on the slab's bottom increased significantly compared with those of the 4 m/s impact. To be more specific, under the impact of the rectangular hammer at the initial velocity of 3 m/s, the effective fracture area on the mortar slab's top was 4872 mm^2; the slab's top fracture area under a 4 m/s impact was 6111 mm^2; and the top fracture area for the 5 m/s impact was 12,474 mm^2. Under the impact of the rectangular hammer at 3 m/s, the effective fracture area on the slab's bottom was 9198 mm^2; the slab's bottom fracture area for the 4 m/s impact was 17,877 mm^2; and the bottom fracture area for the 5 m/s impact was 40,365 mm^2.

Overall, the effective fracture zone of the mortar slab's top was relatively close under the impact of the lower initial velocity, but the effective fracture zone of the slab's top increased significantly under the impact of the larger initial velocity (5 m/s). Moreover, the effective fracture area of the slab's bottom showed a significantly increasing trend with the increase in the impact velocity.

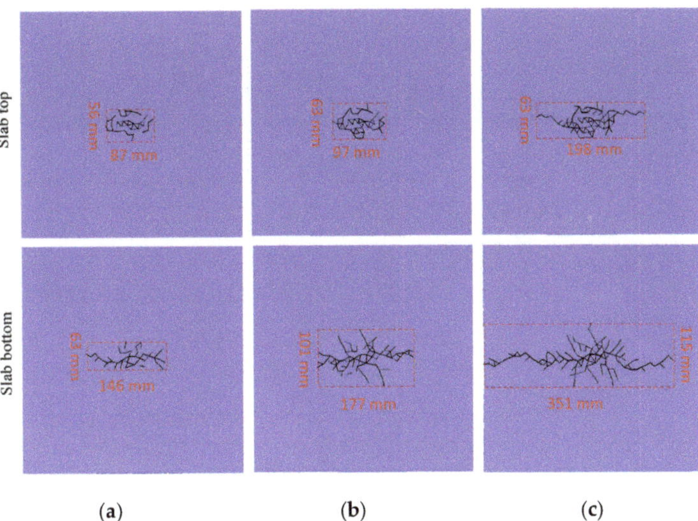

(**a**)   (**b**)   (**c**)

Figure 8. Crack distribution on the slab's top and bottom under different impact velocities: (**a**) 3 m/s impact; (**b**) 4 m/s impact; (**c**) 5 m/s impact.

4.2.2. Impact Reaction and Rupture Degree Analysis of Cement Mortar Slab

To further study the effects of the different impact velocities on the cracking of the mortar slab, the impact-force and rupture-degree curves were also compared and analyzed as shown in Figure 9. The impact processes of the mortar slab at different impact velocities lasted for about 3.75–4.5 ms. The impact-reaction force increased rapidly in the 0–0.5 ms fast-loading stage, but the rupture degree of the mortar slab increased slowly to a lower value. In the slow-loading phase of 0.5–2 ms, the impact reaction increased to a peak at a lower rate, but the rupture degree showed a rapid growth trend. In addition, the impact reaction decreased gradually from the peak to 0 in the unloading stage of 2–4.5 ms. The rupture degree showed a significant increase under the impact loading of 5 m/s, and the rupture degree variation of the 4 m/s impact was relatively slow; however, the rupture degree was basically unchanged under the 3 m/s impact. At the impact-termination stage of 4.5–6 ms, the impact-reaction force remained 0, and the corresponding rupture degree of the mortar slab did not increase. The final rupture degree of the 5 m/s impact was obviously greater than that of the 4 m/s impact, which in turn was obviously greater than that of the 3 m/s impact.

When comparing Figure 9a,b, as the impact velocity increased, the impact duration also increased, as did the ultimate rupture degree of the mortar slab. At the same time, as the impact velocity increased, the increase rate of the impact reaction during the loading stage was larger, and the growth rate of the corresponding rupture degree was also larger. In the stage of impact unloading, although the attenuation rate of the impact force and the growth rate of the rupture degree show an increasing trend with the increase in the impact velocity, the rupture degree increased obviously only when the impact speed was higher.

To further analyze the influence of the impact velocity, the peak values of the impact force and final rupture degree shown in Figure 9 were fitted and analyzed as shown in Figure 10.

As shown in Figure 10, with an increase in the initial impact velocity of the drop hammer, the maximum impact force on the mortar slab showed a linear growth law; the linear fitting degree R^2 was 0.991. The final rupture degree of the mortar slab also increased linearly as the initial impact velocity increased; the linear fitting degree R^2 was 0.9944.

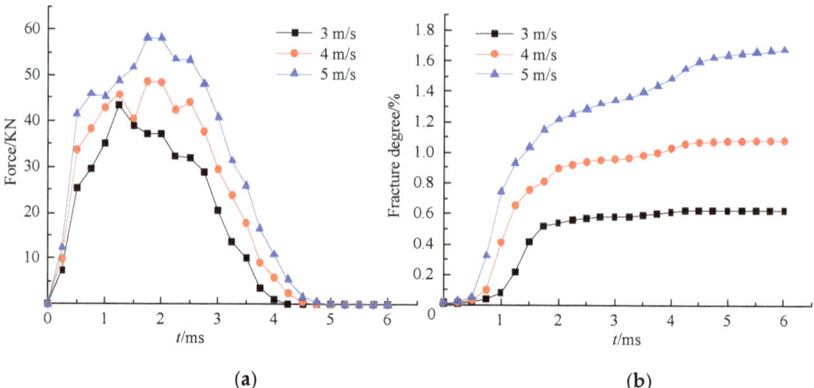

Figure 9. Curves of the impact force and rupture degree under different impact velocities: (**a**) impact-force curves; (**b**) rupture-degree curves.

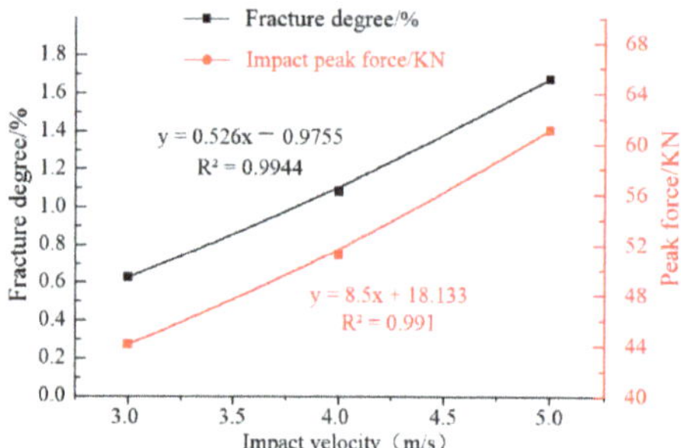

Figure 10. Comparison of the impact force and rupture degree.

5. Discussion

The impact velocity is an important factor in the cracking of concrete-like materials and structures; when the impact velocity is larger, the initial impact energy and the impact force on a cement mortar slab are higher, so the overall rupture degree of the mortar slab is also larger. Moreover, as the impact speed is higher, the change rate of the impact force is larger in the loading and unloading stages, so mortar slabs will fracture not only in the impact-loading stage but also in the impact-unloading stage. Generally, the impact force and overall rupture degree increase with an increase in the impact velocity.

When the initial impact velocity was the same, the stress-concentration degree and stress distribution caused by the differently shaped hammer sections were different, which affected the crack-propagation path, rupture degree, and impact force. Because the impact reaction of the circular and square hammers focused on the slab's middle, the crushing degree of the mortar slab was larger, and the impact force transmitted to the slab's bottom was also larger for the rectangular hammerhead's impact. Therefore, the impact reactions and rupture degrees under the circular and square hammerheads were similar; however, the impact reaction under the rectangular hammerhead's impact was larger than those of the circular and square hammerheads.

Because the stress-concentration degree and stress-distribution difference of the rectangular hammer were larger than those of the circular and square hammerheads. This corresponded to the section characteristics of the rectangular hammer, and the compression–shear-fracture zone and tension-fracture zone were easier to develop along the long axis. The mortar slab generated the entire fracture under the rectangular hammerhead impact, while the dispersion cracks and local crushing occurred in the slab's middle under the circular and square hammerheads. Therefore, the rupture degree of the mortar slab under the rectangular hammerhead was lower than those of the circular and square hammerheads.

Further, the specific impact-cracking mode could effectively fracture the concrete material and structure under artificially contorted impact loading; therefore, the rectangular hammer's impact can be more suitable for application in the renovation of old pavement structures.

6. Conclusions

In this paper, the cracking simulations of a cement mortar slab under low-velocity impact were carried out via the continuum–discontinuum element method (CDEM). Then, the influences of the impact velocity and the hammerhead shape on the mortar slab's cracking were discussed by comprehensively analyzing the slab's crack distribution, fracture-area size, rupture degree, hammerhead-impact stress, and impact-reaction force of the supporting plate. The following conclusions were drawn:

1. The shapes of the impact hammerheads had a significant effect on the crack model of the mortar slab. The crack model of the slab's top showed a circular, square, or rectangular distribution that corresponded to the impact-head shapes. The mortar slab's bottom produced symmetrical dispersion cracks under the circular and square hammerheads. However, the impact cracks mainly propagated along the longitudinal direction of the rectangular hammer. Under the differently shaped hammerheads, the effective fracture area of the mortar slab's top was significantly smaller than that of the slab's bottom; under the circular and square hammers, the sizes and shapes of the effective fracture zones on slab's top and bottom were close, while the effective fracture zone of the rectangular hammer's impact was significantly greater than those under the circular- and square-section hammers.

2. Under the same initial velocity and different hammerhead shapes, the impact reactions and rupture degrees of the mortar slab showed an opposite trend. The ability to transmit the impact force to the slab's bottom decreased as the rupture degree increased, so the impact reaction generated by the supporting plate was also smaller. The peak value (45.5 KN) of the impact force under the circular hammer was slightly smaller than the peak value (45.9 KN) of the impact force under the square hammer, which in turn was significantly smaller than the peak value (48.7 KN) of the impact force under the rectangular hammer. However, the final value (1.096%) of the rupture degree under the circular hammer was significantly larger than that under the rectangular hammer (1.083%), which in turn was larger than that under the square hammer (1.075%). The impact velocity was also an important factor that affected the mortar slab cracking. With an increase in the initial impact velocity, the effective fracture area of the mortar slab increased significantly. The maximum impact force showed a linear growth trend and a linear fitting degree R^2 of 0.991; the final rupture degree also increased linearly and showed a linear fitting degree R^2 of 0.9944.

3. When the impact velocities and impact areas of the hammerheads were constant, the stress concentration (maximum stress 147.3 MPa) caused by the rectangular hammer's impact was greater than that (145.5 MPa) of the square hammer's impact, which in turn was significantly greater than that (87.5 MPa) of the circular hammer's impact. Therefore, the rupture area, rupture degree, crack form, and impact force of the mortar slab under the rectangular hammer's impact were significantly different from those of the circular and square hammers. Moreover, corresponding to the section characteristics of the rectangular hammer, the mortar slab was prone to directional rupture; therefore, the rectangular hammer should be more suitable for reconstruction projects involving mortar slab structures.

7. Outlook

Slab structures composed of rock and concrete materials are widely used in road, airport, and various municipal and hydraulic engineering. To ensure their normal use, this investigation aimed to provide a theoretical reference for understanding the generation and treatment mechanisms of pavement defects. The interactive relationships of a mortar slab structure and impact drop hammers were discussed, the impact-fracturing processes of the mortar slab under different conditions were simulated, and the crack propagation and failure degree were analyzed. To verify the investigation results based the continuum–discontinuum element method (CDEM), corresponding impact tests will be carried out in the future.

Author Contributions: Data curation, Q.Z., D.W. and J.Y.; writing—original draft, Q.Z., D.W. and J.Y.; formal analysis, Q.Z., D.W. and J.Y.; writing—review and editing, Q.Z., D.W. and J.Y.; software, C.F.; methodology, C.F.; writing—review, D.W., C.F. and R.Y; funding acquisition, D.W., C.F. and R.Y. All authors have read and agreed to the published version of the manuscript.

Funding: This research was supported by the National Key Research and Development Program of China (No. 2018YFC1505504); the National Natural Science Foundation of China (No. 52174109); the Key Research and Development Project of Henan Province (No. 222102320407); the Program for Innovative Research Team (in Science and Technology) in University of Henan Province (No. 22IRT-STHN005); and the Science and Technology Collaborative Innovation Special Project of Zhengzhou, Henan Province in 2021.

Institutional Review Board Statement: Not applicable.

Informed Consent Statement: Not applicable.

Data Availability Statement: Not applicable.

Conflicts of Interest: The authors declare no conflict of interest.

References

1. Ma, J.; Sun, S.Z.; Rui, H.T. Review on China's road construction machinery research progress: 2018. *China J. Highw. Transp.* **2018**, *31*, 136–139.
2. Li, W.; Zhang, Q.; Zhi, Z.; Feng, C.; Cai, Y.; Yue, J. Investigation on the fracture mechanism of homogenized micro-crack crushing technology for portland cement conc rete pavement rehabilitation. *AIP Adv.* **2019**, *9*, 075113. [CrossRef]
3. Fu, J.; Yang, J.; Yin, L.; Liu, W.; Wang, J.; Chen, Z. Dynamic Properties of Zirconia Ceramic Bullets under High-Speed Impact. *J. Chin. Ceram. Soc.* **2016**, *44*, 346–352.
4. Zu, Z.; Yuan, T.; Tang, S.; Dai, X. Repeated Low Velocity Impacts on Honeycomb Sandwich Panels and Residual Strength after Impacts. *Sci. Technol. Eng.* **2019**, *19*, 101–109.
5. Chen, Y.; May, I.M. Reinforced concrete members under drop-weight impacts. *Proc. Inst. Civ. Eng.-Struct. Build.* **2009**, *162*, 45–56. [CrossRef]
6. Dey, V.; Bonakdar, A.; Mobasher, B. Low-velocity flexural impact response of fiber-reinforced aerated concrete. *Cem. Concr. Compos.* **2014**, *49*, 100–110. [CrossRef]
7. Yoo, D.Y.; Yoon, Y.S. Influence of steel fibers and fiber-reinforced polymers on the impact resistance of one-way concrete slabs. *J. Compos. Mater.* **2014**, *48*, 695–706. [CrossRef]
8. Zhang, W.M.; Chen, S.H.; Zhang, N.; Zhou, Y. Low-velocity flexural impact response of steel fiber reinforced concrete subjected to freeze–thaw cycles in NaCl solution. *Constr. Build. Mater.* **2015**, *101*, 522–526. [CrossRef]
9. Zhang, W.M.; Chen, S.H.; Liu, Y.Z. Effect of weight and drop height of hammer on the flexural impact performance of fiber-reinforced concrete. *Constr. Build. Mater.* **2017**, *140*, 31–35. [CrossRef]
10. Anil, O.; Kantar, E.; Yilmaz, M.C. Low velocity impact behavior of RC slabs with different support types. *Constr. Build. Mater.* **2015**, *93*, 1078–1088. [CrossRef]
11. Radnić, J.; Matešan, D.; Grgić, N.; Baloević, G. Impact testing of RC slabs strengthened with CFRP strips. *Compos. Struct.* **2015**, *121*, 90–103. [CrossRef]
12. Sakthivel, P.B.; Ravichandran, A.; Alagamurthi, N. Impact strength of hybrid steel mesh-and-fiber reinforced cementitious composites. *KSCE J. Civ. Eng.* **2015**, *19*, 1385–1395. [CrossRef]
13. Yahaghi, J.; Muda, Z.C.; Beddu, S.B. Impact resistance of oil palm shells concrete reinforced with polypropylene fibre. *Constr. Build. Mater.* **2016**, *123*, 394–403. [CrossRef]
14. Elavarasi, D.; Mohan, K.S.R. On low-velocity impact response of SIFCON slabs under drop hammer impact loading. *Constr. Build. Mater.* **2018**, *160*, 127–135. [CrossRef]
15. Fu, Y.; Dong, X. An experimental study on impact response and failure behavior of reinforced concrete beam. *Sci. Sin. Technol.* **2016**, *46*, 400–406.
16. Mei, F.; Dong, X.; Yu, X. On failure behavior of concrete and RC beam to different velocity impact. *J. Ningbo Univ. Nat. Sci. Eng. Ed.* **2017**, *30*, 83–88.
17. Gu, S.; Peng, F.; Yu, Z.; Li, J. An experimental study on the damage effects of the concrete slabs under low-velocity impact. *J. Vib. Shock.* **2019**, *38*, 107–114.

18. Nia, A.A.; Hedayatian, M.; Nili, M.; Sabet, V.A. An experimental and numerical study on how steel and polypropylene fibers affect the impact resistance in fiber-reinforced concrete. *Int. J. Impact Eng.* **2012**, *46*, 62–73.

19. Gopinath, S.; Ayashwarya, R.; Kumar, V.R.; Prem, P.R. Low velocity impact behaviour of ultra high strength concrete panels. *Sadhana-Acad. Proc. Eng. Sci.* **2014**, *39*, 1497–1507. [CrossRef]

20. Othman, H.; Marzouk, H. An experimental investigation on the effect of steel reinforcement on impact response of reinforced concrete plates. *Int. J. Impact Eng.* **2016**, *88*, 12–21. [CrossRef]

21. Othman, H.; Marzouk, H. Impact response of ultra-high-performance reinforced concrete plates. *Aci. Struct. J.* **2016**, *113*, 1325–1334. [CrossRef]

22. Othman, H.; Marzouk, H. Finite-element analysis of reinforced concrete plates subjected to repeated impact loads. *J. Struct. Eng.* **2017**, *143*, 04017120. [CrossRef]

23. Xiao, Y.; Li, B.; Fujikake, K. Experimental study of reinforced concrete slabs under different loading rates. *Aci Struct. J.* **2016**, *113*, 157–168. [CrossRef]

24. Xiao, Y.; Li, B.; Fujikake, K. Predicting response of reinforced concrete slabs under low-velocity impact. *Mag. Concr. Res.* **2017**, *69*, 996–1010. [CrossRef]

25. Zhao, W.Y.; Guo, Q.Q.; Dou, X.Q.; Zhou, Y.; Ye, Y. Impact response of steel-concrete composite panels: Experiments and FE analyses. *Steel Compos. Struct.* **2018**, *26*, 255–263.

26. Zhang, Q.; Zhi, Z.; Feng, C.; Cai, Y.; Pang, G.; Yue, J. Investigation of concrete pavement cracking under multi-head impact loading via the continuum-discontinuum element method. *Int. J. Impact Eng.* **2020**, *135*, 103410. [CrossRef]

27. ACI Committee 544. *ACI 544.2R-89: Measurement of Properties of Fiber Reinforced Concrete*; American Concrete Institute: Indianapolis, IN, USA, 1999; pp. 6–7.

Article

The Typical Damage Form and Mechanism of a Railway Prestressed Concrete Sleeper

Ruilin You [1],*[iD], Jijun Wang [1], Na Ning [1], Meng Wang [1] and Jiashuo Zhang [2]

[1] Railway Engineering Research Institution, China Academy of Railway Sciences Corporation Limited, Beijing 100081, China
[2] School of Civil Engineering, Universiti Sains Malaysia, Pinang 14300, Malaysia
* Correspondence: youruilin0731@126.com

Abstract: Prestressed concrete sleepers are an important track component that is widely used in railway ballast track. Prestressed concrete sleepers have high strength, strong stability, and good durability; thus, their operation and use in railways are beneficial. However, in different countries and regions, certain damage to sleepers typically appears. Existing research on concrete sleepers focuses primarily on the structural design method, the application of new materials, theoretical analysis, and bearing strength test research, while ignoring sleeper damage. There are a few sleeper damage studies, but they look at only one type of damage; thus, there is no comprehensive study of prestressed concrete sleeper damage. The damage forms of prestressed concrete sleeper damage are thus summarized in this study, and the theory of the causes of prestressed concrete sleepers is analyzed based on the limit state method for the first time. The findings indicate that sleeper structure design is the primary cause of its operation and use status, and that special measures should be considered based on sleeper use conditions. In addition to meeting design requirements, materials, curing systems, product inspection, and other factors must be considered during manufacturing to improve the sleepers' long-term performance. Keeping the track in good condition, including but not limited to the state of fasteners, ballast bed, and track geometry is also an important aspect of preventing sleeper damage. The outcomes of this study provide better insights into the influences of damage to railway prestressed concrete sleepers and can be used to improve track maintenance and inspection criteria.

Keywords: concrete sleeper; damage form; damage mechanisms; longitudinal crack; transversal crack; limit state method

Citation: You, R.; Wang, J.; Ning, N.; Wang, M.; Zhang, J. The Typical Damage Form and Mechanism of a Railway Prestressed Concrete Sleeper. *Materials* **2022**, *15*, 8074. https://doi.org/10.3390/ma15228074

Academic Editors: Dan Huang, Lisheng Liu, Zhanqi Cheng and Liwei Wu

Received: 2 October 2022
Accepted: 12 November 2022
Published: 15 November 2022

Publisher's Note: MDPI stays neutral with regard to jurisdictional claims in published maps and institutional affiliations.

1. Introduction

Railways are one of the most important modes of transportation for both passengers and freight worldwide. Railway traffic is increasing due to the widespread use of high-speed lines, and new railway lines are being built in many countries. Parallel to these developments, the production and the use of railway sleepers are increasing. Railway sleepers are key components in railway tracks that carry the loads transferred from vehicles to the rails, support the rails, protect the gauge, and withstand horizontal and vertical rail movement [1–4]. Sleepers can be made from a variety of materials, including wood, concrete, steel, and composites [5–9].

The use of prestressed concrete sleepers has grown in recent years due to their high quality. Furthermore, prestressed concrete sleepers are more environmentally friendly than creosote-treated wooden sleepers [5,8,10]. Currently, approximately 500 million prestressed concrete sleepers are required every year in railway networks all over the world [11].

Most prestressed concrete sleepers are in good operating condition, but some are damaged in different ways [12,13]. Many studies have been performed on prestressed concrete sleepers, primarily focusing on dynamic load [14,15], bearing capacity [10,16], and structural design [17,18]. However, systematic research on the damage form and mechanism of railway prestressed concrete sleepers is currently limited.

Berntsson and Chandra investigated calcium chloride damage in concrete sleepers and concluded that calcium chloride is harmful to concrete [19]. Pawluk et al. researched the durability of prestressed concrete sleepers [20]. Ravindrarajah and White investigated the effect of non-delayed heat application on prestressed concrete sleeper strength [21]. Shojaei et al. conducted a study on the application of alkali-activated slag (AAS) concrete in the production of prestressed reinforced concrete sleepers [22]. Rezaie et al. performed a study on the factors affecting longitudinal crack propagation in prestressed concrete sleepers [23]. Zeman et al. investigated the mechanism of rail-seat abrasion of prestressed concrete sleepers in North America [24]. Zakeri et al. study on the variation of loading pattern of concrete sleeper due to ballast sandy contamination in sandy desert areas and the failures of railway concrete sleepers during service life [25,26]. These studies only analyze a single type of damage in prestressed concrete sleepers, but the damage mechanism and analysis methods were not investigated. to date. Thus, the various types of damage in prestressed concrete sleeper damage are summarized in detail in this study. In addition, for the first time, the theory of the causes of prestressed concrete sleepers is analyzed using the limit state method. The findings of this study provide a clearer picture of the effects of damage to railway prestressed concrete sleepers and improve track maintenance and inspection criteria.

2. Typical Form and Influence of Prestressed Concrete Sleeper Damage

2.1. Transverse Cracks

Transverse cracks are perpendicular to the long axis of the concrete sleeper. Because of the change in the ballasted bed support state and the randomness of the trainload, the concrete sleeper may experience a load-bending moment that exceeds its strength during service, resulting in transverse cracks. According to the appearance position, the transverse crack of the concrete sleeper can be classified into two types: a transverse crack of the rail seat section and a transverse crack of the central section.

The transverse crack of the rail seat section can be divided into the lower part crack and upper part crack, as shown in Figure 1; these cracks are caused by the positive and negative bending moments of the lower section of the sleeper exceeding its bearing strength, respectively. Typically, the width of the transverse crack is narrow, but if the load bending moment is too large, a crack will develop, and the sleeper will fail.

(a)   (b)

Figure 1. Transverse crack in a rail seat section: (**a**) lower part crack in the rail seat section; (**b**) upper part crack in the rail seat section.

A transverse crack of the central section can be divided into a lower part crack and upper part crack, as shown in Figure 2. These cracks are caused by the positive and negative bending moments of the lower section of the sleeper exceeding its bearing strength, respectively. Typically, cracks in the upper part of the central section of the concrete sleeper due to the negative moment being too large are more common.

(a)   (b)

Figure 2. The transverse crack of the central section: (**a**) lower part crack of the central section; (**b**) upper part crack of the central section.

2.2. Longitudinal Crack

Cracks along the long axis of the sleeper are collectively referred to as longitudinal cracks. For the prestressed concrete sleeper, stress was applied to the concrete in the design, manufacturing, construction and maintenance links of improper treatment, which may lead to the occurrence and development of longitudinal cracks. A longitudinal crack in a sleeper is generally categorized as an end surface crack, a longitudinal crack on the upper surface, a horizontal longitudinal crack on the side, a longitudinal crack at the embedded parts and a longitudinal crack through the sleeper (as shown in Figure 3).

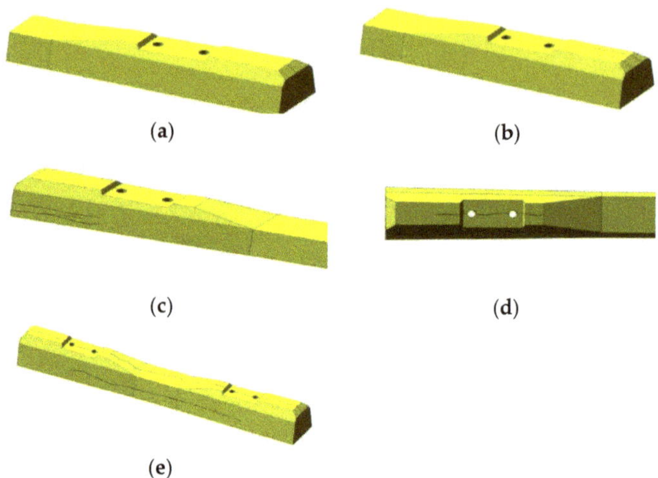

Figure 3. Longitudinal crack: (**a**) end surface crack; (**b**) longitudinal cracks on the upper surface of the end part (**c**) longitudinal crack at the embedded parts; (**d**) horizontal and longitudinal crack on the side; (**e**) the through longitudinal crack of the sleeper.

Longitudinal cracks in prestressed concrete sleepers have long been a source of concern. After the longitudinal crack appears, the sleeper's strength and durability decrease, as does its ability to maintain track geometry; thus, longitudinal cracks must be prioritize during maintenance work.

2.3. Inclined Crack

Inclined cracks on the sleeper's surface become oblique cracks and typically appear at the bottom corner of the retaining shoulder or on the surface of the sleeper, as shown in Figure 4. The former is generally caused by the excessive shear load of the sleeper, while the latter may be caused by improper tamping and repair operations in the maintenance process.

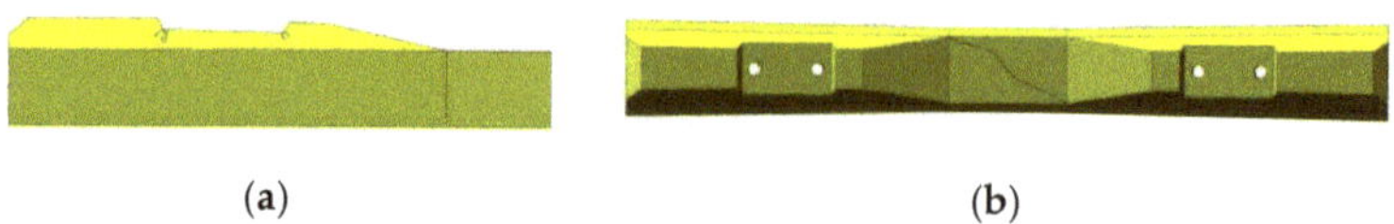

Figure 4. Inclined crack: (**a**) inclined crack at the base corner of the sleeper shoulder; (**b**) inclined crack in the middle of the sleeper.

2.4. Map-Like Crack

An irregular mesh of cracks on the surface of a sleeper is also called sleeper map-like cracks, as shown in Figure 5. Typically, a sleeper map-like crack is related to the materials, the manufacturing process and the environmental conditions used of the concrete sleeper. When a map-like crack appears in the early stage of sleeper curing process, the crack width is small and generally appears on the sleeper's surface. With the extension of the width and range of the crack, it will continue to develop and eventually lead to the sleeper's life being markedly reduced.

Figure 5. Map-like crack of a concrete sleeper.

2.5. Rail Seat Abrasion

Abrasion refers to the use of a sleeper during interaction with the connected rail parts for a long time, resulting in damage to the connection site. The abrasion form of the sleeper is primarily the wear of the rail seat area, as shown in Figure 6, which is primarily caused by the long-term impact friction between the under-rail pad and the bearing surface. The geometry of the track is altered and the bearing capacity of the sleeper is weakened by the abrasion of the rail seat region, which worsens the state of the railway track.

Figure 6. Rail seat abrasion of the concrete sleeper.

2.6. Break Damage and Block Dropped

Break damage and drop blocks generally appear in the shoulder and the upper surface of the concrete sleeper, as shown in Figure 7. Typically, breaks and drops are caused by external loads that are too large, such as in the process of transportation and unloading and the process of trackbed tamping. The accidental improper external loads will also lead to the sleeper knock block in the small radius curve section because the large lateral load of the train will also lead to the phenomenon of shoulder damage. The stress condition of the sleeper, which is vulnerable to stress concentration and other damage, will get worse with loss and block drop, and the stability of the track construction will also suffer.

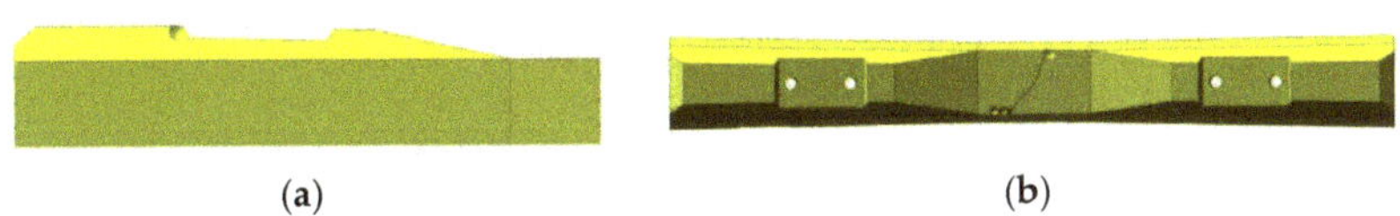

(a)   (b)

Figure 7. Break damage and block dropped: (**a**) block dropped off the shoulder; (**b**) break damage in the middle part of the sleeper.

2.7. Comprehensive Injury

Sleeper damage to a railroad track often consists of two or more types of damage occurring at the same time rather than just one. For instance, the simultaneous appearance of a transverse fracture and collapse, a longitudinal crack and a surface crack, and a shoulder oblique crack and knock are frequent. Different types of harm will result from the sources of these simultaneous incidences of injury, or the "inducing variables." On the other hand, one kind of damage might potentially cause another kind of harm to occur. The bearing capacity of the sleeper and the stability of the track structure are more severely affected when various types of damage take place at the same time. It is important to identify the precise causes of damage during maintenance and to suggest early preventive solutions.

3. Analysis of the Damage Mechanism of Prestressed Concrete Sleepers

3.1. Analysis Method

The essence of the damage of a prestressed concrete sleeper is that the external load effect exceeds its resistance; thus, damage analysis can be evaluated using the limit state method.

The limit state method is commonly used in the structure design. This method uses failure probability or a reliable index to measure structure reliability and to establish a relationship between the structure limit equation of state and the probability theory of structural reliability [27]. When designed by the limit state method, the structural resistance uses the strength R of the material multiplied by a load factor; this value shall not be any less than the total load effect of each load effect S of the structure multiplied by the respective load factor S, as shown in Equation (1):

$$\Sigma(\gamma \cdot S) \leq \Phi R \tag{1}$$

The damage phenomenon of prestressed concrete sleepers is contrary to the working condition considered in the design process due to the external load effect exceeding its resistance; however, the concept of its analysis is the same. The probability distribution curve of the total load effect, structural resistance and sleeper damage failure is shown in Figure 8. Therefore, the damage mechanism of prestressed concrete sleepers can be analyzed from the total load effect, internal resistance and comprehensive factors.

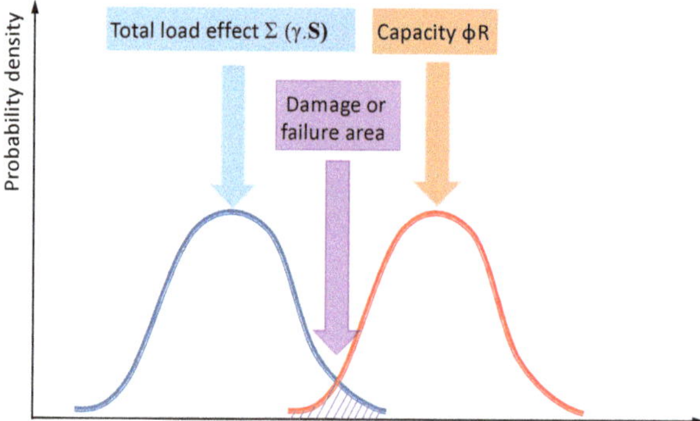

Figure 8. Schematic diagram of the distribution curve of sleeper damage or failure probability.

3.2. External Load Effect

3.2.1. Design Load Effect

When designing a prestressed concrete sleeper, the load moment produced by the train dynamic load in the sleeper is primarily considered. The designed load moment primarily includes the lower section of the sleeper and the positive load bending moment of the section in the sleeper.

Because the load bending moment of the sleeper is directly related to the dynamic load of the train and the support state of the sleeper itself, if the above factors exceed the design range, an excessive load bending moment will lead to sleeper damage.

Queensland University of Technology in Australia tested the impact force of two separate sites, Braeside and Raglan [10,17]. At these two sites, the maximum static axle load of the operational vehicles is 28 tons. The field measurement data in Braeside is shown in Figure 9. The trainload of the sleeper has a wide fluctuation range.

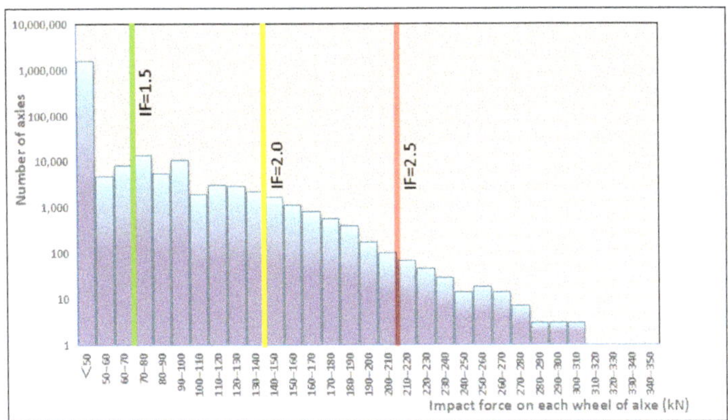

Figure 9. Typical impact force statistical data on the track at Braeside. Note: Impact factor (IF) = 1+ (impact force)/(static wheel load).

The bottom compression stress of the sleeper in various bed states is varied according to the measured findings of the compression stress at the bottom of the sleeper in those beds [26,28]; thus, the load-bending moment of the sleeper will differ appropriately., as shown in Figure 10.

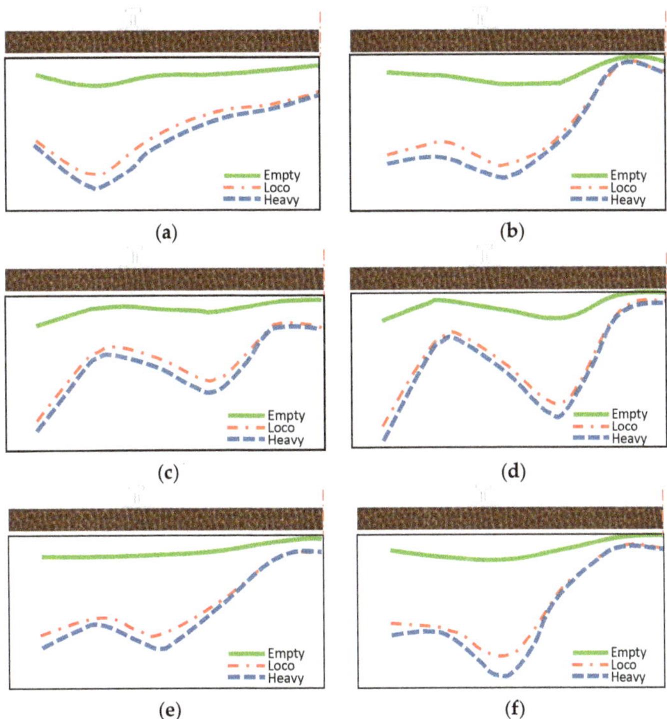

Figure 10. Support conditions experimentally measured in the field. (**a**) moderate ballast 1; (**b**) moderate ballast 2; (**c**) new ballast 1; (**d**) new ballast 2; (**e**) fouled ballast 1; (**f**) fouled ballast 2.

3.2.2. Environmental Effect

Environmental utility primarily refers to the external environmental effects of moisture, temperature, chemical erosion, and freezing and thawing during the operation of the concrete sleeper. The AAR and DEF that often occur in concrete sleepers are typical damages caused by environmental effects.

Alkali-Aggregate Reaction

The alkaline-aggregate reaction (AAR) refers to the concrete hole solution by cement or alkali admixture, mineral admixture, and the environment released Na^+, K^+, OH^- and the aggregate of harmful active minerals in the expansion reaction, resulting in concrete expansion and cracking phenomenon [29]. Figure 11 shows the concrete sleeper's AAR damage.

Figure 11. AAR damaged the concrete sleeper. Note: These images also show other factors in addition to AAR.

There are three conditions for AAR occurrence:

- The aggregate in the concrete has alkali activity;
- There is a certain amount of soluble alkali in concrete;
- Sufficient water or wet environmental conditions are present.

The alkali-aggregate reaction can primarily be divided into two types of reaction types: alkali-silicic acid reaction (ASR) and alkali-carbonate reaction (ACR).

ASR is the chemical reaction between the alkali in concrete and the active SiO_2 in the aggregate to form the alkali silicate gel. The gel volume is greater than the SiO_2 volume before the reaction (as shown in Equation (2)). The gel water absorption causes concrete expansion and cracking, and the internal reaction mechanism is shown as follows. The ASR generally has a large swelling area that accounts for cracking within 10–20 years after the construction of the concrete structure and can further develop damage to the entire structure:

$$Na^+(K^+) + SiO_2 + OH^- \rightarrow Na(K)\text{-}Si-Hgel \tag{2}$$

ACR is the reaction of the alkali in the concrete with the dolomite crystals contained in the active carbonate aggregate, producing concrete expansion and cracking, as shown in Equations (2) and (3). Studies have shown that LiOH can be used to distinguish ASR from ACR. The ACR reaction development speed is fast, and the general concrete project is built in 2~3 years of expansion and cracking, and cannot be repaired and reinforced:

$$CaMg(CO_3)_2 + 2ROH = Mg(OH)_2 + CaCO_3 + R_2CO_3 \tag{3}$$

$$R_2CO_3 + Ca(OH)_2 = 2ROH + CaCO_3 \tag{4}$$

Delayed Ettringite Formation

The Delayed Ettringite Formation (DEF) is a form of sulfate erosion in cement concrete with sulfate ions from the inside of cement concrete. Therefore, the definition of delayed alum rock can be considered as follows: in hardened cement concrete, not from the process of sulfate outside the cement concrete, the harm caused by delayed alum rock often appears only months or years later. The delay of slurry expansion caused by ettringite formation will cause cracks in the interface between the cement slurry and aggregate-cement slurry [30], which leads to cracks in the concrete sleeper (as shown in Figure 12).

Figure 12. DEF damaged the concrete sleeper. Note: These images also show other factors in addition to DEF.

Results show that the primary characteristics of delayed calcium generation are as follows:

- Typically, delaying the formation of ettringite occurs when the cement concrete experiences a temperature of 70 °C and causes serious expansion cracking in an environment of high relative humidity. Between 70 °C and 100 °C, the higher the temperature is, the more severely inflated;
- The expansion and development of cement concrete in wet air are slower than those in water, but its macrocracks are larger;
- The expansion caused by delayed ettringite begins from the outside of the cement concrete and gradually expands to the inside;
- The expansion caused by delayed calcium ettringite is affected by the nature and size of the aggregate used;
- When the concrete structure has been damaged, delayed calcium damage is typically accompanied by an AAR.

There are still some disputes about the mechanism of DEF caused by concrete structure damage, which primarily includes the following aspects. First, a real concrete structure engineering environment is diverse, and the causes of concrete damage are complex. For example, AAR consumes the alkali in the liquid phase, which causes the solubility of calcium in the liquid phase to decline and precipitate out. Therefore, when concrete experiences cracking failure, it is often considered the joint action of AAR and DEF, and the reaction of AAR occurs before DEF. In addition, other factors can also affect DEF, which cannot show whether the primary cause of concrete destruction is DEF. Second, calcium ettringite is widely used as an indispensable component of concrete shrinkage compensation and cement early strong and fast hard, even if the discovery of calcium ettringite cannot prove its real impact on the performance of concrete.. Third, laboratory small size test blocks can be observed in the water after several days of the new formation of ettringite, marginally larger mortar test blocks can see the delayed formation of ettringite, and the real engineering of large size concrete to several years to see this phenomenon. Therefore, the action mechanism and destruction process must be investigated in more detail.

3.2.3. Accidental Load Effect

The load effect under accidental conditions must be taken into account in the sleeper design in addition to bearing the operating load of the design. In rare circumstances, these unintended load effects can harm the sleeper. When a train is derailed [31], the sleeper receives a very large impact load and may become damaged or even fail, as shown in Figure 13. Excessive bolt torque on the fasteners can also cause damage during sleeper laying and maintenance [32] (as shown in Figure 14). During sleeper laying and maintenance, the sleeper can also be knocked and damaged, as shown in Figure 15.

Figure 13. Damaged concrete sleeper in the derailment.

Figure 14. Excessive torque of the buckling bolt causes sleeper damage. Note: These images also show other factors in addition to the torque of the buckling bolt.

Figure 15. Impact load caused damage during construction of the sleeper.

For sleeper damage with an accidental load effect, the proportion of damaged sleepers is not large, and the characteristics are important. Combined with the situation on the site, it is easier to judge the cause of the damage.

3.2.4. Poor Track Structure Status

Sleepers are a track component in the ballast track structure when the track structure is in poor condition (e.g., the line appears empty hanging phenomenon, trackbed hardening, trackbed frost boiling) will also lead to sleeper damage or even failure, as shown in Figures 16 and 17.

Figure 16. Track bed hardening damages the concrete sleeper.

Figure 17. Track bed frost boiling damages the concrete sleeper.

Sleeper damage and poor railway conditions have an interaction. Inadequate railroad conditions can cause sleeper damage, and in the other direction, poor track conditions can be made worse by sleeper damage [33]. Examples of this include the mutual impact and aggravation between the track's geometric condition and the wear surface of the sleeper, as shown in Figure 18. Therefore, sections with poor track conditions should be rapidly repaired during track structure operation to prevent further degradation of the track state and escalation of track component damage.

3.3. Structural Self-Resistance

3.3.1. Structural Design Strength

When prestressed concrete sleepers are used in a track design, if the strength is insufficient, damage will occur during its service life. For one type of sleeper, the same damage form usually occurs in different lines, different manufacturers and under similar operating conditions. The design strength of the sleeper structure is insufficient; typically, the reasons for this phenomenon are a small concrete sleeper section, an unreasonable prestressed steel wire configuration, insufficient configuration or no stirrups.

Typical sleeper damage caused by insufficient structural design strength is a transverse crack at the rail seat section, as shown in Figure 19, where a transverse crack in the sleeper (Figure 20) and a horizontal crack on the side of the sleeper occur (Figure 21).

Figure 18. Concrete sleeper rail seat abrasion.

Figure 19. Transverse crack in the rail seat section of the concrete sleeper.

Figure 20. Transverse crack in the central section of the concrete sleeper.

Figure 21. Horizontal crack on the side of the concrete sleeper.

3.3.2. Manufacturing Quality

In addition to the structural design leading to the insufficient structural resistance of prestressed concrete sleepers, the production quality also has an important impact on the structural resistance of sleepers, and the general production quality of sleeper damage will appear in different external load superposition forms. In the production and manufacturing process, the resistance of sleepers is primarily affected by the material quality and curing process.

(1)   Raw material quality

Under the action of the sleeper mentioned above, the AAR will lead to damage, which is closely related to the raw material quality of the sleeper. An important indicator of high-quality materials is not to produce or cause AAR damage to concrete. Steps to prevent AAR are typically required to eliminate the reaction:

- Select the inactive aggregate;
- Control the alkali content of cement ($Na_2O+0.658K_2O$), which should not exceed 0.6%;
- Appropriate incorporation of mineral fine admixture reduces the amount of cement to reduce the temperature difference stress in the concrete caused by the heat of cement hydration.

(2)   Curing process

A prestressed concrete sleeper is typically made using steam curing, as shown in Figure 22. Steam curing is a type of hardening process that accelerates the development of concrete strength. During hardening, the hydration reaction of cement produces a large amount of hydration heat. Sleeper curing in the early stage is the most important period of health preservation. If the temperature in the curing pool is high, the temperature in the pool is higher than that inside the sleeper, and thermal hydration accumulation inside the pool is not easy to distribute. The gas and moisture inside the concrete mixture are heated and expanded. At this time, the concrete strength is low, and there is no resistance. However, internal stress causes concrete deformation and can even produce microfine cracks. This situation will lead to lower sleeper structural resistance. Thus, damage during operation affects a sleeper's service life. A reasonable sleeper curing system includes the following points:

- Determine the relationship between the temperature of the sleeper concrete and the surrounding air temperature throughout the process of the sleeper life cycle, and avoid a large temperature difference between the temperature of the sleeper concrete and the surrounding air temperature;
- The temperature change of the sleeper core should not exceed 15 °C/h, and the cooling rate should not exceed 15 °C/h;
- The difference between the temperature of the sleeper surface and the outside environment when leaving the curing pool should not exceed 15 °C.

Figure 22. Steam curing of the concrete sleeper.

3.4. Multiple Factors

The external load effect and the structure's resistance are used to analyze the primary causes of prestressed concrete sleeper damage. However, the causes of sleeper damage during operation are frequently caused by multiple factors rather than a single factor. For example, the quality control of sleeper manufacturing lax alkali activity of materials exceeds the standard. This sleeper is used in a humid environment and certain acidic medium, and it is likely to exhibit sleeper surface cracking damage caused by AAR, which is the comprehensive factor of manufacturing quality and environmental effect.

In addition, the comprehensive sleeper damage mentioned above can also be caused by the comprehensive superposition of multiple factors.

4. Conclusions

This paper summarizes the damage forms of concrete sleepers and analyzes the damage mechanism of prestressed concrete sleepers from the three aspects of the external load effect, structural self-resistance and comprehensive factors. Based on the results of this paper, prestressed concrete sleepers can prevent damage and improve the quality level from the following aspects:

(1)　Structural design guarantee of prestressed concrete sleeper

Sleeper structure design is the fundamental cause of its operation and use status. An excellent structural design can avoid damage to a large extent and should consider the appearance size, reinforcement design, manufacturing, operation, maintenance systems and environmental conditions of the sleeper. Therefore, the prevention of concrete sleeper damage starts with improving the design quality of the sleeper structure.

(2)　Manufacturing quality control

In addition to structural design, the production quality is also an important factor to prevent concrete sleeper damage. During manufacturing, in addition to meeting design requirements, materials, curing systems, product inspection and other aspects must be considered to improve the long-term sleeper performance.

(3)　Use of environmental condition adaptability

Temperature, humidity and chemical erosion in the use environment are also important factors that can lead to sleeper damage. Therefore, in the sleeper design process, special measures should be considered based on these use conditions. For example, when used in cold areas, sleeper design must consider the ability to resist freezing and thawing; in acid rain or sea regions, sleeper design must consider resistance to chemicals.

(4)　Track status improvement

A concrete sleeper is a portion of the railway track, and the deterioration of the overall state of the track will inevitably lead to a sleeper bearing a greater load effect, which can lead to damage. Therefore, an important aspect of preventing sleeper damage is to keep the track in good condition, including but not limited to the state of fasteners, the ballast bed and track geometry.

Author Contributions: Methodology, J.W.; Data curation, R.Y.; Writing—original draft, R.Y.; Writing—review & editing, N.N., M.W. and J.Z. All authors have read and agreed to the published version of the manuscript.

Funding: The first author wishes to acknowledge the financial support of China State Railway Group Co., Ltd [J2021G001] and China Academy of Railway Sciences Co., Ltd [2020YJ031].

Institutional Review Board Statement: Not applicable.

Informed Consent Statement: Not applicable.

Data Availability Statement: Not applicable.

Conflicts of Interest: The authors declare no conflict of interest.

References

1. Jing, G.; Siahkouhi, M.; Edwards, J.R.; Dersch, M.S.; Hoult, N. Smart railway sleepers—A review of recent developments, challenges, and future prospects. *Constr. Build. Mater.* **2020**, *271*, 121533. [CrossRef]
2. Farnam, S.M.; Rezaie, F. Experimental Analysis of Fracture and Damage Mechanics of Pre-Stressed Concrete Sleepers B70: Part B-Analysis. *Int. J. Transp. Eng.* **2017**, *15*, 17–29.
3. Silva, R.; Silva, W.V.; de Farias, J.Y.; Santos, M.A.A.; Neiva, L.O. Experimental and Numerical Analyses of the Failure of Prestressed Concrete Railway Sleepers. *Materials* **2020**, *13*, 1704. [CrossRef] [PubMed]
4. Li, D.; Kaewunruen, S.; You, R. Time-dependent behaviours of railway prestressed concrete sleepers in a track system. *Eng. Fail. Anal.* **2021**, *127*, 105500. [CrossRef]
5. Bolin, C.A.; Smith, S.T. Life Cycle Assessment of Creosote-Treated Wooden Railroad Crossties in the US with Comparisons to Concrete and Plastic Composite Railroad Crossties. *J. Transp. Technol.* **2013**, *3*, 149–161. [CrossRef]
6. You, R.; Wang, J.; Kaewunruen, S.; Wang, M.; Ning, N. Comparative Investigations into Environment-Friendly Production Methods for Railway Prestressed Concrete Sleepers and Bearers. *Sustainability* **2022**, *14*, 1059. [CrossRef]
7. Gamage, E.K.; Kaewunruen, S.; Remennikov, A.M.; Ishida, T. Reply to Giannakos, K. Comment on: Toughness of Railroad Concrete Crossties with Holes and Web Openings. *Infrastructures 2017, 2, 3. Infrastructures* **2017**, *2*, 5. [CrossRef]
8. Wang, M.; Han, X.; Jing, G.; Wang, H. Experimental and numerical analysis on mechanical behaviour of steel turnout sleeper. *Constr. Build. Mater.* **2022**, *329*, 127133. [CrossRef]
9. Jing, G.; Zong, L.; Ji, Y.; Aela, P. Optimization of FFU synthetic sleeper shape in terms of ballast lateral resistance. *Sci. Iran.* **2021**, *286*, 3046–3057. [CrossRef]
10. You, R.; Kaewunruen, S. Evaluation of remaining fatigue life of concrete sleeper based on field loading conditions. *Eng. Fail. Anal.* **2019**, *105*, 70–86. [CrossRef]
11. Ruiz, A.E.C.; Edwards, J.R.; Qian, Y.; Dersch, M.S. Probabilistic framework for the assessment of the flexural design of concrete sleepers. *Proc. Inst. Mech. Eng. Part F J. Rail Rapid Transit* **2019**, *234*, 691–701. [CrossRef]
12. Siahkouhi, M.; Li, X.; Han, X.; Jing, G. Improving the Mechanical Performance of Timber Railway Sleepers with Carbon Fabric Reinforcement: An Experimental and Numerical Study. *J. Compos. Constr.* **2022**, *26*, 04021064. [CrossRef]
13. Siahkouhi, M.; Li, X.; Markine, V.; Jing, G. Experimental and numerical study on the mechanical behavior of Kunststof Lankhorst Product (KLP) sleepers. *Sci. Iran.* **2021**, *28*, 2568–2581. [CrossRef]
14. Freudenstein, S.; Concrete Ties designed for high dynamic loads. American Railway Engineering and Maintenance-of-Way Association. 2007, pp. 1–18. Available online: https://www.arema.org/files/library/2007_Conference_Proceedings/Concrete_Ties_Designed-High_Dynamic_Loads_2007.pdf (accessed on 1 October 2022).
15. Grassie, S. Dynamic modelling of concrete railway sleepers. *J. Sound Vib.* **1995**, *187*, 799–813. [CrossRef]
16. Janeliukstis, R.; Clark, A.; Rucevskis, S.; Kaewunruen, S. Vibration-based damage identification in railway concrete sleepers. In Proceedings of the 4th Conference on Smart Monitoring, Assessment and Rehabilitation of Civil Structures, Zurich, Switzerland, 13–15 September 2017; pp. 1–9.
17. Leong, J. Development of a Limit State Design Methodology for Railway Track. Ph.D. Thesis, Queensland University of Technology, Brisbane, Australia, 2007; p. 228.
18. Ruiz, Á.E.C. Analysis of the Design of Railroad Track Superstructure Components for Rail Transit Applications. Ph.D. Thesis, University of Illinois Urbana-Champaign, Champaign, IL, USA, 2018; p. 122.
19. Berntsson, L.; Chandra, S. Damage of concrete sleepers by calcium chloride. *Cem. Concr. Res.* **1982**, *12*, 87–92. [CrossRef]
20. Pawluk, J.; Cholewa, A.; Kurdowski, W.; Derkowski, W. 2014. Some problems with prestressed concrete sleepers durability. *Int. J. Res. Eng. Technol.* **2014**, *3*, 153–157.
21. Ravindrarajah, R.S.; White, S.R. Non-Delayed Heat Application Effects on the Strength of Concrete for Railway Sleepers. *Concr. Res. Lett.* **2014**, *5*, 7.
22. Shojaei, M.; Behfarnia, K.; Mohebi, R. Application of alkali-activated slag concrete in railway sleepers. *Mater. Des.* **2015**, *69*, 89–95. [CrossRef]
23. Rezaie, F.; Farnam, S.M. Sensitivity analysis of pre-stressed concrete sleepers for longitudinal crack prorogation effective factors. *Eng. Fail. Anal.* **2016**, *66*, 385–397. [CrossRef]
24. Edwards, J.R.; Dersch, M.S.; Kernes, R.G. *Improved Concrete Crosstie and Fastening Systems for US High Speed Passenger Rail and Joint Corridors: Volume 2*; Federal Railroad Administration: Washington, DC, USA, 2017; p. 410.
25. Zakeri, J.A.; Abbasi, R. Field investigation of variation of loading pattern of concrete sleeper due to ballast sandy contamination in sandy desert areas. *J. Mech. Sci. Technol.* **2012**, *12*, 3885–3892. [CrossRef]
26. Zakeri, J.A.; Rezvani, F.H. Failures of Railway Concrete Sleepers During Service Life. *Int. J. Constr. Eng. Manag.* **2012**, *1*, 1–5.
27. Silva, É.A.; Pokropski, D.; You, R.; Kaewunruen, S. Comparison of structural design methods for railway composites and plastic sleepers and bearers. *Aust. J. Struct. Eng.* **2017**, *18*, 160–177. [CrossRef]
28. Mchenry, M.T. Pressure Measurement at The Ballast-Tie Interface of Railroad Track Using Matrix Based Tactile Surface Sensors. Master's Thesis, University of Kentucky, Lexington, KY, USA, 2013; p. 123.
29. Mayville, R.A.; Jiang, L.; Sherman, M.; Performance Evaluation of Concrete Railroad Ties on the Northeast Corridor. Federal Railroad Administration US Transportation Collection. 2014; pp. 1–65. Available online: https://rosap.ntl.bts.gov/view/dot/28258 (accessed on 1 October 2022).

30. Sundaram, R.; Matsumoto, K.; Nagai, K.; Awasthi, A. Visual investigation method and structural performance evaluation for DEF induced damaged Indian Railway PC sleepers. *J. Asian Concr. Fed.* **2018**, *4*, 103–115. [CrossRef]
31. Marquis, B.; Yu, H.; Jeong, D. *CSX Derailment on Metro-North Tracks in Bronx, NY July 19, 2013*; U.S. Department of Transportation, Federal Railroad Administration, Office of Research: Washington, DC, USA, 2014.
32. Tsoukantas, S.; Tzanakakis, K.; Spyropoulou, D.; Panopoulos, P.; Mintzia, M. Investigation on the Causes of Longitudinal Cracks on Prestressed Monoblock Railway Sleepers of Metric Gauge of the Greek Railway Network. In Proceedings of the International Conference on Case Histories in Geotechnical Engineering, Arlington, VA, USA, 11–16 August 2008.
33. Lutch, R.; Harris, D.; Ahlborn, T. Prestressed Concrete Ties in North America. In Proceedings of the AREMA 2009 Annual Conference, Chicago, IL, USA, 20–23 September 2009.

Thermo-Mechanical Coupling Model of Bond-Based Peridynamics for Quasi-Brittle Materials

Haoran Zhang [1,2], Lisheng Liu [1,2,*], Xin Lai [1,2,*], Hai Mei [1,2] and Xiang Liu [1,2]

1 Hubei Key Laboratory of Theory and Application of Advanced Materials Mechanics, Wuhan University of Technology, Wuhan 430070, China
2 Department of Engineering Structure and Mechanics, Wuhan University of Technology, Wuhan 430070, China
* Correspondence: liulish@whut.edu.cn (L.L.); laixin@whut.edu.cn (X.L.)

Abstract: The mechanical properties of quasi-brittle materials, which are widely used in engineering applications, are often affected by the thermal condition of their service environment. Moreover, the materials appear brittle when subjected to tensile loading and show plastic characteristics under high pressure. These two phenomena manifest under different circumstances as completely different mechanical behaviors in the material. To accurately describe the mechanical response, the material behavior, and the failure mechanism of quasi-brittle materials with the thermo-mechanical coupling effect, the influence of the thermal condition is considered in calculating bond forces in the stretching and compression stages, based on a new bond-based Peridynamic (BB-PD) model. In this study, a novel bond-based Peridynamic, fully coupled, thermo-mechanical model is proposed for quasi-brittle materials, with a heat conduction component to account for the effect of the thermo-mechanical coupling. Numerical simulations are carried out to demonstrate the validity and capability of the proposed model. The results reveal that agreement could be found between our model and the experimental data, which show good reliability and promise in the proposed approach.

Keywords: bond-based peridynamics theory; quasi-brittle materials; thermo-mechanical coupling; crack propagation

Citation: Zhang, H.; Liu, L.; Lai, X.; Mei, H.; Liu, X. Thermo-Mechanical Coupling Model of Bond-Based Peridynamics for Quasi-Brittle Materials. *Materials* **2022**, *15*, 7401. https://doi.org/10.3390/ma15207401

Academic Editor: Danuta Barnat-Hunek

Received: 31 July 2022
Accepted: 18 October 2022
Published: 21 October 2022

Publisher's Note: MDPI stays neutral with regard to jurisdictional claims in published maps and institutional affiliations.

1. Introduction

Thanks to their unique brittle characteristics, materials such as ceramics, rocks, and concrete are widely used in various engineering fields, including the aerospace, armor protection, and construction industries. However, temperature, as an essential environmental factor, can impact the mechanical properties and service performance of the above materials during operation [1]. It may even cause serious accidents associated with the thermal cracking of surrounding rocks upon the underground storage of nuclear waste [2] and heat-induced spalling of building concrete [3]. Therefore, to ensure the safety and load-bearing capacity of quasi-brittle engineering materials, their mechanical properties and damage behavior must be understood under the thermos-mechanical coupling.

To study the behavior of quasi-brittle materials undergoing thermal damage, numerous experimental facilities have been developed and utilized, such as X-ray computed tomography (X-CT) [4], acoustic emission (AE) [5], scanning electron microscopy (SEM) [6], and thermal stress devices (TSD) [7]. These approaches have enabled researchers to significantly improve their knowledge about the hermos-mechanical properties of the materials. However, experimental studies are not only time-consuming and laborious, but it is also challenging to observe the sprouting and extension of micro-cracks inside the material in situ. Thus, by providing a detailed and cost-effective prediction, numerical methods shed insight into the mechanical failure processes in quasi-brittle materials.

Extensive numerical investigations have been devoted to understanding the thermal damage behavior of quasi-brittle materials, which are generally based on continuum mechanics approaches. Various numerical approaches have been developed on the basis of the Finite Element Method (FEM) [8–10], Extended Finite Element Method (X-FEM) [11], Finite Difference Method (FDM) [12], and Boundary Element Method (BEM), etc. Since the numerical approaches mentioned above are based on continuum mechanics, in which partial differential equations need to be solved to find the numerical solution, the ability to deal with the problem of cracks and fractures is often limited, even after the introduction of special-made shape functions in X-FEM. Moreover, special treatment is often needed to maintain the stability of the system, determine when the nucleation and crack happens, and keep track of the crack propagation path, such as the remeshing and level-set method. In this regard, those methods are highly dependent on numerical modeling and discretization.

Peridynamics (PD) theory [13], a nonlocal continuum theory proposed in recent years, is a theory in which spatial differential equations are replaced by integral equations, which provide a uniform framework for both continuities and discontinuities. Unlike the partial differential equations in classical theory, the controlling equations of PD still hold at geometric discontinuities, such as crack discontinuities, making it possible to model crack nucleation and expansion along arbitrary paths. Hence, PD can efficiently deal with fracture problems concerning brittle fractures in solids, complex fracture morphologies, crystal dislocations, and high-speed impacts of geological materials under explosive action.

Until recently, PD has been employed extensively to investigate the thermal impact damage mechanisms of brittle materials such as rocks, ceramics, and concrete. For instance, Chen et al. [14] proposed the refined thermo-mechanical, fully coupled PD approach by applying the PD differential operator to a classical thermal differential equation. This method is suitable for studying the heat conduction and thermal deformation of, and damage to, concrete structures. Shou and Zhou [15] added the thermal expansion coefficient of a solid material to the thermal coupling equation of the non-ordinary state-based Peridynamics (SB-PD). They first used the temperature field to calculate the deformation gradient tensor, then the latter was introduced into the non-ordinary SB-PD motion equation to realize the thermo-mechanical coupling process. Moreover, this method was subsequently employed to simulate the thermal cracking of rocks, and the simulation results showed good convergence with experimental results. Bazazzadeh et al. [16] developed a thermo-mechanical coupling PD model for simulating crack extension in ceramics using an adaptive mesh. Specifically, this model was adaptively transformed based on a stretching control criterion for mesh discretization and then applied in the desired finite region, thereby enabling the prediction of complex cracking forms. Yang et al. [17] proposed a new method of characterizing the mineralogical composition and distribution of heterogeneous rock materials using fully coupled conventional thermo-mechanical equations of PD. The model under this method has the ability of describing the thermal-force damage behavior of granite after thermal cycling treatment. Taking the study [18] as an example, Liu et al. [19] further increased the tangential bond force by considering the influence of the bond on the rotation effect so that the ceramic model developed by Chu et al. could break through the Poisson's ratio limit and be adopted to more types of ceramic materials. Wang et al. [20] derived a microscopic thermal conductivity parameter that links various micro- and macro-geometric conditions based on a weakly coupled thermoelastic, non-ordinary, state-based Peridynamics (OSB-PD) model by analyzing the temperature distribution. Furthermore, they proposed a tensile damage criterion that takes into account the softening effect of stretched parts. However, the above study focuses only on the tensile damage of quasi-brittle materials, ignoring the nonlinear mechanical behavior caused by the generation of micro-cracks in the compression phase. As a result, the simulated results of quasi-brittle materials under thermal coupling deviated from the experimental data.

In this study, a model suitable for the failure of quasi-brittle materials under thermo-mechanical coupling is proposed, and the effect of temperature within the elastic and plastic stages is considered. To verify the reliability and validity of this model, numerical simulations of ceramics under a heating load and the pre-cracked Brazilian disk under uniaxial compression were conducted. Finally, two-dimensional granite plates were exposed to cold, uniaxial compression experiments after heat treatment were conducted, and the coincidence between the simulated and experimental results were analyzed.

2. Thermo-Mechanical Coupling Model

In this section, the classic fully coupled thermo-mechanical BB-PD model, in which the bonds remain elastically deformed during deformation until broken and are not applicable to quasi-brittle materials, is first introduced. Then, the mechanical behavior of quasi-brittle materials in tension and compression is presented. The Peridynamics model for quasi-brittle materials is presented in detail in Section 2.3, and the substance of this paper, i.e., the study of thermal effects in the tensile and compression phases, is presented. Finally, the numerical discretization and time integration methods of the proposed model are presented.

2.1. Fully Coupled Thermo-Mechanical Equation

Unlike the partial derivative of deformation with respect to the spatial coordinate in continuum mechanics, the BB-PD theory adopts the spatial integral equation, which can be applied to discontinuous bodies [21]. As shown in Figure 1, each material point x in a region **R** of an object interacts with all other points within its neighborhood radius H_x. For the domain x', the material point is called a neighborhood particle of the material point x. When a solid is deformed under an external load, the matter points x and x' arrive at the post-deformation positions y and y' through displacements u and u', respectively, and $|\xi|$ is the distance between the two particles before the deformation and $|\xi + \eta|$ is the distance between the two particles after the deformation.

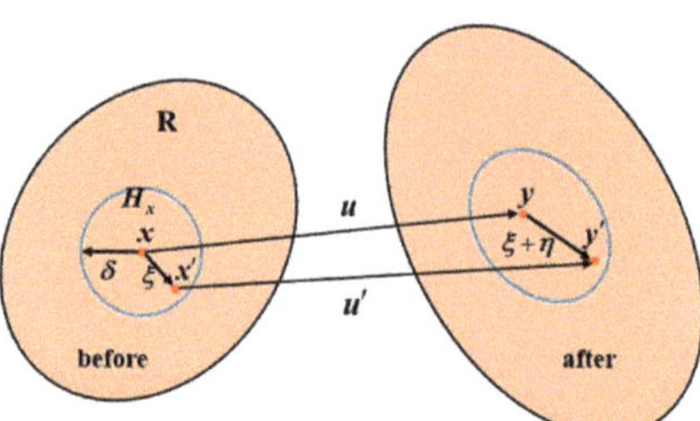

Figure 1. Schematic diagram of the BB-PD model.

In contrast to the classical derivation of the heat equation [22], PD laws are derived based on irreversible thermo-mechanicals, i.e., energy conservation and free energy density functions. The fully coupled thermo-mechanical equations for BB-PD are as follows [23–25]:

$$\rho c_v \dot{T}(x,t) = \int_H \left(\kappa \frac{\tau}{|\xi|} - T_0 \frac{c\alpha}{2} \dot{e} \right) dV' + h_s(x,t) \tag{1}$$

$$\rho \ddot{u}(x,t) = \int_H \frac{\xi + \eta}{|\xi + \eta|} csdV' + b(x,t) \tag{2}$$

Here, Equation (1) is the PD thermal diffusion equation with a structural coupling term, where c_v is the specific heat capacity, κ is the thermal conductivity of the bond in the PD system, $h_s(x,t)$ is the rate of heat production per unit volume at the point of matter x at time t. The temperature difference between the substance points x and x' can be expressed as $\tau = T(x,t) - T(x',t)$, where $T_0(c\alpha/2)\dot{e}$ are the deformation terms caused by heating and cooling, T_0 is the reference temperature, and $\dot{e}$ is the rate of change of bond lengths, denoted as:

$$\dot{e} = \frac{\xi + \eta}{|\xi + \eta|} \cdot \dot{\eta} \tag{3}$$

Equation (2) is the equation of motion for the BB-PD system with a temperature coupling term. In this equation, ρ is the mass density; $\ddot{u}(x,t)$ is the acceleration of the matter point x at time t; H is the neighborhood range of the matter point x; and V' is the volume of the matter point x'; c is the bond constant; α is the coefficient of thermal expansion; $b(x,t)$ is the force density of the matter point x at time t; s is the stretch rate between x and x'; and T_{avg} is the average temperature of the matter points x and x', respectively, expressed as:

$$s = \frac{\|\xi + \eta\| - \|\xi\|}{\|\xi\|} - \Delta T_{avg} \cdot \alpha \tag{4}$$

$$\Delta T_{avg} = \frac{\Delta T + \Delta T'}{2} \tag{5}$$

2.2. The Characterization of the Mechanical Behavior of Quasi-Static Brittle Materials

For quasi-brittle materials such as ceramics and concrete, the damage models under compression and tension are distinct. Under tension, the behavior is brittle, while a more ductile behavior can be observed under compression. The brittle behavior of materials under tensile loading is attributed to macro-crack formation. On the other hand, the ductile behavior of ceramics under compression can be explained by micro-crack formation and plasticity [26–28].

Under tensile loading, the quasi-brittle materials undergo direct, brittle damage at the end of elastic deformation, with essentially no plastic deformation. Figure 2 depicts a typical stress–strain diagram of the quasi-brittle material model in the tensile phase, showing a linear relationship between stress and strain, where the material breaks down and loses its load-bearing capacity after reaching the tensile-strength limit of the material.

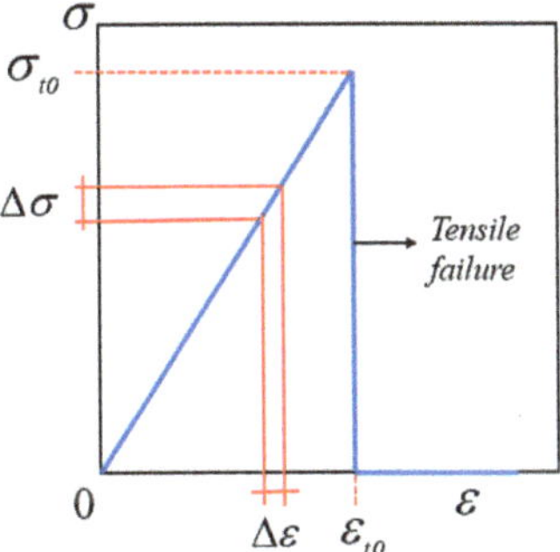

Figure 2. Stress–strain diagram of tensile behavior.

When a quasi-brittle material is subjected to a compressive load, the model remains intact at the initial stage, and as the load continues to increase, structural crushing occurs internally due to the generation of micro-cracks and lattice plasticity, and the interaction of these cracks leads to a decrease in the compressive strength of the quasi-brittle material, a phenomenon described as plastic-softening behavior. The quasi-brittle material is treated as elastic material before damage occurs, and after damage occurs, it is treated as material that remains intact but whose strength decreases with the accumulation of damage.

Figure 3 shows a typical stress–strain diagram for the ideal quasi-brittle material which is typical of ultra-high-strength concrete [29]. Before reaching the elastic limit of the material, for a sufficiently small segment, the relation is close to linear. After exceeding the elastic limit, micro-cracks appear in the concrete, resulting in an inelastic response that differs from the plastic-flow behavior of ductile materials, as shown by the fact that the strength of the material decreases with the accumulation of plastic damage, i.e., the material begins to soften. Finally, the concrete is completely broken and loses its load-bearing capacity.

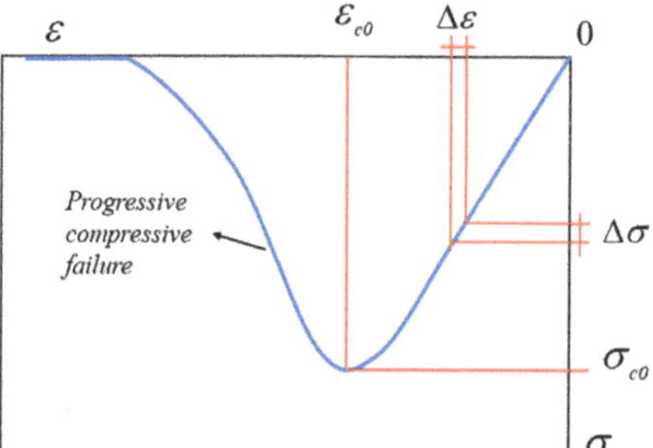

Figure 3. Stress-strain diagram of compression behavior.

2.3. Quasi-Brittle Peridynamics Model

2.3.1. Description of the Stretching Stage

The mechanical behavior of quasi-brittle materials in the tensile phase, as described in Section 2.2, has been known to exhibit mainly brittle characteristics. When using PD to describe the behavior of brittle materials in the tensile phase, the bond forces are considered to be related only to the relative elongation of the bonds and can be described as:

$$f(\eta,\xi) = cs(\xi,\eta)\frac{\xi+\eta}{|\xi+\eta|} \tag{6}$$

where c is the micro-elastic modulus, obtained from the consistency between the strain energy density of the isotropic material and the theoretical strain energy density of the continuum mechanics. The micro-elastic modulus of the isotropic material in the plane stress state is $c = 9E/(\pi h\delta^3)$, where E is the elastic modulus and δ is the neighborhood radius.

2.3.2. Description of the Compression Stage

In the compression stage, the quasi-brittle material can be divided into two phases: the elastic phase before reaching the elastic limit and the plastic phase after exceeding the elastic limit. For the elastic stage, the bond force is calculated in the same way as for the tensile stage. For the plastic stage, the quasi-brittle material exhibits plastic softening, and the influence of plastic deformation must be considered in the calculation of the bond force, which can be expressed as:

$$f(\eta,\xi) = c\left[s(\xi,\eta) - s_p(\xi,\eta)\right]\frac{\xi+\eta}{|\xi+\eta|} \tag{7}$$

where s_p indicates the amount of plastic deformation after the bond enters the plastic phase.

2.3.3. Yield Criteria

In compression, once the bond force exceeds the critical force of the elastic limit of the bond, plastic deformation occurs in the bond, resulting in the accumulation of damage. The bond force decreases due to the accumulation of plastic deformation, and the critical force of the bond decreases due to the accumulation of damage until the bond force is less than the critical force of the bond.

In order to accurately calculate the true bond force for quasi-brittle materials undergoing plastic softening behavior in the compression phase, the bond strengths need to be known. Chu et al. [18], in their work, expressed the bond strengths as follows by emulating the strength expressions in the JH-2 model in the classical framework [30]:

$$p = p_i - D\left(p_i - p_f\right) \tag{8}$$

where p_i is the critical force when the bond is undamaged, p_f is the critical force when the bond is fully damaged, expressed as:

$$p_f = \beta p_i \tag{9}$$

and D represents the cumulative damage from plastic deformation of the bond:

$$D = \frac{\sum \Delta s_p}{s_1 - s_e} \tag{10}$$

where Δs_p is the plastic deformation in one time step, $\sum \Delta s_p$ is the accumulated plastic deformation, s_e is the elastic compression limit, and s_1 is the plastic-compression deformation limit.

In order to characterize the relationship of the bond in the compressive plastic softening phase, it is necessary to define a function to determine whether the bond force has reached the maximum allowed value; referring to the theory of continuum-media mechanics, the yield criterion of the bond is expressed as:

$$\varphi\left(s, s_p, \dot{s}\right) = f(s) - p\left(s_p, \dot{s}\right) = f - [1 - (1 - \beta)D]p_i \tag{11}$$

where $\dot{s}$ is the compressive deformation rate of the bond, which is the time derivative of the relative stretch of the bond, expressed as follows:

$$\dot{s}(\xi, \eta) = \frac{\dot{\eta}(\eta + \xi)}{|\eta||\eta + \xi|} \tag{12}$$

2.3.4. Flow Rule

Similar to the consistency condition in elastoplastic mechanics, the consistency condition for the bond is defined by deriving Equation (11) as follows:

$$\dot{\varphi}\left(s, s_p, \dot{s}\right) = \frac{\partial \varphi}{\partial f}\dot{f} - \frac{\partial \varphi}{\partial s_p}\dot{s}_p = \dot{f} + \frac{(1 - \beta)p_i}{s_1 - s_e}\dot{s}_p = 0 \tag{13}$$

where the second-order derivative of the relative deformation of the bond is neglected. According to Equation (13), the bond force remains equal to the elastic limit value of the bond. The derivation of Equation (7) is as follows:

$$\dot{f} = c\left(\dot{s} - \dot{s}_p\right) \tag{14}$$

Based on Equations (13) and (14), the relative compression plastic deformation rate of the bond can be solved as:

$$\dot{s}_p = \frac{c}{c - H}\dot{s} \tag{15}$$

where $H = \frac{(1 - \beta)p_i}{s_1 - s_e}$.

2.3.5. Consideration of Thermal Effects

Since the mechanical behavior of the material in the elastic phase is not affected by the loading rate, the thermo-mechanical coupling bond-force function in this phase can be simply expressed in a linear form as follows:

$$f(\eta, \xi, T) = c(T)\left[s - \Delta T_{avg}\alpha(T)\right]\frac{\xi + \eta}{|\xi + \eta|} \tag{16}$$

where $\alpha(T)$ is the coefficient of thermal expansion under different temperatures; and $c(T)$ is the bond constant considering the effect of temperature, which is a material-dependent constant. This can be obtained by making the elastic strain energy density in the theory of elastic mechanics equal to the deformation energy density at the material point. The relation between $c(T)$ and the elastic modulus of the material under a plane-stress state can be expressed as:

$$c(T) = \frac{9E(T)}{(\pi h \delta^3)} \tag{17}$$

where the $E(T)$ is Young's modulus of the material at different temperatures.

When the bond is in the plastic phase, the bond-force function can be expressed as follows due to the plastic softening behavior and temperature effects:

$$f(\xi, \eta, \dot{\eta}, T) = c(T)\left[s(T) - s_p(\xi, \eta, \dot{\eta}, T)\right] \tag{18}$$

Meanwhile, the yield function, Equation (13), which determines whether the bond force reaches the maximum permissible value, can be expressed as:

$$\varphi(s, s_p, \dot{s}, T) = f(s, T) - p(s_p, \dot{s}, T) = f - \left[1 - (1 - \beta)D(T)\right]p_i \tag{19}$$

where $D(T)$ is the damage that gradually accumulates, considering the thermal effect, could be expressed as:

$$D(T) = D(T) + \frac{\sum \Delta s_p(T)}{s_1 - s_e} \tag{20}$$

Here, $\Delta s_p(T)$ is the plastic compression deformation, considering the thermal effect, at each time step Δt, which can be expressed as:

$$\Delta s_p(\xi, \eta, \dot{\eta}, T) = \dot{s}(\xi, \eta, \dot{\eta}, T)\Delta t - \frac{\Delta f}{c(T)} \tag{21}$$

To obtain the change rate of bond stretching $\dot{s}(\xi, \eta, \dot{\eta}, T)$, we derived Equation (12):

$$\dot{s}(\xi, \eta, \dot{\eta}, T) = \frac{\dot{\eta}(\eta + \xi)}{|\eta||\eta + \xi|} - \frac{\Delta T_{avg}}{\Delta t}\alpha \tag{22}$$

It should be noted that the consistency condition also needs to be satisfied, i.e., the time derivative of Equation (13):

$$\dot{\varphi}(s, s_p, \dot{s}, T) = \frac{\partial \varphi}{\partial f}\dot{f} - \frac{\partial \varphi}{\partial s_p}\dot{s}_p = \dot{f}(s, T) + \frac{(1 - \beta)p_i}{s_1 - s_e}\dot{s}_p(s, \dot{s}, T) = 0 \tag{23}$$

The relationship between $\dot{s}(\xi, \eta, \dot{\eta}, T)$ and $\dot{s}_p(\xi, \eta, \dot{\eta}, T)$ can be similarly obtained from the above equation and Equation (15) and is expressed as:

$$\dot{s}_p(\xi, \eta, \dot{\eta}, T) = \frac{c(T)}{c(T) - H}\dot{s}(\xi, \eta, \dot{\eta}, T) \tag{24}$$

Therefore, the BB-PD thermo-mechanical coupling equations applicable to quasi-brittle materials are as follows:

$$\begin{cases} \rho c_v \dot{T}(x,t) = \int\limits_{H} \left(\kappa \frac{\tau}{|\xi|} - T_0 \frac{c\alpha}{2} \dot{e} \right) + h_s(x,t) \\ \rho\, \ddot{u}(x,t) = \int\limits_{H} \frac{\xi+\eta}{|\xi+\eta|} f(\xi,\eta,T) dV_{x'} + b(x,t) \end{cases} \tag{25}$$

2.4. Numerical Discretization and Time Integration

For Equation (25), the motion equation and heat conduction can be replaced by a discretized form, as given below:

$$\begin{cases} \rho_i c_i \dot{T}_i(t) = \sum\limits_{j=1}^{N_i} \mu_{ij} \left(h_{cij}(t)\kappa_{ij} \frac{T_i - T_j}{|\xi_{ij}|} - T_{i,0} \frac{c_{ij}\alpha_{ij}}{2} \dot{e}_{ij} \right) V_j + h_{s,i}(t) \\ \rho_i \ddot{u}_i(t) = \sum\limits_{j=1}^{N_i} f_{ij} \frac{\xi_{ij}+\eta_{ij}}{|\xi_{ij}+\eta_{ij}|} V_j + b_i(t) \end{cases} \tag{26}$$

where f_{ij} is the bond force function between the matter points i and j. This can be calculated using Equation (16) in the elastic phase and Equation (18) in the plastic phase.

In this study, in order to describe the thermo-mechanical coupling in the framework of the PD model of quasi-brittle materials, an interleaved scheme is used to approximate the solution. This means that the coupled system equations are solved separately, and different time-explicit algorithms are employed to solve the heat conduction equation and the kinetic equation. In particular, the explicit integration algorithm with the first-order forward difference is used to solve the heat conduction equation, and the temperature of the next time step is obtained as:

$$T_{(i)}(t + \Delta t) = T_{(i)}(t) + \Delta t^T \cdot \dot{T}_{(i)}(t) \tag{27}$$

where Δt^T is the thermo-mechanical time step.

In addition, similar to the quasi-static problem, virtual inertia and damping terms are introduced to solve the dynamics equations, which can be expressed as:

$$D\ddot{u}(x,t) + cD\dot{u}(x,t) = f(u_i, u_j, x_i, x_j) \tag{28}$$

where D is the virtual diagonal density matrix and c is the damping coefficient, obtained from Reschgorin law and Rayleigh quotient [31], respectively. Using the central difference algorithm, the displacement and velocity for the next time step are defined as:

$$\dot{u}^{n+1/2} = \frac{(2 - c_n \Delta t)\dot{u}^{n-1/2} + 2\Delta t^M D^{-1} F^n}{(2 + c^n \Delta t^M)} \tag{29}$$

$$u^{n+1} = u^n + \Delta t^M \cdot \dot{u}^{n+1/2} \tag{30}$$

where Δt^M is the kinetic time step.

In thermo-mechanical coupled problems, the kinetic characteristic time scale depends on the propagation velocity of the stress wave in the material, and the heat conduction characteristic time scale depends on the thermal diffusivity of the material. In general media, the time scale of heat transfer characteristics is usually much larger than that of kinetic characteristics. Therefore, the whole thermal coupling solution can be divided into the following three steps.

Step 1: The heat conduction equation is solved and the temperature field distribution of the whole model is calculated.

Step 2: The motion equations are solved until the whole model reaches a steady state.

Step 3: The heat conduction equation is solved for the next thermo-mechanical time step.

The above steps are repeated to obtain the entire thermo-mechanical coupling solution. It should be noted that the convergence criterion is also needed to determine the steady state of the kinetic iterative solution. When the whole model reaches the steady state, the displacement increment of each kinetic iteration step tends to 0. Chen et al. [14] have previously defined the parametric number, **Re**, as shown in Equation (32), and provided a minimal value, Ω. When $\mathbf{Re} \leq \Omega$, the system reaches the steady state and can be solved in the next thermo-mechanical time step. Otherwise, the iteration continues until it converges with the formula below:

$$\mathbf{Re} = \sqrt{\frac{\sum\limits_{m=1}^{M} \left(u_m^{i,j} - u_m^{i,j-1}\right)^2}{M}} \tag{31}$$

where M is the entire number of model particles.

3. Model Verification and Convergence Analysis

In this section, the BB-PD model proposed in Section 2 is implemented in Fortran code, and two typical cases are applied to verify the efficiency of the model. The convergence of the numerical model is also analyzed in the following section.

3.1. Ceramic Plates Subjected to Heating Loads

Figure 4 shows the computational model of the ceramic plates subjected to heating loads. A flat directional plate with a side length $L = W = 1$ m is adiabatically constrained to the normal directional displacement with respect to three sides, except for the top. The initial temperature of the whole plate is $0\,^\circ$C and the temperature of $T = 1.0\,^\circ$C is applied to the top boundary. The peridynamic, mechanical, and thermo-mechanical parameters used in the numerical simulation are listed in Table 1. Three points, referred to as A, B, and C on the vertical symmetry axis at the center of the plate, and located on the top boundary, positive center, and bottom boundary of the plate, respectively, were selected as reference points.

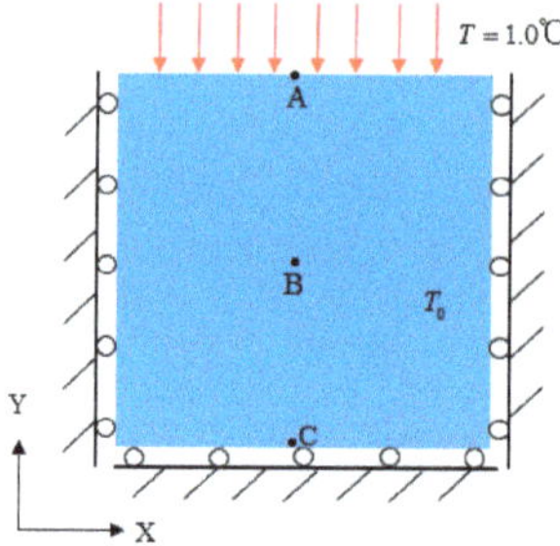

Figure 4. The two-dimensional flat plate subjected to heating loading.

Table 1. Parameters involved in the numerical simulation.

	Parameter	Value
PD parameters	Number of discrete points in the xy direction	200×200
	Material point spacing Δx (m)	0.005
	non-locality parameter m	3
	Heat transfer time step Δt^{TH} (s)	1×10^{-5}
Mechanical parameters	Young's modulus E (GPa)	1
	Poisson's ratio ν	0.33
	Density ρ (kg/m^3)	1
Thermal parameters	Thermal conductivity k_T (W $\cdot$ m^{-1}K^{-1})	1
	Coefficient of thermal expansion α (1/K)	0.02
	Specific heat capacity c_v (J $\cdot$ Kg^{-1}K^{-1})	1

An identical finite element model with the same material properties was built using the commercial ABAQUS finite element software. The model was discreted into 200×200 grids with a time step of 1×10^{-5} s, and a direct thermo-mechanical coupling method was used. Additionally, the theoretical calculation formulas for the temperature and vertical displacement of three reference points were provided by Timoshenko [32] and Carslaw [33]:

$$T(y,t) = 1 - \frac{4}{\pi} \sum_{n=0}^{\infty} \frac{(-1)^n}{2n+1} \exp\left(-\frac{(2n+1)\pi^2 kt}{4L^2}\right) \cos\left(\frac{(2n+1)\pi y}{2L}\right) \tag{32}$$

$$u_y(y,t) = (1+v)\alpha \int_0^y T(y,t)dy \tag{33}$$

In both simulations, in spite of the temperature loads applied, the temperature change causes deformation inside the plate due to the thermo-mechanical coupling effect, and the reference points are displaced at the same time. The simulation results of different reference points according to both models are shown in Figure 5, the calculation results of PD and ABAQUS are consistent with those of analytical results, thus, demonstrating the reliability of the proposed approach in solving the thermo-mechanical coupling problem.

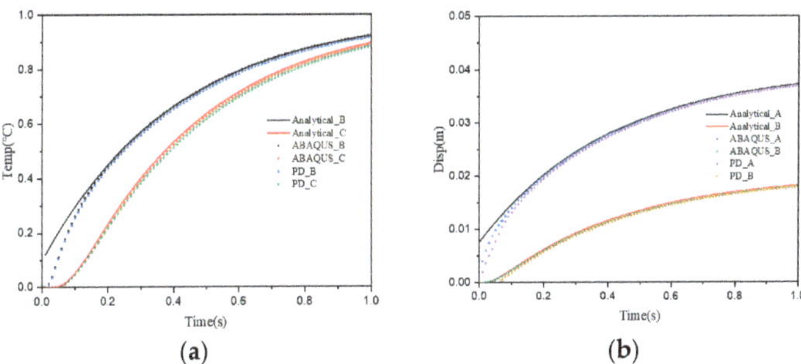

Figure 5. Comparison of calculation results of different methods. (**a**) Temperature; (**b**) Vertical displacement.

3.2. Pre-Cracked Brazilian Disk under Uniaxial Compression

The Ayatollahi's experiments on the brittle fracture of polycrystalline graphite were simulated [34]. A modified version of the cracked Brazilian disk (CBD) specimen called the V-notched Brazilian disk (VBD) specimen was used in this experiment. As shown in Figure 6, the specimen is a circular disk of diameter D containing a central rhombic hole with an opening angle 2α and length d for the VBD specimen. The disk diameter and the notch depth were 60 mm and 15 mm, respectively, and the angles used in the experiment were $2\alpha = 30°$, $\beta = 15°$. The basic material properties of polycrystalline graphite are as follows: density of 1710 kg/m^3, Young's modulus of 8.05 GPa, Poisson's ratio of 0.33, and the fracture toughness of 1.0 MPa m$^{0.5}$. In the experiment, the fracture test was performed by using a universal tension–compression test machine under displacement conditions with a loading rate of 0.05 mm/min.

Using the proposed model to simulate the crack extension of the VDB specimen, the splitting damage process is shown in Figure 7. At 70 s, the tips of both sides of the pre-existing crack begin to accumulate damage due to stress concentration exceeding the strength limit, and crack initiation occurs here. Then at 80 s, cracks appear at the tips of both sides of the pre-existing crack and begin to propagate outward. Next, at 110 s, the cracks on both sides remain symmetrical and propagate to both sides of the loading. Finally, at 170 s, cracks penetrate the tips of both ends of the precast crack and both sides of the loading point, at which point the Brazilian disc specimen fails.

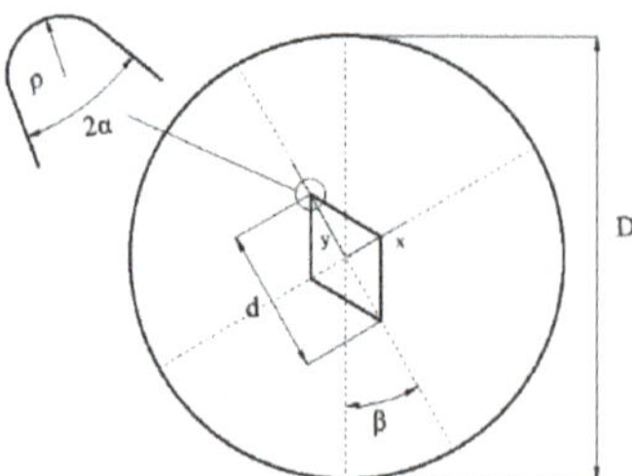

Figure 6. The VBD specimen used in experiments.

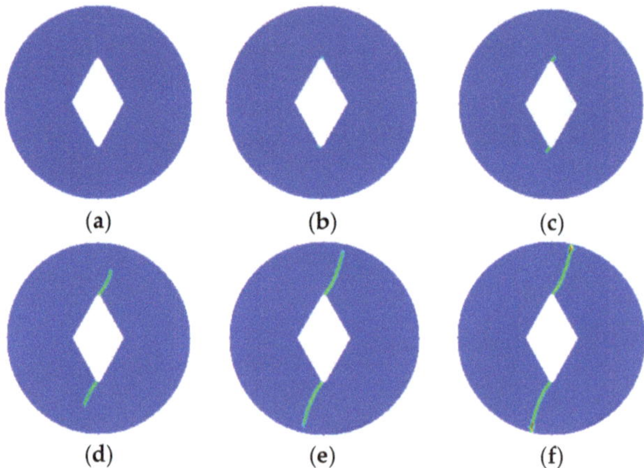

Figure 7. Splitting and destruction process of the VBD specimen. (**a**) 0 s; (**b**) 70 s; (**c**) 80 s; (**d**) 110 s; (**e**) 130 s; (**f**) 170 s.

The comparison of the prefabricated cracked Brazilian disc splitting damage before and after the experiment with the PD simulation results is shown in Figure 8. It can be observed that the simulated results are in high agreement with the experimental results.

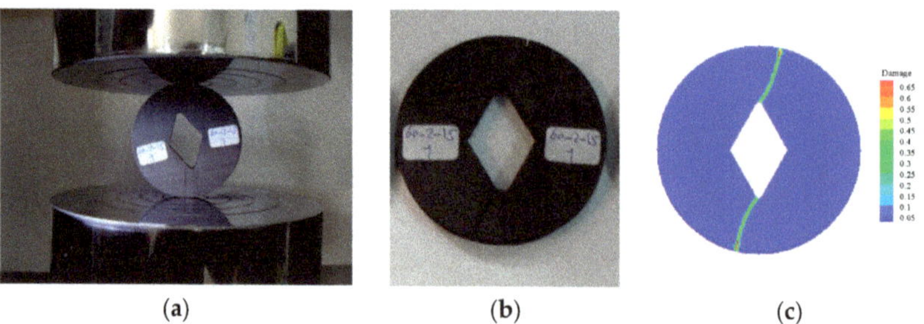

Figure 8. (**a**) Specimen before experiment; (**b**) Specimen after experiment; (**c**) The PD simulation result.

3.3. Convergence Analysis

Furthermore, the numerical convergence of the model was also analyzed using the case study of Section 3.1. Currently, the convergence analysis for PD consists of two main types: m-convergence and δ-convergence [35].

In the m-convergence, horizon δ is kept constant as $\delta = 5.0 \times 10^{-3}$ m throughout the computation, while m and Δx are taken as 2, 3, and 4, and 2.5×10^{-3} m, 1.66×10^{-3} m, and 1.25×10^{-3} m, respectively, as shown in Figure 9. The vertical displacements of the reference point obtained usinig the proposed model, the analytical solution, and the FEM simulation are shown in Figure 10. For a fixed horizon, as the value of m increases, the error rate between the simulation results and the analytical results becomes smaller. Although the error can be captured when $m = 3$ or 4, $m = 3$ is preferred, considering the effect of computational efficiency.

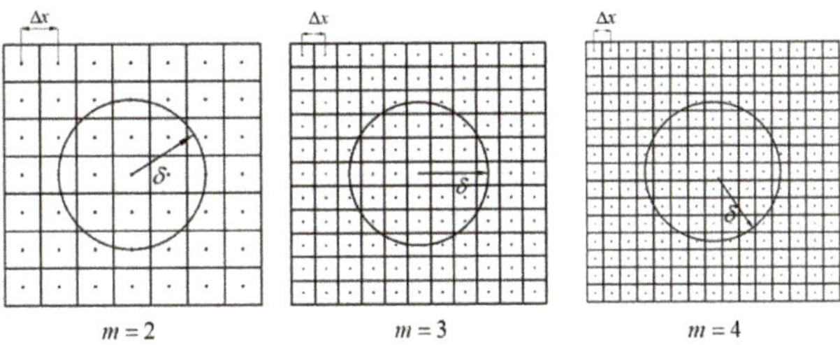

Figure 9. *m*-convergence with a fixed horizon size.

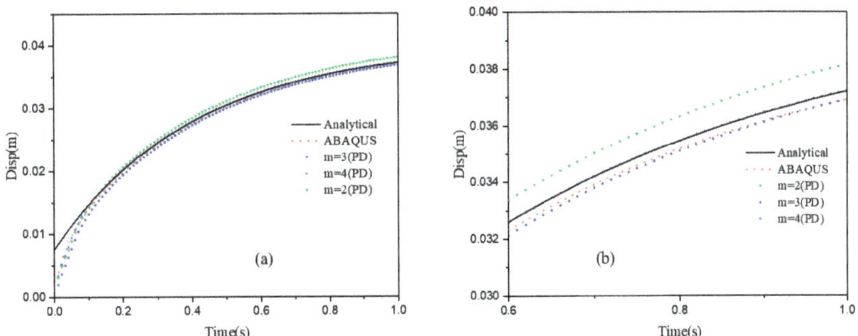

Figure 10. (**a**) The vertical displacement of point A with a different non-locality parameter m; (**b**) an enlarged detail from (**a**).

In the δ-convergence, the non-locality parameter m is kept as a constant, i.e., $m = 3$. Three different horizon sizes are chosen as $\delta = 1.5 \times 10^{-2}$ m, 3×10^{-2} m, 6×10^{-2} m, and $\Delta x = 5 \times 10^{-3}$ m, 1×10^{-2} m, 2×10^{-2} m (see Figure 11). The vertical displacements calculated at reference points using the different δ are compared to those obtained from the finite element method, as shown in Figure 12. For a fixed non-locality parameter, with a decrease in the horizon sizes, the error calculation will also decrease, but it will lead to an increase in the calculation efficiency. Therefore, choosing the right value of horizon sizes requires special consideration.

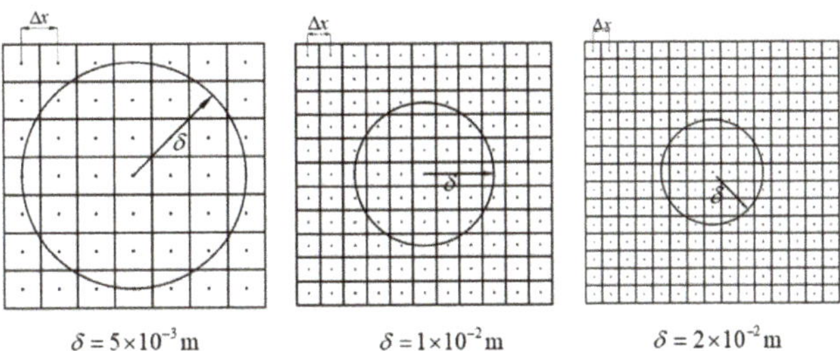

Figure 11. δ-convergence with a fixed parameter m.

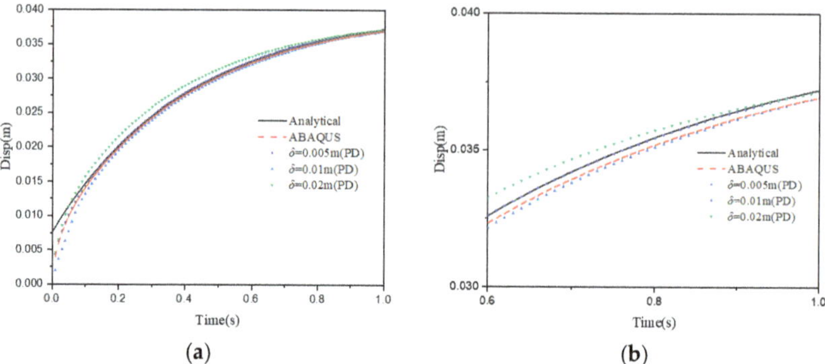

Figure 12. (**a**) The vertical displacement of point A with different horizon sizes δ; (**b**) an enlarged detail from (**a**).

4. Numerical Applications

The previous studies have shown that the proposed model is able to accurately simulate the mechanical behavior of quasi-brittle materials under thermal loading (in Section 3.1) and static loading (in Section 3.2). For the purpose of clarifying the applicability of the model, this section applies it to two more complex coupled thermal-force processes: the ceramic quenching process (in Section 4.1) and the process of compressing the rock after heat treatment (in Section 4.2).

4.1. Ceramic under Cold Shock

Referring to the quenching experiments of ceramic plates at different temperatures conducted by Jiang et al. [36], an alumina ceramic plate with the dimensions of 50 mm × 10 mm is heated to 873 K and subsequently allowed to freefall into the water at 293 K. Considering the symmetry of the load and boundary conditions, a 1/4 model, shown in Figure 13, could be established to perform the calculation. In such a model, the left and lower boundaries are constrained during the normal displacements, and a uniform and constant-cold impact load is applied to the upper and right boundaries. The convective heat transfer coefficient $h = 70,000 \ \text{W}/(\text{m}^2 \cdot \text{K})$ is taken when the ceramic plate is dropped into the water. The PD and thermo-mechanical parameters involved in the numerical simulations are listed in Table 2.

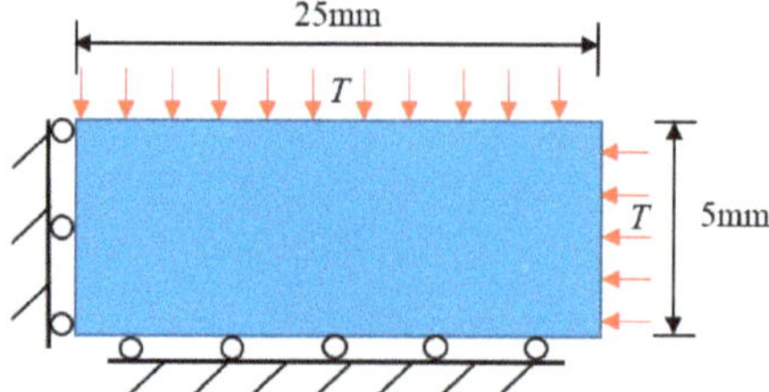

Figure 13. Schematic diagram of the geometry and boundary condition of the ceramic subjected to cold shock.

Table 2. Parameters involved in the PD model of the ceramic under cold shock.

	Parameter	Value
PD parameters	Number of discrete points in the xy direction	500×100
	Material point spacing Δx (m)	0.00005
	Non-locality parameter m	3
	Heat transfer time step Δt^{TH} (s)	1×10^{-4}
Mechanical parameters [36]	Young's modulus E (GPa)	370
	Poisson's ratio ν	0.33
	Density ρ (kg/m^3)	3980
	Fracture energy G_0 (J/m^2)	24.3
Thermal parameters [36]	Thermal conductivity k_T (W $\cdot$ m^{-1}K^{-1})	31
	Coefficient of thermal expansion α (1/K)	7.5×10^{-6}
	Specific heat capacity c_v (J $\cdot$ Kg^{-1}K^{-1})	880

When the high-temperature ceramic plate (873 K) enters the room-temperature water (293 K), the heat energy of the plate spreads rapidly to the surrounding ambient medium. The surface temperature of the ceramic plate decreases sharply, as shown in Table 3, forming a vast temperature gradient with the interior. This temperature gradient from the inside to the outside (i.e., hot inside and cold outside) causes the ceramic surface to undergo tensile stress while the interior is subjected to compressive stress. When the tensile stress on the ceramic surface exceeds that of the interior of the ceramics, damage occurs, and cracks propagate throughout the material. Since the temperature bond also breaks due to the thermal effect, heat conduction through the crack is blocked and the temperature on both sides of the bending crack exhibits a significant temperature jump.

Table 3. The temperature field and crack evolution in high-temperature ceramics under cold shock loading.

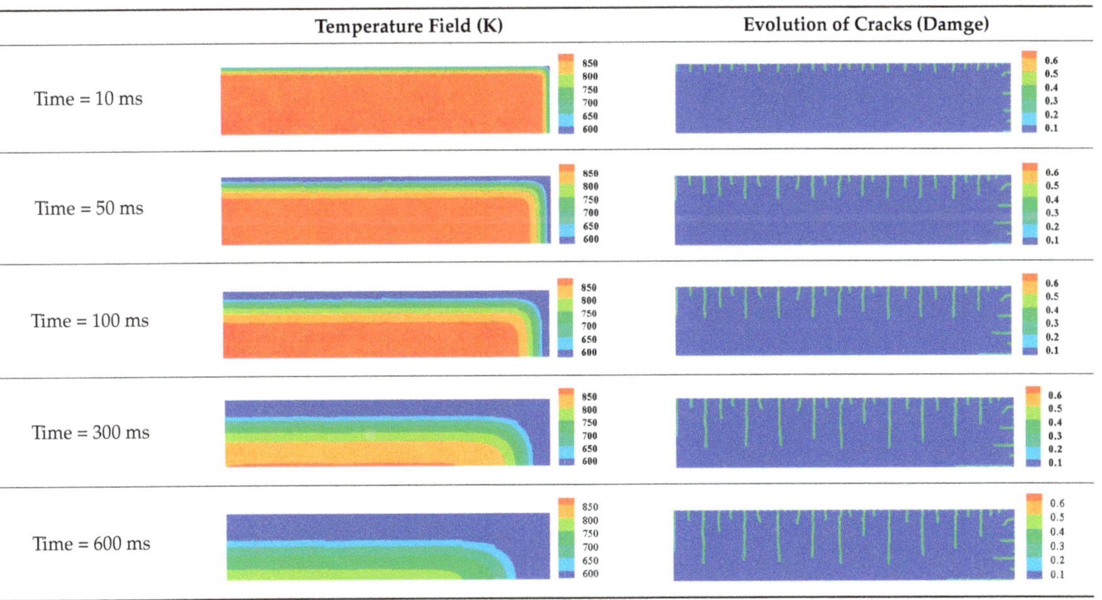

For cracks due to thermal shock, all cracks are distributed in a parallel manner at equal distances on the outer surface of the ceramic (upper and right side) and extend from the outside to the inside. As the cold shock continues, some of the initial cold shock cracks stop growing, while other thermal shock cracks continue to grow.

During the first 10 ms, the cracks are uniformly distributed at intervals of roughly 0.001 mm, and the length of each crack remains consistent. As the cold shock continues, some initial cold shock cracks stop growing at 50 ms, while other heat shock cracks continue to grow. Thereafter, the ceramic plate temperature gradient decreases, the thermal stress becomes smaller, the crack expansion slows down, and the crack stops growing at 600 ms, reaching the maximum length.

In the work of Jiang et al. [36], the area within 10 mm of the ends of the specimen was excluded in order to eliminate the effect of the end boundaries. The average dimensionless crack spacing $\bar{s}$ and dimensionless crack length $\bar{p}$ were proposed, denoted as $\bar{s} = s/L_C$ and $\bar{p} = p/L_C$, respectively, where s is the crack spacing, p is the crack length, and L_C is the specimen width. The average dimensionless crack spacing in the simulation results is 0.112, compared with 0.12 in experiments, and the dimensionless crack length in the simulation results is 0.715, compared with 0.79 in experiments [36]. The thermal impact cracks show a clear spacing distribution, i.e., there are short cracks in the middle of long cracks. The comparison between the simulated and experimental results is shown in Figure 14, where the thermal impact cracks remain similar in terms of spacing, length, length hierarchy, and periodicity. However, since the model used in the experiments is not an ideal model, the ceramic plate is a non-homogeneous material and there are small gaps in the structure, which cannot be consistent with the simulated results, as evidenced by the asymmetry of the thermal cracks in the experimental results.

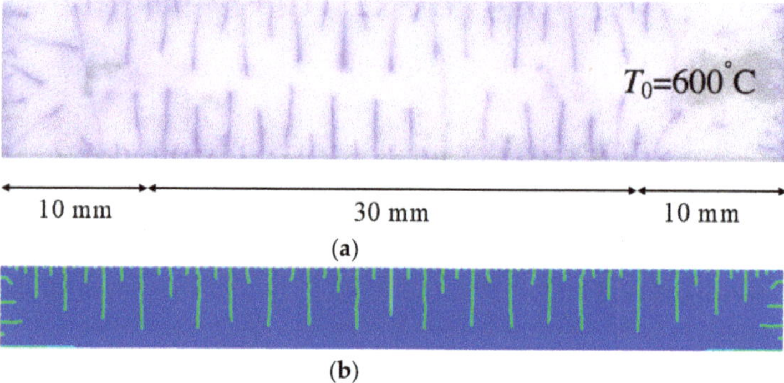

Figure 14. Comparison of ceramic plate thermal impact cracking results: (**a**) Specimens after thermal shock [36]; (**b**) PD simulation results for the 1/2 model.

4.2. Granite under Uniaxial Compression after Heat Treatment

The granite specimen with prefabricated cracks was first subjected to thermocycling and then compressed uniaxially, as performed by Yang et al. [37]. The dimensions of the granite specimen were 80 mm × 160 mm (see Figure 15a), and there was a prefabricated crack with a length of 20 mm, width of 1.5 mm, and inclination angle of 30° in the center of the specimen. At the thermal cycling stage, Yang et al. first heated the granite specimen to 573 K and then kept the temperature constant to make the inside and outside of the sample converge to the same temperature. Subsequently, the sample was placed in the open air and cooled down naturally to room temperature (293 K). At the uniaxial compression stage, the top and bottom ends of the specimens were loaded in compression using a loading speed of 0.1m/s, and the crack nucleation and expansion were observed. The mechanical and thermo-mechanical parameters in the PD simulation are listed in Table 4.

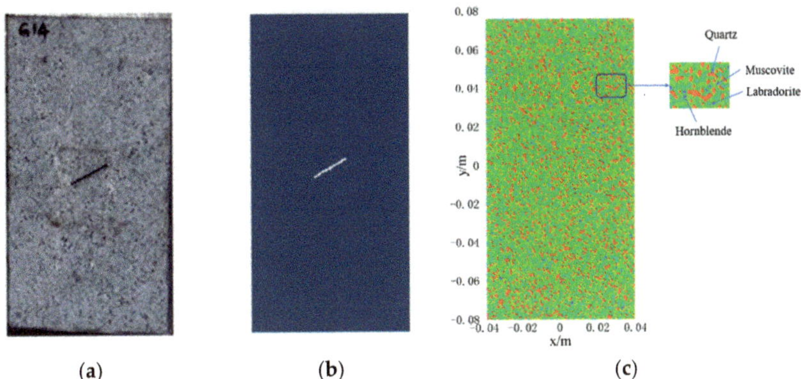

Figure 15. (**a**) Granite samples containing prefabricated cracks; (**b**) PD model; (**c**) composition distribution of granite.

Table 4. Peridynamic, mechanical, and thermal parameters of the PD numerical model.

	Parameter	Value
PD parameters	Number of discrete points in the xy direction	100×200
	Material point spacing Δx (m)	0.00008
	Non-locality parameter m	3
	Heat transfer time step Δt^{TH} (s)	2×10^{-4}
Mechanical parameters [37]	Mechanical time step during single-axis compression Δt^{ME} (s)	5×10^{-8}
	Young's modulus E (GPa)	36
	Poisson's ratio ν	0.33
	Density ρ (kg/m^3)	2790
	Fracture energy G_0 (J/m^2)	50
Thermal parameters [37]	Thermal conductivity k_T (W $\cdot$ m^{-1}K^{-1})	3.5
	Specific heat capacity c_v (J $\cdot$ Kg^{-1}K^{-1})	900

Before the numerical simulation, the same geometric model as that for the granite specimen was established, as shown in Figure 15b. Due to the non-homogeneous properties of rock materials, the Weibull distribution is often introduced to describe the statistical distribution of the characteristic parameters [38], such as elastic modulus, Poisson's ratio, and thermal expansion coefficient in the PD simulation. However, the Weibull distribution does not accurately reflect the properties of granite due to the variety of mineral components and contents of rocks and their vastly different material properties. Therefore, this study consists of reconstructing the PD calculation model of the non-homogeneous granite with the non-uniform and discontinuous thermal expansion coefficients using the Knuth–Durstenfeld stochastic algorithm proposed by Yang et al. [17] (see Figure 15c). The proportions of mineral compositions and thermal expansion coefficients of granite materials are listed in Table 5.

Table 5. Proportions of mineral compositions and thermal expansion coefficients of granite materials.

Type of Mineral 1	Proportion (%)	Coefficient of Thermal Expansion (10^{-6}K^{-1})
Quartz	17.73	24.3
Muscovite	36.33	17.3
Labradorite	39.32	14.1
Hornblende (rock-forming mineral, type of amphibole)	6.62	8.7

After 1000 s of thermal loading, the temperature of the granite specimen increased from 293 K to 573 K. Subsequently, after 3200 s, the sample naturally cooled down to room temperature (293 K). The simulation of the whole heat treatment process is shown in Figure 16a. Due to the slow temperature rise at the warming stage, the temperature difference between the inside and outside of the granite is small, and the non-uniform thermal stress caused by the temperature gradient is minor. Therefore, only a tiny amount of discontinuous thermal cracks is generated inside the granite during the entire heating-up stage. In addition, the higher compression strength of the granite also suppresses the crack generation at the warming stage. Unlike the warming stage, the outer surface of the granite decays sharply to room temperature during the natural cooling stage. This drastic heat transfer behavior leads to the formation of a vast temperature gradient inside and outside the granite, which provokes a rapid increase of the tensile (thermal) stress on the surface of the specimen under tensile strength, causing more discontinuous cracks to occur on both sides of the sample. These cracks continue to expand in the course of the cooling process and then gradually penetrate and fall off. Moreover, due to the inconsistency between the thermal expansion coefficients of different mineral components inside the specimen, there are more and more cracks induced by uneven thermal expansion.

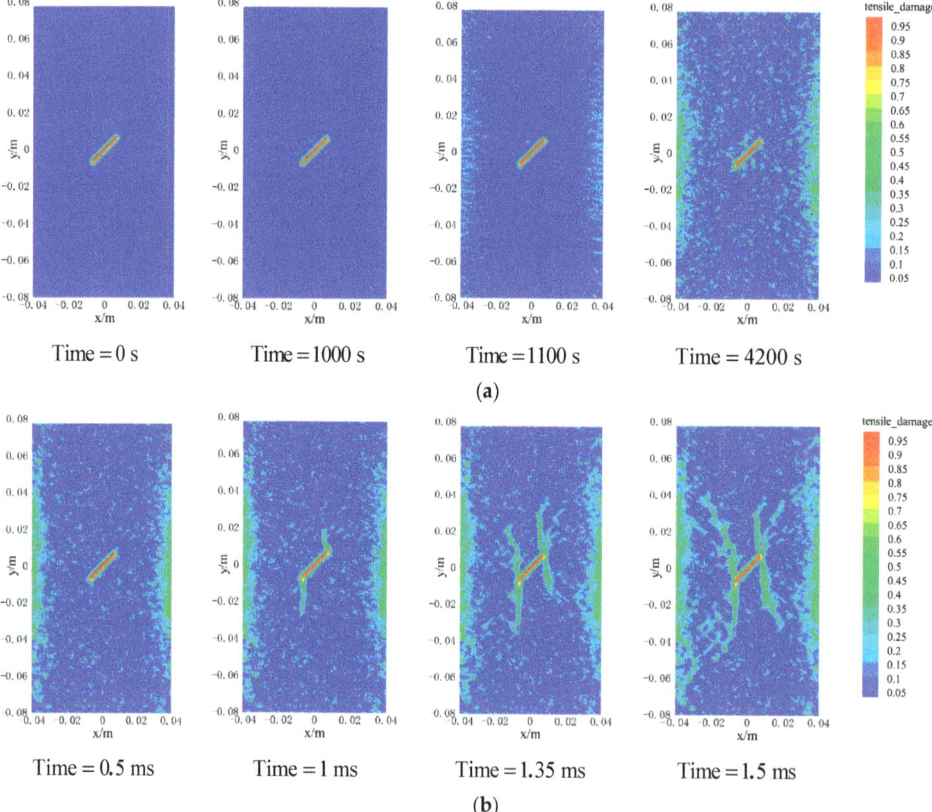

Figure 16. PD simulation of damage in granite under uniaxial compression after thermal cycling. (**a**) Thermal cycle stage; (**b**) Uniaxial compression stage.

The crack initiation and propagation process of the granite specimens containing prefabricated cracks under uniaxial compression simulated through the PD model is shown in Figure 16b. In a future study, the crack types will be analyzed according to the classification of crack types proposed by Yang et al. [38]. At the initial stage of loading, the main strain concentrations are found from the tips of the pre-existing fissure; the granite specimens had no macroscopic crack generation except for a large number of thermal micro-cracks on both sides, due to the thermal cycling process. With the increase of load, when the time reached 1 ms, the main strain concentrations develop obviously, the secondary tensile crack appeared at both the upper and lower ends of the prefabricated crack of the specimen, but the development of the secondary tensile crack was not symmetrical due to the uneven distribution of thermal micro-cracks inside the model. Subsequently, at 1.35 ms, a downward expanding tensile wing crack appeared at the upper end of the precast crack, while an upward expanding anti-shear crack appeared at the lower end of the precast crack. At the same time, thermal micro-cracks can be observed developing into secondary tensile cracks on both sides of the prefabricated cracks that afterward form web-like cracks. The macroscopic cracks in the final granite specimens were classified as secondary tensile cracks and anti-shear cracks.

The comparison between the experimental and PD simulation results of uniaxial compression after thermal cycling of precast cracked granite is shown in Figure 17. Secondary tensile cracks and anti-shear cracks existed at the tips of both sides of the precast cracks and were approximately the same in the form of extension. However, both tensile wing cracks and anti-shear cracks exist on the right side of the precast crack in the experimental results [37], while only anti-shear cracks exist in the simulation process. This is caused by the existence of tiny voids inside the granite, which is a non-homogeneous material, and the presence of a large number of non-uniformly distributed thermal micro-cracks during the thermal cycling process, which prevent the anti-tensile cracks that should appear along the axial stress direction; hence, only the anti- shear cracks mainly caused by shear damage appeared. The inconsistency of crack forms on both sides in the experimental results also indicates the non-homogeneous nature of granite.

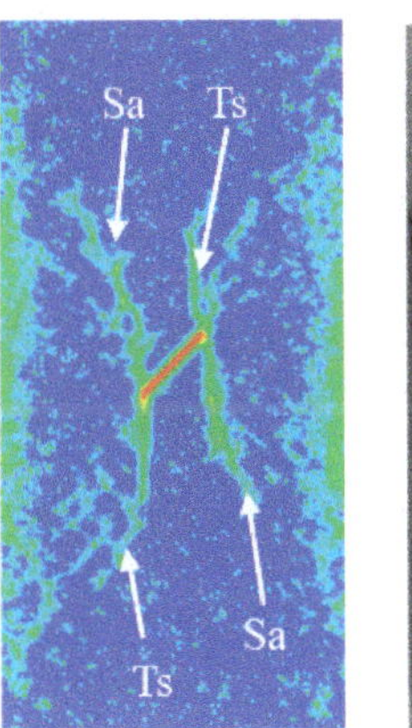

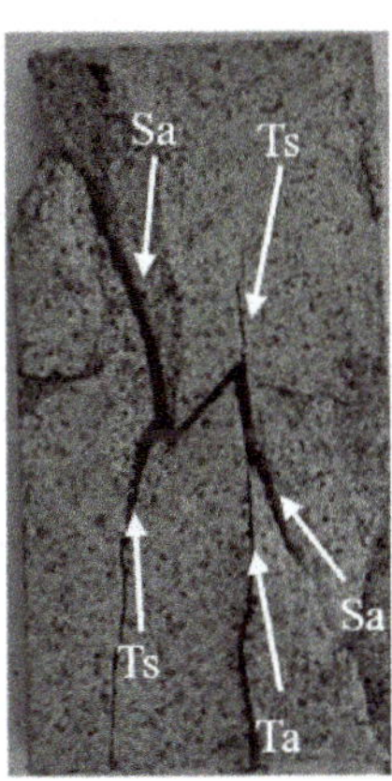

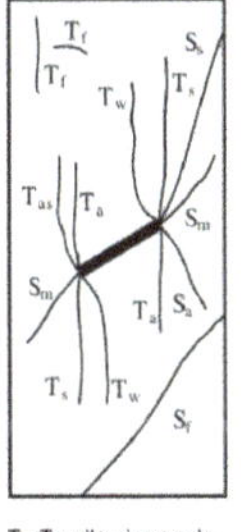

T_w=Tensile wing crack;
T_s=Secondary tensile crack;
S_s=Secondary shear crack;
S_m=Main shear crack;
T_a=Anti-tensile wing crack;
T_{as}= Anti-tensile secondary crack;
S_a=Anti-shear crack;
F_t=Far-field tensile crack;
F_s=Far-field shear crack;

Crack types, Yang et al, 2017

Figure 17. Comparison of PD simulated cracks extension with experiment [37,38].

Existing studies have shown that the deformation process of concrete is very complex due to its heterogeneity and involves progressive damage, such as the generation, propagation, and coalescence of microcracks [39]. While we have considered the parameters of the different components of the granite in an effort to construct a macroscopic heterogeneous model subjected to thermo-mechanical coupling loads, we neglected the mechanical characteristics at the microscopic scale. Through the references [40–44], it may be observed that the study of thermo-mechanical coupling under microstructures tends to focus on the mechanical properties of microstructures using the theory of nonlocal elasticity, by combining intermolecular or interatomic bonds into their specific intrinsic structural relationships. Further study in this area would increase the validity of the model.

5. Conclusions

In this paper, a coupled model capable of simulating the thermal-force damage behavior of quasi-brittle materials was proposed based on the bond-based PD theory using the fully thermodynamic coupling equation. The model consists in describing different mechanical properties of quasi-brittle materials in tensile and compressive states, constructing bond force functions in the tensile and compressive phases, and introducing the role of temperature terms in the bond-based peridynamic model. The simulations of the thermal expansion process of ceramics and the static compression damage of polycrystalline graphite were applied to the proposed model, and the numerical model results showed agreement with the experimental results in the references. In addition, the model was also used to simulate thermal damage processes in ceramics and in homogeneous rocks, revealing the potential capacity of the model in analyzing the post-thermal damage behavior of quasi-brittle materials.

Author Contributions: Conceptualization, L.L. and H.Z.; methodology, L.L.; software, H.Z.; validation, H.Z., L.L. and X.L. (Xin Lai); formal analysis, H.Z.; investigation, H.Z.; resources, H.Z.; data curation, H.Z.; writing—original draft preparation, H.M. and X.L. (Xin Lai); writing—review and editing, H.Z., X.L. (Xin Lai) and X.L. (Xiang Liu); visualization, H.Z.; supervision, L.L.; project administration, L.L.; funding acquisition, L.L. All authors have read and agreed to the published version of the manuscript.

Funding: This work was supported by the National Natural Science Foundation of China (Nos. 11972267, 11802214, 12102313 and 51932006) and the Fundamental Research Funds for the Central Universities (WUT:223114010).

Institutional Review Board Statement: Not applicable.

Informed Consent Statement: Not applicable.

Data Availability Statement: The data used to support the findings of this study are available upon request from the corresponding author.

Conflicts of Interest: The authors declare no conflict of interest.

References

1. Gautam, P.K.; Verma, A.K.; Jha, M.K. Effect of high temperature on physical and mechanical properties of Jalore granite. *J. Appl. Geophys.* **2018**, *159*, 460–474. [CrossRef]
2. Xie, K.; Jiang, X.; Jiang, D. Change of crackling noise in granite by thermal damage: Monitoring nuclear waste deposits. *Am. Mineral.* **2019**, *104*, 1578–1584. [CrossRef]
3. Ripani, M.; Etse, G.; Vrech, S. Thermodynamic gradient-based poroplastic theory for concrete under high temperatures. *Int. J. Plast.* **2014**, *61*, 157–177. [CrossRef]
4. Fan, L.F.; Gao, J.W.; Wu, Z.J. An investigation of thermal effects on micro-properties of granite by X-ray CT technique. *Appl. Therm. Eng.* **2018**, *140*, 505–519. [CrossRef]
5. Li, B.Q.; Gonçalves da Silva Einstein, H. Laboratory hydraulic fracturing of granite: Acoustic emission observations and interpretation. *Eng. Fract. Mech.* **2019**, *209*, 200–220. [CrossRef]
6. Kumari WG, P.; Beaumont, D.M.; Ranjith, P.G. An experimental study on tensile characteristics of granite rocks exposed to different high-temperature treatments. *Geomech. Geophys. Geo-Energy Geo-Resour.* **2019**, *5*, 47–64. [CrossRef]
7. Chu, I.; Lee, Y.; Amin, M.N. Application of a thermal stress device for the prediction of stresses due to hydration heat in mass concrete structure. *Constr. Build. Mater.* **2013**, *45*, 192–198. [CrossRef]
8. Bou Jaoude, I.; Novakowski, K.; Kueper, B. Identifying and assessing key parameters controlling heat transport in discrete rock fractures. *Geothermics* **2018**, *75*, 93–104. [CrossRef]
9. Fu, Y.; Wang, Z.; Ren, F. Numerical model of thermo-mechanical coupling for the tensile failure process of brittle materials. *AIP Adv.* **2017**, *7*, 105023. [CrossRef]
10. Tang, S.B.; Tang, C.A. Crack propagation and coalescence in quasi-brittle materials at high temperatures. *Eng. Fract. Mech.* **2015**, *134*, 404–432. [CrossRef]
11. Jiang, W.; Spencer, B.W.; Dolbow, J.E. Ceramic nuclear fuel fracture modeling with the extended finite element method. *Eng. Fract. Mech.* **2020**, *223*, 106713. [CrossRef]
12. Kwon, S.; Cho, W.J. The influence of an excavation damaged zone on the thermal-mechanical and hydro-mechanical behaviors of an underground excavation. *Eng. Geol.* **2008**, *101*, 110–123. [CrossRef]

13. Silling, S.A. Reformulation of elasticity theory for discontinuities and long-range forces. *J. Mech. Phys. Solids* **2000**, *48*, 175–209. [CrossRef]
14. Chen, W.; Gu, X.; Zhang, Q. A refined thermo-mechanical fully coupled peridynamics with application to concrete cracking. *Eng. Fract. Mech.* **2021**, *242*, 107463. [CrossRef]
15. Shou, Y.; Zhou, X. A coupled thermomechanical nonordinary state-based peridynamics for thermally induced cracking of rocks. *Fatigue Fract. Eng. Mater. Struct.* **2020**, *43*, 371–386. [CrossRef]
16. Bazazzadeh, S.; Mossaiby, F.; Shojaei, A. An adaptive thermo-mechanical peridynamic model for fracture analysis in ceramics. *Eng. Fract. Mech.* **2020**, *223*, 106708. [CrossRef]
17. Yang, Z.; Yang, S.Q.; Chen, M. Peridynamic simulation on fracture mechanical behavior of granite containing a single fissure after thermal cycling treatment. *Comput. Geotech.* **2020**, *120*, 103414. [CrossRef]
18. Chu, B.; Liu, Q.; Liu, L. A rate-dependent peridynamic model for the dynamic behavior of ceramic materials. *Comput. Modeling Eng. Sci.* **2020**, *124*, 151–178. [CrossRef]
19. Liu, Y.; Liu, L.; Mei, H. A modified rate-dependent peridynamic model with rotation effect for dynamic mechanical behavior of ceramic materials. *Comput. Methods Appl. Mech. Eng.* **2022**, *388*, 114246. [CrossRef]
20. Wang, Y.; Zhou, X.; Zhang, T. Size effect of thermal shock crack patterns in ceramics, Insights from a nonlocal numerical approach. *Mech. Mater.* **2019**, *137*, 103133. [CrossRef]
21. Silling, S.A.; Epton, M.; Weckner, O. Peridynamic states and constitutive modeling. *J. Elast.* **2007**, *88*, 151–184. [CrossRef]
22. Nowinski, J.L. *Theory of Thermoelasticity with Applications*; Sijthoff & Noordhoff International Publishers: Alphen aan den Rijn, The Netherlands, 1978.
23. Oterkus, S.; Madenci, E. Peridynamics for fully coupled thermomechanical analysis of fiber reinforced laminates. In Proceedings of the 55th AIAA/ASME/ASCE/AHS/ASC Structures, Structural Dynamics, and Materials Conference, National Harbor, MD, USA, 13–17 January 2014; American Institute of Aeronautics and Astronautics: Reston, VA, USA, 2014.
24. Oterkus, S.; Madenci, E.; Agwai, A. Fully coupled peridynamic thermomechanics. *J. Mech. Phys. Solids* **2014**, *64*, 1–23. [CrossRef]
25. Oterkus, S. *Peridynamics for the Solution of Multiphysics Problems*; The University of Arizona: Tucson, AZ, USA, 2015.
26. Lankford, J., Jr.; Anderson, C.E.; Nagy, A.J.; Walker, J.D. Inelastic response of confined aluminium oxide under dynamic loading conditions. *J. Mater. Sci.* **1998**, *33*, 1619–1625. [CrossRef]
27. Wade, J.; Robertson, S.; Wu, H. Plastic deformation of polycrystalline alumina introduced by scaled-down drop-weight impacts. *Mater. Lett.* **2016**, *175*, 143–147. [CrossRef]
28. Bhattacharya, M.; Dalui, S.; Dey, N.; Bysakh, S. Kumar Mukhopadhyay A. Low strain rate compressive failure mechanism of coarse grain alumina. *Ceram. Int.* **2016**, *42*, 9875–9886. [CrossRef]
29. Nguyen, T.T.; Thai, H.T.; Ngo, T. Optimised mix design and elastic modulus prediction of ultra-high strength concrete. *Constr. Build. Mater.* **2021**, *302*, 124150. [CrossRef]
30. Johnson, G.R.; Holmquist, T.J. An improved computational constitutive model for brittle materials. *Am. Inst. Phys.* **1994**, *309*, 981–984.
31. Belytschko, T.; Hughes, T.J. Computational method for transient analysis. *Amsterdam* **1986**, *1*, 245–263. [CrossRef]
32. Timoshenko, S.P.; Goodier, J.N. *Theory of Elasticity*; Mcgraw-Hill: New York, NY, USA, 1970.
33. Carslaw, H.S.; Jaeger, J.C. *Conduction of Heat in Solids*; Clarendon Press: Oxford, UK, 1959.
34. Ayatollahi, M.R.; Berto, F.; Lazzarin, P. Mixed mode brittle fracture of sharp and blunt V-notches in polycrystalline graphite. *Carbon* **2011**, *49*, 2465–2474. [CrossRef]
35. Bobaru, F.; Yang, M.; Alves, L.F. Convergence, adaptive refinement, and scaling in 1D peridynamics. *Int. J. Numer. Methods Eng.* **2009**, *77*, 852–877. [CrossRef]
36. Jiang, C.P.; Wu, X.F.; Li, J. A study of the mechanism of formation and numerical simulations of crack patterns in ceramics subjected to thermal shock. *Acta Mater.* **2012**, *60*, 4540–4550. [CrossRef]
37. Yang, S.Q.; Huang, Y.H.; Tian, W.L. Effect of High Temperature on deformation failure behavior of granite specimen containing a single fissure under uniaxial compression. *Rock Mech. Rock Eng.* **2019**, *52*, 2087–2107. [CrossRef]
38. Yang, S.Q.; Huang, Y.H. An experimental study on deformation and failure mechanical behavior of granite containing a single fissure under different confining pressures. *Environ. Earth Sci.* **2017**, *76*, 1–22. [CrossRef]
39. Li, G.; Tang, C.A. statistical meso-damage mechanical method for modeling trans-scale progressive failure process of rock. *Int. J. Rock Mech. Min. Sci.* **2015**, *74*, 133–150. [CrossRef]
40. Kiani, K.; Wang, Q. Nonlocal magneto-thermo-vibro-elastic analysis of vertically aligned arrays of single-walled carbon nanotubes. *Eur. J. Mech. A/Solids* **2018**, *72*, 497–515. [CrossRef]
41. Kiani, K.; Pakdaman, H. Nonlocal vibrations and potential instability of monolayers from double-walled carbon nanotubes subjected to temperature gradients. *Int. J. Mech. Sci.* **2018**, *144*, 576–599. [CrossRef]
42. Khanchehgardan, A.; Shah, M.A.; Rezazadeh, G. Thermo-elastic damping in nano-beam resonators based on nonlocal theory. *Int. J. Eng.* **2012**, *26*, 1505–1514. [CrossRef]
43. Ansari, R.; Gholami, R. Size-dependent nonlinear vibrations of first-order shear deformable magneto-electro-thermo elastic nanoplates based on the nonlocal elasticity theory. *Int. J. Appl. Mech.* **2016**, *8*, 1–33. [CrossRef]
44. Liu, C.; Ke, L.L.; Wang, Y.S.; Yang, J. Thermo-electro-mechanical vibration of piezoelectric nanoplates based on the nonlocal theory. *Compos. Struct.* **2013**, *106*, 167–174. [CrossRef]

Article

Identification of Relatively Weak Areas of Planar Structures Based on Modal Strain Energy Decomposition Method

Dongwei Wang, Kaixuan Liang and Panxu Sun *

School of Civil Engineering, Zhengzhou University, Zhengzhou 450001, China
* Correspondence: panxusun@zzu.edu.cn

Abstract: Identifying relatively weak areas is of great significance for improving the seismic reliability of structures. In this paper, a modal strain energy decomposition method is proposed, which can realize the decoupling of the comprehensive modal strain energy of a planar structure into three basic modal strain energies. According to the decomposition results, the modal strain energy decomposition diagram and the modal strain energy cloud diagram can be drawn so as to realize the quantitative and visual analysis of the vibration modes. The method is independent of load cases and can identify relatively weak areas of a structure from the perspective of inherent characteristics. The comparison with the shaking table test results of the two-story shear wall shows that the modal strain energy decomposition method can effectively identify the type of the relatively weak area of a structure and locate the position of the relatively weak area. Finally, the 6-story shear wall is analyzed by the modal strain energy decomposition method, and the relatively weak areas under the first two vibration modes are identified.

Keywords: planar structures; vibration mode; modal strain energy; relatively weak areas; quantification and visualization

check for
updates

Citation: Wang, D.; Liang, K.; Sun, P. Identification of Relatively Weak Areas of Planar Structures Based on Modal Strain Energy Decomposition Method. *Materials* **2022**, *15*, 6391. https://doi.org/10.3390/ma15186391

Academic Editor: Francesco Fabbrocino

Received: 11 August 2022
Accepted: 12 September 2022
Published: 14 September 2022

Publisher's Note: MDPI stays neutral with regard to jurisdictional claims in published maps and institutional affiliations.

1. Introduction

The weak areas of a structure may be the to be damaged in an earthquake. Several examples show that, in earthquakes [1–3], the damage to a structure is mainly caused by the domino effect, which starts with localized damage in weak areas and eventually leads to overall damage or collapse. The method of effectively identifying weak areas and taking targeted strengthening measures is of great significance to the analysis and design of structures.

Generally speaking, the identification of structurally weak areas needs to be carried out under certain load conditions. For example: (1) based on the accumulation of a large number of engineering cases (including earthquake damage examples), important identification experience is formed [4]; (2) the structural details are checked by mechanical tests [5]; (3) through theoretical analysis, the parameter indices of key parts are rechecked (under specific load conditions) to identify weak areas [6]. However, different load cases will result in different identification results, and it is often difficult to perform a complete analysis of load cases.

According to the d'Alembert equation, the frequency and mode shape are inherent, characteristic parameters of a structure, related only to mass and stiffness, not to load cases. Some scholars have thus used frequency to identify the weak areas of a structure [7,8]. At present, modal eigenvalues have been widely used in structural damage identification [9,10], health monitoring [11,12], and dynamic analysis [13,14]. However, the frequency of a structure represents the distribution characteristics of the absolute stiffness and absolute mass of the structure. Structural analysis using eigenvalues usually only yields quantitative thresholds (such as the critical load, the strength of key interfaces of a structure, etc.), and it is not very effective for identifying relatively weak areas of the structure.

The vibration mode vector is made up of the distribution characteristics of the relative stiffness and relative mass of a structure, and it is an ideal index for analyzing the relative stiffness of a structure [15]. Guan et al. [16] analyzed the dynamic response results of the temporary middle wall of a tunnel and found that this area was easily damaged under impact loading. Using modal analysis, Mazanoglu and Kandemir-Mazanoglu [17] found that a frame structure with column cracks had local vibration modes, and that this area was vulnerable to damage by earthquakes. At present, the vibration mode participation mass coefficient method [18] and the animation observation method are mainly used to roughly identify vibration modes. However, it is difficult to use traditional methods to quantitatively analyze vibration modes. Thus, it is difficult to achieve the purpose of effectively identifying the relatively weak areas of a structure.

Hence, a modal strain energy decomposition method is proposed in this paper. The proposed method does not involve load cases, and it can analyze a structure from the perspective of inherent characteristics. It can be applied to the vibration mode analysis of a structure and decompose the comprehensive modal strain energy so as to realize the quantitative visualization analysis of the vibration mode and the identification of the weak areas. The body of this paper is divided into five main sections. Section 2 presents the modal strain energy decomposition method of a planar square element. Section 3 uses the modal strain energy decomposition method to draw the modal strain energy decomposition diagram and the modal strain energy cloud diagram of a two-story shear wall with openings, and then it compares that with the crack diagram of the shear wall in a shaking table test. Section 4 takes a 6-story shear wall with openings as an example to decompose the modal strain energy of the first two vibration modes. Section 5 concludes the presentation of the research on which this paper is based and proposes topics for future research.

2. Modal Strain Energy Decomposition of a Planar Square Element

The free vibration equation of an undamped planar structure with N nodes is

$$\mathbf{M} \cdot \frac{\mathrm{d}^2 u(t)}{\mathrm{d}t^2} + \mathbf{K} \cdot u(t) = 0 \tag{1}$$

where $\mathbf{M}$ and $\mathbf{K}$ are the mass matrix and the stiffness matrix of the structure, respectively; $u(t) = [x_1(t),\ \ y_1(t),\ \ x_2(t),\ \ y_2(t),\ \ \ldots,\ \ x_N(t),\ \ y_N(t)]$ is the nodal displacement vector, which contains $2N$ components, that is, the displacements of N nodes in the X and Y directions; $\frac{\mathrm{d}^2 u(t)}{\mathrm{d}t^2}$ is the nodal acceleration vector; and 0 is a $2N$-dimensional zero vector.

The corresponding eigenvector is expressed as

$$\mathbf{\Phi} = \begin{bmatrix} \phi_1 & \phi_2 & \cdots & \phi_N \end{bmatrix} (j = 1,\ 2,\ \ldots,\ N) \tag{2}$$

where ϕ_j is the j-th eigenvector.

Element k is extracted from the structure. The modal strain and modal stress of the j-th eigenvector satisfy

$$\varphi_{k,j}^{\mathrm{T}} B_k^{\mathrm{T}} D_k B_k \varphi_{k,j} = \sigma_x \varepsilon_x + \sigma_y \varepsilon_y + \tau_{xy} \gamma_{xy} \ (j = 1,\ 2,\ \ldots,\ N) \tag{3}$$

where $\varphi_{k,j}$ is the displacement vector of the element k for the j-th eigenvector; B_k and D_k are the strain matrix and elastic matrix of element k, respectively; σ_x, σ_y, and τ_{xy} are the X-direction modal normal stress, Y-direction modal normal stress, and modal shear stress at any point of the element, respectively; ε_x, ε_y, and γ_{xy} are the corresponding modal strains at this point.

The X-direction modal normal strain energy W_x can be calculated by

$$W_x = \frac{b}{2} \int_x \int_y \sigma_x \varepsilon_x dx dy \tag{4}$$

The Y-direction modal normal strain energy W_y can be calculated by

$$W_y = \frac{b}{2} \int_x \int_y \sigma_y \varepsilon_y dx dy \qquad (5)$$

The modal shear strain energy W_{xy} can be calculated by

$$W_{xy} = \frac{b}{2} \int_x \int_y \gamma_{xy} \tau_{xy} dx dy \qquad (6)$$

where b is the thickness of the element.

The main modal strain energy of the element can be obtained by comparing the values of W_x, W_y, and W_{xy}. To facilitate the analysis, each modal strain energy corresponds to a color (as shown in Table 1), and this operation is applied to all of the elements of the structure. Thus, the modal strain energy decomposition diagram of the entire structure can be obtained. In addition, a cloud diagram of specific modal strain energy can be drawn according to the magnitude of the energy.

Table 1. Colors corresponding to the modal strain energy.

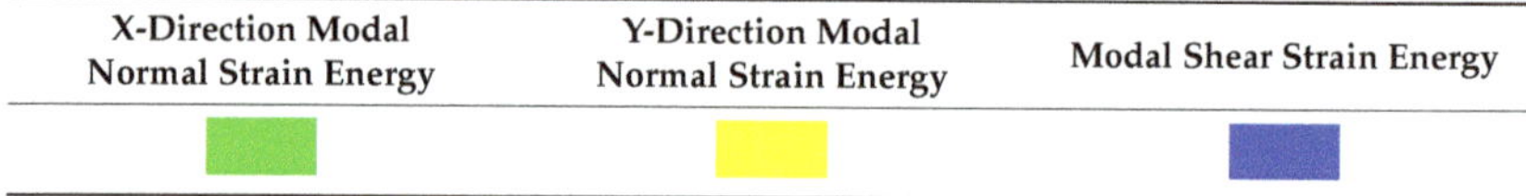

X-Direction Modal Normal Strain Energy	Y-Direction Modal Normal Strain Energy	Modal Shear Strain Energy

3. Experimental Verification

Two-story shear walls A1 and A2 with openings [19] are taken as examples. The dimensional parameters and reinforcement layout of A1 and A2 are shown in Figure 1. The design concrete strength grade of A1 was C30. The elastic modulus of the concrete was 24.43 GPa, and the density was 23.25 kN/m^3. The walls of A1 had a T-shaped cross-section and were 50 mm thick. The simplified boundary elements around the openings were 2Ø6 longitudinal bars, while the simplified boundary elements for the limbs were columns concealed with triangular stirrups, with 2Ø6 and 1Ø4 longitudinal bars placed [19,20]. The coupling beam and the first-story wall limbs of A2 were additionally provided with 75°-inclined steel bars, and the diameters of the steel bars were Ø6 and Ø4, respectively. The other parameters of A2 were the same as A1. The mechanical properties of the steel bars are shown in Table 2.

Table 2. Mechanical properties of steel bars [19].

Steel Reinforcement Diameter (mm)	Yield Strength (MPa)	Ultimate Strength (MPa)	Elastic Modulus (GPa)
Ø4	730.3	903.0	190.6
Ø6	394.5	578.3	220.7

An additional mass of 7.485 tons was applied to the top of A1 and A2 by using a load trough. The specific test setup is shown in Figure 2.

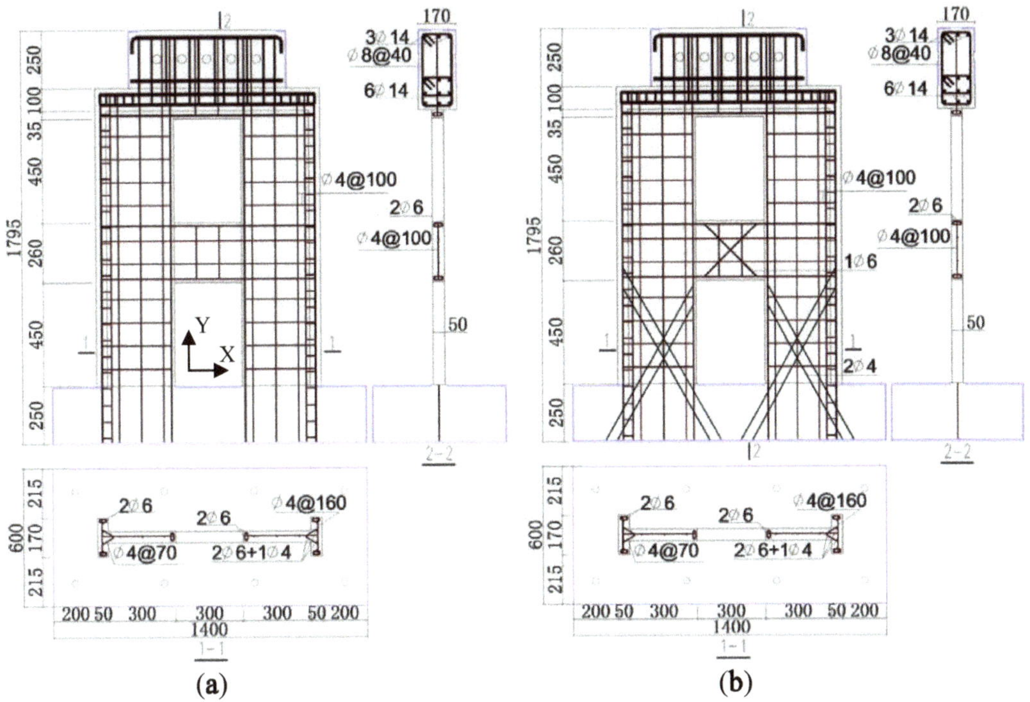

Figure 1. Dimensions and reinforcement of (**a**) A1; (**b**) A2 (unit: mm) [19].

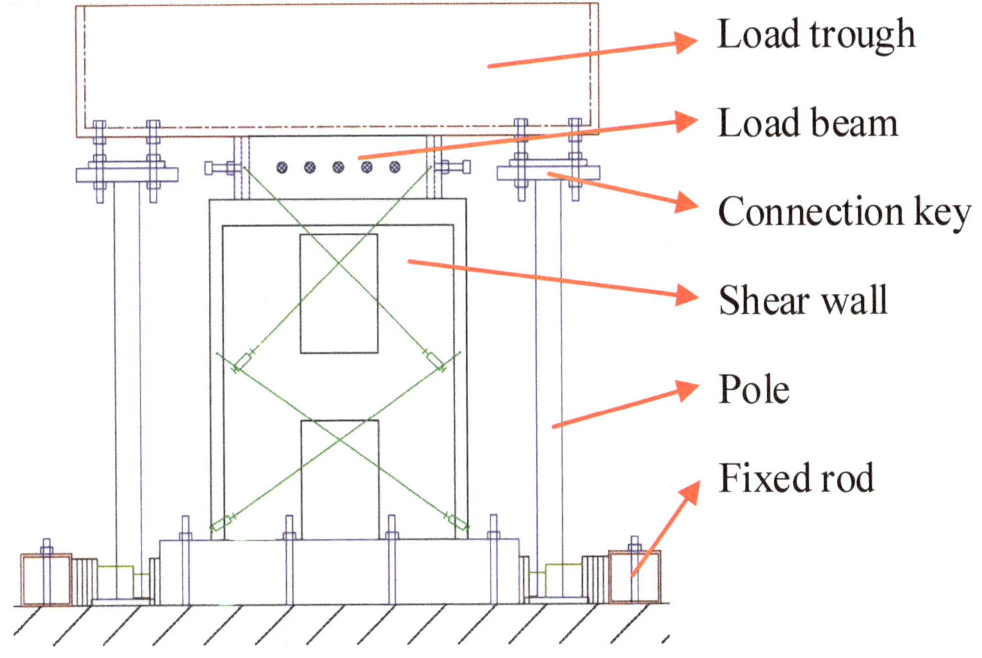

Figure 2. Schematic diagram of test setup [19].

In the shaking table test in reference [19], the El Centro (1940) N-S wave was used. The scale values of the time interval and duration of the earthquake motion were $0.02 \times 0.5 = 0.01$ s and $53.74 \times 0.5 = 26.87$ s, respectively. The time history acceleration and response spectrum of the scaled earthquake motion are shown in Figure 3.

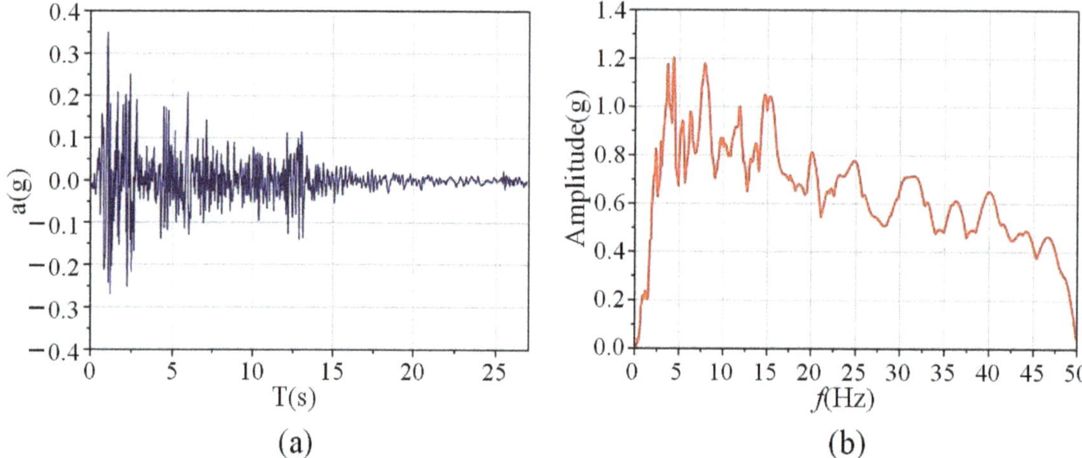

Figure 3. Inputted earthquake: (**a**) time history acceleration; (**b**) response spectra [19].

A total of 7 seismic hazard levels were selected and named T-1–T-7. The actual PGA obtained on the shaking table surface during the experiment is shown in Table 3. Using white noise, the initial natural frequencies of shear walls A1 and A2 were 8.25 and 8.30, respectively.

Table 3. Test procedure [19].

Shear Walls	PGA (g)						
	T-1	T-2	T-3	T-4	T-5	T-6	T-7
A1	0.189	0.388	0.524	0.615	0.815	1.155	1.219
A2	0.191	0.330	0.492	0.650	0.782	0.875	1.126

After the T-6 test, the crack diagrams and failure modes of A1 and A2 are shown in Figures 4 and 5, respectively.

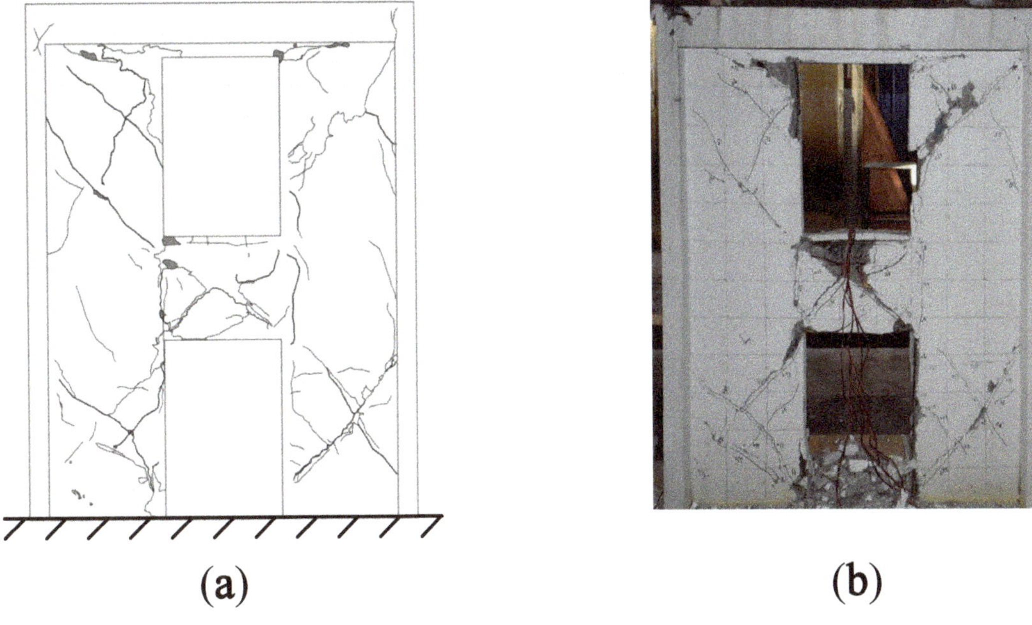

Figure 4. (**a**) crack diagram; (**b**) failure mode of A1 [19].

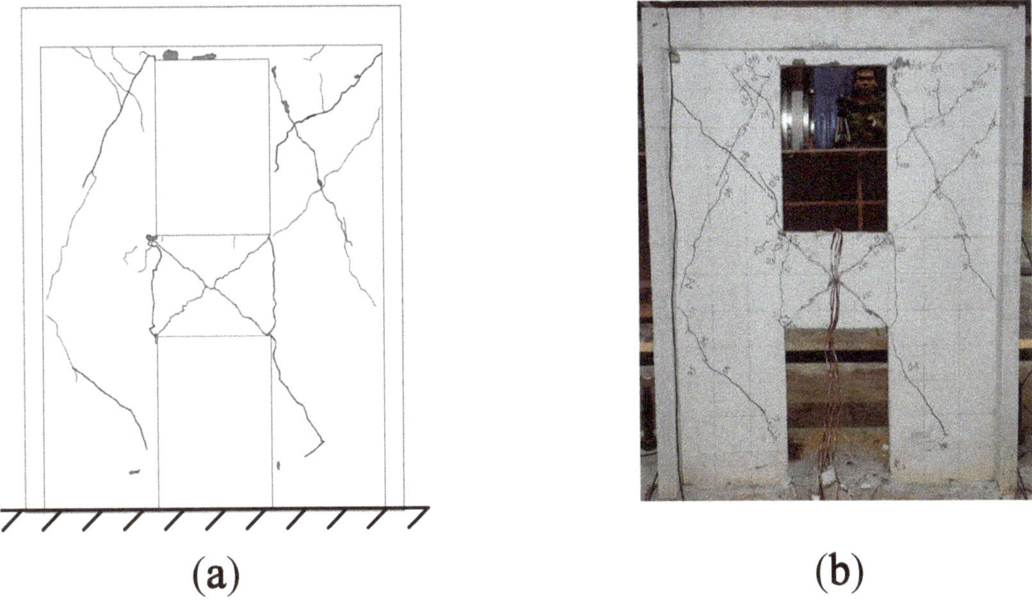

Figure 5. (**a**) crack diagram; (**b**) failure mode of A2 [19].

It can be seen from Figure 4 that vertical cracks appeared at the joint areas between the coupling beam and the wall limbs of A1, while the inclined shear cracks were mainly located in the middle of the coupling beam and the wall limbs on both sides of the openings. The middle part of the coupling beam and the corner of the openings were seriously damaged, and there was obvious concrete spalling. Horizontal cracks appeared at the bottom of the wall limb after the T-3 test, and the concrete at the corner of the lower wall limbs was crushed after the T-7 test.

It can be seen from Figure 5 that the locations and types of cracks in A2 were basically the same as those in A1. Due to the constraint effect of the inclined steel bars, the number of inclined cracks in A2 was fewer than that in A1, and the degree of concrete spalling was lesser.

According to the parameters given in reference [19], the finite element models of A1 and A2 were established by ANSYS software. The concrete was established with SOLID65 elements, and the steel bars were established with LINK8 elements. The modeling process used a combination of integral and individual modeling, and the additional mass at the top of the shear wall model was simulated by prestressing. The modal solution was carried out. The first-order natural frequencies of A1 and A2 were 8.54 and 8.66, respectively, which were less than 5% from the measured results. Modal strain energy decomposition was performed on the first-order vibration modes of A1 and A2, and the energy decomposition diagrams are shown in Figure 6.

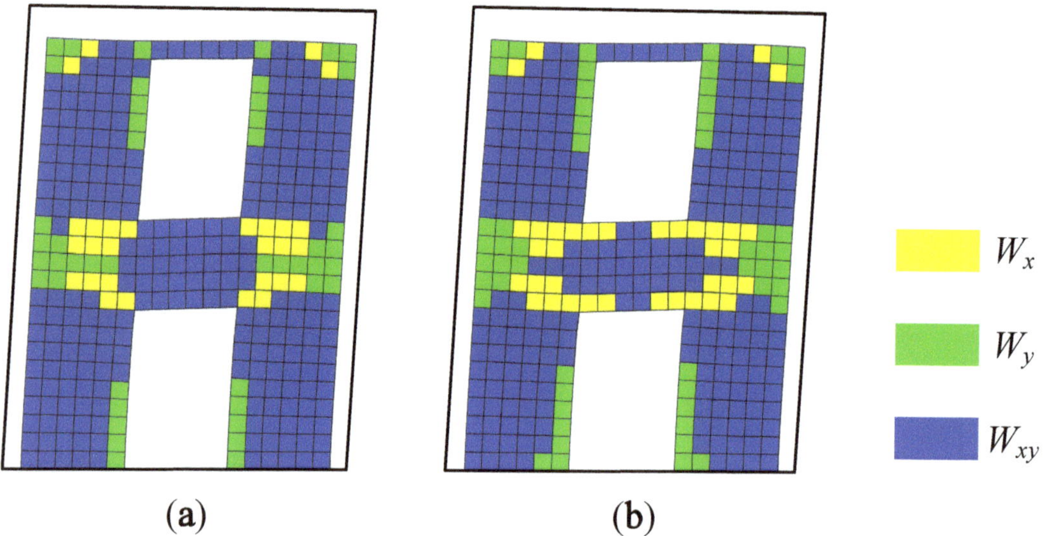

Figure 6. Modal strain energy decomposition diagrams of (**a**) A1; (**b**) A2.

The modal strain energy decomposition diagram was compared with the crack diagram. The yellow areas (dominated by the X-direction modal normal strain energy) in Figure 6 correspond to the vertical cracks in Figures 4 and 5, while the green areas (dominated by the Y-direction modal normal strain energy) correspond to horizontal cracks, and the blue areas (dominated by the modal shear strain energy) correspond to inclined cracks. The comparison results show that the cracking mode of the structure can be effectively obtained by using the modal strain energy decomposition diagram, which further proves the correctness of the modal strain energy decomposition method.

Strain, stress, and strain energy can be used to solve the problem of absolute stiffness, such as judging whether a structure will crack under external force and studying the development process of fracturing [21]. The vibration mode vector is the relative value of the nodal displacement. Therefore, the modal strain energy can be used to solve the relative stiffness problems of a structure, such as predicting the location of the first crack and the sequence of cracking so as to realize the purpose of identifying relatively weak areas of a structure. The maximum value of the X-direction modal normal strain energy is regarded as 1.00 J, and each modal strain energy is normalized separately. The modal strain energy cloud diagram of A1 and A2 can be drawn, as shown in Figures 7 and 8.

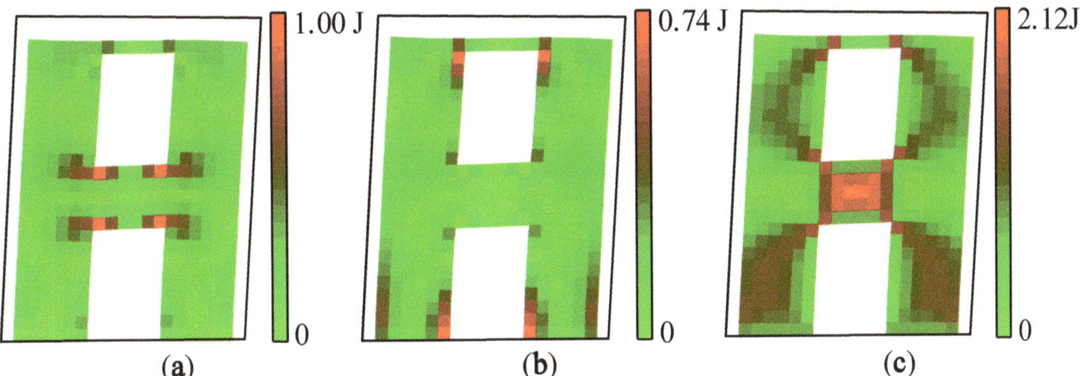

Figure 7. Modal strain energy cloud diagrams of A1: (**a**) X-direction modal normal strain energy, (**b**) Y-direction modal normal strain energy, and (**c**) modal shear strain energy.

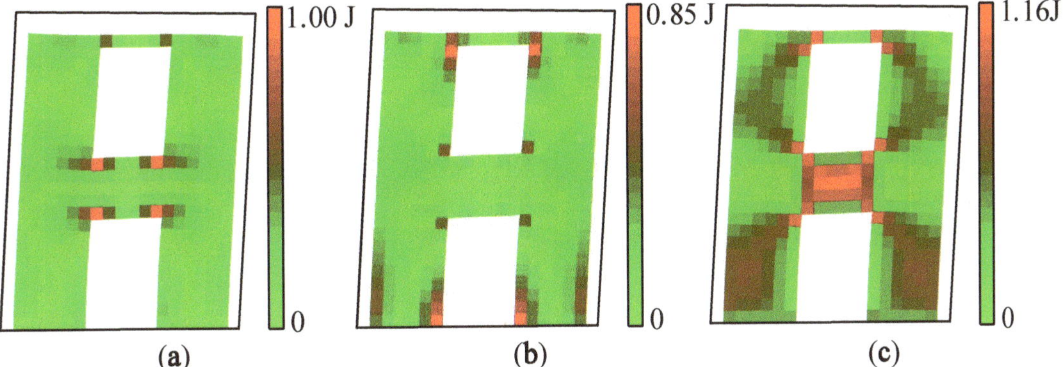

Figure 8. Modal strain energy cloud diagrams of A2: (**a**) X-direction modal normal strain energy, (**b**) Y-direction modal normal strain energy, and (**c**) modal shear strain energy.

It can be seen from Figures 7 and 8 that the modal strain energy cloud diagrams of A1 and A2 are similar. The maximum X-direction modal normal strain energy appears at the joint areas between the coupling beam and the wall limbs, and the maximum Y-direction modal normal strain energy appears at the lower wall limbs and the corner of the upper opening. The shear modal strain energy is mainly concentrated in the middle area of the coupling beam, and there is also a relatively obvious energy concentration in the middle of the wall limbs. It is worth noting that the relative value of the modal shear strain energy of A2 was significantly smaller than that of A1.

During the shaking table test, vertical cracks first appeared at the corner of the coupling beam of A1 and gradually increased, and then horizontal cracks appeared at the bottom of the wall limbs. When the PGA was further increased, the coupling beam began to crack obliquely, and the cracks of the wall limbs gradually developed into the inclined shear cracks. Finally, under a seismic excitation with a PGA of 1.219 g, the concrete of the coupling beam of A1 fell off, and the concrete at the corner of the lower wall limbs was crushed.

The cracking sequence of A2 is similar to that of A1. The first crack appeared in the coupling beam, and the wall limbs continued to crack. The damage in the coupling beam was more severe than that in the wall limbs. The difference is that the cracks of A1 were more concentrated in the lower part of the wall limbs, while the cracks of A2 were mainly concentrated in the upper part of the wall limbs, and the number of shear cracks was relatively small in general.

In conclusion, the modal strain energy decomposition diagram and modal strain energy cloud diagram are consistent with the cracking sequence of shear walls A1 and A2, which proves that the modal strain energy decomposition method is an effective way to identify relatively weak areas of a structure. In addition, the process of the modal strain energy decomposition method is in the elastic range and does not involve nonlinear analysis. It is simpler and more convenient than traditional elastic–plastic analytic methods.

4. Identification of Relatively Weak Areas of a Structure Based on Different Order Vibration Modes

Taking 6-story shear wall B with a thickness 0.5 m as an example, its plane size is shown in Figure 9. The elastic modulus of shear wall B is 31.5 GPa, and the density is 2250 kg/m^3. The finite element model of shear wall B is established with SOLID65 elements, and the number of elements used in the X- and Y-directions are 33 and 78, respectively. According to the principle that "the number of vibration modes can generally be taken as the number required for the participating mass of the vibration modes to reach 90% of the total mass", the modal strain energy decomposition was carried out for the first two transverse vibration modes of shear wall B.

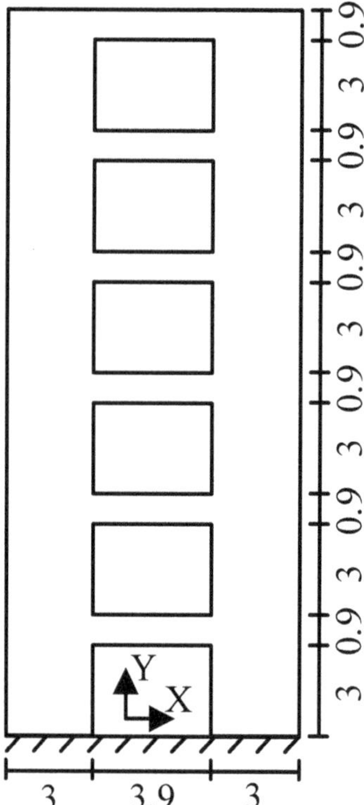

Figure 9. Dimension diagram of shear wall B (unit: m).

4.1. Identification Result Based on First-Order Vibration Mode

Modal strain energy decomposition was performed on the first-order vibration mode of shear wall B, and the obtained modal strain energy decomposition diagram is shown in Figure 10.

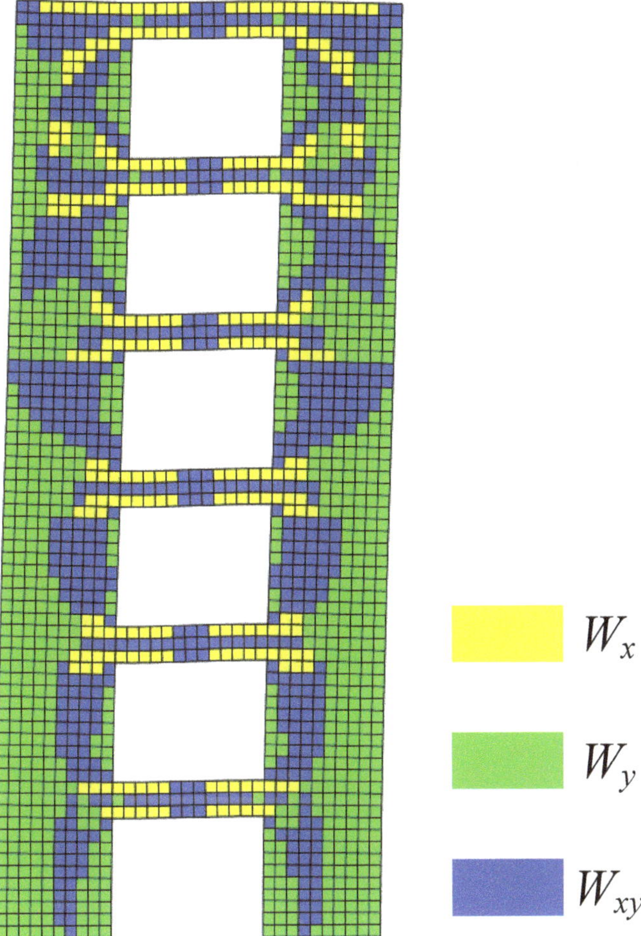

Figure 10. Modal strain energy decomposition diagram of first-order vibration mode.

It can be inferred from Figure 10 that the X-direction modal normal strain energy dominated the joint areas between the coupling beams and the wall limbs, while the Y-direction modal normal strain energy was the main energy of the outer side of the 1st–4th stories. In addition, the inner side of the wall limbs of the 1st–4th stories, the middle areas of the wall limbs of the 5th–6th stories, and the middle areas of the coupling beams were dominated by the modal shear strain energy, and these areas were prone to oblique cracking. In order to further identify the relatively weak areas of shear wall B, the modal strain energy cloud diagram, as shown in Figure 11, was drawn according to the decomposition results.

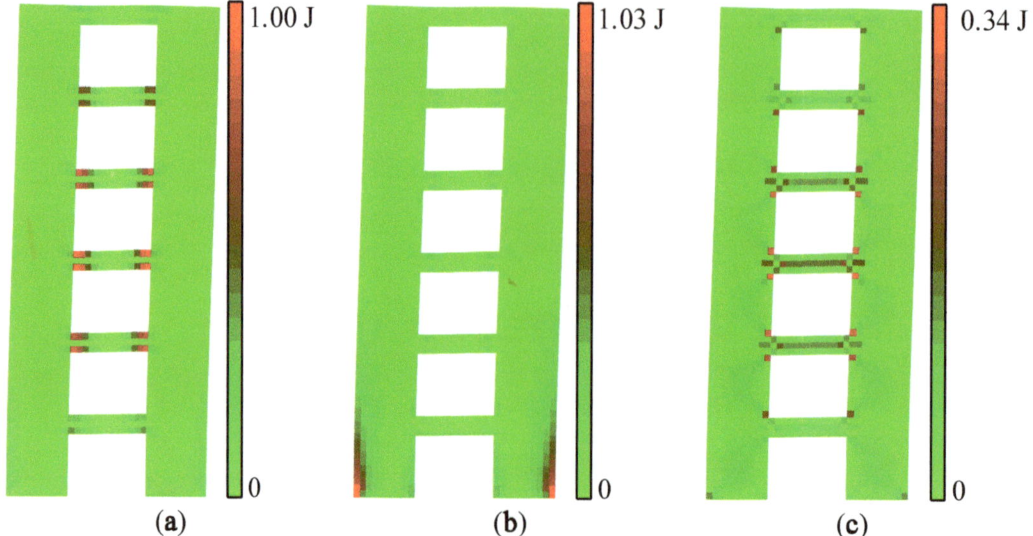

Figure 11. Modal strain energy cloud diagram of first-order vibration mode: (**a**) X-direction modal normal strain energy, (**b**) Y-direction modal normal strain energy, and (**c**) modal shear strain energy.

Figure 11 visualizes the concentration areas of different modal strain energies. It can be seen that the joint areas between the coupling beam and the wall limbs of the 2nd–4th stories are the relatively weak areas of X-direction tension and compression, and vertical cracks will first appear in these areas. The bottom of the wall limbs is a relatively weak area of Y-direction tension and compression; that is, the first horizontal crack will appear at the bottom of shear wall B. There is an obvious modal shear strain energy concentration in the middle of the coupling beams of the 2nd–4th stories, and there is also a certain degree of energy concentration in the middle of the wall limbs of the 1st–3rd stories. It can be inferred that the inclined cracks will first appear at the coupling beams of the 2nd–4th stories, and then gradually develop in the wall limbs.

4.2. Identification Results Based on Second-Order Vibration Mode

The first-order vibration mode is the most important vibration mode in a structural analysis. However, sometimes the structure is excited to a higher order vibration mode under the action of high-frequency earthquakes. The modal strain energy decomposition of the second-order mode shape of shear wall B was carried out, and the decomposition results are shown in Figure 12.

It can be seen from Figure 12 that the X-direction modal normal strain energy was dominant at both ends of the coupling beam of the B shear wall. The Y-direction modal normal strain energy was the main energy for the 3rd–4th-floor wall. The remaining areas were governed by the modal shear strain energy. According to the decomposition results, the modal strain energy cloud diagrams of shear wall B were drawn, as shown in Figure 13.

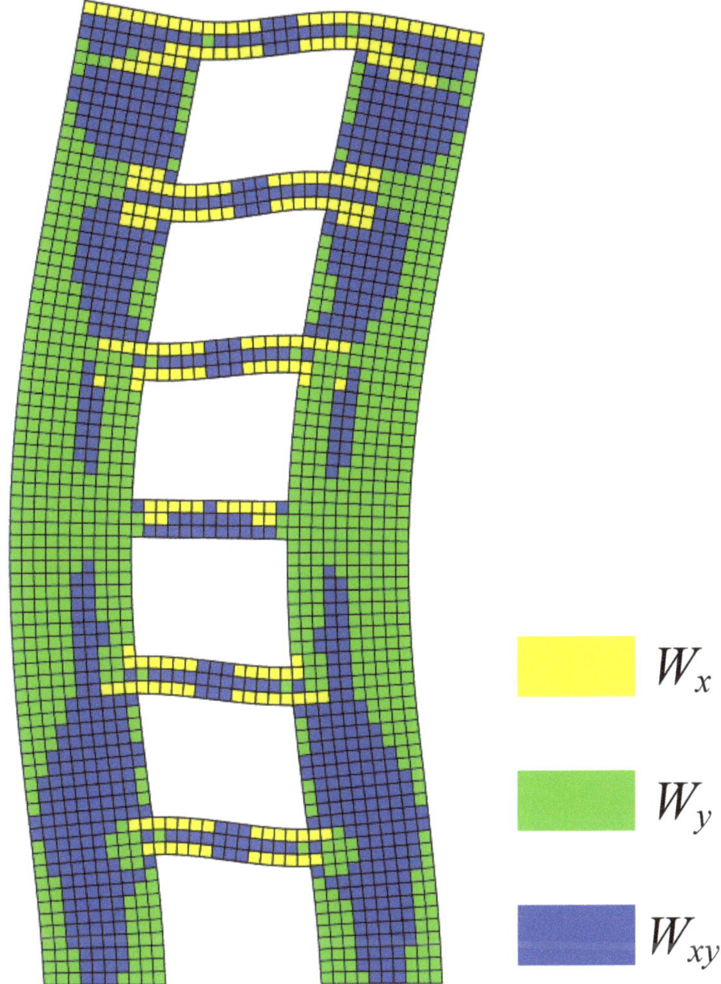

Figure 12. Modal strain energy cloud diagram of second-order vibration mode.

For the second-order vibration mode, it can be inferred from Figure 13 that vertical cracks will first appear at the joint areas between the coupling beam and the wall limbs of the 5th story. Horizontal cracks will first appear at the bottom of the wall limbs, and horizontal cracking is also prone to occur on the outer side of the 3rd–4th stories. The modal shear strain energy was mainly concentrated in the coupling beams of the 5th story, and there was also obvious energy concentration in the wall limbs of the 1st–2nd stories, which are prone to oblique shear cracks.

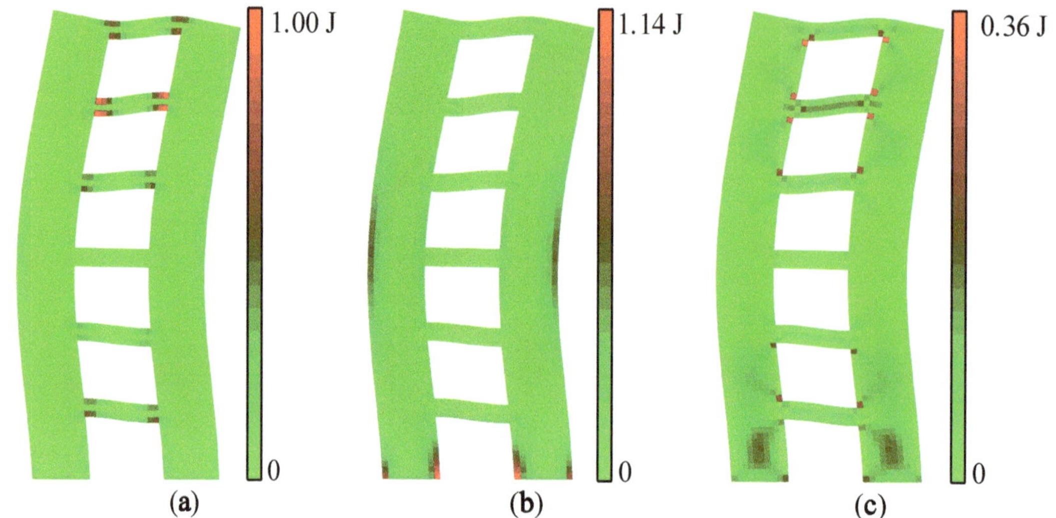

Figure 13. Modal strain energy cloud diagram of second-order vibration mode: (**a**) X-direction modal normal strain energy, (**b**) Y-direction modal normal strain energy, and (**c**) modal shear strain energy.

The relatively weak areas of shear wall B under the first two modes are shown in Table 4.

Table 4. Relatively weak areas of shear wall B under the first- and second-order vibration modes.

Order	Tension and Compression		Shear
	X-Direction	**Y-Direction**	
1st	joint areas between coupling beam and wall limbs of the 2nd–4th stories	bottom of the wall limbs	coupling beams of the 2nd–4th stories, and the wall limbs of the 1st–3rd stories
2nd	joint areas between coupling beam and wall limbs of the 5th story	bottom of the wall limbs, and outer side of the 3rd–4th stories	coupling beams of the 5th story, and the wall limbs of the 1st–2nd stories

In summary, the modal strain energy decomposition diagram and modal strain energy cloud diagram can intuitively and effectively locate the relatively weak areas of the shear wall structure, predict which parts are prone to cracking, and evaluate the cracking type, which is beneficial to the targeted reinforcement of the structure.

5. Conclusions

A modal strain energy decomposition method of planar square elements is proposed and applied to a finite-element analysis of structures. Based on this method, the modal strain energy of a structure can be decomposed, and the quantitative basic modal strain energy information can be obtained. The following conclusions can be drawn:

1. By decomposing the comprehensive modal strain energy of the planar structure into X- and Y-direction modal normal strain energy and modal shear strain energy, the quantitative visual analysis of the vibration mode can be realized.
2. The modal strain energy decomposition diagram can visualize the areas dominated by different modal strain energies, and the cracking mode of a structure can be predicted. The modal strain energy cloud diagram can quantify the concentration degree of different modal strain energies, thereby predicting which areas are prone to cracking and evaluating the cracking type.

3. The modal strain energy decomposition method does involve load condition information and can effectively identify the relatively weak areas of a structure and guide the implementation of targeted reinforcement.

There are still some limitations to this study. In our follow-up work, a strain energy analysis method in the elastic–plastic stage is being developed to realize the real-time evaluation of the deformation state of plastic structures.

Author Contributions: Conceptualization, D.W. and P.S.; methodology, D.W. and P.S.; software, K.L.; data curation, K.L.; formal analysis, P.S.; writing—original draft, K.L.; writing—review and editing, D.W. and P.S.; visualization, K.L.; supervision, P.S.; funding acquisition, D.W. and P.S. All authors have read and agreed to the published version of the manuscript.

Funding: This research was funded by the National Natural Science Foundation of China (Grant No. 51878621), the Natural Science Foundation of Henan Province (Grant No. 222300420316), the China Postdoctoral Science Foundation (Grant No. 2022M712905), and the Key Research Project of Henan Higher Education Institutions (Grant No. 22A560005).

Institutional Review Board Statement: Not applicable.

Informed Consent Statement: Not applicable.

Data Availability Statement: Some or all data that support the findings of this study are available from the corresponding author upon reasonable request.

Conflicts of Interest: The authors declare no conflict of interest.

References

1. Seyedkhoei, A.; Akbari, R.; Maalek, S. Earthquake-Induced Domino-Type Progressive Collapse in Regular, Semiregular, and Irregular Bridges. *Shock Vib.* **2019**, *2019*, 8348596. [CrossRef]
2. Göçer, C. Structural Evaluation of Masonry Building Damages during the April 24, 2014 Gökçeada Earthquake in the Aegean Sea. *Bull. Earthq. Eng.* **2020**, *18*, 3459–3483. [CrossRef]
3. Nale, M.; Minghini, F.; Chiozzi, A.; Tralli, A. Fragility Functions for Local Failure Mechanisms in Unreinforced Masonry Buildings. *Bull. Earthq. Eng.* **2021**, *2021*, 6049–6079. [CrossRef]
4. Cimmino, M.; Magliulo, G.; Manfredi, G. Seismic Collapse Assessment of New European Single-Story Rc Precast Buildings with Weak Connections. *Bull. Earthq. Eng.* **2020**, *18*, 6661–6686. [CrossRef]
5. Gong, M.S.; Zuo, Z.X.; Wang, X.M.; Lu, X.T.; Xie, L.L. Comparing Seismic Performances of Pilotis and Bare Rc Frame Structures by Shaking Table Tests. *Eng. Struct.* **2019**, *199*, 109442. [CrossRef]
6. Li, S.; Tian, J.B.; Liu, Y.H. Performance-Based Seismic Design of Eccentrically Braced Steel Frames Using Target Drift and Failure Mode. *Earthq. Struct.* **2017**, *13*, 443–454.
7. Guo, T.N.; Meng, L.J.; Zhou, C.; Hua, X. Research on the Dynamic Hammering Method for Identifying Weak Parts in Cantilever Structures. *Sci. Prog.* **2021**, *104*, 1–22. [CrossRef]
8. Zhang, A.L.; Xie, Z.Q.; Zhang, Y.X.; Lin, H.P. Shaking Table Test of a Prefabricated Steel Frame Structure with All-Bolted Connections. *Eng. Struct.* **2021**, *248*, 113273. [CrossRef]
9. Sheng, E.F.; Ji, Y.H. Statics-Based Model-Free Damage Detection Under Uncertainties Using Modal Interval Analysis. *Materials* **2020**, *13*, 1567.
10. Deng, T.F.; Huang, J.W.; Cao, M.S.; Li, D.Y.; Bayat, M. Seismic Damage Identification Method for Curved Beam Bridges Based on Wavelet Packet Norm Entropy. *Sensors* **2022**, *22*, 239. [CrossRef]
11. Mccrory, J.P.; Pearson, M.R.; Pullin, R.; Holford, K.M. Optimisation of Acoustic Emission Wavestreaming for Structural Health Monitoring. *Struct. Health Monit.* **2020**, *19*, 1475–9217. [CrossRef]
12. Cao, P.Y.; Niu, K.M.; Sun, J.H.; Zhao, S.L.; Liu, Y.P. Reordering of Vertical Vibration Modes in an Axially Compressed Beam on a Winkler Foundation. *J. Sound Vib.* **2022**, *526*, 116841. [CrossRef]
13. Cong, P.H.; Van Thom, D.; Duc, D.H. Phase Field Model for Fracture Based on Modified Couple Stress. *Eng. Fract. Mech.* **2022**, *269*, 108534. [CrossRef]
14. Wang, Z.Y.; Shi, Y.Z.; You, X.; Jiang, R.J.; Gai, W.M. Dynamic Behaviour of Bridge Girders with Trapezoidal Profiled Webs Subjected to Moving Loads. *Materials* **2021**, *14*, 38. [CrossRef]
15. Treyssède, F. Finite Element Modeling of Temperature Load Effects on the Vibration of Local Modes in Multi-Cable Structures. *J. Sound Vib.* **2018**, *413*, 191–204. [CrossRef]
16. Guan, X.M.; Yang, N.; Zhang, W.J.; Li, M.G.; Liu, Z.L.; Wang, X.H.; Zhang, S.L. Vibration Response and Failure Modes Analysis of the Temporary Support Structure Under Blasting Excavation of Tunnels. *Eng. Fail. Anal.* **2022**, *136*, 106188. [CrossRef]

17. Mazanoglu, K.; Kandemir-Mazanoglu, E.C. A New Simplified Method for Anti-Symmetric Mode in-Plane Vibrations of Frame Structures with Column Cracks. *J. Vib. Control* **2018**, *24*, 5794–5810. [CrossRef]
18. Ibrahimbegovic, A.; Wilson, E.L. Automated Truncation of Ritz Vector Basis in Modal Transformation. *J. Eng. Mech.* **1990**, *116*, 2506–2520. [CrossRef]
19. Zhang, J.W.; Zheng, W.B.; Yu, C.; Cao, W.L. Shaking Table Test of Reinforced Concrete Coupled Shear Walls with Single Layer of Web Reinforcement and Inclined Steel Bars. *Adv. Struct. Eng.* **2018**, *21*, 2282–2298. [CrossRef]
20. Zhang, J.W.; Cao, W.L.; Yin, W.S. Study on Seismic Performance of Mid-Rise RC Shear Wall with Single Row of Steel Bars and Simplified Boundary Elements. *Chin. Civ. Eng. J.* **2009**, *42*, 99–104. (In Chinese)
21. Tho, N.C.; Ta, N.T.; Thom, D.V. New Numerical Results from Simulations of Beams and Space Frame Systems with a Tuned Mass Damper. *Materials* **2019**, *12*, 1329. [CrossRef]

 *materials*

Article

Numerical Investigation of Prefabricated Utility Tunnels Composed of Composite Slabs with Spiral Stirrup-Constrained Connection Based on Damage Mechanics

Qinghua Wang [1,2], Guobin Gong [3,*], Jianli Hao [3] and Yuanfeng Bao [4]

[1] Department of Structural Engineering, Tongji University, Shanghai 200092, China
[2] School of Architectural Engineering, Nantong Vocational University, Nantong 226007, China
[3] Department of Civil Engineering, Xi'an Jiaotong-Liverpool University, Suzhou 215123, China
[4] Suzhou Integrated Infrastructure Technology Research Institute (SITRI), Suzhou 215000, China
* Correspondence: guobin.gong@xjtlu.edu.cn

Abstract: This paper investigates prefabricated utility tunnels composed of composite slabs with a spiral stirrup-constrained connection, considering material nonlinearity with concrete damage. An experiment was set up based on the prototype of a practical utility tunnel project, and the results were compared with finite element (FEM) simulation results with reasonable agreement obtained. The parametric analysis was carried out considering variations of seam location, haunch height and reinforcement, and embedment depth, using FEM simulations. It is found that, as with the increase in seam distance above haunch, the load capacity increases slightly, while the ductility does not vary much. The haunch height is not found to have an apparent effect on stiffness, load capacity or ductility. The increase in the embedment depth can enhance both the yield and peak loads while decreasing the ductility. A simplified method is proposed for evaluating the seismic performance in terms of deformation coefficient considering ductility demand, based on three different methods for calculating interaction coefficients considering soil–structure interactions. The findings from this investigation provide theoretical and practical guidance for underground engineering design of prefabricated utility tunnels.

Keywords: precast utility tunnel; concrete damage; spiral stirrup; finite element; ductility

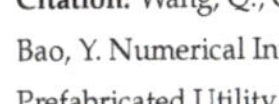 check for updates

Citation: Wang, Q.; Gong, G.; Hao, J.; Bao, Y. Numerical Investigation of Prefabricated Utility Tunnels Composed of Composite Slabs with Spiral Stirrup-Constrained Connection Based on Damage Mechanics. *Materials* **2022**, *15*, 6320. https://doi.org/10.3390/ma15186320

Academic Editors: Dan Huang, Lisheng Liu, Zhanqi Cheng and Liwei Wu

Received: 25 August 2022
Accepted: 8 September 2022
Published: 12 September 2022

Publisher's Note: MDPI stays neutral with regard to jurisdictional claims in published maps and institutional affiliations.

1. Introduction

Underground utility tunnels integrate various engineering pipelines, such as electricity, communication, gas, heating, water supply and drainage, which have special inspection and hoisting ports to implement unified planning and management [1–3]. These tunnels have become critical infrastructure to ensure the smooth operation of modern cities. In recent years, with the development of China's economy and process of urbanization, the construction of utility tunnels has developed rapidly throughout the country [4,5].

According to the fabrication materials, utility tunnels can be divided into those built of concrete and those built of steel [6]. Concrete utility tunnels are more popular and widely used, and they are either cast-in-place (CIP) or prefabricated tunnels. The main components of a prefabricated utility tunnel are cast in a factory and assembled onsite to create a whole utility tunnel. The construction period for prefabricated utility tunnels is relatively short, with good and easy quality control and remarkable environmental protection and energy saving [7,8]. Prefabricated concrete utility tunnels, which are the trend in China, are mainly of three types according to the assemblage methods: integrated prefabricated utility tunnels (IPUT), prefabricated utility tunnels composed of composite slabs (PUTCCS), and prefabricated utility tunnels composed of groove-shaped elements (PUTCGE) [9].

The focus of this investigation is PUTCCS, which comprises double-sided composite sidewalls, a composite top slab, and a cast-in situ bottom slab or sometimes composite bottom slab, all of which are then connected to form a monolithic structure by post-casting concrete on site. Researchers have extensively investigated the seismic performance of PUTCCS. Wei et al. [10] took the volume ratio of stirrups and anchorage length of the longitudinal reinforcement as parameters and carried out seismic performance tests of the joints of PUTCCS. They found that adoption of stirrups in the core region of the joint can enhance both the shear and moment capacities of the members. Yang et al. [11] carried out low-cycle repeated loading tests of scaled models of the PUTCCS. The test parameters were the height of the haunch and the depth of the overburden soil layer. The results revealed that bending failure occurred in all the specimens, with the limit drift ratio between 1/27 and 1/20, and with relatively good ductility. Wang [12] subjected two full-scale PUTCCS specimens to cyclic loadings, one with an exterior precast top joint and the other with a CIP top joint. It was found that the bearing capacity of the precast specimen was 9.5% lower than that of the CIP specimen.

It should be mentioned that the top and bottom slabs of PUTCCS in the aforementioned research are composite slabs; in actual engineering practice, the CIP bottom slab is more convenient in terms of construction. The connection between the sidewall and the CIP bottom slab of PUTCCS is generally formed by reserving a certain anchorage length with overlapping of the reinforcements between the bottom slab and sidewall. However, this type of connection for PUTCCS is different from the traditional overlapping connection for steel, because, for the former type, the vertical steel bars in the prefabricated sidewalls are not in contact with the steel bars extended upward from the CIP bottom slab. It is therefore difficult to form a valid connection. It is for this purpose that spiral stirrups are used to restrain the overlapping connection, as shown in Figure 1. During prefabrication of the sidewalls, distributed spiral stirrups are imbedded along the vertical steel bars on the inner and outer leaves of the sidewalls, and a 25 mm-depth groove is reserved so that the steel bars extended upward from the bottom slab can be put in place. In this way, the vertical reinforcements for the sidewall and the extended reinforcements from the bottom slab are all inside the spiral stirrups, forming an effective overlapping connection, and so the overlapping length can be shortened to some degree, due to the restraint effect of the stirrup, as demonstrated in [13,14]. Imai [13] showed that the pull-out capacity of the overlapping connection can be improved by increasing the thickness of the cover and the volume ratio of the stirrups. Jiang et al. [14] conducted overlapping tests of 108 specimens, considering the influences of factors such as steel bar diameter, concrete strength, and overlapping length, and established relationships among these factors for the failure mode of the overlapping connection. They found that the overlapping length can be shortened to one standard anchorage length, subject to such factors as concrete strength, steel strength, steel bar diameter, and type of steel; see also DB23/T1813-2016 [15].

Little research can be found on the seismic performance of the spiral stirrup-constrained connection for PUTCCS, except Tian [16], who performed a few joint static performance tests and found that a constrained spiral stirrup improved the connection performance. However, there have been no reported studies on the seismic performance of the over-all structure of PUTCCS with a spiral stirrup-constrained connection, especially when soil–structure interactions are involved.

In addition, there are not many design specifications for utility tunnels. Many structural design codes or guidelines [17–20] do not include specifications of design or construction requirements of precast concrete utility tunnels. Some detailed regulations and descriptions of the static and seismic design of the utility tunnel are provided in [21], but mainly for the CIP concrete utility tunnel and do not include that of the precast utility tunnel. GB50838-2015 [22] proposes a design method only for precast utility tunnels composed of groove-shaped elements. It can be concluded that there is a lack of studies on the structural design of precast utility tunnels composed of composite slabs.

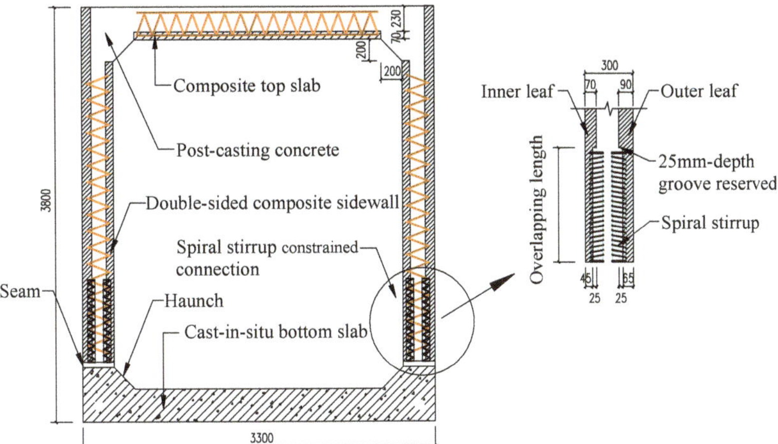

Figure 1. Schematic diagram of composite precast utility tunnel with spiral stirrup-constrained connection.

This investigation takes a practical underground utility tunnel project, located on Hongtu Street in Harbin, China, as the prototype. The ABAQUS software is used to establish a nonlinear finite element model of PUTCCS with a spiral stirrup-constrained connection. Parametric analysis is performed with a focus on the influences of seam location, haunch height and reinforcement and embedment depth on the overall structural behavior of PUTCCS, with exploration of the seismic performance in terms of ductility demand considering soil–structure interactions.

2. Overview of the Experiment

The overall dimension of the concrete specimen (C40) is 3800 mm high, 3300 mm wide, and 800 mm thick. The reinforcement layout of the test specimen is shown in Figure 2. The reinforcement mainly comprises longitudinal bars, truss bars and spiral stirrups. HRB400 steel is used for the longitudinal bars and HPB300 steel is used for the spiral stirrups. The outer sides for the vertical walls and top and bottom slabs have a cover thickness of 50 mm, with a diameter of 20 mm for the longitudinal bars. The inner sides for the vertical walls and top and bottom slabs have a cover thickness of 30 mm, with a diameter of 16 mm for the longitudinal bars. A truss bar system comprises one straight bar (diameter of 10 mm) at the top, two straight bars (diameter of 10 mm) at the bottom, and a V-shaped tie bar (diameter of 6 mm) connecting the straight bars. For the truss bar system, HRB400 steel is used for the straight bars and HPB300 steel is used for V-shaped tie bars. Two truss bar systems with a spacing of 400 mm are arranged longitudinally in each sidewall, as well as in the top slab. There are two types of spiral stirrups. One type is with a diameter of 6 mm, a spacing of 50 mm, a height of 600 mm, and an overall section diameter of 70 mm, which is for the overlapping of the outer longitudinal bars. The second type is with a diameter of 4 mm, a spacing of 40 mm, a height of 480 mm, and an overall section diameter of 70 mm, which is for the overlapping of the inner longitudinal bars. The mechanical properties of concrete and steel reinforcements, were tested in the structural laboratory of Tongji University according to the standard test method specified in Chinese code GB/T 50081-2019 [23] and GB/T 2281-2010 [24].

Cyclic horizontal loadings were applied at the top slab via a hybrid-controlled loading mode as prescribed by the Chinese Specification for Seismic Test of Buildings (JGJ/T101-2015) [25]. The pin supports are set on both the left and right edges of the bottom slab. Some details about the loading can be found in [26]. The loading setup is shown in Figure 3, with an attempt to model PUTCCS imbedded in the ground under the action of an earthquake [27,28]. Figure 4 shows the failure mode of the specimen, which indicates a bending failure as characterized by crushing of concrete observed mainly at the bottom joint connecting the outer sides of the side wall and bottom slab, with the bending bars exposed. The general procedure for experimental data processing and analysis can be found in [29,30].

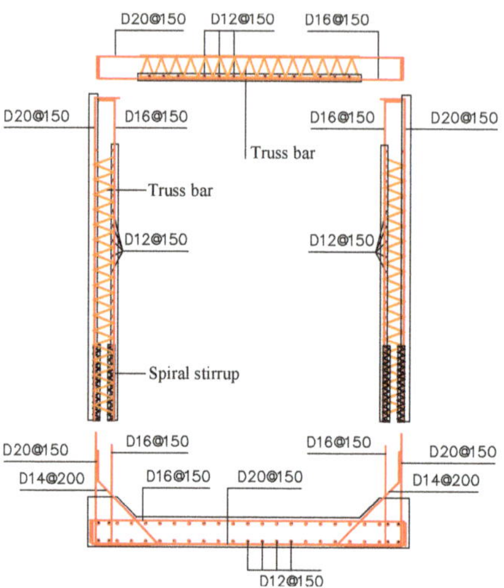

Figure 2. Reinforcement details of specimen.

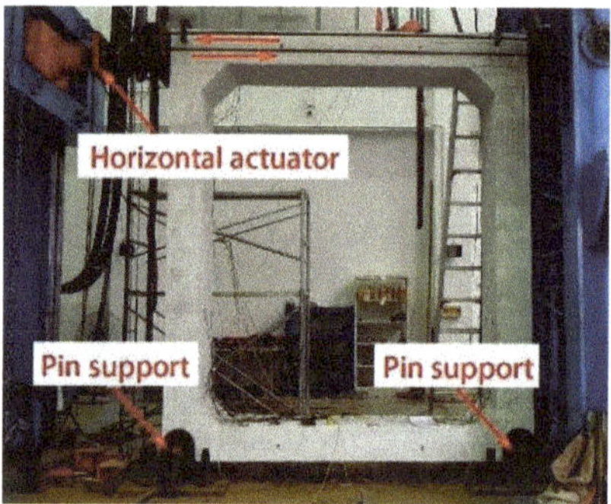

Figure 3. Test setup.

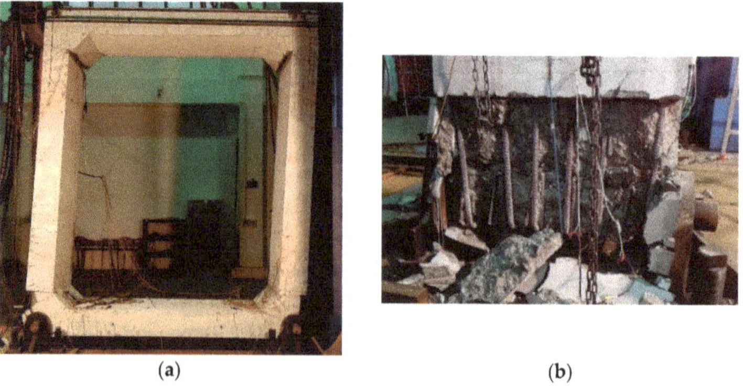

(**a**)   (**b**)

Figure 4. Failure pattern of specimen. (**a**) The whole structure; (**b**) Bottom joint.

3. Finite Element Modeling and Verification

3.1. Finite Element Modeling

Finite element method is a powerful numerical method for solving differential equations in an approximate manner [31,32]. Based on the above testing prototype, a finite element model of PUTCCS with a spiral stirrup-constrained connection using ABAQUS is established, as shown in Figure 5.

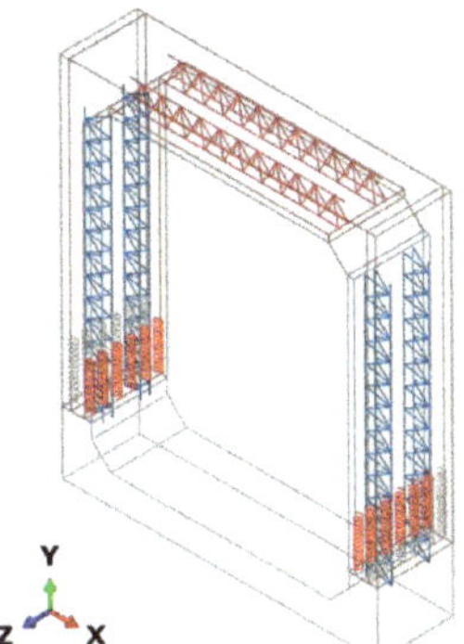

Figure 5. Finite element model.

The damage-plastic material model is adopted to simulate the constitutive stress–strain relationship of concrete [33,34]. The bilinear elastic-plastic material model is adopted to simulate the constitutive relationship of the reinforcements. The constitutive relationship curve of concrete under uniaxial compression is shown in Figure 6, where $\varepsilon_{,i}^{el}$ corresponds to the elastic strain without damage during unloading, ε_i^{el} is the elastic strain with damage during unloading, ε_i^{pl} is the plastic strain after unloading, and ε_i^{in} is the inelastic strain (part of inputs in Abaqus). In the figure, E_0 indicates the initial elastic modulus with a value of 32,500 MPa, d_i indicates plastic-damage factor ranging from 0 to 1, wherein "0" indicates no damage, while "1" indicates full damage (strength completely lost). It should be mentioned that the inelastic strain ε_i^{in} is not equal to plastic strain ε_i^{pl} when d_i is not zero. In this discussion concerning Figure 6, uniaxial tension could also be imposed, with a similar but different pattern, details of which can be found in [33].

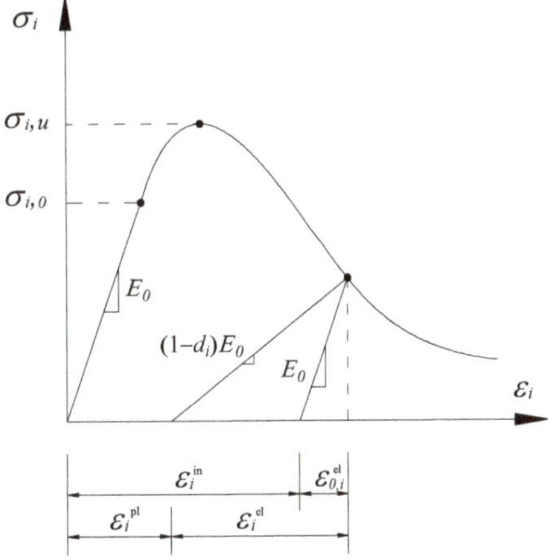

Figure 6. Constitutive relation of concrete.

The Abaqus software automatically converts the inelastic strain values to the plastic strain values using the following Equation (1) [33]:

$$\varepsilon_i^{pl} = \varepsilon_i^{in} - \left(\varepsilon_i^{el} - \varepsilon_{0,i}^{el}\right) = \varepsilon_i^{in} - \frac{d_i \sigma_i}{(1 - d_i)E_0} \tag{1}$$

where the subscript "i" refers to compression or tension.

According to Sidoroff's principle of energy equivalence, the elastic residual energy W_d^e produced by the stress acting on the damaged material is assumed to be the same in form as the elastic residual energy W_0^e acting on an imagined non-damaged material.

For the imagined non-damaged material:

$$W_{0,i}^e = \frac{\sigma_i^2}{2E_0} \tag{2}$$

For the damaged material which has the same form as Equation (2):

$$W_{d,i}^e = \frac{\overline{\sigma}_i^2}{2E_d} \tag{3}$$

where $\overline{\sigma}_i$ is the "effective" tensile and compressive cohesion stress, and defined as in Equation (4), where E_d is the secant modulus:

$$\overline{\sigma}_i = \frac{\sigma_i}{1 - d_i} \tag{4}$$

From Equations (2)–(4), we obtain:

$$E_d = E_0(1 - d_i)^2 \tag{5}$$

Substituting

$$\sigma_i = E_d \varepsilon_i \tag{6}$$

into Equation (5) leads to:

$$d_i = 1 - \sqrt{\frac{\sigma_i}{E_0 \varepsilon_i}} \tag{7}$$

Table 1 shows the input parameters for the concrete damage model used in the simulations, based on the concrete's stress–strain relationship as specified for C40 in the Chinese concrete code GB50010 [35].

Table 1. Input parameters for the concrete damage model in Abaqus.

x	0.4	0.6	0.8	1.0	1.2	1.4	1.6	1.8	2.0	2.2	2.4	2.6	2.8	3.0	5.0	8.0	10.0
$\varepsilon_c^{el}/\mu\varepsilon$	0	232	469	766	1115	1499	1892	2280	2661	3032	3395	3752	4103	4449	7778	12,622	15,826
$\sigma_c/$MPa	18.16	23.47	26.10	26.80	25.79	23.64	21.22	18.93	16.91	15.18	13.71	12.45	11.39	10.47	5.65	3.28	2.56
d_c	0	0.130	0.206	0.280	0.355	0.428	0.494	0.549	0.596	0.635	0.668	0.696	0.720	0.740	0.852	0.911	0.930
$\varepsilon_t^{el}/\mu\varepsilon$	—	—	—	0	58	88	116	143	169	194	218	241	264	287	505	822	1032
$\sigma_t/$MPa	—	—	—	2.39	2.18	1.88	1.63	1.42	1.26	1.14	1.03	0.95	0.88	0.82	0.50	0.34	0.28
d_t	—	—	—	0	0.267	0.369	0.451	0.516	0.568	0.609	0.643	0.672	0.696	0.716	0.828	0.888	0.909

Note: x represents the ratio of the compressive or tensile strain of concrete to the strain corresponding to the peak stress.

The linear truss element type of T3D2 is used to model the steel bars (reinforcements) and the solid element type of C3D8R is used to model the concrete. The automatic meshing is used with an element division of 50 mm. The numbers of truss and solid elements are 9172 and 26,208, respectively. Automatic time step in Abaqus is adopted, which is sufficient for the simulations reported in this work since only material nonlinearity is involved. Good convergency of the results is obtained. The reinforcement elements (T3D2) are coupled to the concrete elements (C3D8R) through "embedded" constraints, ignoring the bond-slip between reinforcement and concrete.

The experimental results indicate obvious cracking on the seam interface between the bottom of the side wall and the top of the bottom slab, while no obvious cracking or slip was observed in the double-sided sidewalls and the composite top slab combined with the post-casting layers. Therefore, contact surface interaction with "surface-to-surface contact" is adopted for the horizontal seam surface at the bottom of the sidewall. In terms of the contact properties, "hard contact" is adopted along the normal contact direction with separation allowed after contact, and "friction contact" is adopted along the tangential contact direction with a friction coefficient of 0.6 between the rough concrete surfaces.

3.2. Model Validation

In the simulations performed in this investigation, monotonic loading with displacement control is applied to the utility tunnel, which is different from the cyclic loading in the experiments as described in Section 2. However, it is well known that hysteretic loop curves can be obtained from the cyclic loading with a well-defined skeleton curve (response envelope) connecting the peak points of each loading in the same direction on the hysteretic loop curve. Such a skeleton curve can clearly reflect the strength and deformation of the structure. It is on this basis that monotonic loading is adopted in the simulations.

Figure 7a shows the contour of the Mises stress invariant (equal to the axial stress under uniaxial loading) for the reinforcement, where the inner and outer longitudinal reinforcement and the haunch reinforcement at the corner of the utility tunnel all yield, which conforms to the experimental observations. Figure 7b shows the contour of damage factors (describing the damage degree) for the concrete, where it can be seen the main damage occurs in the corners of the utility tunnel with a damage factor up to 0.859, which is also consistent with experimental observations.

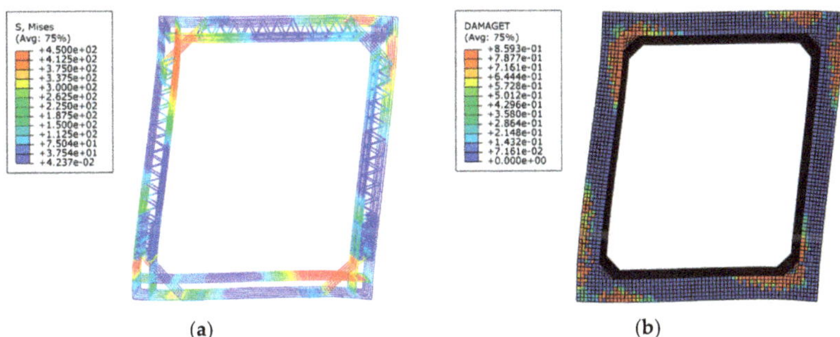

(a)   (b)

Figure 7. Failure pattern of specimen in FEM. (**a**) Reinforcement stress (MPa); (**b**) Concrete damage factor.

Figure 8 shows comparison of load–displacement curves for PUTCCS between FEM and experimental results, wherein the curve for the experimental results is a skeleton curve of the hysteretic loop curves. The positive and negative peak loads, and ductility from the FEM and experimental values are listed and compared in Table 2, where it can be seen that good agreement is obtained between the FEM and experimental results in terms of the peak load values and ductility.

Figure 9a,b show comparisons of load-strain curves between FEM and experimental results, for the outer reinforcements at the upper and lower ends of the sidewall, respectively, where it can be seen that the load-strain curves from FEM can well represent the skeleton curves (response envelopes) from the experiments, which is another way of showing the agreement between the FEM and experimental results.

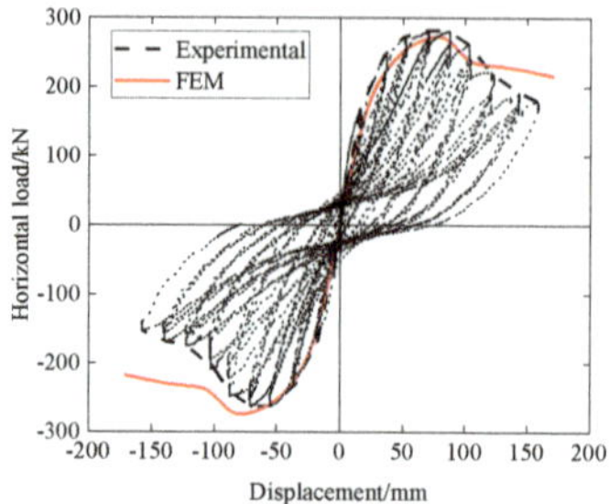

Figure 8. Comparison of load–displacement curves for PUTCCS between FEM and experimental results.

Table 2. Comparison of peak load and ductility for PUTCCS between FEM and experimental results.

Items	Experimental Value	FEM Value	(FEM Value Experimental Value)/Experimental Value
Positive peak load/kN	281.84	277.22	−1.64%
Negative peak load/kN	263.56	277.22	5.18%
Ductility	2.92	2.88	−1.37%

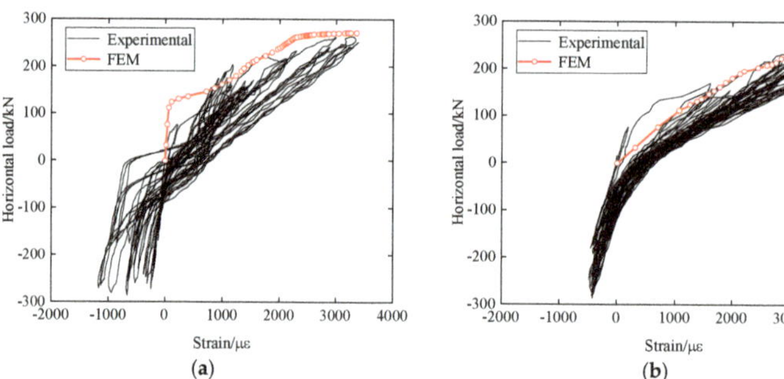

Figure 9. Comparison of load-strain curves for reinforcement between FEM and experimental results. (**a**) Outer reinforcement at the upper end of sidewall; (**b**) Outer Reinforcement at the lower end of sidewall.

4. Parametric Analysis

Parametric analysis is performed using Abaqus considering the variations of the seam location, haunch height and reinforcement, and embedment depth. The load–displacement curve of the specimen is obtained by FEM simulations, and the structural performance indicators include yield load (P_y), peak load (P_{max}), yield displacement (Δ_y), ultimate displacement (Δ_u), and ductility (μ) of the specimen. The yield load and yield displacement are calculated based on the energy method proposed by Park [36]. The basic principle of this energy method is that the curved area A0F is equal to the curved area ACD in Figure 10 and the points C and E can be determined accordingly, the details of which can be found in [36]. The ultimate displacement is taken as the displacement corresponding to 85% of the peak load during the softening stage, and the ductility, i.e., the ductility ratio, is taken as the ratio of the ultimate displacement to the yield displacement.

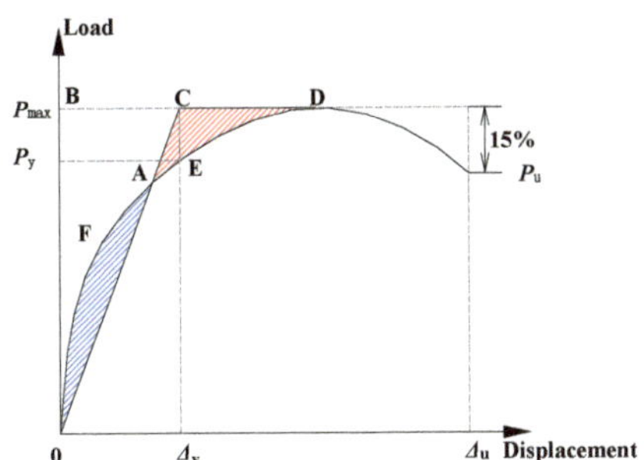

Figure 10. Method used to define yield and ultimate displacements.

4.1. Seam Location

In order to study the effects of seam locations, four sets of seam locations are used for this investigation, which vary according to the vertical distance between the top of the haunch (for the bottom slab) and the seam. Such vertical distances are 0, 0.5, 1.0, and 1.5 times the wall thickness (300 mm), respectively, as shown in Table 3.

Table 3. Analysis parameters and calculation results of seam location.

Specimen	Seam Location	P_y/kN	P_{max}/kN	Δ_y/mm	Δ_u/mm	μ	Remark
SL-0	0 mm above haunch	237.35	277.22	41.62	119.90	2.88	Test conditions
SL-150	150 mm above haunch	245.25	284.46	41.49	105.04	2.53	
SL-300	300 mm above haunch	247.65	289.44	41.27	105.96	2.57	
SL-450	450 mm above haunch	249.03	289.92	42.10	106.29	2.52	

The simulation results in terms of the horizontal load–displacement relationship are plotted in Figure 11, and the results in terms of values of the yield load, peak load, yield displacement, ultimate displacement, and ductility are shown in Table 3. It should be noted that the scenario for the SL-0 sample in Figure 11 corresponds to the actual experimental situation with loading in one direction, and the FEM result for SL-0 is also plotted in Figure 8. It can be seen from Figure 11 that all the four samples (with different distances above haunch) have similar load–displacement patterns. All the four samples share the same initial slope of the load–displacement curve before the yield load is reached. As with the increase in the distance above haunch, both the yield and peak loads are found to increase slightly. It can be seen from Table 3 that when the distance above haunch is 0 mm, i.e., when the top of the haunch and the seam coincide, the ductility of the system is the largest, while the ductility does not vary much when the distance above the haunch is from 150 mm to 450 mm.

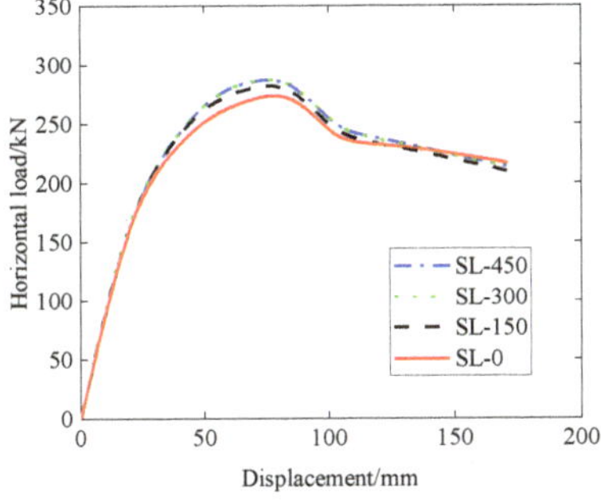

Figure 11. Load–displacement curve (effect of seam location).

4.2. Haunch Height and Reinforcement

In order to study the effects of haunch height and reinforcement, five sets of haunch height and reinforcement are used for this investigation, which vary according to the haunch height and/or haunch reinforcement (four corners for both the top and bottom slabs). The setting of the haunch height and reinforcement is based on common practice in engineering projects. The haunch height ranges from 0 (indicating no haunch) to 150 mm to 200 mm. The haunch reinforcement ranges from D12@200 to D14@200 to D16@200, where D12@200 refers to reinforcements (HRB400 steel) with a diameter of 12 mm distributed at 200 mm center-to-center distances along the longitudinal direction of the utility tunnel.

The simulation results in terms of the horizontal load–displacement relationship are plotted in Figure 12, and the results in terms of values of the yield load, peak load, yield displacement, ultimate displacement, and ductility are shown in Table 4. It should be noted that the scenario for the HD-200-14 sample in Figure 12 corresponds to the actual experimental reinforcement situation, and the FEM result for HD-200-14 is also plotted in Figure 8. It can be seen from Figure 12 that the four samples with haunches have similar load–displacement patterns, while the sample without haunches (HD-0-0) has an obviously lower initial stiffness, obviously lower yield and peak loads, and lower ductility (refer to Table 4). Comparing the results of the sample HD-200-14, the peak load capacity for HD-0-0 is found to decrease by 29.6%, and the ductility for HD-0-0 is found to decrease by 16.7%. A comparison of the results of the samples HD-150-14 and HD-200-14 indicates that the increase in the haunch height (from 150 mm to 200 mm) can slightly enhance the load capacity in terms of both yield and peak loads as well as the ductility. A comparison of the results of the samples HD-200-12, HD-200-14 and HD-200-16 indicates that the increase in the reinforcement size from 12 mm to 14 mm to 16 mm (without changing other aspects) can slightly enhance both the yield and peak loads, but slightly decrease the ductility.

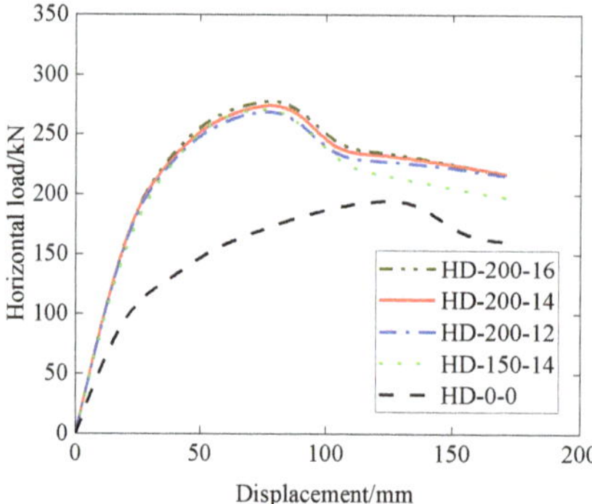

Figure 12. Load–displacement curve (effect of haunch size and reinforcement).

Table 4. Analysis parameters and calculation results of haunch height and reinforcement.

Specimen	Haunch Height/mm	Haunch Reinforcement	P_y/kN	P_{max}/kN	Δ_y/mm	Δ_u/mm	μ	Remark
HD-0-0	0	None	161.77	195.26	63.96	153.26	2.40	
HD-150-14	150	D14@200	233.72	272.10	41.38	102.98	2.49	
HD-200-12	200	D12@200	232.36	271.33	40.75	120.57	2.96	
HD-200-14	200	D14@200	237.35	277.22	41.62	119.90	2.88	Test conditions
HD-200-16	200	D16@200	241.74	281.28	42.03	115.74	2.75	

4.3. Embedment Depth

The embedment depth for a prefabricated utility tunnel in practice is usually not more than 10 m, corresponding to shallow embedment depth, especially when the open-cut method is used for the construction of such tunnels. The change in the embedment depth only involves the change in earth pressure on the top slab of the utility tunnel in the current investigation. The effect of the embedment depth is thus reflected by the loading on the top slab. In order to study the effects of such embedment depths, four sets of embedment depths are considered, which range from 0 m to 2 m to 5 m to 10 m, as shown in Table 5.

Table 5. Analysis parameters and calculation results of embedment depth.

Specimen	Embedment Depth/m	P_y/kN	P_{max}/kN	Δ_y/mm	Δ_u/mm	μ	Remark
DB-0	0	237.35	277.22	41.62	119.90	2.88	Test conditions
DB-2	2	243.44	284.93	40.76	113.45	2.78	
DB-5	5	254.39	295.49	41.53	106.84	2.57	
DB-10	10	263.78	298.56	42.12	103.08	2.45	

The simulation results in terms of the horizontal load–displacement relationship are plotted in Figure 13, and the results in terms of values of the yield load, peak load, yield displacement, ultimate displacement, and ductility are shown in Table 5. It should be noted that the scenario for the DB-0 sample in Figure 13 corresponds to the actual experimental situation with loading in one direction, and the FEM result for DB-0 is also plotted in Figure 8. It can be seen from Figure 13 that all four samples have similar load–displacement patterns, and that they share the same initial stiffness in the load–displacement curve before the yield load is reached. As with the increase in the embedment depth, both the yield and peak loads are found to increase gradually, as can be seen in Table 5, while the ductility is found to decrease gradually. Comparison of the results of the samples DB-0 and DB-10 indicates that the increase in the embedment depth from 0 m to 10 m can enhance the load capacity in terms of the peak loads by 7.7% and decrease the ductility by 14.9%. Physically, the increase in the embedment depth in this investigation is equivalent to the increase in the axial/vertical loading of the sidewall, which is like the behavior of a column with combined axial compression and bending, leading to an increase in the horizontal load capacity and decrease in ductility.

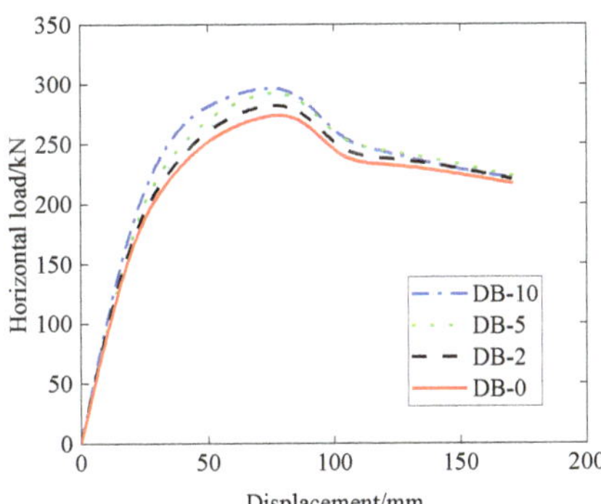

Figure 13. Load–displacement curve (effect of embedment depth).

5. Seismic Performance Evaluation Considering Soil–Structure Interaction

The utility tunnel is a large-scale underground structure, and the soil–structure interaction effect should not be ignored under earthquake actions. The flexibility coefficient method, also known as the soil–structure interaction coefficient method, is a commonly used simplified design method for underground structures under earthquake conditions [37–39]. The main core of the calculation steps of this method is:

$$\Delta_s = R\Delta_{free-field} \tag{8}$$

where Δ_s is the racking deformation of the structure between the top and bottom slabs under earthquake action considering soil–structure interactions, $\Delta_{\text{free-field}}$ is the free-field deformation under earthquake action, and R is the interaction coefficient or racking coefficient. The free-field deformation refers to the deformation of the soil with no structure and no opening in the ground.

The free-field deformation under an earthquake can be calculated according to GB50909-2014 [40]:

$$\Delta_{free-field} = U(z_1) - U(z_2) \tag{9}$$

where $U(z_1)$ and $U(z_2)$ are the horizontal displacements of the soil stratum corresponding to the locations of the top and bottom slabs of the structure under free-field deformation, respectively.

The horizontal displacement of the soil stratum under free-field deformation $U(z)$ can be obtained from the following:

$$U(z) = \frac{1}{2} U_{max} \cos \frac{\pi z}{2H} \tag{10}$$

where z is the depth of the soil stratum, $U_{\max}$ is the maximum horizontal displacement of the site surface, which can be found with reference to GB50909-2014 [40], and H is the vertical distance from the ground surface to the seismic action datum.

There are several methods to calculate the interaction coefficient R in the literature, which are all related to the relative stiffness ratio of the soil and the structure and/or Poisson's ratio of the soil. Wang [41], Penzien [42], and Nishioka [43] proposed Equations (11)–(13), respectively, for calculating R, as follows:

$$R = \frac{8K_r(1-\nu)}{2K_r + (5-6\nu)} \ (\text{Wang's method}) \tag{11}$$

$$R = \frac{4K_r(1-\nu)}{K_r + (3-4\nu)} \ (\text{Penzien's method}) \tag{12}$$

$$R = \frac{2K_r}{K_r + 1} \ (\text{Nishioka's method}) \tag{13}$$

where ν is the Poisson's ratio of the soil, and K_r is the ratio of the soil stiffness (k_s) to the structural stiffness (k_{st}) for the utility tunnel. Setting a value of 0.5 for ν in Equations (11) and (12) leads to Equation (13). In other words, Nishioka's method is a special case of either Wang's method or Penzien's method by setting ν equal to 0.5.

The soil stiffness is calculated as follows:

$$k_s = \frac{LG}{B} \tag{14}$$

where G is the shear modulus of the soil, L is the width of the cross section of the utility tunnel, and B is the height of the cross section of the utility tunnel.

The structural stiffness k_{st} can be obtained based on FEM. Pinned supports are set at the lower ends of the sidewall of the structure, and the load–displacement curve can be obtained by applying a concentrated lateral force to the top of the structure and the elastic stiffness is taken as the structural stiffness, see also Wang [41]. However, the method provided by Wang [41] has a limitation in that the structure needs to be elastic. To extend the applicability of the above methodology to the structure with plastic deformation, a simplified method is proposed for the utility tunnel in this investigation, where the load–displacement curve is idealized as an elastic rigid-plastic bilinear line, as shown in Figure 14. In this simplified method, the yield load and yield displacement are determined by using the energy method proposed by Park [36] in relation to the discussion of point A in Figure 10. In Figure 14, K_y corresponds to the stiffness in the elastic phase in the idealized elastic rigid-plastic model, K_{eq} is the secant stiffness corresponding to the plastic phase.

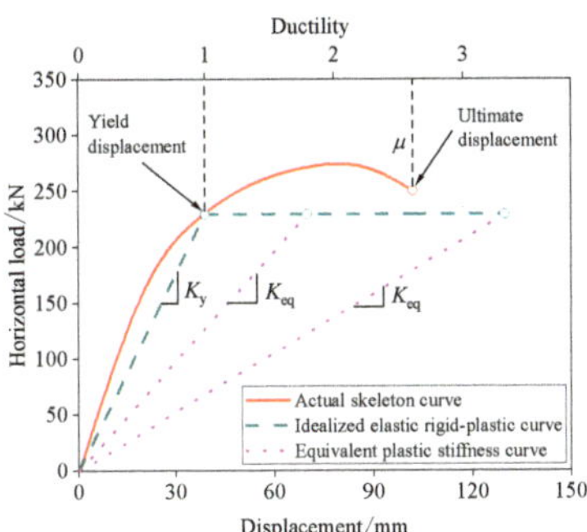

Figure 14. Idealized elastic rigid-plastic curve.

The calculation procedure of the racking deformation can be outlined as follows:

Determine the basic parameters such as site characteristic, dimensions of the utility tunnel, and seismic conditions.

1. Input the initial value D_{ini} (can be assumed to be a small trial value for the first attempt) of the racking deformation of the utility tunnel.
2. Calculate the structural stiffness k_{st} using the simplified elastic rigid-plastic model.
3. Calculate the soil stiffness k_{s} based on Equation (14).
4. Calculate the interaction coefficient R by using one of Equations (11)–(13).
5. Calculate the racking deformation Δ_{s} of the structure under earthquake action using Equation (8).
6. If the difference between Δ_{s} in Step 5 and D_{ini} in Step 2 is relatively large, meeting $|D-\Delta_{\mathrm{s}}|/D \geq 0.001$, increase the initial displacement D_{ini} and repeat steps 3 to 6 iteratively until the updated values of Δ_{s} and D_{ini} meet $|D-\Delta_{\mathrm{s}}|/D < 0.001$.
7. The racking deformation Δ_{s} can then be obtained from the final value of Δ_{s} in Step 6.

The above procedure can be illustrated as a flow chart shown in Figure 15.

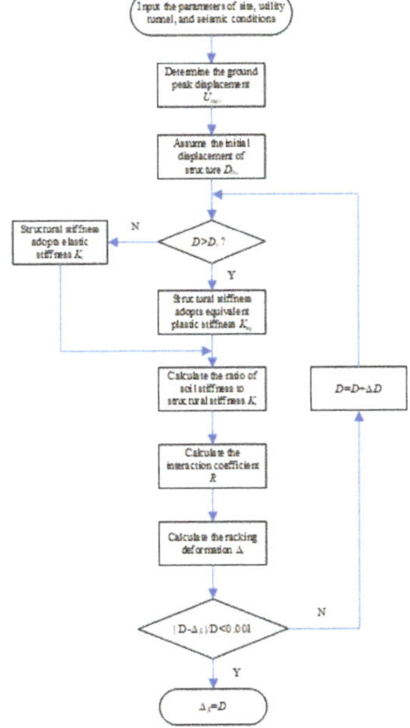

Figure 15. Flow chart of the procedure to obtain racking deformation.

The site characteristic, dimensions of the utility tunnel, and seismic conditions are listed in Table 6. It should be noted that according to seismic precautionary intensity of 8 (0.3 g), the maximum displacement of the site surface is set to 490 mm for analysis [40].

Table 6. Value of the input parameters.

Parameters	U_{max}/mm	ν	G/MPa	L/mm	B/mm	Embedment Depth/mm	H/m
Value	490	0.3	100	3300	3800	3000	15

For the convenience of discussion, we define a deformation coefficient μ_s as the ratio of the racking deformation Δ_s to the yield displacement Δ_y. Three methods for calculating R are used to obtain the final Δ_s, respectively, following the calculation procedure in Figure 15, and the results are compared considering effects of seam location, haunch height and reinforcement, and embedment depth. The results for the seismic performance evaluation (in terms of the deformation coefficient) considering soil–structure interaction are shown in Figure 16. A summary of the discussions is listed below.

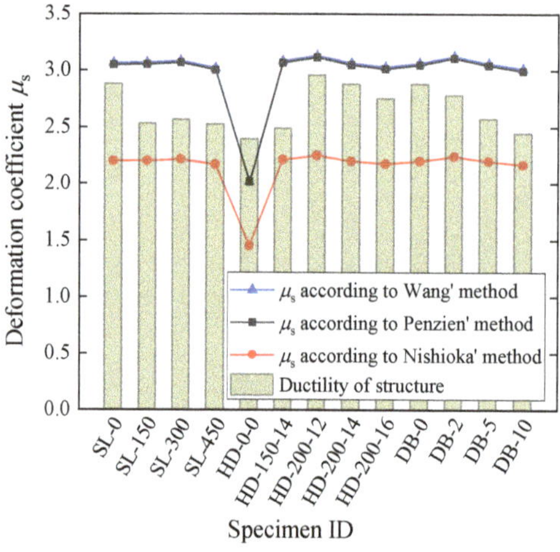

Figure 16. Seismic performance evaluation considering soil–structure interaction.

1. The results of the seismic performance in terms of deformation coefficient indicate that Penzien's and Wang's methods (with a traditional value of 0.3 for Poisson's ratio) are almost identical, giving a larger value of μ_s than that using Nishioka's method for each sample. This is mainly because Nishioka's method does not consider the influence of the changes in Poisson's ratio, which is equivalent to the case where a fixed value of Poisson's ratio of 0.5 is used in Penzien's and Wang's methods. It can be easily verified that from Equations (11) and (12), the interaction coefficient R decreases with the increase in Poisson's ratio. Table 7 shows the interaction coefficient R using Wang's, Penzien's and Nishioka's methods, which confirms the above discussion in relation to the effects of change in Poisson's ratio, since μ_s is in positive proportion to R.

2. The smallest deformation coefficient occurs for the haunch-free specimen HD-0-0 for each of the three methods (Penzien's, Wang's, and Nishioka's), respectively. This is mainly due to the large yield displacement (63.96 mm) of this specimen, which is significantly larger than that of the other specimens ranging from 40.75 mm to 42.12 mm, as shown in Tables 3–5. It should be mentioned that specimen HD-0-0's interaction coefficient R is the largest for each of the methods, respectively, due to the small structural stiffness. However, the variation of R across the whole specimen list

is not much, as can be seen in Table 7. Therefore, the dominant factor affecting the deformation coefficient is the yield displacement for the cases under consideration.

3. If the ductility obtained from the load–displacement curve without considering soil–structure interactions is treated as a limit of the deformation coefficient and applied to the seismic conditions (where soil–structure interactions are normally considered), we can compare the value of deformation coefficient under different scenarios. It can be seen from Figure 16 that the deformation coefficient of each specimen based on Nishioka's method falls below that corresponding to the structural ductility histogram, while the deformation coefficient for each specimen using Penzien's and Wang's methods, lies above that corresponding to the structural ductility histogram, except the very special specimen HD-0-0. In terms of the limit value of the deformation coefficient (ductility), the comparison of the results for the specimen HD-0-0 indicates that a smaller loading capacity and lower ductility without considering soil–structure interactions may correspond to a lower value of deformation coefficient under seismic conditions with soil–structure interactions, and may therefore exhibit a better seismic performance in terms of ductility demand.

4. It should also be mentioned that different methods for calculating R may lead to different conclusions in terms of an evaluation of the seismic performance, and further investigations should be made as regards which method provides realistic results.

Table 7. Interaction coefficient R using Wang's, Penzien's, and Nishioka's methods.

Specimen ID	SL-0	SL-150	SL-300	SL-450	HD-0-0	HD-150-14	HD-200-12	HD-200-14	HD-200-16	DB-0	DB-2	DB-5	DB-10
Wang's	2.68	2.68	2.68	2.68	2.72	2.68	2.69	2.68	2.68	2.68	2.68	2.67	2.67
Penzien's	2.67	2.66	2.66	2.66	2.71	2.67	2.67	2.67	2.67	2.67	2.67	2.66	2.65
Nishioka's	1.93	1.92	1.92	1.92	1.95	1.93	1.93	1.93	1.93	1.93	1.92	1.92	1.92

6. Conclusions

This investigation focused on numerical investigations of the seismic performance of prefabricated utility tunnels composed of composite slabs (PUTCCS) with a spiral stirrup-constrained connection. An experiment was set up based on the prototype of a practical utility tunnel, the results of which were compared with FEM simulation (considering material nonlinearity with concrete damage) results with reasonable agreement obtained. The parametric analysis without considering soil–structure interactions was conducted in terms of seam location, haunch height and reinforcement, and embedment depth, based on FEM simulations. A simplified method was proposed for evaluating the seismic performance in terms of deformation coefficient considering ductility demand, based on three different approaches of obtaining interaction coefficient (racking coefficient) considering soil–structure interactions. The main conclusions are as follows.

1. All the samples with varying seam locations share the same initial slope of the load–displacement curve before the yield load is reached. As with the increase in seam distance above haunch, a slight increase in both the yield and peak loads is observed, while the ductility does not vary much.

2. The haunch-free sample has an obviously lower initial stiffness and lower yield and peak loads, and lower ductility in comparison with those for the samples with haunches. The increase in the reinforcement size can enhance both the yield and peak loads, but decrease the ductility.

3. The increase in the embedment depth can enhance both the yield and peak loads while decreasing the ductility.

4. The seismic performance in terms of deformation coefficient considering soil–structure interactions indicates that Penzien's and Wang's methods are almost identical, giving a larger deformation coefficient than that using Nishioka's method.

5. The smallest deformation coefficient occurs for the haunch-free specimen for each of the three methods (Penzien's, Wang's, and Nishioka's), respectively. The dominant factor affecting the deformation coefficient is the yield displacement.

Author Contributions: Q.W.: conceptualization, funding acquisition, model development, validation, and original drafting; G.G.: funding acquisition, original drafting, writing, review, and editing; J.H.: supervision, writing, review and editing; Y.B.: supervision, writing, review and editing. All authors have read and agreed to the published version of the manuscript.

Funding: This research was funded by the Science and Technology Program Projects in Nantong, China (grant number: MSZ21046), and by Xi'an Jiaotong-Liverpool University (grant numbers: KSF-E-19, RDF 18-01-23, PGRS1906002 and REF-20-01-01).

Institutional Review Board Statement: Not applicable.

Informed Consent Statement: Not applicable.

Data Availability Statement: The data used to support the findings of this study are available upon request from the corresponding author.

Acknowledgments: The authors acknowledge the support provided by the Science and Technology Program Projects in Nantong, China (MSZ21046), and by Xi'an Jiaotong-Liverpool University (KSF-E-19, RDF 18-01-23, PGRS1906002 and REF-20-01-01).

Conflicts of Interest: The authors declare that they have no known competing interests that could have appeared to influence the work reported in the paper.

References

1. Luo, Y.; Alaghbandrad, A.; Genger, T.; Hammad, A. History and recent development of multi-purpose utility tunnels. *Tunn. Undergr. Space Technol.* **2020**, *103*, 103511. [CrossRef]
2. Canto-Perello, J.; Curiel-Esparza, J.; Calvo, V. Criticality and threat analysis on utility tunnels for planning security policies of utilities in urban underground space. *Expert Syst. Appl.* **2013**, *40*, 4707–4714. [CrossRef]
3. Phillips, J. A quantitative evaluation of the sustainability or unsustainability of three tunnelling projects. *Tunn. Undergr. Space Technol.* **2016**, *51*, 387–404. [CrossRef]
4. Wang, T.; Tan, L.; Xie, S.; Ma, B. Development and applications of common utility tunnels in China. *Tunn. Undergr. Space Technol.* **2018**, *76*, 92–106. [CrossRef]
5. Yang, C.; Peng, F.-L. Discussion on the Development of Underground Utility Tunnels in China. *Procedia Eng.* **2016**, *165*, 540–548. [CrossRef]
6. Sun, D.; Liu, C.; Wang, Y.; Xia, Q.; Liu, F. Static performance of a new type of corrugated steel-concrete composite shell under mid-span loading. *Structures* **2022**, *37*, 109–124. [CrossRef]
7. Hao, J.L.; Chen, Z.; Zhang, Z.; Loehlein, G. Quantifying construction waste reduction through the application of prefabrication: A case study in Anhui, China. *Environ. Sci. Pollut. Res. Int.* **2021**, *28*, 24499–24510. [CrossRef]
8. He, H.; Zheng, L.; Zhou, G. Linking users as private partners of utility tunnel public–private partnership projects. *Tunn. Undergr. Space Technol.* **2022**, *119*, 104249. [CrossRef]
9. Wang, Q.; Gong, G.; Hao, J.L. Double-Cell Prefabricated Utility Tunnel Composed of Groove-Shaped Elements: An Extended Study of Stiffness Reduction Method. *Appl. Sci.* **2022**, *12*, 5982. [CrossRef]
10. Wei, Q.; Wang, Y.; Wang, Y.; Pi, Z.; Liao, X.; Wang, S.; Zhang, H. Experiment study on seismic performance of joints in prefabricated sandwich structures of utility tunnels. *J. Build. Struct.* **2019**, *40*, 246–254. (In Chinese)
11. Yang, Y.; Tian, X.; Liu, Q.; Zhi, J.; Wang, B. Anti-seismic behavior of composite precast utility tunnels based on pseudo-static tests. *Earthq. Struct.* **2019**, *17*, 233–244.
12. Wang, Q. Experimental study of the structural performance of the top joints of precast concrete utility tunnel composed of composite slabs. In Proceedings of the fib Symposium (Concrete Structures: New Trends for Eco-Efficiency and Performance), Lisbon, Portugal, 14–16 June 2021.
13. Adajar, J.C.; Imai, H. Behavior of spirally confined lap splice for precast concrete structural walls under tension. *J. Struct. Constr. Eng.* **1997**, *62*, 157–165. [CrossRef]
14. Jiang, H.; Zhang, H.; Liu, W.; Yan, H. Experimental study on plug-in filling hole for steel bar lapping of precast concrete structure. *J. Harbin Inst. Technol.* **2011**, *43*, 18–23. (In Chinese)
15. *DB23/T1813-2016*; Technical Specification for Concrete Shear Wall Structure Assembled with Precast Components. China Building Materials Press: Beijing, China, 2016. (In Chinese)
16. Tian, Z. *Experimental Research on Force Performance of Precast Concrete Underground Comprehensive Municipal Tunnel*; Harbin Institute of Technology Harbin: Harbin, China, 2016. (In Chinese)
17. ACI. *318–19*; Building Code Requirements for Structural Concrete and Commentary. American Concrete Institute: Farmington Hills, MI, USA, 2019.
18. PCI Design Handbook. *Precast and Prestressed Concrete*, 8th ed.; Precast/Prestressed Concrete Institute: Chicago, IL, USA, 2017.
19. *EN-1992-1-1*; Eurocode 2, Design of Concrete Structures–Part 1-1: General Rules and Rules for Buildings. CEN European Committee for Standardization: Brussels, Belgium, 2004.

20. *NZS3101*; The design of concrete structures. Standards Association of New Zealand: Wellington, New Zealand, 2006.
21. Japan Road Association. *Design guidelines of Common Tunnel*; Tokyo Press: Tokyo, Japan, 1991.
22. *GB50838-2015*; Technical Code for Urban Utility Tunnel Engineering. China Planning Press: Beijing, China, 2015. (In Chinese)
23. *GB/T 50081-2019*; Standard for Test Methods of Concrete Physical and Mechanical Properties. China Building Industry Press: Beijing, China, 2019. (In Chinese)
24. *GB/T 228.1-2010*; Metallic Materials-Tensile Testing Part 1: Method of Test at Room Temperature. China Standards Press: Beijing, China, 2010. (In Chinese)
25. *JGJ/T101-2015*; Specification for Seismic Test of Buildings. China Building Industry Press: Beijing, China, 2015. (In Chinese)
26. Yan, H.; Xue, W.; Hu, X.; Liu, W. Seismic performance, key technologies of construction and prefabrication of precast concrete utility tunnel consisting of composite slabs-taking the YUHUI precast concrete utility tunnel as an example. *Spec. Struct.* **2021**, *38*, 85–90. (In Chinese)
27. Hashash, Y.M.; Hook, J.J.; Schmidt, B.; John, I.; Yao, C. Seismic design and analysis of underground structures. *Tunn. Undergr. Space Technol.* **2001**, *16*, 247–293. [CrossRef]
28. Takeda, T.; Ishikawa, H.; Adachi, M.; Honda, K.; Yamaya, A.; Iizuka, K. Study of performance evaluation around a strong earthquake of important reinforced concrete structure at a nuclear power plant. In Proceedings of the 7th International Conference on Nuclear Engineering, Tokyo, Japan, 19–23 April 1999.
29. Zaghloul, M.M.Y.; Zaghloul, M.Y.M.; Zaghloul, M.M.Y. Experimental and modeling analysis of mechanical-electrical behaviors of polypropylene composites filled with graphite and MWCNT fillers. *Polym. Test.* **2017**, *63*, 467–474. [CrossRef]
30. Fuseini, M.; Zaghloul, M.M.Y. Investigation of Electrophoretic Deposition of PANI Nano fibers as a Manufacturing Technology for corrosion protection. *Prog. Org. Coat.* **2022**, *171*, 107015. [CrossRef]
31. Abbasand, I.A.; Abo-Dahab, S.M. On the numerical solution of thermal shock problem for generalized magneto-thermoelasticity for an infinitely long annular cylinder with variable thermal conductivity. *J. Comput. Theor. Nanosci.* **2014**, *11*, 607–618. [CrossRef]
32. Mohamed, R.A.; Abbas, I.A.; Abo-Dahab, S.M. Finite element analysis of hydromagnetic flow and heat transfer of a heat generation fluid over a surface embedded in a non-Darcian porous medium in the presence of chemical reaction. *Commun. Nonlinear Sci. Numer. Simul.* **2009**, *14*, 1385–1395. [CrossRef]
33. ABAQUS. *ABAQUS User Subroutines Reference Manual*; ABAQUS: Pawtucket, RI, USA, 2014.
34. Liu, Y.; Zha, X.; Gong, G. Study on recycled-concrete-filled steel tube and recycled concrete based on damage mechanics. *J. Constr. Steel Res.* **2012**, *71*, 143–148. [CrossRef]
35. *GB50010-2010*; Code for Design of Concrete Structures. China Architecture & Building Press: Beijing, China, 2015. (In Chinese)
36. Park, R. Evaluation of ductility of structures and structural assemblages from laboratory testing. *Bull. New Zealand Soc. Earthq. Eng.* **1989**, *22*, 155–166. [CrossRef]
37. Debiasi, E.; Gajo, A.; Zonta, D. On the seismic response of shallow-buried rectangular structures. *Tunn. Undergr. Space Technol.* **2013**, *38*, 99–113. [CrossRef]
38. Tsinidis, G. Response characteristics of rectangular tunnels in soft soil subjected to transversal ground shaking. *Tunn. Undergr. Space Technol.* **2017**, *62*, 1–22. [CrossRef]
39. Tsinidis, G.; Pitilakis, K. Improved RF relations for the transversal seismic analysis of rectangular tunnels. *Soil Dyn. Earthq. Eng.* **2018**, *107*, 48–65. [CrossRef]
40. *GB50909-2014*; Code for Seismic Design of Urban Rail Transit Structures. China Planning Press: Beijing, China, 2014. (In Chinese)
41. Wang, J.N. *Seismic Design of Tunnels: A Simple State-Of-Art Design Approach*; Parsons Brinckerhoff Quade & Douglas Inc.: New York, NY, USA, 1993.
42. Penzien, J. Seismically induced racking of tunnel linings. *Earthq. Eng. Struct. Dyn.* **2000**, *29*, 683–691. [CrossRef]
43. Nishioka, T.; Unjoh, S. A simplified evaluation method for the seismic performance of underground common utility boxes. In Proceedings of the Pacific Conference on Earthquake Engineering, Christchurch, New Zealand, 13–15 February 2003.

Article

Influence and Mechanism of Curing Methods on Mechanical Properties of Manufactured Sand UHPC

Chengfang Yuan [1,2], Shiwen Xu [1,*], Ali Raza [1], Chao Wang [1] and Di Wang [1]

[1] College of Water Resources and Civil Engineering, Zhengzhou University, Zhengzhou 450001, China
[2] Yellow River Laboratory, Zhengzhou University, Zhengzhou 450001, China
* Correspondence: shiwen1125@126.com

Abstract: The mechanical properties of ultra-high performance concrete (UHPC) made of manufactured sand (MS) under four curing methods (steam, standard, sprinkler and saturated $Ca(OH)_2$) were investigated via compressive, flexural and uniaxial tensile tests, and the bond strength of steel fiber and manufactured sand UHPC (MSUHPC) matrix. Based on the analysis of the microstructure, the influence mechanism of curing methods on the mechanical properties of materials was explored. The results showed that the early compressive strength of MSUHPC under steam curing (SM) is much higher than that of the other three curing methods, but the difference gradually decreases with the increase of age. The compressive strength of MSUHPC under SM is higher than that of river sand UHPC (RSUHPC). The bending strength and compressive strength of MSUHPC under different curing methods are similar, and the bending strength of 28 days steam cured samples is the highest. The uniaxial tensile properties of MSUHPC did not show significant difference under standard curing (SD), sprinkler curing (SP) and saturated $Ca(OH)_2$ curing (CH), and the uniaxial tensile properties of MSUHPC under SM are slightly better than RSUHPC. The ultimate bond strength and fiber pullout energy of steel fiber and MSUHPC increase with the development of age. The bond strength and fiber pullout work of SM is higher than those of the other three curing methods, but there are lower increases in the later stage than that of the other three curing methods.

Keywords: manufactured sand; UHPC; curing methods; mechanical properties; bond strength

Citation: Yuan, C.; Xu, S.; Raza, A.; Wang, C.; Wang, D. Influence and Mechanism of Curing Methods on Mechanical Properties of Manufactured Sand UHPC. *Materials* **2022**, *15*, 6183. https://doi.org/10.3390/ma15186183

Academic Editors: Francesco Fabbrocino, Dan Huang and Zhanqi Cheng

Received: 22 July 2022
Accepted: 2 September 2022
Published: 6 September 2022

Publisher's Note: MDPI stays neutral with regard to jurisdictional claims in published maps and institutional affiliations.

1. Introduction

Since the second half of the 20th century, with the rapid development of engineering structures and economic growth, people also have new requirements for the performance of engineering materials, and have begun to explore and research high performance and even ultra-high performance engineering materials. In 1994, the concept of ultra-high performance concrete (UHPC) was proposed [1]; then the research on reactive powder concrete was published and had a warm response in the field of civil engineering, which marked the entry of concrete into the era of ultra-high performance [2].

UHPC is a high density cement-based composite material with high strength and high ductility, which has been applied in maintenance and reinforcement, bridge engineering and other fields [3–9]. It is well known that the manufacturing of UHPC requires a large quantity of natural river sand (RS) as fine aggregates. However, natural RS is increasingly scarce due to the exhaustive exploitation. Therefore, it is urgent to find other raw materials to replace natural sand in producing UHPC. Manufactured sand (MS) is a kind of artificial fine aggregate from natural stone based on a series of breaking and grinding techniques, which could be processed in larger amount and lower cost [10–16]. Studies have shown that MS can replace natural sand to prepare UHPC and be used in practical engineering [17–23]. Yoo et al. [24] studied the mechanical properties of UHPC under various curing methods and found that thermal curing can promote hydration reaction and improve the compressive strength of UHPC. Zdeb [25] found through microstructural studies that with the increase of temperature, C-S-H will produce a new crystalline hydrate, which makes the structure of concrete denser, the temperature increases by 10 °C, and the compressive strength of concrete increases by 16 MPa. Yazıcı et al. [26] studied the influence of auxiliary cementitious materials and curing conditions on the compressive strength of UHPC, and found the optimal dosage of different auxiliary cementitious materials and optimal curing conditions. Wu et al. [27] investigated the mechanical properties of UHPC under three curing methods. The results showed that when the curing time is same, hot water curing and SM resulted in higher compressive strength and bending strength of UHPC than that of standard curing. However, Shi et al. [28] conducted studies on hydration products, microscopic morphology, pore structure and interfacial transition zone (ITZ), and the results showed that SM can adversely affect the long-term performance of concrete by changing the internal temperature and humidity field. Liu et al. [29] found through the study of pore structure that although SM can adversely affect the long-term performance of concrete, proper subsequent wet curing can effectively reduce the adverse effects of SM on concrete. At the same time, Ballim Y. [30] measured the oxygen permeability and water absorption at different depths in the concrete, and the result showed that wet curing had a protective and sustainable effect on the durability of concrete.

The curing method of concrete has an important influence on the formation of its strength and later development. The commonly used curing method for UHPC is high-temperature steam curing, which is relatively complex and more suitable for preparing prefabricated components in fixed places such as prefabricated yards [31]. For some components that need to be poured on site, it is generally difficult to achieve the conditions of SM due to site factors. Therefore, it is also very important to study the mechanical properties of UHPC under other normal temperature curing conditions.

Gao H et al. [32] studied the case of concrete CO_2 erosion; by using a $Ca(OH)_2$ solution, wetting concrete can improve its density and compressive strength, and this method can solve the problem of CO_2 erosion of cement and concrete materials.

Existing research results have shown that different curing conditions have a great influence on the hydration reaction rate and physical and mechanical properties of UHPC [33,34]. The object of this study is to gain insights into the mechanical and microstructural properties of manufactured sand UHPC (MSUHPC) under different curing methods. The effects of four curing methods (steam curing (SM), standard curing (SD), sprinkler curing (SP) and saturated $Ca(OH)_2$ curing (CH)) on mechanical and microstructural properties of concrete are investigated. It is greatly significant to promote the application of MS in UHPC preparation.

2. Materials and Methods

2.1. Material

This study adopted ordinary Portland cement (P·O52.5) produced by Mengdian Cement Industry (Huixian, Henan, China). The physical properties of the cement are shown in Table 1. Fly ash produced by Hengnuo Filter Material (Gongyi, China) and silica fume (SiO_2 content 97%) produced by Hengyuan Material (Xining, Qinghai, China) were used. The steel fiber was made of copper-plated steel fiber produced by Sida Steel Fiber Industry (Tengzhou, China). The shape diagram is shown in Figure 1, and the specifications and performance indicators are shown in Table 2. Two types of fine aggregate (MS and RS) were applied here. The test reports are shown in Tables 3 and 4. The test equipment, methods and calculations are in accordance with the specification JGJ 52-2006 [35]. Polycarboxylic superplasticizer with water-reducing of 26% was produced by Chenqi Industy (Shanghai, China). Common tap water was used for mixing and maintenance.

Table 1. Chemical and physical properties of cement.

	Properties	Value	Range [1]
Physical properties	Specific surface area (m^2/kg)	386	$\geq$300
	Initial set (min)	170	$\geq$45
	Final set (min)	222	$\leq$600
Compressive strength	28 days (MPa)	55.6	$\geq$52.5
Flexural strength	28 days (MPa)	7.5	$\geq$7.0
	Stability	Qualified	-
	Loss on ignition (%)	2.30	$\leq$5.0
Chemical properties	SO_3 (%)	2.33	$\leq$3.5
	Cl^- (%)	0.036	$\leq$0.06
	Mixed admixture (%)	6.0	>5 and $\leq$20

[1] GB 175-2007 [36].

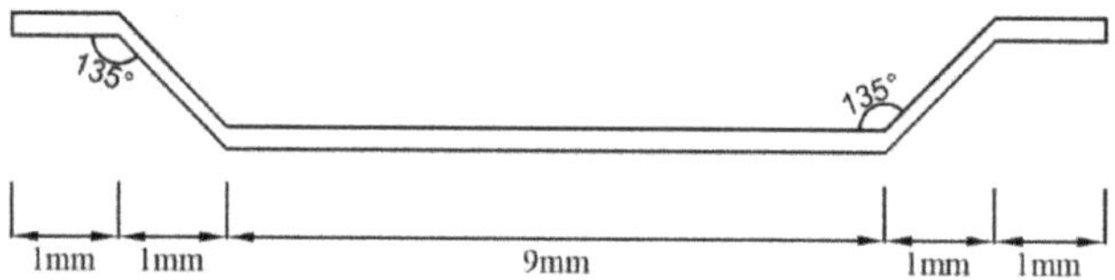

Figure 1. Shape diagram of steel fiber.

Table 2. Physical properties of steel fiber.

Index	Length (mm)	Diameter (mm)	Density (g/cm^3)	Tensile Strength (MPa)
Unit value	13	0.2	7.8	$\geq$2850

Table 3. Physical properties of natural river sand.

	Properties	Value
	Fineness modulus	2.10
	Bulk density (kg/m^3)	1570
Physical properties	Tight density (kg/m^3)	1740
	Apparent density (kg/m^3)	2550
	Water absorption of the dry saturation surface (%)	1.3
	Dust content (%)	0.5

Table 4. Physical properties of manufactured sand.

	Properties	Value
	Fineness modulus	2.10
	Bulk density (kg/m^3)	1720
	Tight density (kg/m^3)	1900
	Apparent density (kg/m^3)	2560
Physical properties	Water absorption of the dry saturation surface (%)	1.0
	Crash value index (%)	5.4
	Crusher dust content (%)	4.5
	Methylene blue value (g/kg)	0.5
	Soundness (%)	6

2.2. Sample Preparation

The workability of concrete is characterized by the slump of the specification GB/T 50080-2016 [37]. The test process is shown in Figure 2, and the mix proportion and slump spread are listed in Table 5. UHPC mixture was mixed by SJD60 forced mixer. First, the inner wall of the mixer was wiped with wet cloth. During mixing, dry powders, including cement, silica fume, fly ash, superplasticizer and expansive agent were mixed for 1–2 min. Water was then added and mixed for approximately 5 min. MS or RS was then added and mixed for approximately 3 min. Steel fibers were added uniformly and mixed for 3 min after forming flowing mortar. Finally, the mixture was cast in the mold and compacted with a mechanical vibration table until there was no bubble and covered with a layer of film. After 48 h in a room, the samples were removed from the molds and subjected to four different curing conditions.

(a) MS

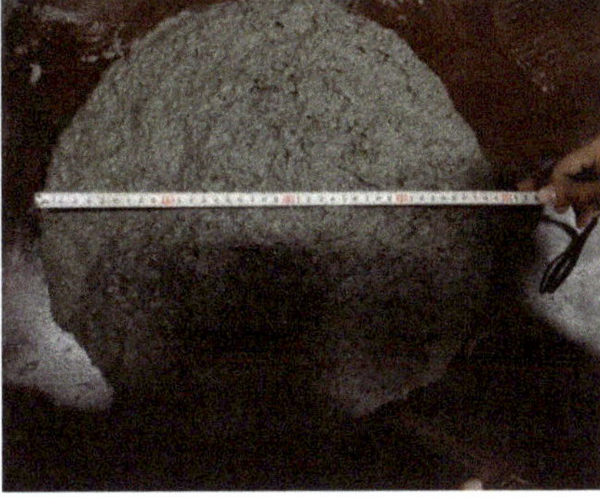

(b) RS

Figure 2. Concrete slump spread test.

Table 5. Mixture proportions of UHPC (kg/m^3).

Group	Cement	Fly Ash	Silica Fume	Superplasticizer	MS	RS	Steel Fiber	Water	Slump Spread (mm)
MS	1000	100	200	40.2	1200	-	235.5	221	425
RS	1000	100	200	40.2	-	1200	235.5	221	395

2.3. Test Methods

2.3.1. Mechanical Tests

The mechanical properties of the specimen are characterized according to the compressive, tensile and bending strength of specification T/CECS 864-2021 [38]. The compressive strength was performed at a loading rate of 1.2 MPa/s on 100 mm × 100 mm × 100 mm specimens. The samples were tested by YAW-3000 (microcomputer controlled electrohydraulic pressure testing machine) at 3, 7, 14, 28 and 56 days for four different curing conditions; the loading process is shown in Figure 3a. The uniaxial tensile strength was performed at a loading rate of 0.2 mm/min on dog-bone specimens. The samples were tested by WDW-100 (microcomputer controlled electronic universal testing machine) at 28 days for four different curing conditions; Lh-S05 tension sensor, YWC-100 strain displacement sensor and DN3816N static stress-strain system are used to collect and record the information of load and displacement, respectively; the loading process is shown in Figure 3b. The bending strength was performed at a loading rate of 0.1 mm/min on 100 mm × 100 mm × 400 mm specimens. The samples were loaded by four-point bending and tested by WAW-600B (microcomputer controlled electrohydraulic pressure testing machine) at 28 days for four different curing conditions; the loading process is shown in Figure 3c.

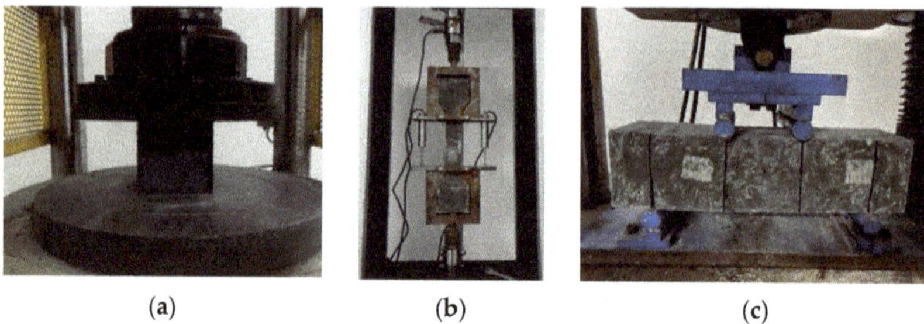

(a)   **(b)**   **(c)**

Figure 3. Mechanical property loading test of manufactured sand UHPC. (**a**) Compressive test. (**b**) Uniaxial tensile test. (**c**) Bending test.

2.3.2. Steel Fiber Pullout Test

A steel fiber pullout test was used to study 3, 7, 14 and 28 days bond strength between steel fiber and matrix according to specification CECS 13-2009 [39]. The 8-shaped specimens were used for the pullout test, which was placed a 0.5–1 mm thick plastic diaphragm at the minimum section with four round holes. Four steel fibers were placed in the round holes and fixed with a spacing of 15 mm. The steel fiber on both sides of the diaphragm was at the embedded end and the fixed end, respectively; the length of the embedded end was 5 mm, and five specimens were tested in each group. The test was performed at a loading rate of 0.5 mm/min by WDW-20 (microcomputer controlled electronic universal tensile testing machine), and stopped when steel fiber slip exceeded 2.5 mm, the loading process is shown in Figure 4.

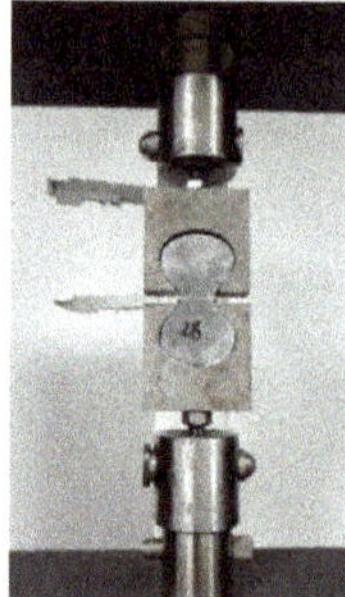

Figure 4. Steel fiber pullout test.

The bond strength between steel fibers and the UHPC cement matrix, f_{fb}, can be calculated according to Equation (1). The pullout work, W_{con}, which can be reflected by the area enclosed by the Load-slip curves and x axes, can be calculated according to Equation (2).

$$f_{fb} = \frac{F_{max}}{4u_f l_{em}} \tag{1}$$

$$W_{con} = \int F dx \tag{2}$$

In Equation (1), F_{max} is the maximum pullout load, μ_f is fiber diameter and l_{em} represents the embedment length of the target fiber in the concrete matrix.

2.3.3. Microscopic Test

The samples used for the microscopic test were taken from UHPC specimens with four different curing methods and 28 days age. Scanning electron microscopy (SEM, Hitachi S4800, Tokyo, Japan) was used to analyze the morphology and the internal microstructure of concrete under different curing conditions. Small pieces of samples (about 3 mm thick at least) were selected, and the surface of the samples was sprayed with Au to make them better conductive. The test condition was in a vacuum and the test voltage was 15 kV. X-ray diffraction (XRD, Rigaku Ultima IV, Tokyo, Japan) using a diffractometer at 40 kV and 35 mA. UHPC samples were ground to 200 mesh in a mortar before the test. The pore structure of the UHPC samples was analyzed using mercury intrusion porosimetry (MIP, Micromeritics AutoPore IV-9500, Atlanta, GA, USA). Firstly, the specimens were broken into approximately 5 mm pieces, and then to stop hydration the specimens were soaked in ethyl alcohol. The samples were dried at 50 °C in a drying oven for 20 h before experiment. The specimen tests were then carried out under high pressure of 413.70 MPa and low pressure of 0.28 MPa, respectively.

3. Results and Discussion

3.1. Mechanical Properties

3.1.1. Compressive Strength

The results of the UHPC compressive strength test are shown in Figure 5. As can be seen, under SM, the compressive strength of the MSUHPC developed rapidly, and the early compressive strength was much higher than that under the other three curing methods [40]. At the curing age of 3 days, the compressive strength of the MSUHPC under SM reached 149.7 MPa, while the compressive strength under SD, SP and CH were 80.9 MPa, 107.0 MPa and 105.0 MPa, respectively. Under SM, the later compressive strength of UHPC increased less, and the compressive strength even decreased slightly at the age of 56 days. This is because early hydration rate of UHPC is faster under the SM, and the hydration products quickly wrap the cement particles, which leads to the limited development of the later strength. Compared with normal temperature curing, SM promotes the formation of early hydration product C-S-H gel, which has longer molecular chain. However, when the molecular chain of C-S-H gel is too long, it will hinder the continuous formation of later hydration products and affect the later strength development of MSUHPC [41,42].

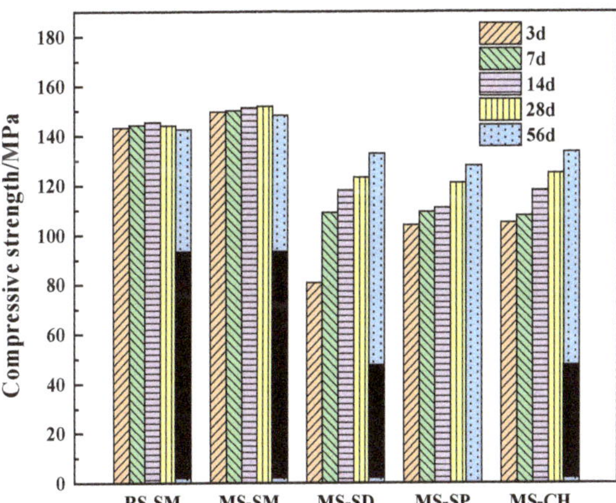

Figure 5. Compressive strength of manufactured sand UHPC at different ages under different curing methods.

With the increase of age, the gap between the compressive strength of MSUHPC with other three curing methods and SM is gradually narrowed. CH can better improve the compressive strength of MSUHPC. Among them, the size rule of 28 days compressive strength of the material is as follows: CH > SD > SP, reaching 82.2%, 81.1% and 79.6% of the compressive strength of SM at the same age, respectively. When the curing age was 56 days, the compressive strength of CH, SD and SP reached 90.2%, 89.6% and 86.3% of that of SM, respectively.

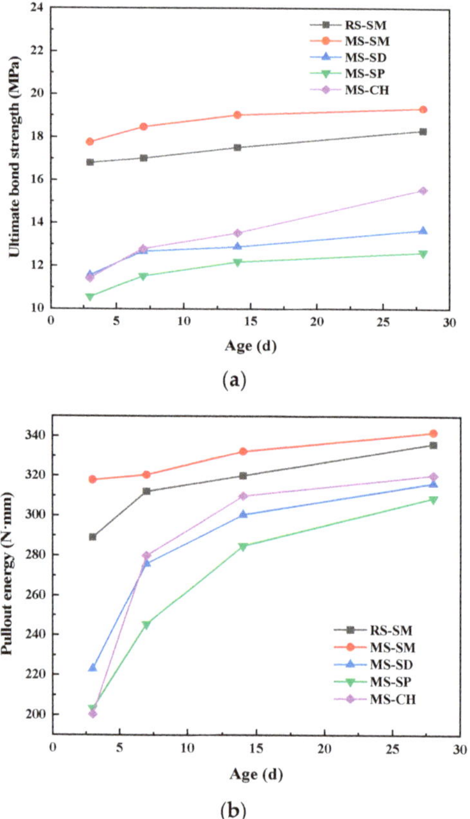

Figure 9. Bonding properties of manufactured sand UHPC at different ages under different curing methods. (**a**) Ultimate bond strength. (**b**) Pullout energy.

The bonding properties and force failure characteristics of steel fibers and UHPC matrix are mainly related to three factors: steel fiber, UHPC matrix and the bonding force between them [45]. The curing method can change the strength and shrinkage characteristics of the UHPC matrix itself and then affect its bonding force to the steel fiber. It can be seen from Figures 8 and 9 that under different curing methods, when the curing age increases from 3 days to 28 days, the load-slip curves of steel fiber and matrix are basically consistent. The ultimate load of the load-slip curve of steel fiber and MSUHPC matrix and the size of the area enclosed by the curve and the abscissa are: SM > CH > SD > SP; this shows that under SM, the fiber—matrix pullout energy is the highest, and the bonding performance of steel fiber and MSUHPC is the strongest. This also shows that the fiber—matrix pullout energy increases with the strength of the cemented matrix [46]. Under SM, compared with RSUHPC, steel fiber and MSUHPC matrix have higher ultimate load and better fiber—matrix bonding performance.

Under the four curing methods, the ultimate bond strength and fiber pullout energy of steel fiber and MSUHPC increased with the increase of curing age. Under SM, when the curing age is 3 days, the bond strength and fiber pullout energy of steel fiber and MSUHPC matrix are significantly higher than those of the other three curing methods, and the later strength and energy growth level are relatively flat. This is because SM can promote the hydration reaction at an early stage, thereby increasing the strength of the cementitious matrix. Under SD, SP and CH, with the increase of age, the hydration degree of the MSUHPC matrix is also increasing, and its ultimate bond strength with steel fiber and fiber pullout energy have a high increase. Under SM, the ultimate bond strength and fiber pullout energy of steel fiber and MSUHPC matrix increased slightly, and the 28 days age increased by 8.9% and 7.68% compared with the 3 days age, respectively. However, the ultimate bond strength of the other three curing methods increased by more that 16%, and the fiber pullout energy increased by at least 41%. From 3 days to 28 days, the ultimate bond strength and pullout energy increased the most under CH mode, by 36.2% and 57.8%, respectively.

3.3. Mechanism Analysis

3.3.1. SEM Experiment

The microscopic morphologies of MSUHPC under different curing methods are shown in Figure 10. Under SM, a large amount of acicular substance can be observed in RSUHPC, which is the acicular ettringite (AFt) formed by the further reaction of calcium aluminate hydrate (C_3AH_6) produced in the cement hydration process with gypsum in the cement (Figure 10a). The MSUHPC hydration products are tightly wrapped with steel fibers and have compact microstruture (Figure 10b). Therefore, MSUHPC exhibits better mechanical properties and stronger fiber—matrix adhesion than RSUHPC. Under CH and SD, a large amount of C-S-H gel and acicular ettringite (AFt) were generated in the MSUHPC, and many incompletely hydrated particles were also observed (Figure 10c,d), indicating that the hydration reaction of the cementitious material is insufficient. Under SP, the MSUHPC has many pores, relatively loose structure and relatively weak bonding force with fiber. $Ca(OH)_2$ crystal and more spherical fly ash particles can be obviously observed in the matrix, indicating that the degree of secondary hydration reaction is low (Figure 10e,f).

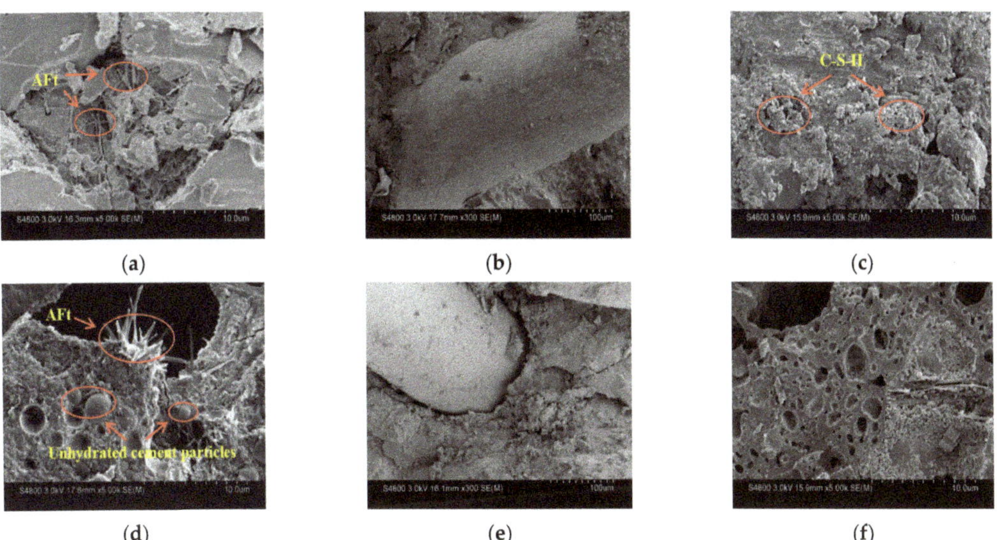

Figure 10. SEM images of manufactured sand UHPC samples under different curing methods. (a) RS-SM; (b) MS-SM; (c) MS-CH; (d) MS-SD; (e) MS-SP; (f) MS-SP.

3.3.2. XRD Experiment

The XRD patterns of MSUHPC under different curing methods are shown in Figure 11. The crystal phases measured mainly include SiO_2, $CaCO_3$, $Ca(OH)_2$, C_3S, C_2S, AFt and C-S-H. Among them, the MSUHPC uses limestone MS as the fine aggregate, so a strong $CaCO_3$ diffraction peak is detected, while the RSUHPC uses the RS as the fine aggregate, so a strong SiO_2 diffraction peak is detected. The $Ca(OH)_2$ diffraction peak under SM is relatively weak, which confirms the promotion effect of SM on secondary hydration reaction of UHPC matrix. C_3S and C_2S diffraction peaks were found in each group of MSUHPC, because the water–binder ratio of MSUHPC was extremely low, the cement hydration reaction was insufficient, and there was unhydrated cement clinker in the sample. Among them, the diffraction peaks of C_3S and C_2S under SP are relatively strong, indicating that due to the relatively low hydration degree of cementitious materials under SP, there are more unhydrated cement clinkers.

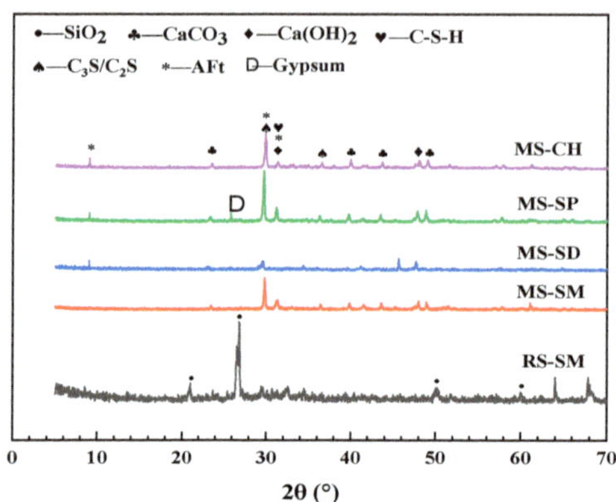

Figure 11. XRD diffraction patterns of manufactured sand UHPC under different curing methods.

3.3.3. MIP Experiment

The pore size distribution curves of MSUHPC under different curing methods are shown in Figure 12. The porosity, the most probable pore size and the pore fractal dimension are shown in Table 7.

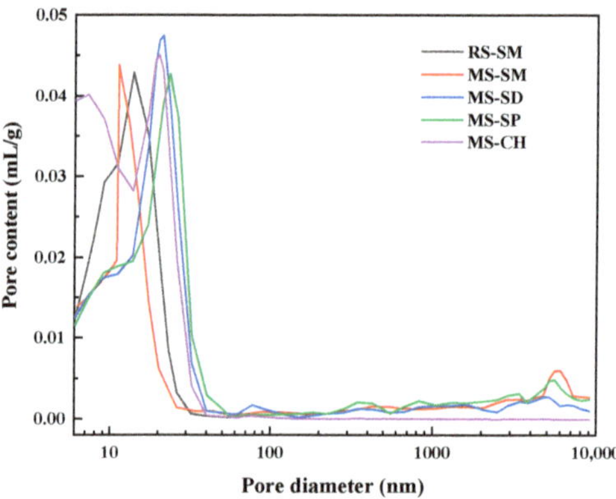

Figure 12. Pore size distribution of manufactured sand UHPC under different curing methods.

Table 7. Pore index of manufactured sand under different curing methods.

Specimen NO.	Porosity (%)	Most Probable Pore Size (nm)	Fractal Dimension
RS-SM	2.97	13.74	2.855
MS-SM	2.48	11.10	2.872
MS-SD	6.99	21.09	2.821
MS-SP	7.59	23.43	2.698
MS-CH	6.79	19.93	2.827

It can be seen from Table 7 that the total porosity of MSUHPC under different curing methods is as follows: SM < CH < SD < SP. During SM, high temperature accelerates cement hydration and promotes pozzolanic reaction to generate more hydration products. The material structure is more compact, the total porosity is the lowest, and the most probable pore size is the smallest. The porosity index under CH and SD is relatively close, and the porosity is 2.74 times and 2.81 times that under SM, respectively. The porosity of SP is the highest, and the porosity is 3.06 times that of SM. This is because SP cannot continuously provide water for the hydration process of MSUHPC, resulting in the slowing down of its hydration rate and the reduction of hydration products.

The fractal dimension can reflect the pore characteristics of concrete and quantitatively describe the roughness and complexity of concrete pore surface [47,48]. The larger the pore fractal dimension, the more complex the spatial geometric features representing the pores [49]. Comparing the pore fractal dimension of MSUHPC under different curing methods, it is found that SM is the largest, followed by CH and SD, and SP is the smallest. This shows that more hydration products are produced during SM, which makes the matrix structure more compact, the macropores decrease, the small pores increase and the specific surface area of the pores increases. The fractal dimension of CH and SD is close. The fractal dimension of SP is the lowest, and there are relatively more macropores in the material, and the pore complexity is low. Under SM, the pore fractal dimension of the MSUHPC is larger than that of the RSUHPC, which indicates that the pore distribution inside the MSUHPC material is more complex, and the proportion of small pores is higher; this is because the RS has a more regular shape and a smoother surface than the MS, and produces less pore complexity when combined with the matrix.

4. Conclusions

In this research, UHPC was prepared by fully replacing RS with MS. The effects of different curing methods on the mechanical properties of MSUHPC and the bonding properties between steel fibers and MS were analyzed through mechanical property tests combined with microscopic tests. The following conclusions were drawn:

(1) Under SM, the compressive strength of MSUHPC increased greatly in the early stage, and the later strength increased less. The 28 days compressive strength increased by only 1.5% compared with the 3 days compressive strength. The 28 days compressive strength of CH, SD and SP increased by 19.1%, 52.3% and 16.3% respectively compared with 3 days compressive strength. In addition, the compressive strength of four curing methods was only negatively increased in SM at 56 days. The development law of bending strength and compressive strength of MSUHPC under different curing methods is similar, but due to the significant influence of steel fiber on the bending properties, the difference in bending strength of the four curing methods is small.

(2) Under SD, SP and CH methods, the tensile properties of MSUHPC do not show significant differences. Compared with the other three curing methods, SM method increased the initial crack strength and tensile strength of MSUHPC by more than 19%. The ratio of tensile strength to initial crack strength of MSUHPC is greater than 1.1 under all four curing methods and shows strain hardening characteristics. The tensile properties of MSUHPC are slightly better than RSUHPC under SM method.

(3) The ultimate bond strength of steel fiber and MSUHPC matrix under SM has a small increase in the later period, and the 28 days ultimate bond strength only increases by 8.9% compared with the 3 days ultimate bond strength. The 28 days ultimate bond strength of CH, SD and SP increased by 36.2%, 18.3% and 16.5% respectively compared with 3 days ultimate bond strength. Moreover, compared with the other three curing methods, SM shows that the increase of fiber pullout energy of MSUHPC is more different. The SM 28 days fiber pullout energy increases by 7.7% compared with 3 days fiber pullout energy, while the other three curing methods all increase by 40%.

(4) Under SM, the porosity and the most propable pore diameter of MSUHPC were lower, and the Ca(OH)$_2$ diffraction peak is relatively weak, indicating that SM promoted the secondary hydration reaction of UHPC matrix. However, SM has an adverse

effect on the long-term performance of concrete, so it needs to be kept wet in the later stage to eliminate its adverse effects. Under SD, CH and SP, the porosity and the most probable pore diameter of MSUHPC are relatively close, but larger than the SM method. The pore fractal dimension of MSUHPC is the largest under SM, indicating that the pore distribution in the material is more complex and the proportion of small pores is higher.

Author Contributions: Conceptualization, C.Y.; data curation, S.X. and D.W.; formal analysis, S.X.; funding acquisition, C.Y.; investigation, D.W.; project administration, C.Y.; resources, C.Y.; supervision, C.Y.; visualization, S.X. and A.R.; writing-original draft, S.X., C.W. and D.W.; writing-review and editing, S.X., A.R. and C.W. All authors have read and agreed to the published version of the manuscript.

Funding: This study was funded by the Natural Foundation of China (grant numbers 52178258), Henan Province key research and development and promotion projects (202102310255), the Henan Province transportation science and technology plan project (2020J-2-7) and the Henan Provincial department of transportation Science and Technology plan project (2021J3).

Institutional Review Board Statement: Not applicable.

Informed Consent Statement: Not applicable.

Data Availability Statement: The general data are included in the article. Additional data are available on request.

Conflicts of Interest: The authors declare no conflict of interest.

References

1. De Larrard, F.; Sedran, T. Optimization of Ultra-High-Performance Concrete by the Use of a Packing Model. *Cem. Concr. Res.* **1994**, *24*, 997–1009. [CrossRef]
2. Richard, P.; Cheyrezy, M.H. Reactive Powder Concretes with High Ductility and 200-800 MPa Compressive Strength. *Spec. Publ.* **1994**, *144*, 507–518. [CrossRef]
3. Gu, C.; Ye, G.; Sun, W. Ultrahigh Performance Concrete-Properties, Applications and Perspectives. *Sci. China-Technol. Sci.* **2015**, *58*, 587–599. [CrossRef]
4. Wang, J.; Wang, Z.; Tang, Y.; Liu, T.; Zhang, J. Cyclic Loading Test of Self-Centering Precast Segmental Unbonded Posttensioned UHPFRC Bridge Columns. *Bull. Earthq. Eng.* **2018**, *16*, 5227–5255. [CrossRef]
5. Akeed, M.H.; Qaidi, S.; Ahmed, H.U.; Faraj, R.H.; Mohammed, A.S.; Emad, W.; Tayeh, B.A.; Azevedo, A.R.G. Ultra-High-Performance Fiber-Reinforced Concrete. Part I: Developments, Principles, Raw Materials. *Case Stud. Constr. Mater.* **2022**, *17*, e01290. [CrossRef]
6. You, W.; Bradford, M.A.; Liu, H.; Zhao, W.; Yang, G. Steel-Alkali Activated Cement Based Ultra-High Performance Concrete Lightweight Composite Bridge Decks: Flexural Behavior. *Eng. Struct.* **2022**, *266*, 114639. [CrossRef]
7. Zhang, X.Y.; Yu, R.; Zhang, J.J.; Shui, Z.H. A Low-Carbon Alkali Activated Slag Based Ultra-High Performance Concrete (UHPC): Reaction Kinetics and Microstructure Development. *J. Clean. Prod.* **2022**, *363*, 132416. [CrossRef]
8. Sun, M.; Yu, R.; Jiang, C.; Fan, D.; Shui, Z. Quantitative Effect of Seawater On the Hydration Kinetics and Microstructure Development of Ultra High Performance Concrete (UHPC). *Constr. Build. Mater.* **2022**, *340*, 127733. [CrossRef]
9. Sobuz, H.R.; Visintin, P.; Mohamed Ali, M.S.; Singh, M.; Griffith, M.C.; Sheikh, A.H. Manufacturing Ultra-High Performance Concrete Utilising Conventional Materials and Production Methods. *Constr. Build. Mater.* **2016**, *111*, 251–261. [CrossRef]
10. Shen, W.; Yang, Z.; Cao, L.; Cao, L.; Liu, Y.; Yang, H.; Lu, Z.; Bai, J. Characterization of Manufactured Sand: Particle Shape, Surface Texture and Behavior in Concrete. *Constr. Build. Mater.* **2016**, *114*, 595–601. [CrossRef]
11. Guan, M.; Wei, C.; Wang, Y.; Lai, Z.; Xiao, Q.; Du, H. Experimental Investigation of Axial Loaded Circular Steel Tube Short Columns Filled with Manufactured Sand Concrete. *Eng. Struct.* **2020**, *221*, 111033. [CrossRef]
12. Zhao, S.; Ding, X.; Zhao, M.; Li, C.; Pei, S. Experimental Study On Tensile Strength Development of Concrete with Manufactured Sand. *Constr. Build. Mater.* **2017**, *138*, 247–253. [CrossRef]
13. Selva Ganesh, M.; Jagadeesh, P. Assessment of Usage of Manufactured Sand and Recycled Aggregate as Sustainable Concrete: A Review. *Mater. Today Proc.* **2022**, *64*, 1029–1034. [CrossRef]
14. Li, B.; Ke, G.; Zhou, M. Influence of Manufactured Sand Characteristics On Strength and Abrasion Resistance of Pavement Cement Concrete. *Constr. Build. Mater.* **2011**, *25*, 3849–3853. [CrossRef]
15. Ji, T.; Chen, C.; Zhuang, Y.; Chen, J. A Mix Proportion Design Method of Manufactured Sand Concrete Based On Minimum Paste Theory. *Constr. Build. Mater.* **2013**, *44*, 422–426. [CrossRef]

16. Kottukappalli, F.G.; Ramkrishnan, R.; Sathyan, D.; Mini, K.M. Strength Properties of Concrete Blocks with Sand Manufacture Sludge as Partial Replacement to Fine Aggregate. *Mater. Today Proc.* **2018**, *5*, 23733–23742. [CrossRef]

17. Han, Z.; Zhang, Y.; Qiao, H.; Feng, Q.; Xue, C.; Shang, M. Study On Axial Compressive Behavior and Damage Constitutive Model of Manufactured Sand Concrete Based On Fluidity Optimization. *Constr. Build. Mater.* **2022**, *345*, 128176. [CrossRef]

18. Mane, K.M.; Joshi, A.M.; Kulkarni, D.K.; Prakash, K.B. Influence of Retempering On Properties of Concrete Made with Manufactured Sand and Industrial Waste. *Clean. Mater.* **2022**, *4*, 100060. [CrossRef]

19. Li, Y.; Liu, Y.; Jin, C.; Mu, J.; Li, H.; Liu, J. Multi-Scale Creep Analysis of River Sand and Manufactured Sand Concrete Considering the Influence of ITZ. *Constr. Build. Mater.* **2022**, *344*, 128175. [CrossRef]

20. Skare, E.L.; Sheiati, S.; Cepuritis, R.; Mørtsell, E.; Smeplass, S.; Spangenberg, J.; Jacobsen, S. Rheology Modelling of Cement Paste with Manufactured Sand and Silica Fume: Comparing Suspension Models with Artificial Neural Network Predictions. *Constr. Build. Mater.* **2022**, *317*, 126114. [CrossRef]

21. Zhang, P.; Kang, L.; Zheng, Y.; Zhang, T.; Zhang, B. Influence of SiO2 /Na2O Molar Ratio On Mechanical Properties and Durability of Metakaolin-Fly Ash Blend Alkali-Activated Sustainable Mortar Incorporating Manufactured Sand. *J. Mater. Res. Technol.* **2022**, *18*, 3553–3563. [CrossRef]

22. Duan, Y.; Wang, Q.; Yang, Z.; Cui, X.; Liu, F.; Chen, H. Research On the Effect of Steam Curing Temperature and Duration On the Strength of Manufactured Sand Concrete and Strength Estimation Model Considering Thermal Damage. *Constr. Build. Mater.* **2022**, *315*, 125531. [CrossRef]

23. Ren, Q.; Tao, Y.; Jiao, D.; Jiang, Z.; Ye, G.; De Schutter, G. Plastic Viscosity of Cement Mortar with Manufactured Sand as Influenced by Geometric Features and Particle Size. *Cem. Concr. Compos.* **2021**, *122*, 104163. [CrossRef]

24. Yoo, D.; Banthia, N. Mechanical Properties of Ultra-High-Performance Fiber-Reinforced Concrete: A Review. *Cem. Concr. Compos.* **2016**, *73*, 267–280. [CrossRef]

25. Zdeb, T. An Analysis of the Steam Curing and Autoclaving Process Parameters for Reactive Powder Concretes. *Constr. Build. Mater.* **2017**, *131*, 758–766. [CrossRef]

26. Yazıcı, H.; Deniz, E.; Baradan, B. The Effect of Autoclave Pressure, Temperature and Duration Time On Mechanical Properties of Reactive Powder Concrete. *Constr. Build. Mater.* **2013**, *42*, 53–63. [CrossRef]

27. Wu, Z.; Shi, C.; He, W. Comparative Study On Flexural Properties of Ultra-High Performance Concrete with Supplementary Cementitious Materials Under Different Curing Regimes. *Constr. Build. Mater.* **2017**, *136*, 307–313. [CrossRef]

28. Shi, J.; Liu, B.; Zhou, F.; Shen, S.; Guo, A.; Xie, Y. Effect of Steam Curing Regimes On Temperature and Humidity Gradient, Permeability and Microstructure of Concrete. *Constr. Build. Mater.* **2021**, *281*, 122562. [CrossRef]

29. Liu, B.; Jiang, J.; Shen, S.; Zhou, F.; Shi, J.; He, Z. Effects of Curing Methods of Concrete After Steam Curing On Mechanical Strength and Permeability. *Constr. Build. Mater.* **2020**, *256*, 119441. [CrossRef]

30. Ballim, Y. Curing and the Durability of OPC, Fly Ash and Blast-Furnace Slag Concretes. *Mater. Struct.* **1993**, *26*, 238–244. [CrossRef]

31. Shi, J.; Liu, B.; Zhou, F.; Shen, S.; Dai, J.; Ji, R.; Tan, J. Heat Damage of Concrete Surfaces Under Steam Curing and Improvement Measures. *Constr. Build. Mater.* **2020**, *252*, 119104. [CrossRef]

32. Gao, H.; Liao, H.; Wang, M.; Cheng, F. Reinforcing the Physicochemical Properties of Concrete through Synergism of CO2 Curing and Ca(OH)2 Solution Drenching. *Constr. Build. Mater.* **2021**, *280*, 122546. [CrossRef]

33. Liu, Y.; Wei, Y. Internal Curing Efficiency and Key Properties of UHPC Influenced by Dry Or Prewetted Calcined Bauxite Aggregate with Different Particle Size. *Constr. Build. Mater.* **2021**, *312*, 125406. [CrossRef]

34. Dong, E.; Yu, R.; Fan, D.; Chen, Z.; Ma, X. Absorption-Desorption Process of Internal Curing Water in Ultra-High Performance Concrete (UHPC) Incorporating Pumice: From Relaxation Theory to Dynamic Migration Model. *Cem. Concr. Compos.* **2022**, *133*, 104659. [CrossRef]

35. *JGJ 52-2006*; Standard for Technical Requirements and Test Method of Sand and Crushed Stone (or Gravel) for Ordinary Concrete. China Construction Industry Press: Beijing, China, 2006.

36. *GB 175-2007*; Common Portland Cement. China Construction Industry Press: Beijing, China, 2007.

37. *GB/T 50080-2016*; Standard for Test Method of Performance on Ordinary Fresh Concrete. China Construction Industry Press: Beijing, China, 2016.

38. *T/CECS 864-2021*; Standard for Test Method of Ultra-High Performance Concrete. China Construction Industry Press: Beijing, China, 2021.

39. *CECS 13-2009*; Standard Test Methods for Fiber Reinforced Concrete. China Planning Press: Beijing, China, 2009.

40. Hiremath, P.N.; Yaragal, S.C. Effect of Different Curing Regimes and Durations On Early Strength Development of Reactive Powder Concrete. *Constr. Build. Mater.* **2017**, *154*, 72–87. [CrossRef]

41. Akeed, M.H.; Qaidi, S.; Ahmed, H.U.; Faraj, R.H.; Mohammed, A.S.; Emad, W.; Tayeh, B.A.; Azevedo, A.R.G. Ultra-High-Performance Fiber-Reinforced Concrete. Part II: Hydration and Microstructure. *Case Stud. Constr. Mater.* **2022**, *17*, e01289. [CrossRef]

42. Kang, S.; Lee, J.; Hong, S.; Moon, J. Microstructural Investigation of Heat-Treated Ultra-High Performance Concrete for Optimum Production. *Materials* **2017**, *10*, 1106. [CrossRef]

43. Yang, R.; Yu, R.; Shui, Z.; Guo, C.; Wu, S.; Gao, X.; Peng, S. The Physical and Chemical Impact of Manufactured Sand as a Partial Replacement Material in Ultra-High Performance Concrete (UHPC). *Cem. Concr. Compos.* **2019**, *99*, 203–213. [CrossRef]

44. Kumar, K.P.; Radhakrishna; Ramesh, P.S.; Aravinda, P.T. Effect of Fines On Strength and Durability of Concrete with Manufactured Sand. *Mater. Today Proc.* 2022; *in press.* [CrossRef]
45. Chan, Y.; Li, V.C. Effects of Transition Zone Densification On Fiber/Cement Paste Bond Strength Improvement. *Adv. Cem. Based Mater.* **1997**, *5*, 8–17. [CrossRef]
46. Ding, X.; Zhao, M.; Li, C.; Li, J.; Zhao, X. A Multi-Index Synthetical Evaluation of Pull-Out Behaviors of Hooked-End Steel Fiber Embedded in Mortars. *Constr. Build. Mater.* **2021**, *276*, 122219. [CrossRef]
47. Lan, X.; Zeng, X.; Zhu, H.; Long, G.; Xie, Y. Experimental Investigation On Fractal Characteristics of Pores in Air-Entrained Concrete at Low Atmospheric Pressure. *Cem. Concr. Compos.* **2022**, *130*, 104509. [CrossRef]
48. Wang, L.; Jin, M.; Wu, Y.; Zhou, Y.; Tang, S. Hydration, Shrinkage, Pore Structure and Fractal Dimension of Silica Fume Modified Low Heat Portland Cement-Based Materials. *Constr. Build. Mater.* **2021**, *272*, 121952. [CrossRef]
49. Han, X.; Wang, B.; Feng, J. Relationship Between Fractal Feature and Compressive Strength of Concrete Based On MIP. *Constr. Build. Mater.* **2022**, *322*, 126504. [CrossRef]

 *materials*

Article

A Promising Mortar Produced with Seawater and Sea Sand

Zhigang Sheng [1,*], Yajun Wang [2,3,*] and Dan Huang [4,*]

1   Key Laboratory of Building Collapse Mechanism and Disaster Prevention,
    Institute of Disaster Prevention, China Earthquake Administration, Sanhe 101601, China
2   Zhejiang Key Laboratory of Offshore Marine Engineering Technology, Zhoushan 316002, China
3   School of Marine Engineering Equipment, Zhejiang Ocean University, Zhoushan 316002, China
4   College of Mechanics and Materials, Hohai University, Nanjing 210098, China
*   Correspondence: shengzg999@sina.com (Z.S.); wlxmwyjup@zju.edu.cn (Y.W.); danhuang@hhu.edu.cn (D.H.)

Abstract: The aim of the study is the deep understanding of the essential reactivity of the environmentally friendly mortar by which its applicability can be justified. Created in the study was the environmentally friendly mortar, which helped relieve the increasing requirements on conventional building materials that are produced from exhausted freshwater and river sand nowadays. Seawater (SW) and sea sand (SS) collected from the Eastern Seas of China were used to produce the mortar at various ages, including 10-day, 33-day, and 91-day. Both the curing and working conditions of the mortar were natural marine ones. The physicochemical-mechanical behaviors were investigated using uniaxial compression tests (UCTs), Energy Dispersive Spectrometer (EDS), X-ray Diffraction (XRD), and thermal-field emission scanning electron microscopy (SEM) analysis to understand the essential reactivity of the mortar with age accumulation. The results indicated that hydration products and favorable components were generated promisingly in the mortar: the C-S-H (xCaO·SiO$_2$·zH$_2$O) development was certainly achieved in the critical environment during the curing and working period; the extensive generation of C-A-S-H (CaO·Al$_2$O$_3$·2SiO$_2$·4H$_2$O) helped densify the C-S-H grid, which caused the promising development of the uniaxial compression strength (UCS); the framework porosity of the mortar was restrained effectively due to the development of Friedel's salt that re-bonded the interfacial cracks between SS and the hydration products with the age accumulation in the critical environment. Consequently, UCS and the resistance against damage of the mortar showed increasing behavior even in the critical environment. The study established Friedel's salt working models and strength and damage models to interpret the physicochemical reactivity of the mortar as: the source of the strength and toughness was the proper polymerization between the native saline components and the hydration product mixture generated throughout the production, curing, and application without the leaching phenomenon. The novel models and interpretation of the physicochemical reactivity ensured the applicability of the mortar produced with SW and SS in the critical environment.

Keywords: environment-friendship; mortar; seawater and sea sand; strength and damage

Citation: Sheng, Z.; Wang, Y.; Huang, D. A Promising Mortar Produced with Seawater and Sea Sand. *Materials* **2022**, *15*, 6123. https://doi.org/10.3390/ma15176123

Academic Editor: F. Pacheco Torgal

Received: 7 July 2022
Accepted: 31 August 2022
Published: 3 September 2022

Publisher's Note: MDPI stays neutral with regard to jurisdictional claims in published maps and institutional affiliations.

1. Introduction

Concrete and mortar are the most popular materials (i.e., cement-matrix materials) in infrastructure. However, the exhausting application of river sand has caused many ecological and environmental problems (Figure 1).

Meanwhile, marine resource exploitation requires an efficient construction method that can reduce the freshwater application. The environment friendliness and sustainability were the main concepts that must be included in the novelty of the cement-matrix material composition. Hence, the applicability of seawater and sea sand has been the focus of construction and building materials, which have shown promising potential in many marine engineering cases. Particularly, without steel bars utilization, the direct application of seawater and sea sand in the production of cement-matrix materials achieved environmentally friendly and sustainable characteristics. Moreover, in China, the construction and building materials that used seawater and sea sand directly on-site showed increasing needs (Figure 2).

(**a**) Yellow river (North-west China)

(**b**) Hanjiang river (Central China)

(**c**) Lijiang river (South-west China)

Figure 1. Wide-spreading disastrous cases due to river sand excavation in China.

Figure 2. Sea embankment covers in the Eastern Seas of China; the blocks that composed the covers were produced with SW and SS without reinforcement.

The work of Wang et al. [1] reported that the application of seawater and sea sand was a promising choice for marine structure construction. Their work also indicated that the cement-matrix materials using seawater and sea sand could offer adjustable performance by composition optimization. Karthikeyan and Nagarajan [2] studied the chloride quantification of sea sand and estimated the corresponding treatment in concrete. Dhondy et al. [3] produced concrete materials using seawater and sea sand under natural and unaltered curing conditions. They also investigated the short-term mechanical properties of the concrete using compressive tests. The results showed that the sea sand and seawater concrete held slightly higher strength at 28 days than conventional concrete. Suraksha et al. [4] presented two numerical approaches to study the durability of basalt fiber-reinforced polymer bars in seawater and sea sand concrete solutions subjected to various temperatures. The main goal of their work was to numerically predict the degradation of a basalt fiber-reinforced polymer bar in the marine environment. Sakthieswaran et al. [5] produced a polymer concrete modified using epoxy resin and sea sand. Their study results showed a significant improvement in the compressive and flexural strength due to the sea sand substitution in the polymer concrete. They also indicated that the application margin was unreinforced concrete. Bazli et al. [6–10] offered serial studies on fiber-reinforced polymer materials that were used in seawater and sea sand concrete; they also introduced glass-fiber-reinforced polymer composites using seawater and sea sand. The durability of the composites using seawater and sea sand was the focus. The damage mechanism of the fiber-reinforced polymer composite was also studied in their work. The serial work of Bazli et al. indicated that the mechanical performance of the concrete produced with seawater and sea sand showed no significant degradation under natural conditions. The proper treatment for the fiber-reinforced polymer materials even strengthened the concrete block. Ahmed et al. [11] also paid attention to the fiber-reinforced polymer material application in seawater and sea sand concrete. They pointed out that the appropriate mix design still achieved the needed strength and workability for the field concrete. Vafaei et al. [12,13] studied the sorptivity and microstructural behaviors of the high-strength fiber-reinforced seawater and sea sand concrete. Their work reported that the total porosity of the concrete showed finer characteristics. El-Khoury et al. [14] studied the effect of salinity, cement type, and specimen geometry on the degradation phenomenon of cement-based materials. Their work indicated that the salinity increase did not result in significant volumetric deformation in the specimens. El-Khoury et al. also reported the differentiation between conventional thoughts and their findings on the saline impact. According to the study, they confirmed that the sulfate attack from seawater environments was not the predominant one, and the pore-blocking effect of brucite and calcium carbonate formation near the surface slowed down the ionic diffusion into the cement matrix. Their study also indicated that the mechanical properties of the seawater-attacked specimens still remained equal to or slightly lower than the properties of the reference specimens. Seoung et al. [15] investigated the effects of seawater exposure on the mechanical, durability, and microstructural properties of Portland cement mortars. Their test results revealed that seawater exposure yielded positive effects, including flexural strength and durability improvements. Their interpretation was that the seawater curing, which filled specific size pores of 50–200 μm, generated the additional hydration, i.e., matrix densification. Weerdt et al. [16] studied the chloride ingress in Portland cement mortars exposed to seawater by comparing them with the ingress in mortars exposed to a NaCl solution. After 180 days of exposure to seawater, only the outer 1mm was enriched in sulfur and magnesium, which had only a limited impact on the chloride ingress. Particularly, the decalcification of C-S-H at the exposed surface was less severe. Palin et al. [17] tried to quantify the crack-healing capacity of seawater-submerged mortar, which was produced with blast furnace slag cement. Their study indicated that the cracks measuring 0.2 mm were fully healed after 28 days of submersion in seawater and the precipitation of the saline minerals from the carbonation in the cracks was principally responsible for healing. The study of Santhanam et al. [18] was focused on understanding the physical, chemical, and microstructural differences in sulfate attack from seawater and groundwater. They immersed the Portland cement mortars completely in solutions of seawater and groundwater. The results from their study reported that Portland cement mortars performed better in seawater in comparison with groundwater. Their study also indicated that the high Cl concentration of seawater played an important role by binding the C_3A to form chloroaluminate compounds, including Friedel's salt. Qu et al. [19] investigated the effects of seawater and undesalted sea sand on the hydration products, mechanical properties, and microstructures of Portland cement mortars. Their study showed that the saline content in sea sand accelerated the hydration reaction and increased the early-stage compressive strength. The addition of seawater could significantly accelerate hydration. The positive effect compensated for the side effect of sea sand and produced the cement mortar with seawater and sea sand with the highest compressive strength. Moreover, their study confirmed that the seawater application also upgraded the chemical property of C-S-H; chloride ions in both seawater and sea sand reacted with the C_3A and CH to form Friedel's/Kuzel's salts, which accelerated the hydration reaction and optimized the microstructure of the mortars. The merit of the study of He et al. [20] was the justification of the positive effect of the carbonation in the seawater environment. The mortar using ordinary Portland cement and ion chelator was produced in their study. The investigation of the self-healing behavior of the mortar immersed in seawater showed that calcium carbonate was the key crack-healing product. Zhang et al. [21] evaluated the performance of alkali-activated mortars using seawater and sea sand. Their study indicated that the existence of seawater and sea sand promoted the formation of the C-S-H gel phase. Their mortar also showed higher compressive and flexural strengths at an early age. Zhang et al. interpreted that the natural internal curing effect and the dense interfacial transition zone in the mortar were the main causes. The study of Li et al. [22] confirmed that the alumina-rich cementitious materials effectively improved the chloride binding capacity of mortar mixed with seawater and sea sand. The study by Qin et al. [23] also verified the positive effect of the carbonation in the seawater environment. The results from their study reported that the products from the carbonation in the seawater cement mortar played a key role in hydration acceleration through the chemical and crystal nuclei mechanisms. There still exists, however, the crucial subject of how the essential reactivity of the marine-cementitious materials develops in the critical environment. The subject is determined to establish the marine epoch of material science and technology. The experimental study in this paper was conducted to explore the essential reactivity of the mortar produced with SW and SS. The physicochemical-mechanical experimental art was established in the study to interpret the mortar performance. UCTs were carried out in the study to explore the macro-strength and macro-damage characteristics of the mortar at various ages, which reported the mechanical properties. Moreover, the physicochemical behaviors of the mortar were investigated by EDS, XRD, and SEM, which helped understand the physicochemical mechanism. The relationship between the mechanical properties and physicochemical mechanism was discussed in the paper. The key novelty of the study is just the deep understanding of the essential reactivity derived from the physicochemical-mechanical behaviors, which ensures the rational application of the cement-matrix materials using SW and SS, including the mortar [24].

2. Materials and Methods

2.1. Mortar Composition

The study ideology and design composition of the mortar (Table 1) were conceptualized at the School of Marine Engineering Equipment of Zhejiang Ocean University (in Zhoushan, China). The mortar specimens in the study were produced in the Zhejiang Key Laboratory of Offshore Marine Engineering Technology of Zhejiang Ocean University (in Zhoushan, China). The key gravity ratios included SW:cement = 0.458:1 and SS:cement = 0.757:1.

The following information was offered to report the characteristics of the parental materials in detail.

SW: The mortar specimens were prepared with SW that was collected from the Eastern Seas of China. The key ions and the saline minerals in SW included: Cl^- at 17,552.126 ppm (i.e., mg/kg), Na^+ at 8704.95 ppm, Mg^{2+} at 962.494 ppm, K^+ at 466.411 ppm, Ca^{2+} at 314.85ppm, SO_4^{2-} at 2650.5534 ppm, $MgCl_2$ at 3774.72 ppm, $CaSO_4$ at 1070.49 ppm, K_2SO_4 at 1040.46 ppm, and Br^- at 32.7351 ppm.

Table 1. Uniaxial compression system parameters.

Extreme Load (kN)	Minimal Load Rate (kN/s)	Maximal Load Rate (kN/s)	Maximal Stroke of Vertical Main-Shaft (mm)	Return Stroke Velocity (mm/s)	Platform Area (mm²)	Power (kW)
300	0.5	30	260	15	1.47×10^4	0.75

SS: SS was drilled from Dongsha shore of the Eastern Seas of China (Figure 3). An SEM image indicated that SS from Dongsha shore achieved superior roughness, which ensured the reliable interfacial cohesion between the SS particle's faces and the components and hydration products mixture of the mortar. Furthermore, the SS used in the study was composed of SiO_2 matrix, Cl^- complex, and the saline minerals of K^+, Al^{3+}, and Na^+. Due to the dense distribution of the Cl^- complex and saline minerals on the particle's faces, anion-cation exchange would happen between the paste and SS, which also contributed to the promising physicochemical reactivity of the mortar, even in the critical environment, i.e., the marine environment. The size distribution of SS was explored by Malvern Mastersizer 2000 analyzer (Malvern Panalytical Ltd., Malvern, UK) (size range of 2×10^{-5} mm–2 mm), and the results on the cumulative percentages are offered in Figure 4.

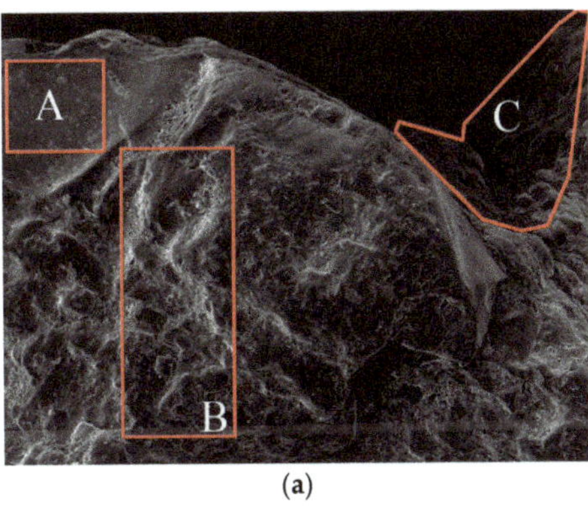

(a)

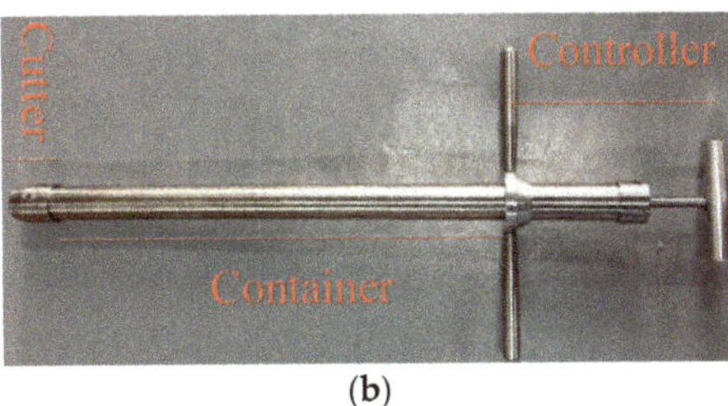

(b)

Figure 3. SS and driller. (**a**) SS image (Zone A was the SiO_2 matrix; Zone B was the Cl^- complex; Zone C was the saline minerals of K^+, Al^{3+}, and Na^+). (**b**) Driller (Cutter helped break the consolidated cover on the shelf; container carried the SS particles; controller was operated by human or machine to press the driller into the deep SS layer).

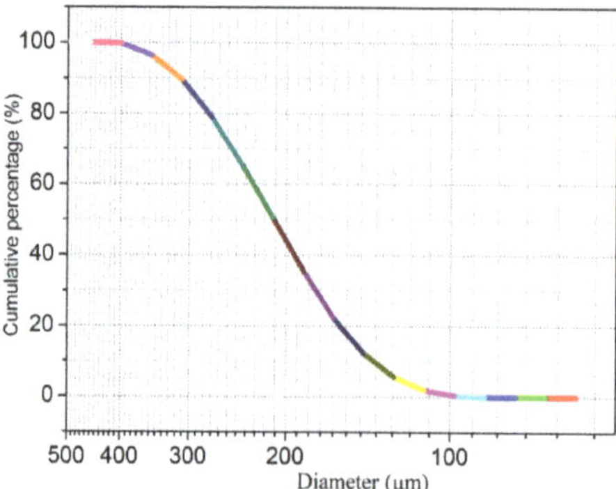

Figure 4. Size distribution of SS.

Based on size distribution, it was proposed that SS from Dongsha shore had uniform and discontinuous characteristics. Therefore, it was deduced that the discontinuousness of SS from Dongsha shore could offer abundant fill space for the components and hydration products mixture.

The study also invited the XRD analysis (BRUKER AXS Inc., Madison, WI, USA) to investigate the SS mineral components, and the results are reported in Figure 5.

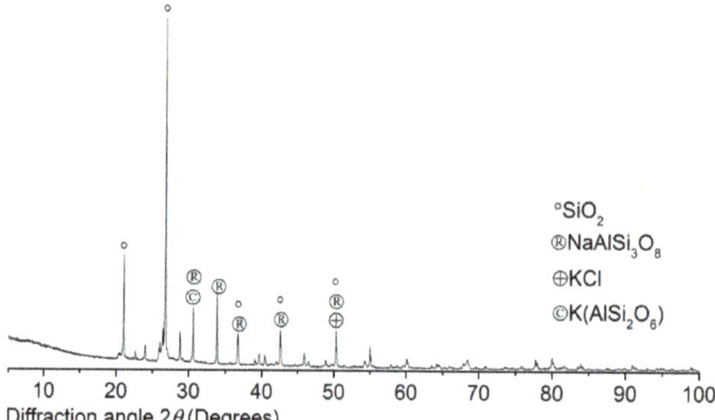

Figure 5. XRD patterns of sea sand.

XRD patterns of the SS from Dongsha shore certainly showed the marine characteristics, and the saline minerals on the particles' faces determined the specific performance of the mortar. The SiO_2 matrix was covered densely by the saline minerals and Cl^- complex. According to the results from XRD, KCl was the predominant Cl^- complex; the main saline minerals included $NaAlSi_3O_8$ and $K(AlSi_2O_6)$. Moreover, the anion-cation exchange between the paste and SS contributes to the physicochemical reactivity of the mortar during the curing and working period in the marine environment.

Cement The composite Portland cement P.C 42.5 R (Hailuo Cement Co. Ltd., Zhangjia-gang City, China) was used in the study to produce the mortar of SW and SS. The specification standards of the composite Portland P.C 42.5 R from CNSMC (Beijing, China) and ASTM were adopted because the two standards had consistency in the physicochemical quality of the composite Portland P.C 42.5 R and inspired the main concept of the study. The chemical composition of the composite Portland P.C 42.5 R was studied by EDS and XRD. EDS investigated the element content (EC) with five spectrums (Figure 6). The spectrums showed obvious consistency in terms of EC. According to the EDS results on the composite Portland cement P.C 42.5 R used in the study, the EC of Al was higher than that of Fe. Hence, it was deduced that the cement was rich in Al^{3+} complexes, including $3CaO·Al_2O_3$ and

$2CaO \cdot Al_2O_3$ and poor in Fe^{2+} and Fe^{3+} complexes, including $4CaO \cdot Al_2O_3 \cdot Fe_2O_3$ and FeO. Based on XRD results (Figure 7), the main components of the composite Portland cement P.C 42.5 R used in the study included CaO, SiO_2, Al_2O_3, FeO, $2CaO \cdot SiO_2$, $3CaO \cdot SiO_2$, $3CaO \cdot Al_2O_3$, and $4CaO \cdot Al_2O_3 \cdot Fe_2O_3$. Therefore, the results of XRD also justified the deduction from EDS. Meanwhile, $3CaO \cdot Al_2O_3$ existed at six dominant peaks in XRD patterns. By contrast, $2CaO \cdot Al_2O_3$ existed at four dominant peaks. Therefore, the predominant chemical component of the composite Portland cement P.C 42.5 is $3CaO \cdot Al_2O_3$, which will accelerate the early 180-day hydration. Hence, it is deduced that the mortar produced with the composite Portland cement P.C 42.5 in the study will reach its compressive strength peak at an early age younger than 180 days. FeO and $4CaO \cdot Al_2O_3 \cdot Fe_2O_3$ lived at one low peak in XRD patterns. Fe^{2+} and Fe^{3+} complexes were not the predominant chemical components. Hence, the results from EDS and XRD showed the exact consistency.

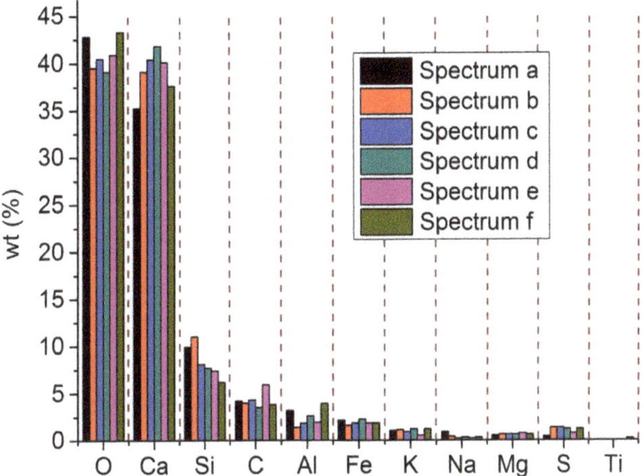

Figure 6. EDS spectrums of composite Portland cement P.C 42.5 R.

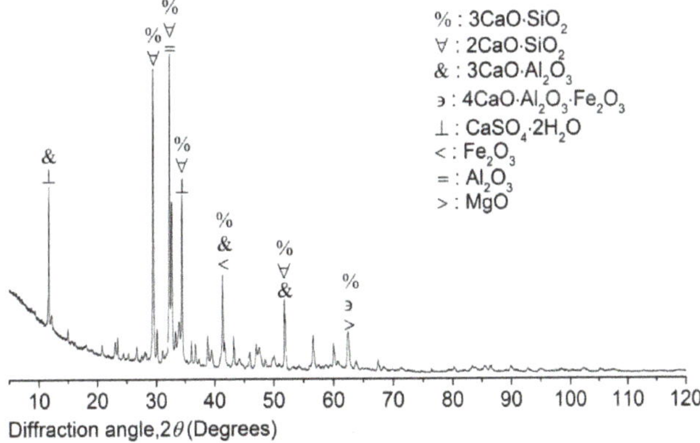

Figure 7. XRD patterns of composite Portland cement P.C 42.5 R.

2.2. Experiments' Standards, Equipment, and Methods

The Sino standards were the predominant references, and the American codes were also invited partly during the composition design and the mortar preparation.

Mixing [25,26]. The mortar paste was prepared in the specific mixer, and the type was UJZ-15 JGJ-90 (maximal power 1.5 kW and capacity 28 L) produced by Xinke company in Shijiazhuang, China. The standard preparation time was 10 min.

Molding [27]. The mortar paste was homogeneously filled into the normal molds (length × width × height = 70.7 × 70.7 × 70.7 mm) that were set on the vibro-machine for vibrating compaction. The key parameters included a vibrating period of 20 min, a vibrating frequency of 45 ± 5 Hz, and a vibrating amplitude of 0.3 mm.

The cubic specimens in the normal molds were kept indoors for 24 h, and the temperature and percentage relative humidity (RH) were 24 ± 5 °C and 85 ± 5%. The cubic specimens were then unloaded from the normal molds when they achieved the initial strength.

Curing [26–28]. The cubic specimens achieving the initial strength were retrieved into the curing chamber where the specific temperature and percentage relative humidity (RH) were 20 ± 5 °C and 92 ± 5%. Furthermore, the cubic specimens were immersed into the tank full of SW that was collected from the Eastern Seas of China, which could imitate the marine environment. The curing ages for the cubic specimens in SW included 9, 32, and 90 days, and the total ages, including the indoor resting period, correspondingly, were 10, 33, and 91 days. The paper referred to the total ages for the following discussion.

Uniaxial compression system [29,30]. The poly-tetra-fluoroethylene (PTFE) plates that were painted with silicon-based grease were used to reduce the shearing friction between the cubic specimens' faces and the apparatus, by which the uniaxial compression load was ensured. The uniaxial compression system parameters are reported in Table 1.

3. Results and Discussion

Based on the mortar composition designed in this paper, the study produced three groups of cubic specimens, the total of which was nine. The physicochemical-mechanical experimental art was established in the study to explore the mortar mechanism, and the key results were reported as follows.

3.1. Mathematics and Discussion on UCT

The study invited the nominal stress in Equation (1) to quantify the UCS, which helped investigate the mechanical-physical performance of the mortar.

$$S = \frac{f_u}{a_0} \tag{1}$$

In Equation (1) f_u represents the maximal load of the uniaxial compression system and a_0 is the non-damaged cross-sectional area of the cubic specimen, the value of which is 70.7 × 70.7 mm^2.

Figure 8 demonstratesthe numerical distribution of S based on Equation (1) and the results from UCTs on nine cubic specimens.

According to the results from UCTs on three groups of the cubic specimens, the S expectation of the mortar with an age of 91 days was above 36 MPa. In terms of the expectation, the S of the cubic specimens with an age of 33 days reached the peak of 36.28 MPa; the S of the cubic specimens with an age of 91 days decreased by 0.7% in comparison with that of the 33-days-old; the S of the cubic specimens with an age of 91 days increased by 49% in comparison with that of the 10-days-old. Hence, the mortar compressive strength in the study generally showed increasing behavior during the working period in the marine environment. Moreover, the mortar's compressive strength reached its peak at an early age of 91 days. Hence, the strength development of the mortar showed obvious consistency with the physicochemical quality of the composite Portland P.C 42.5 R, the predominant chemical component of which was $3CaO \cdot Al_2O_3$, which accelerated the early hydration and strength development. Furthermore, the strength development at an early age partly reduced the damage on the faces of the specimens caused by the salt petering, which impaired the toughness and resistance of the mortar faces. Therefore, the key conclusion of the study was that the mortar produced with SW and SS achieved different physicochemical-mechanical behaviors in comparison with the conventional building materials due to their different physicochemical reactivity.

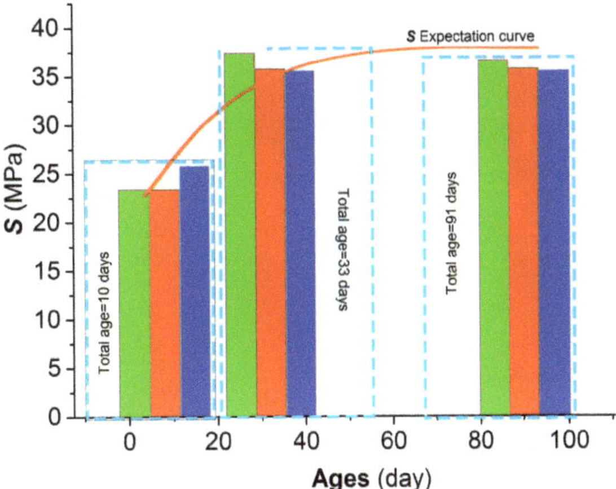

Figure 8. *S* distribution from UCTs.

The salt petering (Figure 9a) generally existed in the mortar during the working period in the marine environment, which was the main cause of why the cubic specimens were damaged on the faces where the saline ions penetrated the framework of the mortar. Particularly, the microscopic young elastic cracks were activated due to the development of Friedel's salt ($3CaO \cdot Al_2O_3 \cdot CaCl_2 \cdot 10H_2O$) in the young specimens.

Friedel's salt in young specimens played the jack role, lifting and breaking the interfacial cracks (Figure 10). Hence, the repulsive force was generated along the interfacial cracks, which activated the development of microscopic young elastic cracks and reduced the toughness and resistance of the mortar faces (Figure 10a); the depth of the young microscopic elastic cracks was below 3 μm, according to SEM results).

The jack-like Friedel's salt and the interfacial cracks formed the statically determinate system. The interfacial cracks would evolve out of control once the statically determinate system was broken, which showed the linear-elastic feature. Consequently, the young specimens failed to generate wider tough cracks except for the microscopic young elastic cracks.

Moreover, in terms of the mortar, the 10-day UCS of which was above 23 MPa in the study, its physicochemical-mechanical behaviors also helped partly constrain the evolution of the microscopic young elastic cracks. The work of Xiao et al. also justified the early strength feature of the cementitious materials produced with SW and SS [31].

On the contrary, the development of Friedel's salt was helpful for the grown specimens because most of the saline ions entered the internal grid of the grown specimens, the interfacial cracks of which were re-bonded due to Friedel's salt's polymerization effect in the marine environment. In the meantime, the open micro-pores in the framework were locked efficiently due to Friedel's salt development in the grown specimens, which upgraded the mortar durability (Figure 9b,c).The specific physicochemical reactivity of the mortar produced in the study with SW and SS promisingly resisted the extreme case of seawater attack in terms of the cement type with >10% $3CaO \cdot Al_2O_3$ content. The key findings on Friedel's salt polymerization effect were also justified partly by Vafaei et al. [12,13].

The glue-like Friedel's salt and the interfacial cracks formed the statically indeterminate system. The interfacial cracks would evolve plastically with the gradual loss of the constraints from the statically indeterminate system, which showed the nonlinear-plastic feature. The grown specimens exactly generated the developed tough cracks during the failure process of the statically indeterminate system.

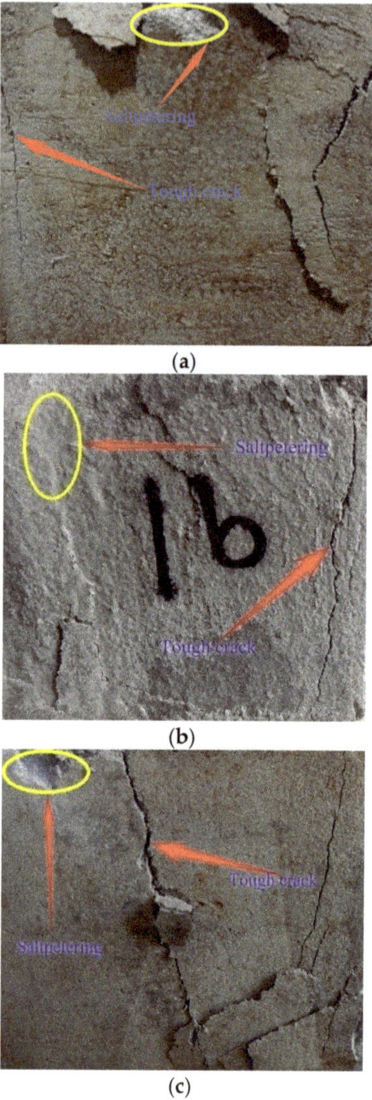

Figure 9. Damaged faces of cubic specimens. (**a**) 10-daysold(maximal width of tough cracks < 1 mm; there lived lower toughness development when the cubic specimen was damaged thoroughly). (**b**) 33-daysold (1 mm < maximal width of tough cracks < 2 mm; the mortar toughness developed certainly when the cubic specimen was damaged thoroughly). (**c**) 91-daysold (2 mm < maximal width of tough cracks < 3 mm; there lived the complete toughness development in the mortar when the cubic specimen was damaged thoroughly).

The typically damaged faces (Figures 9, 10a and 11a,b) from the cubic specimens indicated that the mortar toughness developed with age accumulation. The grown specimens generated wider tough cracks than the young ones under the uniaxial compression condition. The cubic specimens with wider tough cracks offered higher resistance against the load and damage development, which meant that the grown specimens developed superior performance to the young ones. Therefore, the mortar produced in the study achieved promising strength with age accumulation in the marine environment.

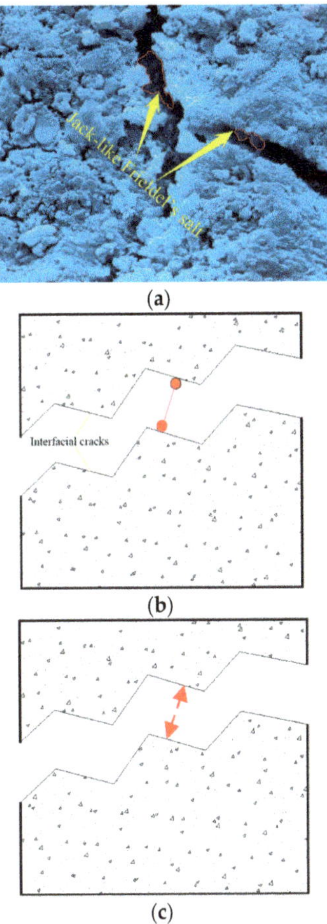

Figure 10. Friedel's salt working model in young specimens. (**a**) SEM image of the 10-day-old specimen. (**b**) Jack working behavior of Friedel's salt; the jack can be represented by a single-direction hinge. (**c**) Repulsive force model of Friedel's salt with jack working behavior.

The physicochemical-mechanical behaviors of the mortar using SW and SS had close similarity with the pressure casting effect. Their key mechanism was the proper polymerization that could ensure the framework of the goal material was as high macroscopic integration and as low damage as possible. In terms of the mortar using SW and SS, the proper polymerization of Friedel's salt in the interfacial cracks contributed promisingly to the toughness and resistance of the mortar without heavy damage to the framework. Meanwhile, the leaching phenomenon of the Ca^{2+} complex in the conventional mortar due to seawater attack did not happen in the mortar using SW and SS (Figure 11a,b). The framework of the hydration products mixture was polymerized properly by the native saline products, including Friedel's salt.

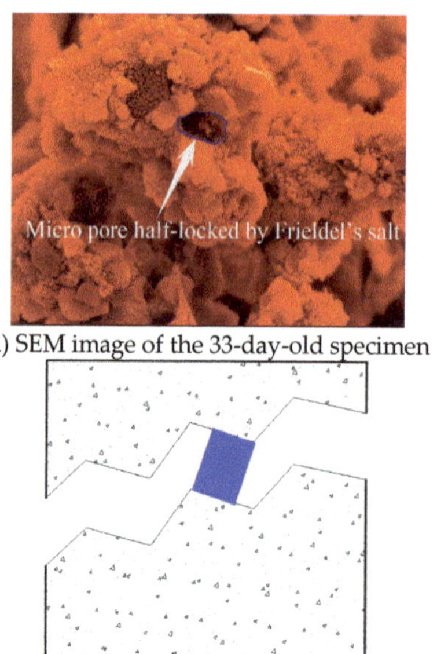

(**a**) SEM image of the 33-day-old specimen

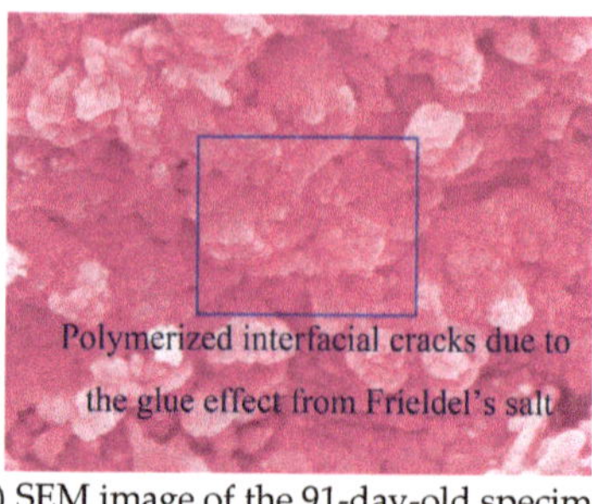

(**b**) SEM image of the 91-day-old specimen

(**c**) Glue working behavior of Friedel's salt; glue can be represented by a built-in zone

(**d**) Attractive force model of Friedel's salt with glue working behavior

Figure 11. Friedel's salt working model in grown specimens. Friedel's salt in grown specimens played the glue role in polymerizing the interfacial cracks (**a**,**b**). Hence, the attractive force was generated along the interfacial cracks (**c**,**d**), which ensured higher toughness and resistance of the mortar under the uniaxial compression condition.

Moreover, the orthogonal damage tensor Ω (Equations (2)–(5)) was created to interpret the mortar strength and toughness developed with age accumulation. Meanwhile, the cubic specimen position in UCTs was expressed in Figure 12.

$$\Omega = \begin{bmatrix} \Omega_1 & & \\ & \Omega_2 & \\ & & \Omega_3 \end{bmatrix} = \begin{bmatrix} \frac{w'_1}{w_{01}} & & \\ & \frac{w'_2}{w_{02}} & \\ & & \frac{w'_3}{w_{03}} \end{bmatrix} \tag{2}$$

$$w'_1 = \max\left(w'_{-1}, w'_{+1}\right) \tag{3}$$

$$w'_2 = \max\left(w'_{-2}, w'_{+2}\right) \tag{4}$$

$$w'_3 = \max\left(w'_{-3}, w'_{+3}\right) \tag{5}$$

where 1, 2, and 3 indicated three principal directions (Figure 12a); uniaxial compression load was applied in direction 1, which was formed by bottom and top faces represented by -1 and $+1$, respectively; w'_{-1} and w'_{+1} designate the maximal widths of the tough cracks on bottom and top faces and w'_1 is the result from the maximizing calculation on w'_{-1} and w'_{+1}; directions 2 and 3 are the free wings of cubic specimen; direction 2 was formed by posterior and anterior faces represented by -2 and $+2$, respectively; w'_{-2} and w'_{+2} designate the maximal widths of the tough cracks on posterior and anterior faces and w'_2 is the result from the maximizing calculation on w'_{-2} and w'_{+2}; direction 3 was formed by on right and left faces represented by -3 and $+3$, respectively; w'_{-3} and w'_{+3} designate the maximal widths of the tough cracks on the right and left faces and w'_3 is the result from the maximizing calculation on w'_{-3} and w'_{+3}; w_{01}, w_{02}, and w_{03} are the side lengths of cubic specimen under the initial state in directions 1, 2, and 3, respectively; the values of w_{01}, w_{02}, and w_{03} in the study are 70.7 mm. All the tough cracks in the study were measured with a vernier caliper (0.1 mm).

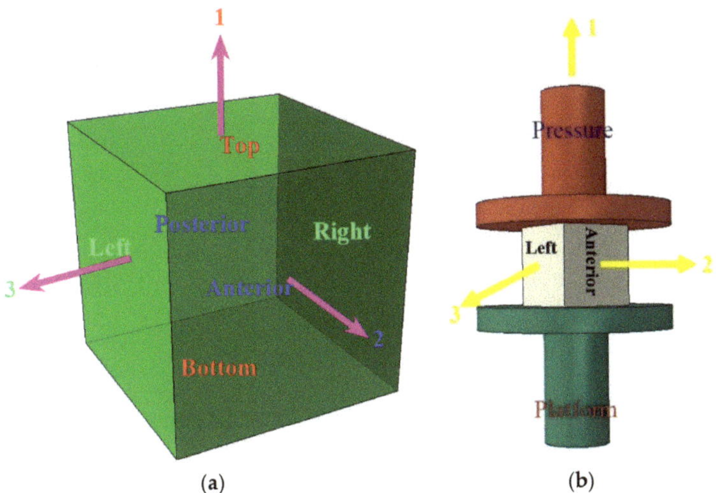

Figure 12. Strength and damage model in UCTs. (**a**) Coordinate system on the orthogonal damage tensor. (**b**) Cubic specimen position. 1 represented the first principal direction where the uniaxial compression load was applied and was formed by bottom and top faces of cubic specimen; 2 was the second principal direction and was formed by posterior and anterior faces; 3 designated the third principal direction and was formed by right and left faces.

The study reported the widths of the tough cracks of nine cubic specimens in Table 2. Meanwhile, the principal damage variables Ω_i (i = 1, 2, and 3) were computed based on Equation (2) and Figure 12b, and the results are offered in Table 2.

Consequently, the resistance tensor against damage $\boldsymbol{R}$ was defined as follows to interpret the macroscopic integration of the mortar developed with age accumulation [32].

$$
\begin{aligned}
\boldsymbol{R} &= \begin{bmatrix} R_1 & & \\ & R_2 & \\ & & R_3 \end{bmatrix} \\
&= \begin{bmatrix} \dfrac{S\varepsilon_0}{(1-\Omega_1)} & & \\ & \dfrac{S\varepsilon_0}{(1-\Omega_2)} & \\ & & \dfrac{S\varepsilon_0}{(1-\Omega_3)} \end{bmatrix} \\
&= \begin{bmatrix} \dfrac{S\varepsilon_0}{\left(1-\left(\frac{w'_1}{w_{01}}\right)\right)} & & \\ & \dfrac{S\varepsilon_0}{\left(1-\left(\frac{w'_2}{w_{02}}\right)\right)} & \\ & & \dfrac{S\varepsilon_0}{\left(1-\left(\frac{w'_3}{w_{03}}\right)\right)} \end{bmatrix} \\
&= S\varepsilon_0 \begin{bmatrix} \dfrac{1}{\left(1-\left(\frac{w'_1}{w_{01}}\right)\right)} & & \\ & \dfrac{1}{\left(1-\left(\frac{w'_2}{w_{02}}\right)\right)} & \\ & & \dfrac{1}{\left(1-\left(\frac{w'_3}{w_{03}}\right)\right)} \end{bmatrix}
\end{aligned} \tag{6}
$$

$$\varepsilon_0 = \frac{D}{w_{01}} \tag{7}$$

where D indicates the maximal displacement of cubic specimen in direction 1; ε_0 is the corresponding strain; R_i ($I = 1, 2,$ and 3) referred to the principal resistance variables against damage in directions 1, 2, and 3, respectively.

Table 2. Mortar damage and resistance characteristics.

Specimens	Direction 1			Direction 2			Direction 3			Ω_1	Ω_2	Ω_3	D_1	ε_{max1}	R_1	R_2	R_3	R_n
	w'_{-1}	w'_{+1}	w'_1	w'_{-2}	w'_{+2}	w'_2	w'_{-3}	w'_{+3}	w'_3									
1	0.2	0.6	0.6	0.6	0.8	0.8	0.7	0.6	0.7	0.008	0.011	0.010	3.3	0.047	1.102	1.105	1.104	3.311
2	0.3	0.6	0.6	0.6	0.9	0.9	0.8	0.6	0.8	0.008	0.013	0.011	3.2	0.045	1.069	1.073	1.072	3.214
3	0.2	0.5	0.5	0.3	0.6	0.6	0.6	0.4	0.6	0.007	0.008	0.008	3.2	0.045	1.177	1.178	1.178	3.533
4	1.0	1.1	1.1	1.2	1.4	1.4	1.2	1.0	1.2	0.016	0.020	0.017	3.0	0.042	1.613	1.619	1.615	4.847
5	1.2	1.3	1.3	1.6	1.8	1.8	1.7	1.4	1.7	0.018	0.025	0.024	3.0	0.042	1.548	1.559	1.557	4.664
6	1.0	1.2	1.2	1.4	1.6	1.6	1.6	1.2	1.6	0.017	0.023	0.023	3.1	0.044	1.589	1.598	1.598	4.783
7	2.0	2.3	2.3	2.1	2.5	2.5	2.5	2.3	2.5	0.033	0.035	0.035	2.8	0.040	1.499	1.503	1.503	4.505
8	2.0	2.6	2.6	2.5	2.9	2.9	2.7	2.2	2.7	0.037	0.041	0.038	2.7	0.038	1.420	1.426	1.422	4.268
9	2.5	2.7	2.7	2.6	3.0	3.0	2.7	2.3	2.7	0.038	0.042	0.038	2.9	0.041	1.519	1.525	1.517	4.563
Unit	mm									/			mm	/	MPa			

Based on the definition of R, the nominal resistance scalar against damage R_n was calculated by Equation (8),

$$R_n = \sum_{i=1}^{3} R_i \tag{8}$$

The study reported the values of the orthogonal damage tensor and resistance tensor against damage of nine cubic specimens in Table 2.

According to the results in Table 2, the grown specimens with wide, tough cracks achieved higher toughness than the young ones with narrow tough cracks. The results showed consistency with the work of El-Khoury et al. [14]. Furthermore, the grown specimens, when reaching their UCS S, had higher resistance against the load and damage than the young ones [33]. Coupled with the physicochemical reactivity, the grown specimens with higher R achieved superior macroscopic integration to the young ones. Therefore, the mortar performance under the uniaxial compression condition was interpreted as the evolution from the microscopic localization of Ω to the macroscopic integration of R. Specifically, the promising macroscopic integration originated from the pressure casting effect due to which the framework of the mortar produced with SW and SS was densified by the proper polymerization of Friedel's salt.

3.2. Results and Discussion on EDS

The study also invited EDS analysis (EDAX Inc., Mahwah, NJ, USA) to explore the mortar composition, and the results are reported in Table 3.

According to the EC of O and C, the oxidization and carbonization were developed during the working period of the mortar. Particularly, the carbonate complex would be polymerized in the study with the hydration products, which formed the framework of the mortar. The EC development of Al, Si, S, and Cl showed the summits at the age of 33 days. The EC of Mg, K, Na, Fe, and Ca decreased with age accumulation. The EC of Si showed the peak feature, which was produced at the age of 33 days. The development of Cl in the mortar produced in the study also created the EC summit at the age of 33 days.

Table 3. Elements' content and distribution feature from EDS analysis.

Ages	C		O		Na	
(Days)	wt (%)	Feature	wt (%)	Feature	wt (%)	Feature
10	9.1		46.6		1.7	
33	8.6		49.3		0.6	
91	8.68		55.7		0.56	

Ages	Mg		Al		Si	
(Days)	wt (%)	Feature	wt (%)	Feature	wt (%)	Feature
10	0.9		1.8		6.8	
33	0.7		2.2		7.3	
91	0.5		0.98		4.47	

Ages	S		Cl		K	
(Days)	wt (%)	Feature	wt (%)	Feature	wt (%)	Feature
10	0.3		0.4		1.5	
33	0.8		1		0.4	
91	0.56		0.23		0.24	

Ages	Ca		Fe	
(Days)	wt (%)	Feature	wt (%)	Feature
10	29.1		1.8	
33	28.2		1	
91	27.69		0.39	

3.3. Resultsand Discussion on XRD

XRD patterns in the study demonstrated that C-S-H was the predominant hydration product (Figure 13). Particularly, the peak height of C-S-H ascended directly from 10-days old to 91-daysold. Therefore, C-S-H developed exactly with age accumulation, which indicated that the physicochemical reactivity of the mortar produced in the study was certainly growing during the working period in the marine environment. The development of CH ($Ca(OH)_2$) showed the saddle feature: XRD reported the upgrowth of CH at 10-day-old and 91-day-olds; the peak height of CH at 33 days old showed a distinct drop. The C-A-S-H was another key hydration product in the mortar. Particularly, XRD patterns in the study exhibited the general existence of C-A-S-H. More peaks of C-A-S-H were identified in XRD patterns than that of CH. The generation of Friedel's salt was also observed. According to XRD patterns, the growth of Friedel's salt created its peak at 33 days old when the Cl EC summit was created consistently. The development of Friedel's salt showed a reduction at 91 days old. The existence of the mixed salt ($Na_2SO_4 \cdot H_2O \cdot 5(CaSO_4) \cdot 2H_2O$) in the mortar came from its low dissolvability. XRD patterns reported that SiO_2 development reached the highest peak at 33 days old, which was consistent with the EDS results that indicated that the mortar generated the corresponding Si EC summit at this age. Furthermore, SS and the clinker were the predominant sources of SiO_2. The mortar also generated the matrix of $CaCO_3$.Correspondingly, the carbonate complex reinforced the framework of the mortar when polymerized with the other hydration products [34]. Based on the results from Xue, the formation of the carbonate complex restrained exactly in the saline environment the side effects of the micro-flaws in the framework of the cementitious materials.

3.4. Resultsand Discussion on SEM

SEM images (Figure 14a, 10-days-old) showed the developing feature of the young mortar at 10 days old: Friedel's salt was growing actively, and the mesoscopic salt petering was generated on the cubic specimens' faces; C-A-S-H with lower density was developing around the C-S-H grid, the porosity of which displayed the loose structure at this age; the maximal size of micro-pores in the framework was 30 μm; the CH platelets were cultivated freely with lower compression from the framework. Therefore, the young mortar achieved only the tender framework, the hydration of which was not completed. With the free development in the micro-pores of the mortar, Friedel's salt did not effectively offer the cohesive and locking function for the framework, except for the superficial salt petering. Correspondingly, the value of the mortar UCS at this age was lower than that at the grown age. However, the cement used in the study was rich in $3CaO \cdot Al_2O_3$. The development of C-A-S-H showed promising potential, which played a key role in the mortar strength increase with age accumulation.

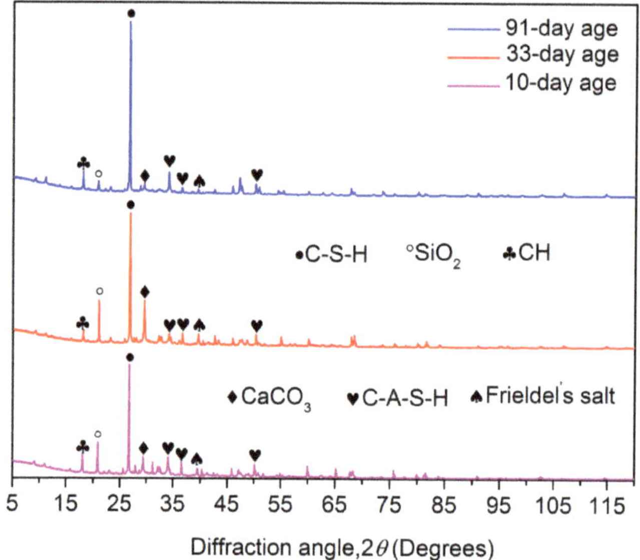

Figure 13. Mortar XRD patterns.

With age accumulation, the gel of C-S-H and C-A-S-H created the mature mortar framework (Figure 14b, 33-day-old) in the marine environment where Friedel's salt stains re-bonded the micro-flaws. The upgrowth of the C-A-S-H crystal clusters densified the C-S-H grid, which caused the promising development of the mortar strength (Figure 8). The maximal size of micro-pores in the framework was compressed to be 2 μm. Consequently, the micro-flaws in the framework were fused and locked by the developed Friedel's salt, and the development of Friedel's salt in the framework of the grown mortar hindered the dilation of the tough macroscopic cracks. Furthermore, the cation exchange, namely, the replacement of Si^{4+} and K^+ on the SS particles faces by Al^{3+} and Ca^{2+} from the paste, raised the interfacial integrity of the mortar. In comparison with the 5μm microscopic young elastic cracks in the mortar at 10 days old, the interfacial cracks in the grown mortar were rehabilitated due to the cation exchange, and the width was below 1 μm. The marine characteristics of the SS used in the study were the main cause of the cation exchange. Therefore, the mortar strength in the study showed a generally increasing behavior during the working period in the marine environment. Meanwhile, the EC summits of Si, Al, and Cl were generated at 33 days old when the development of the hydration products and Friedel's salt reached their peak. Hence, the compaction and cohesion from the hydration products and Friedel's salt were achieved at an intense level. Correspondingly, the expectation of the mortar UCS attained the peak of 36.28 MPa at 33 days old; the *R* of the mortar at this age also achieved the highest level.

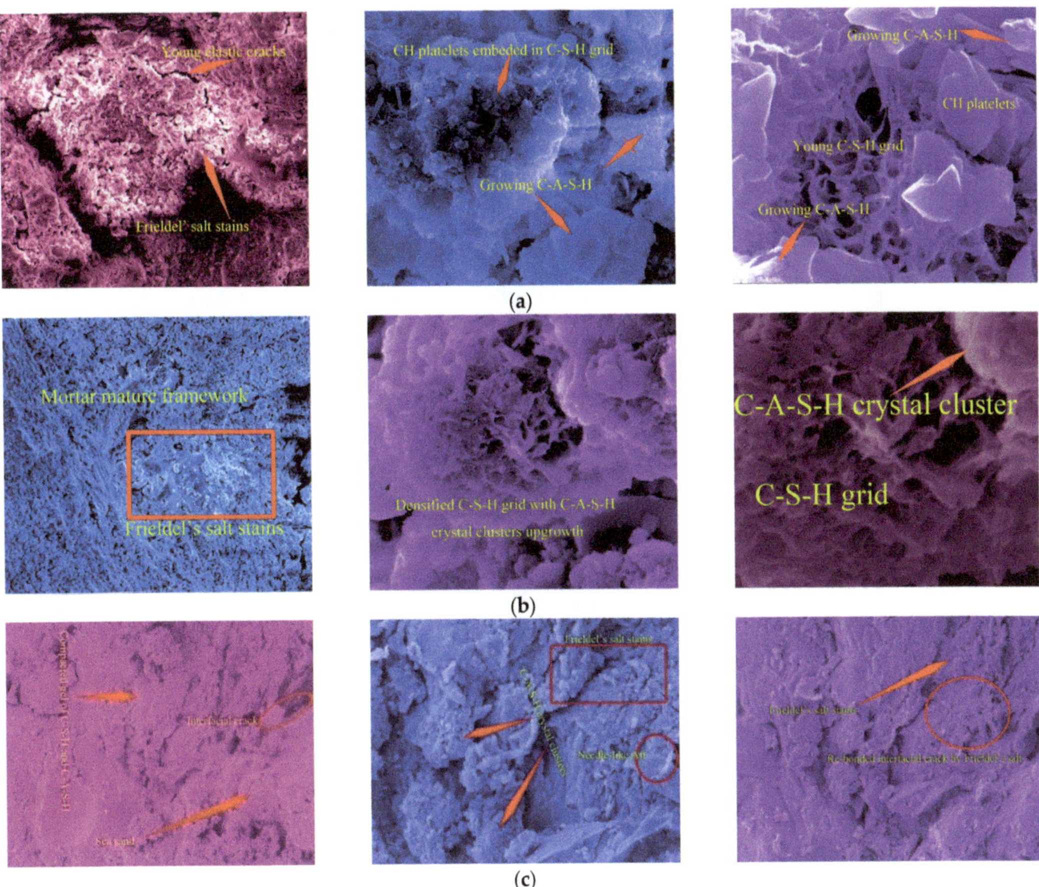

Figure 14. Mortar SEM images. (**a**) Results at 10 days old. (**b**) Results at 33 days old. (**c**) Results at 91 days old.

The SEM images of the mortar at 91 days old (Figure 14c) demonstrated that the gel of C-S-H and C-A-S-H was growing during the working period in the marine environment, and the gel layers were compacted significantly with the hydration development. Particularly, C-A-S-H strengthened the C-S-H grid, and the micro-pores in the mortar were filled with C-A-S-H crystal clusters. Although the interfacial cohesion between the SS particle's faces and the gel of C-S-H and C-A-S-H was reduced partially due to the existence of Aft, the interfacial cracks were compressed with the upgrowth of the gel of C-S-H and C-A-S-H [35]. According to the work by Lu et al., the SW mixture certainly improved the microscopic structures of the cementitious materials. Meanwhile, due to the discontinuousness of the SS used in the study, the abundant fill space in the mortar framework could meet the free upgrowth of the gel of C-S-H and C-A-S-H without inducing destruction in the crystal clusters. Moreover, Friedel's salt re-bonded the interfacial cracks between the SS and the hydration products in the marine environment with age accumulation. The porosity of the mortar was still densified at the grown age. The pressure casting effect from Friedel's salt development offered the proper polymerization in the interfacial cracks without heavy damage to the framework. Therefore, the Cl^- seepage in the mortar was also restricted efficiently [36].

4. Conclusions

In this paper, UCTs were adopted to investigate the mechanical performance of the mortar produced with SW and SS. The physicochemical reactivity of the mortar was also explored with EDS, XRD, and SEM. The mathematical models were established to correspondingly interpret the physicochemical-mechanical experimental art. The following conclusions based on the work can be drawn:

1. The mortar produced in the study achieved the promising development of the components and the hydration products, including C-S-H, C-A-S-H, and CH, even under natural conditions. The physicochemical reactivity of the mortar showed increasing behavior in the marine environment due to the anion-cation exchange between the paste and SS, Friedel's salt polymerization effect and the lasting development of hydration products with age accumulation. Furthermore, the physicochemical reactivity brought out the different physicochemical-mechanical behaviors of the mortar produced using SW and SS in comparison with conventional building materials.

2. Friedel's salt generation caused the salt petering in the young mortar. However, the polymerization from Friedel's salt helped re-bond the micro-flaws in the grown mortar. The cohesive and locking effect from Friedel's salt contributed exactly to the interfacial integrity of the mortar framework. The leaching effect of the Ca^{2+} complex in the conventional mortar due to seawater attack was restrained absolutely by the help of the proper polymerization between the hydration products mixture and the native saline products in the mortar using SW and SS.

3. The interfacial integrity of the mortar produced in the study was also guaranteed by the generally existing cation exchange between the sea sand and the hydration products.

4. The C-A-S-H hydration product helped compact the mortar framework. The mortar was produced with cement rich in $3CaO \cdot Al_2O_3$, which ensured the promising development of the components and hydration products. The development of the components and hydration products reinforced the microscopic structure of the mortar. The early strength feature of the mortar helped constrain the salt petering and reduce the superficial damage. The depth of the microscopic young elastic cracks in the mortar was controlled below 3 µm, which ensured that the strength of the mortar produced in the study increased correspondingly in the marine environment.

5. The findings of the saline ions discussed in the current literature were re-estimated. Not including the addition of the bars, the chemical structure of the cement-matrix materials produced with SW and SS was solidified in the natural marine environment by saline polymerization, the development of which was ensured by the adsorption and adjustment in the hydration products.

6. Existing in the mortar using SW and SS throughout the production, curing and application was the proper polymerization between the hydration product mixture and the native saline products. Being similar to the pressure casting effect, the key mechanism of the proper polymerization was the time, magnitude, and state of the internal force generated by the native saline products in the interfacial cracks.

7. The grown mortar showed promising strength that developed in the marine environment with age accumulation. Coupled with the physicochemical reactivity, the grown mortar, with the increasing macroscopic integration that originated from the pressure casting effect, achieved higher resistance against the load and damage than the young ones. Particularly, the direct application of the mortar under natural conditions was the applicable choice.

8. The study of the direct application of the environmentally friendly mortar was another key job, the applicability of which was justified in this paper. The direct application of the environmentally friendly mortar created in the study reduced the requirement for the river sand, which will help control carbon emissions during the resource-devouring preparation of the river sand.

Author Contributions: Conceptualization, Z.S. and Y.W.; methodology, Z.S.; software, Y.W.; validation, Y.W.; formal analysis, Y.W.; investigation, Z.S.; resources, Y.W.; data curation, Z.S.; writing—original draft preparation, Z.S.; writing—review and editing, D.H.; visualization, Y.W.; supervision, D.H.; project administration, Y.W.; funding acquisition, Z.S. All authors have read and agreed to the published version of the manuscript.

Funding: This research was funded by the National Natural Science Foundations of China, grant numbers 51879236 and 51109118, the Zhejiang Provincial Natural Science Foundation of China, grant number LHY21E090003, and the Teachers' Scientific Research Fund of China Earthquake Administration, grant number 20110104.

Institutional Review Board Statement: Not applicable.

Informed Consent Statement: Not applicable.

Data Availability Statement: Not applicable.

Conflicts of Interest: The authors declare no conflict of interest.

References

1. Wang, Y.J.; Zhang, C.H.; Wang, J.T.; Xu, Y.J.; Jin, F.; Wang, Y.B.; Yan, Q.; Liu, T.; Gan, X.Q.; Xiong, Z. Experimental Study on Foci Development in MortarUsing Seawater and Sand. *Materials* **2019**, *12*, 1799. [CrossRef] [PubMed]
2. Karthkeyan, M.; Nagarajan, V. Chloride Analysis of Sea Sand for Making Concrete. *Natl. Acad. Sci. Lett.* **2017**, *40*, 29–31. [CrossRef]
3. Dhondy, T.; Remennikov, A.; Sheikh, M.N. Properties and Application of Sea Sand in Sea Sand–Seawater Concrete. *J. Mater. Civ. Eng.* **2020**, *32*, 04020392. [CrossRef]
4. Sharma, S.; Zhang, D.; Zhao, Q. Degradation of basalt fiber–reinforced polymer bars in seawater and sea sand concrete environment. *Adv. Mech. Eng.* **2020**, *12*. [CrossRef]
5. Natarajan, S.; Pillai, N.N.; Murugan, S. Experimental Investigations on the Properties of Epoxy-Resin-Bonded Cement Concrete Containing Sea Sand for Use in Unreinforced Concrete Applications. *Materials* **2019**, *12*, 645. [CrossRef]
6. Bazli, M.; Zhao, X.-L.; Jafari, A.; Ashrafi, H.; Raman, R.S.; Bai, Y.; Khezrzadeh, H. Durability of glass-fibre-reinforced polymer composites under seawater and sea-sand concrete coupled with harsh outdoor environments. *Adv. Struct. Eng.* **2020**, *24*, 1090–1109. [CrossRef]
7. Bazli, M.; Zhao, X.-L.; Bai, Y.; Raman, R.S.; Al-Saadi, S. Bond-slip behaviour between FRP tubes and seawater sea sand concrete. *Eng. Struct.* **2019**, *197*. [CrossRef]
8. Bazli, M.; Zhao, X.-L.; Bai, Y.; Raman, R.S.; Al-Saadi, S.; Haque, A. Durability of pultruded GFRP tubes subjected to seawater sea sand concrete and seawater environments. *Constr. Build. Mater.* **2020**, *245*, 118399. [CrossRef]
9. Bazli, M.; Zhao, X.-L.; Raman, R.S.; Bai, Y.; Al-Saadi, S. Bond performance between FRP tubes and seawater sea sand concrete after exposure to seawater condition. *Constr. Build. Mater.* **2020**, *265*, 120342. [CrossRef]
10. Bazli, M.; Heitzmann, M.; Hernandez, B.V. Hybrid fibre reinforced polymer and seawater sea sand concrete structures: A systematic review on short-term and long-term structural performance. *Constr. Build. Mater.* **2021**, *301*, 124335. [CrossRef]
11. Ahmed, A.; Guo, S.; Zhang, Z.; Shi, C.; Zhu, D. A review on durability of fiber reinforced polymer (FRP) bars reinforced seawater sea sand concrete. *Constr. Build. Mater.* **2020**, *256*, 119484. [CrossRef]
12. Vafaei, D.; Hassanli, R.; Ma, X.; Duan, J.; Zhuge, Y. Sorptivity and mechanical properties of fiber-reinforced concrete made with seawater and dredged sea-sand. *Constr. Build. Mater.* **2020**, *270*, 121436. [CrossRef]
13. Vafaei, D.; Ma, X.; Hassanli, R.; Duan, J.; Zhuge, Y. Microstructural behaviour and shrinkage properties of high-strength fiber-reinforced seawater sea-sand concrete. *Constr. Build. Mater.* **2022**, *320*, 126222. [CrossRef]
14. El-Khoury, M.; Roziere, E.; Grondin, F.; Cortas, R.; Chehade, F.H. Experimental evaluation of the effect of cement type and seawater salinity on concrete offshore structures. *Constr. Build. Mater.* **2022**, *322*, 126471. [CrossRef]
15. Choi, S.I.; Kil Park, J.; Han, T.H.; Pae, J.; Moon, J.; Kim, M.O. Early-age mechanical properties and microstructures of Portland cement mortars containing different admixtures exposed to seawater. *Case Stud. Constr. Mater.* **2022**, *16*, e01041. [CrossRef]
16. De Weerdt, K.; Lothenbach, B.; Geiker, M. Comparing chloride ingress from seawater and NaCl solution in Portland cement mortar. *Cem. Concr. Res.* **2018**, *115*, 80–89. [CrossRef]
17. Palin, D.; Jonkers, H.; Wiktor, V. Autogenous healing of sea-water exposed mortar: Quantification through a simple and rapid permeability test. *Cem. Concr. Res.* **2016**, *84*, 1–7. [CrossRef]
18. Santhanam, M.; Cohen, M.; Olek, J. Differentiating seawater and groundwater sulfate attack in Portland cement mortars. *Cem. Concr. Res.* **2006**, *36*, 2132–2137. [CrossRef]
19. Qu, F.; Li, W.; Wang, K.; Tam, V.W.; Zhang, S. Effects of seawater and undesalted sea sand on the hydration products, mechanical properties and microstructures of cement mortar. *Constr. Build. Mater.* **2021**, *310*, 125229. [CrossRef]
20. He, P.; Yu, J.; Wang, R.; Du, W.; Han, X.; Gu, S.; Liu, Q. Effect of ion chelator on pore structure, mechanical property and self-healing capability of seawater exposed mortar. *Constr. Build. Mater.* **2020**, *246*, 118480. [CrossRef]

21. Zhang, B.; Zhu, H.; Shah, K.W.; Dong, Z.; Wu, J. Performance evaluation and microstructure characterization of seawater and coral/sea sand alkali-activated mortars. *Constr. Build. Mater.* **2020**, *259*, 120403. [CrossRef]

22. Li, S.; Jin, Z.; Yu, Y. Chloride binding by calcined layered double hydroxides and alumina-rich cementitious materials in mortar mixed with seawater and sea sand. *Constr. Build. Mater.* **2021**, *293*, 123493. [CrossRef]

23. Qin, Y.; Wang, Q.-K.; Xu, D.-S.; Fan, X.-C. Study of the effects of fine coral powder and salinity on the mechanical behaviour of coral sand-seawater cement mortar. *Constr. Build. Mater.* **2021**, *313*, 125476. [CrossRef]

24. Ebead, U.; Lau, D.; Lollini, F.; Nanni, A.; Suraneni, P.; Yu, T. A review of recent advances in the science and technology of seawater-mixed concrete. *Cem. Concr. Res.* **2021**, *152*, 106666. [CrossRef]

25. China, M. *Standard for Technical Requirements and Test Method of Sand and Crushed Stone (or Gravel) for Ordinary Concrete*; China Architecture& Building Press: Beijing, China, 2006.

26. *ASTM D1141-98*; Standard Practice for the Preparation of Substitute Ocean Water. ASTM International: West Conshohocken, PA, USA, 2013.

27. *ASTM C1084-10*; Standard Test Method for Portland-Cement Content of Hardened Hydraulic-Cement Concrete. ASTM International: West Conshohocken, PA, USA, 2010.

28. *ASTM C1180-18*; Standard Terminology of Mortar and Grout for Unit Masonry. ASTM International: West Conshohocken, PA, USA, 2018.

29. *ASTM C270-19*; Standard Specification for Mortar for Unit Masonry. ASTM International: West Conshohocken, PA, USA, 2019.

30. *ASTM C305-14*; Standard Practice for Mechanical Mixing of Hydraulic Cement Pastes and Mortar of Plastic Consistency. ASTM International: West Conshohocken, PA, USA, 2014.

31. Xiao, J.; Qiang, C.; Nanni, A.; Zhang, K. Use of sea-sand and seawater in concrete construction: Current status and future opportunities. *Constr. Build. Mater.* **2017**, *155*, 1101–1111. [CrossRef]

32. Griffith, A.A. The phenomena of rupture and flow in solids. *Philos. Trans. R. Soc. Lond.* **1921**, *221*, 163–198.

33. Guo, M.; Hu, B.; Xing, F.; Zhou, X.; Sun, M.; Sui, L.; Zhou, Y. Characterization of the mechanical properties of eco-friendly concrete made with untreated sea sand and seawater based on statistical analysis. *Constr. Build. Mater.* **2020**, *234*, 117339. [CrossRef]

34. Xue, C. Cracking and autogenous self-healing on the performance of fiber-reinforced MgO-cement composites in seawater and NaCl solutions. *Constr. Build. Mater.* **2022**, *326*, 126870. [CrossRef]

35. Lu, J.-X.; Shen, P.; Zhang, Y.; Zheng, H.; Sun, Y.; Poon, C.S. Early-age and microstructural properties of glass powder blended cement paste: Improvement by seawater. *Cem. Concr. Compos.* **2021**, *122*, 104165. [CrossRef]

36. Suryavanshi, A.; Scantlebury, J.; Lyon, S. Mechanism of Friedel's salt formation in cements rich in tri-calcium aluminate. *Cem. Concr. Res.* **1996**, *26*, 717–727. [CrossRef]

materials

Article

An Experimental and Analytical Study on a Damage Constitutive Model of Engineered Cementitious Composites under Uniaxial Tension

Dapeng Zhao [1], Changjun Wang [1], Ke Li [1,*], Pengbo Zhang [1], Lianyou Cong [1] and Dazhi Chen [2]

[1] Department of Civil Engineering, Zhengzhou University, Zhengzhou 450001, China
[2] Henan Urban Planning Institute and Corporation, Zhengzhou 450044, China
* Correspondence: irwinlike@163.com

Abstract: Engineered cementitious composites (ECC) exhibit ultra-high ductility and post-cracking resistance, which makes it an attractive material in civil engineering. First, a monotonic uniaxial tensile test was performed, considering the effects of polyvinyl alcohol (PVA) fiber volume content and water-binder ratio. Then, the effects of the above variables on the tensile characteristics including the tensile stress–strain relationship, deformation capacity, and fracture energy were investigated based on test results; and when the water-binder ratio is 0.28 and the fiber volume content is 2%, the deformation performance of ECC is improved most significantly. Next, combined with damage mechanics theory, the damage evolution mechanism of ECC in monotonic uniaxial tension was revealed, based on which the damage factor and damage evolution equation of ECC were developed and the expressions of model parameters were proposed. Moreover, the comparison between the proposed model and test results demonstrated the accuracy of the proposed model. Finally, to further verify the feasibility of the proposed model, a finite element (FE) simulation analysis of the tensile performance of high-strength stainless steel wire rope (HSSWR) reinforced ECC by adopting the proposed model was compared with test results and the simulation analysis results by using anther existing model, the "trilinear model of ECC". The comparison shows that the proposed model in this paper can predict more accurately.

Keywords: engineered cementitious composites; damage constitutive model; stress–strain relationship; tensile performance; monotonic uniaxial tensile test

Citation: Zhao, D.; Wang, C.; Li, K.; Zhang, P.; Cong, L.; Chen, D. An Experimental and Analytical Study on a Damage Constitutive Model of Engineered Cementitious Composites under Uniaxial Tension. *Materials* **2022**, *15*, 6063. https://doi.org/10.3390/ma15176063

Academic Editor: Enrique Casarejos

Received: 28 July 2022
Accepted: 30 August 2022
Published: 1 September 2022

Publisher's Note: MDPI stays neutral with regard to jurisdictional claims in published maps and institutional affiliations.

1. Introduction

Engineered cementitious composites (ECC) is a new type of fiber reinforced cementitious composite material proposed in the 1990s and continuously developed in recent decades [1–3], which mainly consists of cement, sand, fly ash, silica fume, fibers, and water. The fibers used in ECC, which play a crucial role in improving the material performance [4], commonly include polyvinyl alcohol (PVA) [5–7], polypropylene (PE) [8–10], PP [11–13], and hybrid fibers [14–16]. Besides, studies have shown that adding new additives, such as eggshell powder [17] or alkali-activated products composed of industrial waste materials [18], can improve the mechanical properties of cementitious materials and increase environmental protection, which may also be used for improving ECC performance. Due to the presence of fibers, ECC exhibits ultra-high toughness, ductility, and crack-control capacity, which make ECC widely applied to constructing ductile structures and strengthening existing structures [19,20].

Many researchers have focused on the constitutive model of ECC in tension in order to give an accurate analytical calculation or numerical analysis of mechanical properties of structures constructed using ECC [21–26]. A trilinear model for charactering the stress–strain relationship of the ideal strain-hardening material, such as ECC, was first proposed by Li [21], based on which Kanda [22] proposed a simplified bilinear model of the tensile stress–strain curve of ECC. However, these two models did not consider the strain-softening stage of ECC. Then, a simplified trilinear model including a pre-cracking linear elastic section, a strain-hardening section and a strain-softening section has been widely adopted for more accurate FE numerical simulations of the whole-process mechanical behavior of ECC in tension [23–28]. However, this model ignores the nonlinear characteristics of the stress–strain relationship and cannot describe the damage evolution of ECC.

Damage modelling is a new approach to describe the mechanical behavior of materials by using damage mechanics theories [29–31], which attributes the fracture failure of material to the development of microscopic defects in materials in the whole loading process by employing damage factors to express the damage degree. Therefore, in recent years, damage models were applied to fiber-reinforced cementitious composites, such as ECC and fiber reinforced concrete [32–36]. The commonly used methods for defining damage factors include the stiffness method [32–34], the stress method [27], and the energy method [35,36]. However, these methods of defining damage factor do not clearly reveal the damage mechanism of ECC during the whole tension process, nor do they clearly reveal the role played by fibers in the damage process of ECC.

Therefore, in this paper, the damage constitutive model of ECC under uniaxial tension was studied experimentally and analytically. The main contents and purposes of this study include: (1) to evaluate the effects of PVA fiber volume content and water-binder ratio on the tensile properties of ECC under uniaxial tensile test; (2) to investigate the damage evolution law of ECC under uniaxial tensile; (3) to develop the damage constitutive model of ECC under uniaxial tension based on the test results; (4) to validate the acceptability and accuracy of the proposed model through FE analysis of high-strength stainless steel wire rope (HSSWR) reinforced ECC under uniaxial tension by adopting this model and existing trilinear model.

2. Test Procedure

2.1. Test Material and Specimen Production

The component materials of ECC used in this paper are ordinary silicate cement, fly ash, micro-silica fume, silica sand, PVA fiber, superplasticizer, and water. The mix proportions of the test specimens are shown in Table 1. The performance indicators of PVA used in this test are provided by Japan Kuraray Co., Ltd. (Tokyo, Japan), as shown in Table 2.

Table 1. ECC mix proportion.

Group Number	Cement	Silica Sand	Fly Ash	Micro-Silica Fume	Water	PVA Fiber	Superplasticizer
V1R3	1.000	1.500	2.000	0.073	0.860	0.029	0.041
V2R3	1.000	1.500	2.000	0.073	0.860	0.044	0.041
V3R1	1.000	1.500	2.000	0.073	0.738	0.057	0.041
V3R2	1.000	1.500	2.000	0.073	0.768	0.057	0.041
V3R3	1.000	1.500	2.000	0.073	0.860	0.057	0.041

Table 2. Performance indicators of PVA.

Fiber	Length (mm)	Diameter (μm)	Tensile Strength (MPa)	Elastic Modulus (GPa)	Elongation (%)	Density (g/cm^3)
PVA	12	40	1560	41	6.5	1.3

In this test, five groups of specimens with thin plate dimensions of 280 mm (length) × 40 mm (width) × 13 mm (thickness) were used, and each group consisted of 5 identical specimens. The volume content of PVA fibers (v) and the water-binder ratio (r) were considered, as listed in Table 3. It is worth noting that, according to the ready-mix test, when the water-binder ratio is less than 0.24, the matrix is too dry to mix, and when the water-binder ratio is greater than 0.3, the matrix is too dilute to be used normally, so the water-binder ratio of 0.24–0.28 was selected. Considering performance improvement efficiency [2,37–41] and economy [42], the fiber volume content was selected as 1%, 1.5% and 2%. The production process of the test specimen can be briefly summarized as follows: first, weigh the materials according to the mix proportion in Table 1, then mix the dry materials (cement, sand, fly ash, micro-silica fume) for 3 min; next, add water and superplasticizer and mix for 3 min; finally, add the PVA fibers evenly and continue to mix for 3 min. After 28-day curing in saturated lime water, the specimens were taken out and dried. Then, the two ends of the specimens to be clamped by the loading device were reinforced firstly by being externally bonded by carbon fiber reinforced polymer (CFRP) sheet, as shown in Figure 1. Specifically, firstly the two ends of the specimens were polished with sandpaper and cleaned with alcohol. Then, the two ends of the specimen to be clamped were externally bonded by one layer of CFRP sheets using epoxy resin adhesive. Finally, the 0.8-mm thick aluminum plates were bonded to the surface of the CFRP sheets using epoxy resin adhesive. A seven-day curing of the epoxy resin adhesive to make the adhesive reach a certain strength before the ECC tensile test was carried out. The tested tensile strength and elastic modulus of the used CFRP sheet with nominal thickness of 0.111 mm were 4150 MPa and 225 GPa, respectively. As given by the manufacturer, the tensile strength and elastic modulus of the used epoxy resin adhesive were 47.2 MPa and 2860 MPa, respectively.

Table 3. Details of specimens and test results.

Group	Specimen ID	v (%)	r	ε_k/ (%)	σ_k/ MPa	ε_p/ (%)	σ_p/ MPa	ε_u/ (%)	$G_{f,N}$ (kJ/m^2)
V1R3	V1R3-1			0.037	1.347	0.814	3.226	1.697	3.044
	V1R3-2			0.039	2.407	0.274	3.338	2.211	3.950
	V1R3-3	1	0.28	0.015	1.797	0.518	3.438	1.786	3.681
	V1R3-4			0.019	1.960	0.385	2.740	1.652	2.976
	V1R3-5			0.018	2.370	0.563	2.780	2.128	3.913
Average value				0.025	1.976	0.511	3.104	1.895	3.513
Standard deviation				0.011	0.439	0.204	0.324	0.257	0.471
V2R3	V2R3-1			0.072	1.746	1.254	2.867	2.498	5.705
	V2R3-2			0.050	1.259	0.575	3.012	2.485	4.841
	V2R3-3	1.5	0.28	0.055	2.128	0.420	2.792	2.663	5.458
	V2R3-4			0.040	2.166	1.068	2.997	2.461	4.761
	V2R3-5			0.025	2.799	0.663	3.130	2.590	5.313
Average value				0.048	2.020	0.796	2.959	2.539	5.216
Standard deviation				0.017	0.569	0.351	0.132	0.084	0.405
V3R1	V3R1-1			0.033	3.274	0.590	4.392	1.633	7.181
	V3R1-2	2	0.24	0.043	3.508	0.425	4.067	3.050	6.808
	V3R1-3			0.048	3.538	0.495	5.256	2.008	8.270
Average value				0.041	3.440	0.503	4.572	2.230	7.420
Standard deviation				0.008	0.144	0.083	0.613	0.734	0.757
V3R2	V3R2-1			0.090	2.256	1.140	3.680	2.200	6.656
	V3R2-2			0.066	2.960	0.811	3.949	2.682	9.934
	V3R2-3	2	0.25	0.088	3.074	0.665	3.884	3.705	11.000
	V3R2-4			0.040	2.462	1.476	3.769	3.350	10.140
	V3R2-5			0.053	2.574	1.265	3.655	2.940	9.344
Average value				0.067	2.665	1.071	3.787	2.975	9.415
Standard deviation				0.022	0.343	0.331	0.127	0.583	1.653
V3R3	V3R3-1	2	0.28	0.033	2.900	1.755	3.111	4.245	10.166
	V3R3-2			0.092	2.216	1.908	3.477	4.298	12.673

Table 3. *Cont.*

Group	Specimen ID	v (%)	r	$\varepsilon_k/$ (%)	$\sigma_k/$ MPa	$\varepsilon_p/$ (%)	$\sigma_p/$ MPa	$\varepsilon_u/$ (%)	$G_{f,N}$ (kJ/m^2)
	V3R3-3			0.092	2.033	2.317	2.883	3.808	10.053
	V3R3-4			0.075	2.256	2.110	3.309	4.108	11.764
	V3R3-5			0.055	2.827	1.910	3.390	3.929	11.065
Average value				0.069	2.447	2.000	3.234	4.077	11.144
Standard deviation				0.026	0.391	0.217	0.238	0.208	1.104

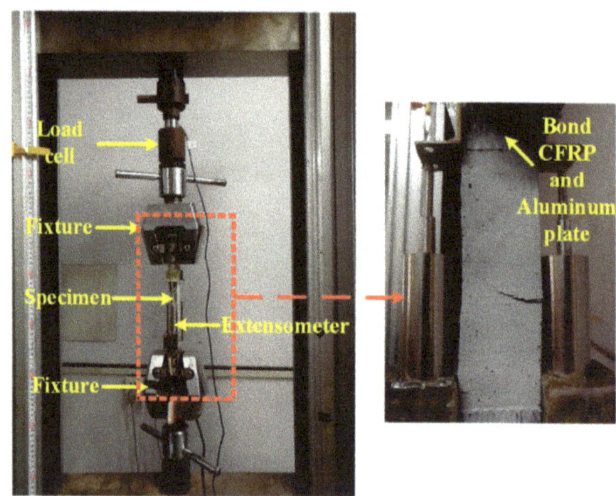

Figure 1. Test setup.

2.2. Test Loading and Measurement

The test setup and loading device is shown in Figure 1. The test was carried out on a 10-ton electro-hydraulic servo material testing machine produced by Jinan Sans Dynamic Testing Technology Co., Ltd. (Jinan, China). The displacement-controlled monotonic loading method with a speed of 0.2 mm/min was adopted. The average temperature and humidity of laboratory are 298.15 K and 95%, respectively. Strain gage was placed on the center of the specimen surface, which was used mainly to measure the strain change of the specimen before cracking. The extensometer was placed at the middle region of the specimen with a measuring range of 120 mm, which was mainly used to obtain the specimen strain after cracking by dividing the measured displacement by the measuring range.

3. Test Results and Discussion

3.1. Analysis of Tensile Mechanical Properties Indicators

The tensile mechanical property indicators including first-cracking strain (ε_k), first-cracking stress (σ_k), peak strain (ε_p), peak stress (σ_p), ultimate strain (ε_u), and fracture energy (G_f) for each group of specimens are shown in Table 3. The relationship between the tensile mechanical property indicators and the volume content of PVA fibers (v) or the water-binder ratio (r) are shown in Figure 2. The test results were summarized in Table 3, and two specimens in Group V3R1 were excluded due to the large deviation of the test specimen results.

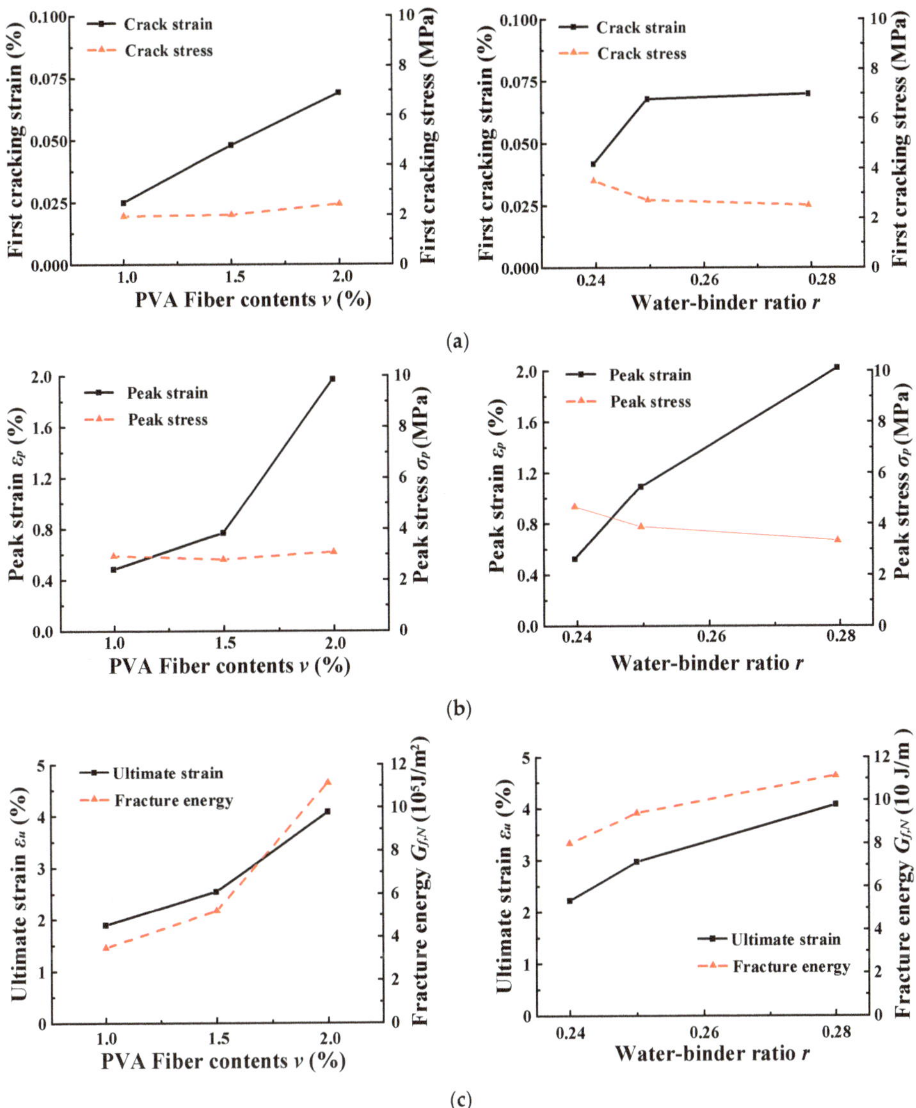

Figure 2. ECC tensile properties: (**a**) First cracking strain and stress; (**b**) Peak strain and stress; (**c**) Ultimate strain and fracture energy.

3.1.1. First-Cracking Strain and Cracking Stress

The first cracking point was detected by the first drop point of the stress–strain curve of ECC under uniaxial tension. As presented in Table 3 and Figure 2a, when the volume content of PVA fibers was increased from 1.0% to 1.5% and 2% respectively, the crack stress increased by 2.2% and 21.1%, respectively, and the first crack strain increased by 92.0% and 43.8%, respectively. This indicates that increasing the volume content of PVA fibers in the range of 1.0–2% can effectively improve the crack resistance of ECC. When the water-binder ratio of ECC was increased from 0.24 to 0.25 and 0.28, respectively, the crack stress decreased by 22.5% and 8.2%, respectively, but the first crack strain increased by 63.4% and 3%, respectively. This indicates that increasing the water-binder ratio would decrease the first-cracking stress, but it can increase the first-cracking strain. This may be because the increase of the water in the matrix would make the interfacial bond strength between the matrix and fibers decrease, which results in the increase of the interfacial relative slip between the fibers and matrix [37].

3.1.2. Peak Strain and Peak Stress

According to Table 3 and Figure 2b, when the volume content of PVA fibers was increased from 1.0% to 1.5% and 2% respectively, the peak stress of ECC increased by −4.7% and 9.3% respectively, and the peak strain increased by 55.8% and 151.3% respectively. When the water-binder ratio of ECC was increased from 0.24 to 0.25 and 0.28 respectively, the peak stresses decreased by 17.2% and14.6% respectively, but the peak strains increased by 112.9% and 86.7% respectively. This indicates that the peak stress of ECC was mainly determined by the water-binder ratio of the matrix, and the increase in fiber content would not significantly increase the peak stress. However, the increase in either PVA fiber content or water-binder ratio could significantly increase the peak strain. Similar phenomenon also appeared in Li's investigation [38]. This may be because the increase of the fiber content in the matrix can make the cracks disperse to more cross sections, and as result the relative slip between fiber and matrix increased [1–3]. In addition, as discussed above, the increase of the water-binder ratio would make the interfacial frictional force between the matrix and fibers decrease, which results in an increase of the interfacial relative slip.

3.1.3. Ultimate Strain

The ultimate strain is the strain at which the ECC specimens rupture, which reflects the maximum deformation capacity of ECC. The ultimate strain of each group of specimens is shown in Table 3. The relationship between the ultimate strain of ECC and the PVA fiber volume content or the water-binder ratio is shown in Figure 2c. According to Table 3 and Figure 2c, the ultimate strain of ECC increases by 34.0% and 60.6% when the volume content of PVA fiber increased from 1.0% to 1.5% and 2%, respectively. The ultimate strain of ECC increased by 33.4% and 37.0%, when the water-binder ratio of ECC increases from 0.24 to 0.25 and 0.28, respectively. This indicates that increasing the content of PVA fibers or the water-binder ratio could enhance the deformation capacity of ECC, and the enhancing effect of increasing the content of PVA fibers is more significant.

3.1.4. Fracture Energy under Tension

Tensile fracture energy is defined as the energy required per unit area of crack extension, and it is a key indicator to describe the crack-control and energy dissipation capabilities of a material [39–41]. According to the calculation method of fracture energy in the literature [41], the total tensile fracture energy ($G_{f,N}$) of a specimen can be expressed as:

$$G_{f,N} = G_{f,A,N} + G_{f,B,N} \tag{1}$$

where

$$G_{f,A,N} = \left[\int_{0}^{\varepsilon_p} \sigma(\varepsilon)d\varepsilon - \frac{1}{2}\frac{\sigma_p^2}{E} \right] L_g \tag{2}$$

$$G_{f,B,N} = \int_{\varepsilon_{res}L_g}^{\varepsilon_p L_g} \sigma(\delta)d\delta \tag{3}$$

$$\varepsilon_{res} = \varepsilon_p - \frac{\sigma_p}{E_0} \tag{4}$$

where L_g is the gauge length and E_0 is initial tensile modulus.

The calculated result of tensile fracture energy of ECC for each specimen was calculated and given in Table 3. The effects of PVA fiber volume content and water-binder ratio on the tensile fracture energy of ECC was presented in Figure 2c. From Table 3 and Figure 2c, it can be noticed that the fracture energy increased by 48.5% and 113.7% when the volume content of PVA fiber increased from 1.0% to 1.5% and 2%, respectively. This indicates that increasing the volume content of PVA fiber in the range of 1.0–1.5% can enhance the fracture energy of ECC, which may be because the increase in fiber content increased the bridging action between the fibers and the matrix at the crack, and as a result more energy can be dissipated from cracking to fracture of the ECC. The fracture energy increased by 26.9% and 18.4% when the water-binder ratio increased from 0.24 to 0.25 and 0.28, respectively. This indicates that the energy dissipation capacity of PVA-ECC can be enhanced with an increase in the water-binder ratio, which was unlike PE-ECC (whose energy dissipation capacity increased with increasing water-binder ratio) [41]. Thus, the energy dissipation capacity of PVA-ECC decreased with the decrease of the water-binder ratio.

3.2. Stress–Strain Curve and Damage Evolution

The uniaxial tensile stress–strain curves of ECC with different fiber volume content and water-binder ratio are shown in Figure 3. As indicated in Figure 3, ECC exhibited obvious strain–hardening characteristic in tension, and after reaching the peak stress, the tensile stress decreased slowly as the strain increased, with a long strain–softening stage, which shows that the ductility of cement mortar matrix can be obviously improved by adding PVA fibers. In addition, the stress–strain curve of the ECC under tension would change significantly when the volume content of fibers or the ECC and the water-binder ratio varied. As shown in Figure 3a,b,e, as the content of PVA fibers increased from 1.0% to 2%, the peak strain, ultimate strain, and the length of the strain–hardening segment increased significantly, but the change of the peak stress is not obvious, which indicates that the PVA fiber mainly plays a role in improving the ductility. As shown in Figure 3c–e, when the water-binder ratio is increased from 0.24 to 0.28, the first cracking stress and the peak stress decrease continuously, while the peak strain and ultimate strain increase significantly, and the length of the strain-hardened section has also increased. This is because the decrease of water-binder ratio would lead to the increase of bond strength between matrix and fibers, which make the fibers vulnerable to rupture rather than being pulled out in tension, and as a result the ductility of specimens reduced [1–3].

The crack pattern and damage evolution process of the specimens under uniaxial tension are shown in Figure 4. Based on the damage mechanism and the characteristics of the stress–strain relationship under tension, the uniaxial tensile damage evolution of the ECC can be described in following three stages, including the first stage named elastic nondestructive stage, the second stage named strain-hardening and stable damage stage, and the third stage named strain-softening and unstable damage stage.

The first stage (elastic nondestructive stage) is from the start of loading to the first-cracking point, where the deformation is elastic and the applied external load is mainly borne by the cement mortar matrix. The tensile stress and strain of ECC conform to a linear relationship at this stage and the specimen is basically in a nondestructive state.

The second stage (strain-hardening and stable damage stage) is from the first-cracking point to the peak point. At this stage the tensile stress of ECC increased slowly with increasing strain and the number of cracks increased, showing clear strain-hardening and multiple cracking characteristics. The damage of ECC began to appear at the first-cracking section, but the damage development at the cracked section was restrained to a certain extent due to the bridging effect of PVA fibers. With the increase of stress, new cracks continuously appeared on different sections of ECC, which reflected that the damage was dispersed to multiple cracked sections. As a result, the rate of damage development could be slowed down. While reaching the peak stress, no new cracks were generated, and the restraint effect of fibers on the damage was maximized and the damage development rate was minimized.

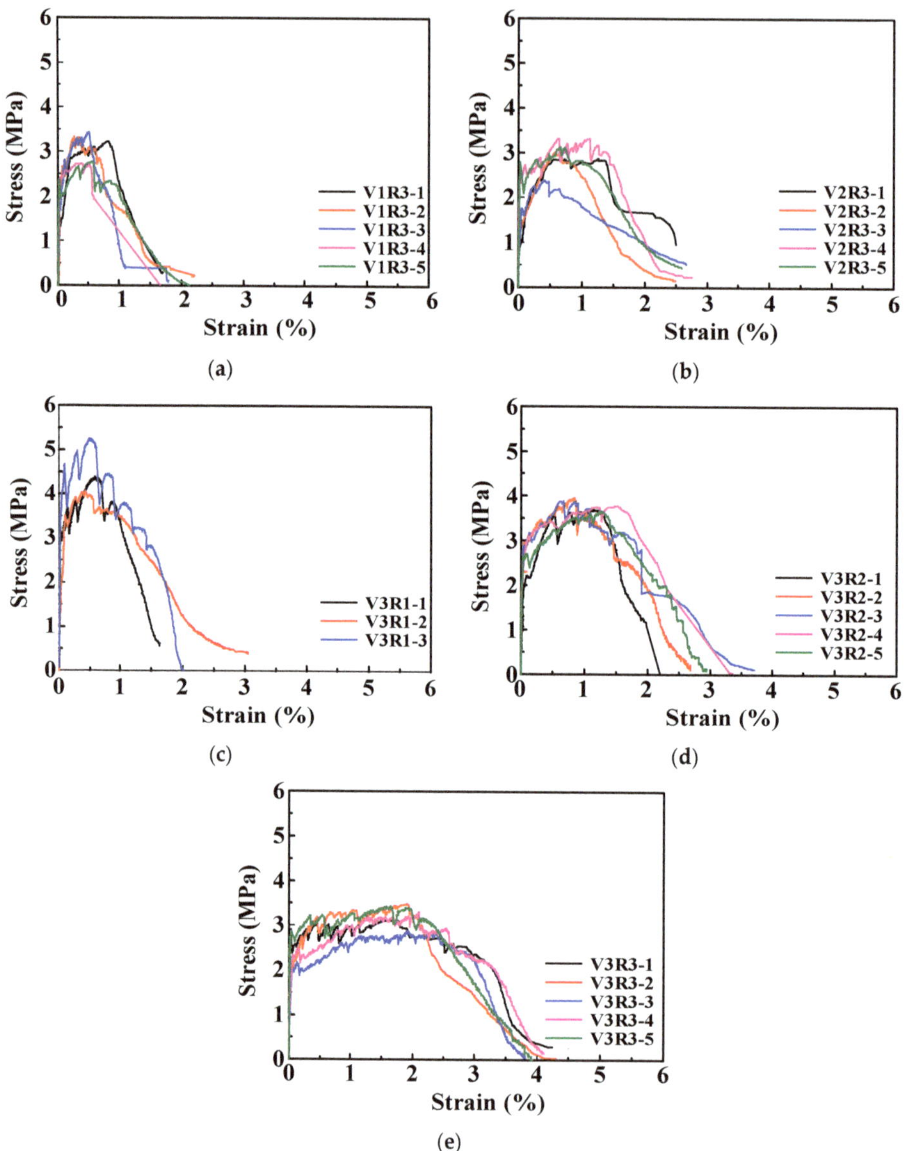

Figure 3. Stress–strain curves of ECC specimens in tension: (**a**) V1R3; (**b**) V2R3; (**c**) V3R1; (**d**) V3R2; (**e**) V3R3.

The third stage (strain-softening and unstable damage stage) is the stress decreasing segment after the peak point. At this stage, as the tensile strain increased, the tensile stress began to decrease, and the fibers ruptured at cracked sections or were continuously pulled out from the matrix. Besides, the width of cracks gradually increased with the increasing strain. Due to losing the bridging effect of the fibers in the matrix, the damage increased rapidly at this stage, making the damage in the increasing state of nonconvergence. Finally, the main crack of most specimens formed close to the clamp because of the stress concentration close to the clamp, and the ECC specimen ruptured at the main crack.

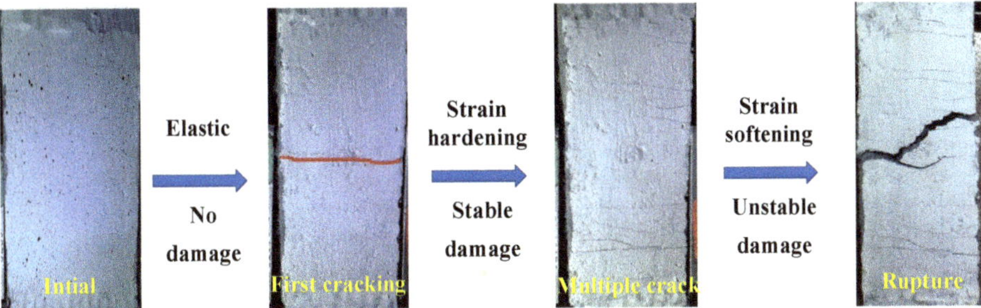

Figure 4. The crack pattern and damage evolution process of the specimens under uniaxial tension.

4. Damage Constitutive Model of ECC in Uniaxial Tension

4.1. The Proposed Damage Constitutive Model of ECC in Uniaxial Tension

In order to reflect the damage development characteristics of ECC in uniaxial tension at different stages, this paper adopts damage factor (d) expressed by the variation of the angle between the secant line of a point of the ECC tensile stress–strain curve and the horizontal axis to represent the degree of damage, as shown in Figure 5.

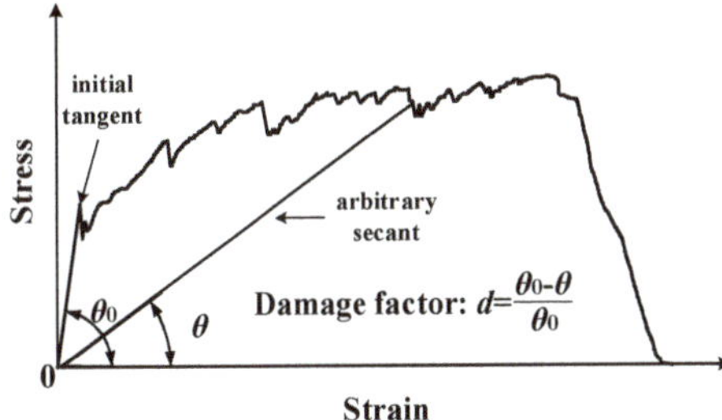

Figure 5. The definition of damage factor of ECC in this paper.

Therefore, the damage factor can be expressed as:

$$d = \frac{\theta_0 - \theta}{\theta_0} = 1 - \frac{\theta}{\theta_0} \tag{5}$$

where θ is the angle between the secant line of a point of the ECC tensile stress–strain curve and the horizontal axis for, θ_0 is the angle between the tensile stress–strain curve of ECC in the elastic nondestructive stage and the horizontal axis. The expressions of θ_0 and θ can be given by:

$$\theta_0 = \arctan(E_0) = \arctan(\sigma_k/\varepsilon_k) \tag{6}$$

$$\theta = \arctan(E) = \arctan(\sigma/\varepsilon) \tag{7}$$

where E_0 is the initial elastic modulus of ECC, and E is the secant modulus of ECC, σ and ε are the tensile stress and strain of ECC under uniaxial tension, respectively. Thus, the stress–strain relationship of ECC under uniaxial tension can be expressed as

$$\sigma = \tan[(1 - d)\arctan(E_0)]\varepsilon \tag{8}$$

Based on the analysis of the evolution process of ECC tensile damage, the damage evolution equation is assumed to be constant zero in the elastic section, a power function in the strain–hardening section, and a quadratic polynomial in the strain–softening section, as follows:

$$d = \begin{cases} 0 & (0 \le \varepsilon \le \varepsilon_k) \\ k(\varepsilon - \varepsilon_k)^b & (\varepsilon_k < \varepsilon \le \varepsilon_p) \\ A\varepsilon^2 + B\varepsilon + C & (\varepsilon_p < \varepsilon \le \varepsilon_u) \end{cases} \tag{9}$$

where k, b, A, B, and C are model coefficients. In fact, any function capable of describing the damage evolution process of ECC under uniaxial tension can also be used.

Besides, the damage factor (d) should satisfy the following boundary conditions:

$$d\big|_{\varepsilon=\varepsilon_k} = 0 \tag{10}$$

$$d\big|_{\varepsilon=\varepsilon_p} = k\left(\varepsilon_p - \varepsilon_k\right)^b = A\varepsilon_p{}^2 + B\varepsilon_p + C \tag{11}$$

$$d\big|_{\varepsilon=\varepsilon_u} = 1 = A\varepsilon_u{}^2 + B\varepsilon_u + C \tag{12}$$

According to Equations (10)–(12), the model coefficients k, B, and C can be obtained by:

$$k = \frac{\arctan(\sigma_k/\varepsilon_k) - \arctan(\sigma_p/\varepsilon_p)}{\left(\varepsilon_p - \varepsilon_k\right)^b \arctan(\sigma_k/\varepsilon_k)} \tag{13}$$

$$B = \frac{\arctan(\sigma_p/\varepsilon_p)}{\left(\varepsilon_u - \varepsilon_p\right)\arctan(\sigma_k/\varepsilon_k)} - A\left(\varepsilon_u + \varepsilon_p\right) \tag{14}$$

$$C = 1 + A\varepsilon_u\varepsilon_p - \frac{\varepsilon_u\arctan(\sigma_p/\varepsilon_p)}{\left(\varepsilon_u - \varepsilon_p\right)\arctan(\sigma_k/\varepsilon_k)} \tag{15}$$

4.2. Determination of Damage Model Coefficients b and A

From the above expressions for the model coefficients (13)–(15), only two coefficients (b and A) are independent unknown coefficients. These two coefficients can be determined by regression analysis of test results. Specifically, Equation (9) was used to fit the test result of damage factor–strain curve of each specimen to obtain the values of model coefficients B and A of each specimen. Then, b and A are substituted into Equations (13)–(15) to obtain the values of coefficients k, B, and C. The obtained values of model coefficients are given in Table 4.

Table 4. Damage model coefficients.

Group	V1R3	V2R3	V3R1	V3R2	V3R3
v (%)	1	1.5	2	2	2
r	0.28	0.28	0.24	0.25	0.28
b	0.997	0.877	0.915	0.844	0.784
A	0.214	0.132	0.197	0.151	0.067
k	0.198	0.199	0.127	0.162	0.203
B	0.138	0.045	0.004	−0.090	−0.090
C	−0.030	0.035	0.011	0.108	0.252

As can be seen from Table 4, the coefficients b and A decrease with increasing PVA fiber volume content or water-binder ratio, indicating that both PVA fiber volume content and water-binder ratio influence damage evolution during the strain-hardening and strain-softening stages. Thus, according to regression of test results, the coefficients b and A can be expressed as a linear function of the PVA fiber volume content (v) and the water-binder ratio (r) with correlation coefficients R^2 of 0.989 and 0.994, respectively, which are given as follows:

$$b = (1.206 - 0.213v)(2.061r + 0.436) \tag{16}$$

$$A = (0.358 - 0.147v)(2.842r + 0.247) \tag{17}$$

where the applicable range of PVA fiber volume content (v) is 1–2%, and the applicable range of ECC water-binder ratio (r) is 0.24–0.28.

In order to further analyze the characteristic of damage evolution in the strain-hardening and strain-softening stages, the first and second order derivatives of the damage factor expression Equation (9) with respect to strain for these stages are derived as follows:

$$\frac{\partial d}{\partial \varepsilon} = \begin{cases} kb(\varepsilon - \varepsilon_k)^{b-1} & (\varepsilon_k < \varepsilon \leq \varepsilon_p) \\ 2A\varepsilon + B & (\varepsilon_p < \varepsilon \leq \varepsilon_u) \end{cases} \tag{18}$$

$$\frac{\partial^2 d}{\partial \varepsilon^2} = \begin{cases} kb(b-1)(\varepsilon - \varepsilon_k) & (\varepsilon_k < \varepsilon \leq \varepsilon_p) \\ 2A & (\varepsilon_p < \varepsilon \leq \varepsilon_u) \end{cases} \tag{19}$$

It can be seen from Equations (16)–(19) that in the strain-hardening stage; the first-order derivative of the damage factor is greater than 0, but the second order derivative is less than 0. This indicates that in the strain-hardening stage, as the tensile strain increased, the damage increased continuously, but the increasing rate of damage decreased due to the presence of fibers, and the damage in a stable development state. In the strain-softening stage, the first and second order derivatives of the damage are both greater than 0. This demonstrates that in the strain-softening stage, as the tensile strain increased, the damage of the ECC increased continuously, and the increasing rate also increased due to the continuous pulling out or rupture of the PVA fibers, and the damage is in an unstable development state. The above analysis is consistent with the test results of the evolutionary mechanism of damage in Section 3.2, which also indicates that the damage factor expressed by Equation (9) proposed in this paper can reasonably characterize the evolutionary law of the damage of ECC in uniaxial tension.

4.3. Model Validation

In order to verify the applicability of the proposed model, existing experimental results in reference [43] will be validated. The average test values [43] of performance indicators of each group were presented in Table 5. In order to validate the proposed calculation formula for damage model coefficients, b, and A, the measured values (b_{expt} and A_{expt}) of the validation group were compared with the calculated values by using Equations (16) and (17) (b_{calc} and A_{calc}), as shown in Table 5. As can be seen from Table 5, the mean values of the ratio of measured value to calculated value for damage model coefficients, b, and A, are 0.971 and 1.155 respectively, with coefficients of variation of 0.023 and 0.208, which indicates that the Equations (16) and (17) are acceptable for calculating the damage model coefficients (b and A).

Table 5. Test results and damage model coefficients of the verification groups.

Group	v (%)	r	ε_k (%)	σ_k (MPa)	ε_p (%)	σ_p (MPa)	ε_u (%)	b_{expt}	b_{calc}	A_{expt}	A_{calc}
V-V1R3	1	0.28	0.018	2.165	0.474	2.760	1.890	0.993	0.997	0.216	0.214
V-V2R3	1.5	0.28	0.033	2.483	0.864	3.063	2.526	0.915	0.877	0.141	0.132
V-V3R2	2	0.25	0.047	2.518	1.371	3.712	3.145	0.880	0.844	0.122	0.151
V-V3R3	2	0.28	0.065	2.542	2.010	3.349	4.018	0.816	0.784	0.046	0.067

In order to verify the accuracy of the damage constitutive model of ECC in uniaxial tension proposed in this paper, the predicted stress–strain curves for ECC under uniaxial tension were obtained by using the proposed model [Equations (9) and (10)] and expressions of model coefficients [Equations (13)–(17)], and were compared with test stress–strain curves and the existing trilinear model [23–25], as shown in Figure 6. The expression of the trilinear model [23–25] is given as follows:

$$\sigma = \begin{cases} \frac{\sigma_k}{\varepsilon_k}\varepsilon & (\varepsilon \le \varepsilon_k) \\ \sigma_k + \left(\frac{\sigma_p - \sigma_k}{\varepsilon_p - \varepsilon_k}\right)(\varepsilon - \varepsilon_k) & (\varepsilon_k < \varepsilon \le \varepsilon_p) \\ \sigma_p - \frac{\sigma_p}{\varepsilon_u - \varepsilon_p}(\varepsilon_u - \varepsilon) & (\varepsilon_p < \varepsilon \le \varepsilon_u) \end{cases} \tag{20}$$

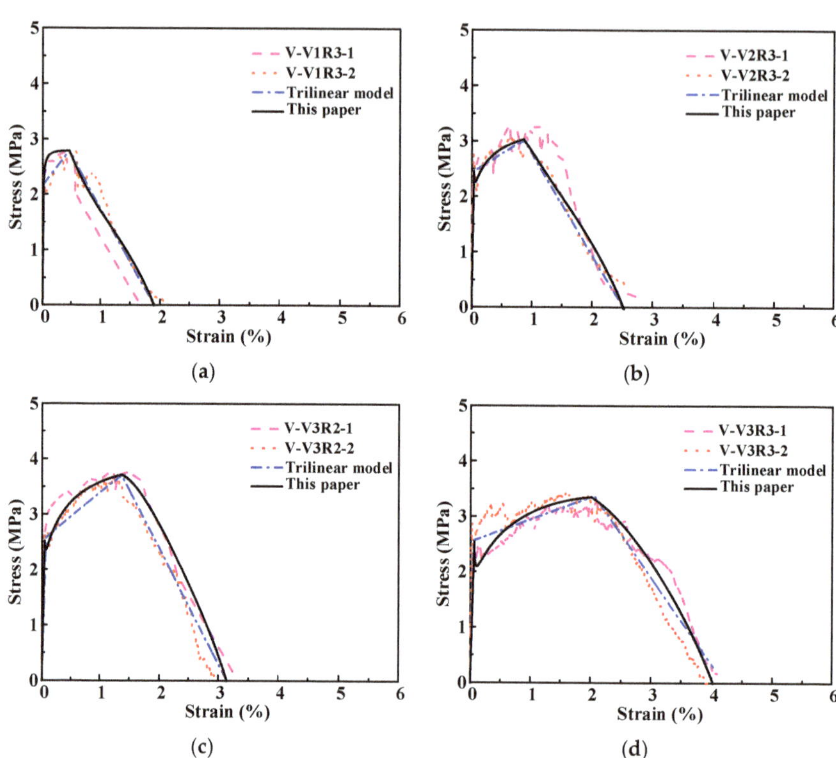

Figure 6. Tensile stress strain curves of: (**a**) Group V-V1R3; (**b**) Group V-V2R3; (**c**) Group V-V3R2; (**d**) Group V-V3R3.

As can be seen from Figure 6, in the elastic nondestructive stage, both the proposed model in this paper and the trilinear model are in good agreement with the test curves. In the strain-hardening stage, the model proposed in this paper can describe accurately the characteristic of the test stress–strain curves, which rise nonlinearly with the reducing slope, whereas the trilinear model [23–25] cannot give accurate prediction of test curves in this stage. In the strain-softening stage (unstable damage), the model proposed this paper can reflect the obvious nonlinear characteristic of test stress–strain curves with fist relatively slow decent rate and then fast decent rate, which cannot be characterized by the trilinear model although the discrepancy between the trilinear model and test curves are relatively small.

5. Application of the Proposed Model to Finite Element Modelling

5.1. Establishment of Nonlinear Finite Element Model

In order to further prove the applicability of the proposed damage constitutive model of ECC in uniaxial tension, this model was incorporated to the FE analysis of the tensile performance of high-strength stainless steel wire ropes (HSSWR) reinforced ECC by using the general FE analysis software ABAQUS. The FE simulation results of by adopting this model would be compared with the results of the uniaxial tension tests on HSSWR reinforced ECC performed by the authors' research group [44] and the FE simulation results by using the trilinear model of ECC [23–25]. The test setup and specimen details are shown in Figure 7 and Table 6, which were detailed in the literature [44]. The volume content of PVA fibers (v), the water-binder ratio (r), first-cracking strain (ε_k), first-cracking stress (σ_k), peak strain (ε_p), peak stress (σ_p), and ultimate strain (ε_u) of the used ECC are 2%, 0.28, 0.02%, 2.45 MPa, 2.79%, 3.53 MPa and 4%, respectively. The maximum tensile stress (σ_{st}) and its corresponding strain (ε_{st}) of the used HSSWR with a nominal diameter of 2.4 mm are 1573 MPa and 2.99%, respectively. The HSSWR reinforced ECC specimen is of the shape of dumbbell, and the HSSWR are symmetrically arranged at the middle height of ECC plate section with equal spacing, as shown in Figure 7a. In order to prevent the end of the specimen from local crushing by the clamp during the loading process, two layers of CFRP sheets were bonded on the surfaces of two ends of the specimen, as shown in Figure 7a.

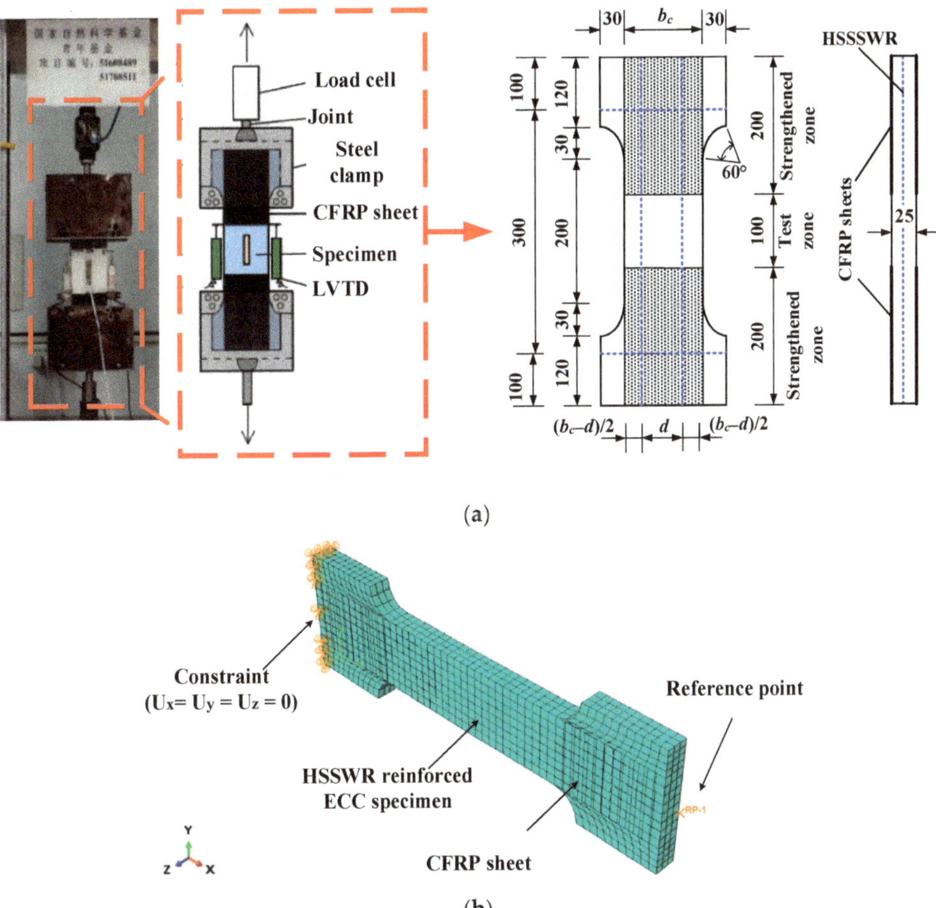

Figure 7. Test setup and 3D FE model of specimens: (**a**) test setup and specimen details (The Chinese project fund in the picture is not related to the research in this paper); (**b**) 3D FE model.

Table 6. HSSWR-ECC Specimen parameter details.

Specimen ID	v (%)	r	Test Section Width, b_c (mm)	HSSWR Spacing, d (mm)	HSSWR Diameter (mm)	HSSWR Reinforcement Ratio (%)
TC1	2	0.28	80	50	2.4	0.28
TC2	2	0.28	70	40	2.4	0.32
TC3	2	0.28	60	30	2.4	0.37
TC4	2	0.28	47	20	2.4	0.48

The three-dimension (3D) nonlinear FE model was used to simulate the tension tests on the HSSWR reinforced ECC specimens, as shown in Figure 7b. In this model, the ECC, HSSSWR and CFRP sheets were assumed to be homogeneous materials and modelled by using eight-node solid elements (C3D8R), two-node truss element (T3D2), and plane shell element, which are available in the FE package of ABAQUS [45].

The mesh size of the elements of ECC and HSSSWR was set to be 10 mm. The interface between ECC and CFRP sheets was assumed to have no relative slip and connected by using the tie constraint feature of ABAQUS [45]. The embedded region constraint feature of ABAQUS [45] was used to prescribe the collaborative deformation between ECC and HSSSWR.

Both the proposed damage constitutive model [Equations (8), (9), and (13)–(17)] and trilinear model [23–25] were adopted to characterize the stress–strain relationship of ECC based on the ABAQUS concrete damage plastic model [45]. Besides, the material parameters dilation angle, flow potential eccentricity, biaxial to uniaxial compression plastic strain ratio, invariant stress ratio, and viscosity were set to, $35°$, 0.1, 1.07, 0.6667, and 0.0005, respectively.

The tensile constitutive model of HSSSWR proposed by the authors' previous studies [44] was adopted to define the stress–strain relationship of HSSSWR in tension based on the ABAQUS Von-Mises plastic model [45], which is expressed as follows:

$$y = 3.33x + 3.66x^2 + 1.33x^3 \tag{21}$$

where $x = \varepsilon_s / \varepsilon_{st}$, $y = \sigma_s / \sigma_{st}$, and σ_s and ε_s indicate the tensile stress and strain of HSSSWR, respectively.

The constraint conditions of the specimen were reproduced by constraining the three translational freedom degrees of one end of the specimen and applying loading by using displacement control at the reference point (RP) created at the section centroid of the other end ECC surface at the loaded end were. The details of 3D nonlinear FE modeling of HSSWR reinforced ECC specimens are shown in Table 6 and Figure 7b.

5.2. Comparison of Simulated Results and Test Results

The predicted load-displacement curves of HSSWR reinforced ECC specimens in tension through FE method by using the proposed model in this paper and the trilinear model [23–25] are compared with test results, as shown in Figure 8. It can be seen in Figure 8 that the predicted results by FE method using the proposed model are in good agreement with test results. Whereas, although before ECC cracking (when the curves are almost linear), the FE simulation results by using the trilinear model [23–25] agree well with test results, after cracking of ECC (when the curves enter the nonlinear phase), the FE simulation results by using the trilinear model [23–25] cannot reflect the nonlinear characteristic of the curves. Besides, the predicted maximum loads by using the trilinear model [23–25] are obviously larger than the test result, while the predicted strains corresponding to the maximum loads are significantly smaller than the test results. The ratio of the predicted maximum loads by using the proposed model and the trilinear model [23–25] to the test result are 1.042 and 1.123, with variation coefficient of 0.026, respectively. The ratio of the predicted displacement corresponding to the maximum load by using the proposed model and the trilinear model [23–25] to the test result are 0.964 and 0.574, with variation coefficient of 0.028 and 0.076, respectively. This demonstrates that the proposed damage constitutive model of ECC in uniaxial tension in this paper can give more accurate prediction of tensile behavior of ECC and can be accepted for characterizing the damage evolution of ECC.

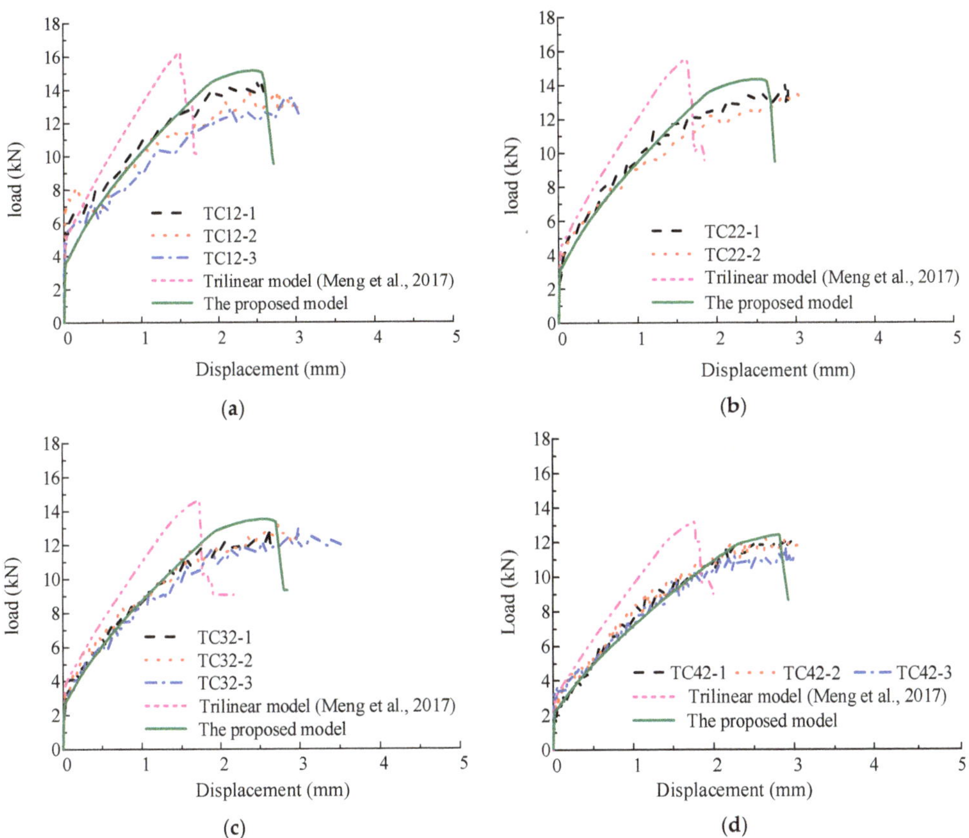

Figure 8. Comparison of tensile load-displacement curves of HSSWR reinforced ECC: (**a**) TC-1 group; (**b**) TC-2 group; (**c**) TC-3 group; (**d**) TC-4 group [23–25].

6. Conclusions

Through the uniaxial tensile test study of ECC specimens with test variables of PVA fiber volume content and water-binder ratio, the effects of the above variables on the uniaxial tensile performance of ECC and the whole process damage mechanism of ECC were revealed, based on which a damage constitutive model of ECC in uniaxial tension was developed. Finally, by applying it to the finite analysis of tensile properties of the new composite material HSSWR-ECC, the acceptability of this model was verified. The following conclusions can be drawn:

(1) An increase in the volume content of PVA fibers could enhance the cracking strain, cracking stress, peak strain, ultimate strain, and fracture energy of ECC. An increase in the water-binder ratio could reduce the cracking stress and peak stress, but would increase the cracking strain, peak strain, ultimate strain, deformation capacity, and fracture energy;

(2) The damage development process of ECC under uniaxial tension can be divided into three stages: (1) elastic undamaged stage, (2) strain hardening and stable damage stage, and (3) strain-softening and nonconverging damage stage. Adding PVA fibers

in ECC could disperse the damage in ECC, and could obviously reduce the damage development rate in the strain hardening and stable damage stage. While reaching the peak stress, the damage development rate was minimized;

(3) By comparing with the experimental results, the damage constitutive model and expressions of model parameters for ECC in uniaxial tension proposed in this paper, which considered the effects of PVA fiber volume content and water-binder ratios, was verified to be able to give accurate prediction of the nonlinear stress–strain relationship of ECC, reflecting the damage evolution characteristic of ECC at different stressing stages;

(4) Based on FE analysis, the model in this paper can be applied to simulate the nonlinear tensile process of HSSWM-ECC. Compared with the trilinear model used by ECC, the model in this paper has higher accuracy and smaller discreteness;

(5) Due to the limitations of the experimental research in this paper, the range of experimental parameters and the number of specimens should be expanded in the future to make the proposed model parameters more accurate. Secondly, the experiments under repeated loads should be explored, and the damage constitutive model of ECC under repeated loads should be proposed. Finally, it may be more meaningful and challenging to apply the proposed model to the analysis of real structures.

Author Contributions: Conceptualization, D.C., and K.L.; methodology, D.Z., and K.L.; investigation, C.W., and L.C.; data curation, P.Z.; writing—original draft preparation, D.Z.; funding acquisition, K.L. All authors have read and agreed to the published version of the manuscript.

Funding: This work was funded by the National Natural Science Foundations of China (No. U U180411663), China Postdoctoral Science Foundation (Grant No. 2020M672236), and Ministry of housing and urban rural development science and technology program of China (Grant Nos. 2019-K-059, K2124).

Institutional Review Board Statement: Not applicable.

Informed Consent Statement: Not applicable.

Data Availability Statement: All the data is available within the manuscript.

Acknowledgments: This work was financially supported by the National Natural Science Foundations of China (No. U U180411663), China Postdoctoral Science Foundation (Grant No. 2020M672236), and Ministry of housing and urban rural development science and technology program of China (Grant Nos. 2019-K-059, K2124).

Conflicts of Interest: The authors declare no conflict of interest.

References

1. Li, V.C. Interface tailoring for strain-hardening PVA-ECC. *ACI Mater. J.* **2002**, *99*, 463–672.
2. Li, V.C. On engineered cementitious composites (ECC)-a review of the material and its applications. *J. Adv. Concr. Technol.* **2003**, *1*, 215–230. [CrossRef]
3. Li, V.C.; Horikoshi, T.; Ogawa, A.; Torigoe, S.; Saito, T. Micromechanics-based durability study of polyvinyl alcohol-engineered cementitious composite. *ACI Mater. J.* **2004**, *101*, 242–248. [CrossRef]
4. Gaverzan, A.; Cadoni, E.; Prisco, M. Tensile behaviour of high-performance fiber-reinforced cementitious composites at high strain rates. *Int. J. Impact. Eng.* **2012**, *45*, 28–38. [CrossRef]
5. Zhang, X.; Liu, S.; Yan, C.; Wang, X.; Wang, H. Effects of Vehicle-Induced Vibrations on the Tensile Performance of Early Age PVA-ECC. *Materials* **2019**, *17*, 2652. [CrossRef]
6. Li, K.F.; Yang, C.Q.; Zhao, Y.B.; Pan, Y.; Wang, G.; Zheng, Y.Y.; Xu, F. Study on the creep behavior of PVA-ECC based on fractional-differential rheological model. *Constr. Build. Mater.* **2020**, *230*, 117064. [CrossRef]
7. Ding, Z.; Wen, J.; Li, X.; Fu, J.; Ji, X. Mechanical behaviour of polyvinyl alcohol-engineered cementitious composites (PVA-ECC) tunnel linings subjected to vertical load. *Tunn. Undergr. Sp. Techol.* **2020**, *95*, 103151. [CrossRef]
8. Tosun, K.; Felekoğlu, B.; Baradan, B. Multiple cracking response of plasma treated polyethylene fiber reinforced cementitious composites under flexural loading. *Cement. Concrete Comp.* **2004**, *26*, 508–520. [CrossRef]
9. Liao, Q.; Yu, J.; Shi, T.; Su, Y. Mechanical behaviors and failure criteria of seawater sea-sand engineered cementitious composites under combined tension and shear. *J. Build. Eng.* **2022**, *54*, 104552. [CrossRef]

10. Yu, K.Q.; Yu, J.T.; Dai, J.G.; Lu, Z.D.; Shah, S.P. Development of ultra-high performance engineered cementitious composites using polyethylene (PE) fibers. *Constr. Build. Mater.* **2018**, *158*, 217–227. [CrossRef]
11. Aizenshtein, E.M.; Efremov, V.N. Production and use of polypropylene fibres and yarn. *Fibre Chem.* **2006**, *38*, 345–350. [CrossRef]
12. Aizenshtein, E.M. World chemical fibre and thread production in 2003. *Fibre Chem.* **2004**, *36*, 467–482. [CrossRef]
13. Zhang, R.; Matsumoto, K.; Hirata, T.; Ishizeki, Y.; Niwa, J. Application of PP-ECC in beam–column joint connections of rigid-framed railway bridges to reduce transverse reinforcements. *Eng. Struct.* **2015**, *86*, 146–156. [CrossRef]
14. Ahmed, S.F.; Maalej, M.; Paramasivam, P. Analytical model for tensile strainhardening and multiple-cracking behavior of hybrid fiber engineered cementitious composites. *ASCE J. Mater. Civ. Eng.* **2007**, *19*, 527–539. [CrossRef]
15. Soe, K.T.; Zhang, Y.X.; Zhang, L.C. Material properties of a new hybrid fibre-reinforced engineered cementitious composite. *Constr. Build. Mater.* **2013**, *43*, 399–407. [CrossRef]
16. Wang, Q.; Yi, Y.; Ma, G.; Luo, H. Hybrid effects of steel fibers, basalt fibers and calcium sulfate on mechanical performance of PVA-ECC containing high-volume fly ash. *Cement. Concrete Comp.* **2019**, *97*, 357–368. [CrossRef]
17. Tosee, S.V.R.; Faridmehr, I.; Bedon, C.; Sadowski, L.; Aalimahmoody, N.; Nikoo, M.; Nowobilski, T. Metaheuristic prediction of the compressive strength of environmentally friendly concrete modified with eggshell powder using the hybrid ANN-SFL optimization algorithm. *Materials* **2021**, *14*, 6172. [CrossRef]
18. Faridmehr, I.; Bedon, C.; Huseien, G.F.; Nikoo, M.; Baghban, M.H. Assessment of mechanical properties and structural morphology of alkali-activated mortars with industrial waste materials. *Sustainability* **2021**, *13*, 2062. [CrossRef]
19. Maalej, M.; Quek, S.T.; Ahmed, S.F.U.; Zhang, J.; Lin, V.W.J.; Leong, K.S. Review of potential structural applications of hybrid fiber Engineered Cementitious Composites. *Constr. Build. Mater.* **2012**, *36*, 216–227. [CrossRef]
20. Shang, X.Y.; Yu, J.T.; Li, L.Z.; Lu, Z.D. Strengthening of RC Structures by Using Engineered Cementitious Composites: A Review. *Sustainability* **2019**, *11*, 3384. [CrossRef]
21. Li, V.C. Performance driven design of fiber reinforced cementitious composites. In *Proceedings of 4th RILEM International Symposium on Fiber*; Swamy, R.N., Ed.; Chapman and Hall: London, UK, 1992; pp. 12–30.
22. Kanda, T.; Lin, Z.; Li, V.C. Tensile stress-strain modeling of pseudo strain hardening cementitious composites. *J. Mater. Civil. Eng.* **2000**, *12*, 147–156. [CrossRef]
23. Deng, M.; Yang, S. Experimental and numerical evaluation of confined masonry walls retrofitted with engineered cementitious composites. *Eng. Struct.* **2020**, *207*, 110249. [CrossRef]
24. Kabir, M.I.; Lee, C.K.; Rana, M.M.; Zhang, Y.X. Flexural and bond-slip behaviors of engineered cementitious composites (ECC) encased steel composite beam. *J. Constr. Steel. Res.* **2019**, *157*, 229–244. [CrossRef]
25. Meng, D.; Huang, T.; Zhang, Y.X.; Lee, C.K. Mechanical behaviour of a polyvinyl alcohol fibre reinforced engineered cementitious composite (PVA-ECC) using local ingredients. *Constr. Build. Mater.* **2017**, *141*, 259–270. [CrossRef]
26. Khan, M.; Lee, C.K.; Zhang, Y.X. Numerical modelling of engineered cementitious composites-concrete encased steel composite columns. *J. Constr. Steel. Res.* **2020**, *170*, 106082. [CrossRef]
27. Kabir, M.I.; Lee, C.K.; Rana, M.M.; Zhang, Y.X. Strength enhancement of high strength steel beams by engineered cementitious composites encasement. *Eng. Struct.* **2020**, *207*, 110288. [CrossRef]
28. Bai, L.; Li, Y.; Hou, C.; Zhou, T.; Cao, M. Longitudinal shear behaviour of composite slabs with profiled steel sheeting and ECC. *Eng. Struct.* **2020**, *205*, 110085.1–110085.20. [CrossRef]
29. Raza, A.; Khan, Q.Z.; Ahmad, A. Numerical investigation of load-carrying capacity of GFRP-reinforced rectangular concrete members using CDP model in ABAQUS. *Adv. Civil. Eng.* **2019**, *3*, 1–21. [CrossRef]
30. Zhuang, H.; Wang, R.; Shi, P.; Chen, G. Seismic response and damage analysis of underground structures considering the effect of concrete diaphragm wall. *Soil. Dyn. Earthq. Eng.* **2019**, *116*, 278–288. [CrossRef]
31. Mao, K.; Soma, Y.; Terada, K. Simulations of cohesive fracture behavior of reinforced concrete by a fracture-mechanics-based damage model. *Eng. Fract. Mech.* **2018**, *206*, 392–407. [CrossRef]
32. Krahl, P.A.; Carrazedo, R.; Debs, M.E. Mechanical damage evolution in UHPFRC: Experimental and numerical investigation. *Eng. Struct.* **2018**, *170*, 63–77. [CrossRef]
33. Behbahani, A.E.; Barros, J.; Ventura-Gouveia, A. Plastic-damage smeared crack model to simulate the behaviour of structures made by cement-based materials. *Int. J. Solids Struct.* **2015**, *73*, 20–40. [CrossRef]
34. Cai, J.M.; Pan, J.L.; Tan, J.W.; Vandevyvere, B. Nonlinear finite-element analysis for hysteretic behavior of ECC-encased CFST columns. *Structures* **2020**, *25*, 670–682. [CrossRef]
35. Chen, X.M.; Duan, J.; Qi, H.; Li, Y.G. Controls on material and mesh for structural nonlinear time-History analysis. *Appl. Mech. Mater.* **2014**, *7*, 580–583; 1564–1569. [CrossRef]
36. Hu, Y.C.; Tan, Y.H.; Xi, F. Failure assessment and virtual scenario reproduction of the progressive collapse of the FIU bridge. *Eng. Struct.* **2021**, *227*, 111423. [CrossRef]
37. Li, V.C.; Leung, C. Steady-state and multiple cracking of short random fiber composites. *J. Eng. Mech.* **1992**, *118*, 2246–2264. [CrossRef]
38. Li, L.; Cai, Z.; Yu, K.; Zhang, Y.X.; Ding, Y. Performance-based design of all-grade strain hardening cementitious composites with compressive strengths from 40 MPa to 120 MPa. *Cement. Concrete Comp.* **2019**, *97*, 202–217. [CrossRef]

39. Zhu, Y.; Zhang, Z.; Yang, Y.; Yao, Y. Measurement and correlation of ductility and compressive strength for engineered cementitious composites (ECC) produced by binary and ternary systems of binder materials: Fly ash, slag, silica fume and cement. *Constr. Build. Mater.* **2014**, *68*, 192–198. [CrossRef]
40. Hoang, N.; Carvelli, V.; Adesanya, E.; Kinnunen, P.; Illikainen, M. High-performance cementitious composite from alkali-activated ladle slag reinforced with polypropylene fibers. Cement. *Concrete Comp.* **2018**, *90*, 150–160. [CrossRef]
41. Yu, K.; Ding, Y.; Liu, J.; Bai, Y. Energy dissipation characteristics of all-grade polyethylene fiber-reinforced engineered cementitious composites (PE-ECC). *Cement. Concrete Comp.* **2020**, *106*, 103459. [CrossRef]
42. Zhang, D.; Yu, J.; Wu, H.L.; Jaworska, B.; Ellis, B.R.; Li, V.C. Discontinuous micro-fibers as intrinsic reinforcement for ductile Engineered Cementitious Composites (ECC). *Compos. Part B Eng.* **2020**, *184*, 107741. [CrossRef]
43. Li, K.; Zhao, D.P.; Liu, W.K.; Fan, J.J. Research on damage constitutive model of engineered cementitious composites under uniaxial tension. *Eng. Mech.* **2022**. (In Chinese) [CrossRef]
44. Wang, X.L.; Yang, G.H.; Qian, W.W.; Li, K.; Zhu, J.T. Tensile stress-strain relationship of engineered cementitious composites reinforced by high-strength stainless steel wire mesh. *Acta Mater. Compos. Sin.* **2020**, *37*, 3220–3228. (In Chinese) [CrossRef]
45. ABAQUS. *ABAQUS Standard User's Manual, Version 6.14*; Dassault Systems Similia Corp.: Johnston, RI, USA, 2014.

Article

Design of Environmentally Friendly Ca-Alginate Beads for Self-Healing Cement-Based Materials

Xiaohang Zhang * and Yonggang Ding

School of Civil Engineering and Architecture, Henan University of Technology, Zhengzhou 450001, China
* Correspondence: zhangxiaohanghaut@163.com

Abstract: Ca-alginate beads have strong hygroscopicity, which have been used for the self-healing and internal curing of cement-based materials. However, ca-alginate beads in cement will chelate with calcium ions, which decreases the swelling rate of ca-alginate beads in the healing environment and is detrimental to the self-healing of cement-based materials. In this paper, the mechanism and steps for preparing ca-alginate beads with a lower ability to chelate with calcium ions were proposed based on protonation theory. In addition, the molecular structure and the swelling rates in cement filtrate and healing environment of ca-alginate beads prepared by the proposed method were characterized. The results showed that the ca-alginate beads prepared by the proposed method had higher molecular density and a lower ability to chelate with calcium ions. The swelling rate in the healing environment is not decreased. Furthermore, the equilibrium swelling rate in cement filtrate can satisfy the need for internal curing of cement-based materials.

Keywords: ca-alginate; protonation theory; cement-based materials; internal curing; self-healing

Citation: Zhang, X.; Ding, Y. Design of Environmentally Friendly Ca-Alginate Beads for Self-Healing Cement-Based Materials. *Materials* **2022**, *15*, 5844. https://doi.org/10.3390/ma15175844

Academic Editor: Miguel Ángel Sanjuán

Received: 14 July 2022
Accepted: 22 August 2022
Published: 24 August 2022

Publisher's Note: MDPI stays neutral with regard to jurisdictional claims in published maps and institutional affiliations.

1. Introduction

Cement is one of the most important building materials. However, the manufacturing process of cement releases a large amount of carbon dioxide, which accounts for 5–7% of carbon dioxide emissions [1,2]. The carbon dioxide emissions and the consumption of cement can be reduced by extending the service life of the structure. However, cement-based materials are brittle and prone to cracks [3]. The generation of cracks accelerates the intrusion of harmful ions and, thus, reduces the service life of cement-based materials [4]. Therefore, it has become an urgent issue for governments around the world to extend the service life of the structure.

Abrams found that cracks in cement-based materials could self-healing in 1925. The basic principle is that the unhydrated cement particles in the cement matrix undergo secondary hydration under a certain humidity to form calcium silicate hydrate and calcium carbonate. However, only the small cracks of 30–50 μm can be healed in cement-based materials [5].

Superabsorbent polymer (SAP) has been used to strengthen the self-healing ability of cement-based materials [6]. In order to ensure self-healing efficiency, SAP equal to 1% of the cement quality should be added to the cement [7]. The cracks smaller than 50 μm in cement-based materials mixed with SAP are completely healed, and the cracks smaller than 150 μm are partially healed [5]. In addition, SAP is also used for the internal curing of cement-based materials [8–10]. The amount of entrained water required for internal curing is 0.18 times the water–cement ratio [8]. Furthermore, SAP can absorb water dozens of times its weight during mixing, which can usually meet the needs of internal curing. However, the excess water is entrained due to the high absorbent ability, which forms a large number of large pores in the cement-based materials after release. The formation of large pores will decrease the strength of cement-based materials [7].

Sodium alginate is a water-soluble anion polysaccharide in brown algae. It has excellent water absorption and moisture retention, which is also environmentally friendly and widely available. Primarily, it is a polysaccharide polymer $((C_6H_7O_6Na)_n)$ composed of 1–4 poly β-D-mannuronic acid (M) and α-L-guluronic acid (G) monomers. Here, ca-alginate beads are formed by the combination of the carboxyl group on sodium alginate and calcium ions by an ionic bond and chelation bond. The ca-alginate beads have a low swelling rate and stable structure. The strength of cement-based materials with 1 wt% ca-alginate beads is 13% higher than that of cement-based materials mixed with 1 wt% SAP under the water-cement ratio of 0.5 when tested at 28 days [11]. In addition, ca-alginate beads have strong hygroscopicity under the relative humidity of the atmosphere, which can be used for self-healing of cement-based materials in the region where rain is scarce [11]. The ca-alginate beads can also act as a precursor of calcium carbonate crystals to promote cement hydration [12–15]. Therefore, the ca-alginate bead is a promising material for cement-based materials.

Enough water needs to be entrained when ca-alginate beads are used for the internal curing of cement-based materials as an absorbent material, which means that the amount of entrained water is 0.18 times the water–cement ratio. In addition, the swelling rate of ca-alginate beads in the healing environment should be much greater than that in the cement paste to ensure the efficiency of the self-healing [6]. However, ca-alginate beads in the cement filtrate chelate with calcium ions, which increases the crosslinking density of ca-alginate beads [11]. When the cement-based materials cracks, the swelling rate of ca-alginate beads will decrease due to the increase in crosslinking density, which is not conducive to self-healing. Therefore, the ability to chelate with calcium ions in the cement filtrate needs to be reduced for the ca-alginate beads applied for the self-healing of cement-based materials.

The ca-alginate bead is a kind of hydrogel with a controllable three-dimensional structure [16]. The ability to chelate with calcium ions can be decreased by increasing the concentration of sodium alginate [17]. The concentration of sodium alginate solution has a limited value due to the viscosity of the sodium alginate solution. The maximum concentration of sodium alginate solution is 8% in the study by Stewart [17]. However, the existing preparation methods all need to prepare a sodium alginate solution first, then disperse the sodium alginate solution into beads and expose them to the calcium salt solution to chelate with calcium ions to form ca-alginate beads [16,17]. Therefore, it is difficult to increase the concentration of sodium alginate solution using the existing methods.

Sodium alginate has a reversible pH sensitivity. Part of –COO– in sodium alginate will be protonated and become –COOH in an acidic solution [18]. On the one hand, this will reduce the hydrophilicity of sodium alginate but, on the other hand, hydrogen bonds can be formed between –COOH, which makes the protonated segments become insoluble in water and precipitate out of the solution [18]. Additionally, Cao makes the sodium alginate solution self-assemble to form sodium alginate micelles in acidic solution based on protonation theory, and then makes the micelles exchange and chelate with calcium ions to form ca-alginate beads.

In this paper, the mechanism and steps of the preparation method for ca-alginate beads with a higher concentration of sodium alginate were proposed based on protonation theory. The molecular structure, content of the chelation bond, morphology, and pore structure of the ca-alginate beads prepared by this method were characterized. In addition, its potential for cement-based materials was discussed based on its swelling behavior in the cement filtrate and the healing environment.

2. Materials and Methods

2.1. Materials

2.1.1. Cement

Portland cement with 42.5 strength grade (P·I 42.5) produced by the China Building Materials Institute was used. The chemical composition of the cement was identified by the X-ray fluorescence test. The cement mainly consists of CaO, SiO_2, SO_3, Fe_2O_3, Al_2O_3, MgO, and K_2O. The content is 64.20 wt%, 22.37 wt%, 2.39 wt%, 3.18 wt%, 4.07 wt%, 2.06 wt%, and 0.45 wt%, respectively.

2.1.2. Alginate

The sodium alginate was provided by Macklin, Shanghai, China. The ratio between mannuronic and guluronic acid was about 1.5. The particle size of the sodium alginate was below 200 μm. The viscosity of a 10% solution was 200 ± 20 mPa·s at 20 °C.

2.1.3. Chemicals

The deionized water and anhydrous ethanol were provided by Macklin, China. The calcium chloride with a purity of 99.99% was provided by Macklin, China.

2.2. The Preparation Method of Ca-Alginate Beads

2.2.1. Preparation Mechanism and Steps of Ca-Alginate Beads with a Higher Concentration of Sodium Alginate

Sodium alginate powder is directly protonated in an acid solution to obtain calcium alginate beads (PCA) with a higher concentration of sodium alginate. The preparation mechanism of ca-alginate beads (PCA) is based on the protonation theory (shown in Figure 1a). The sodium alginate powder will be dissolved and protonated at the same time when it is put into an acid solution, whose PH value is lower than the pKa value of sodium alginate. The molecular chain of the outer layer of sodium alginate stretches gradually, and the –COONa in the molecular chain dissociates into –COO– and Na^+. Part of –COO– in sodium alginate is protonated and becomes -COOH. Additionally, hydrogen bonds can also be formed between –COOH, which reduces the surface charge of sodium alginate and, thus, decreases the charge repulsion between the powders [18]. The hydrophobic shell is formed in the surface layer of sodium alginate. The protonated sodium alginate powder will precipitate out of the solution when the protonation reaches a certain degree. Then, the protonated sodium alginate powder is mixed with calcium salt and the pH value of the acid solution is increased at the same time. In the process, ionic bonds (Figure 1c) and chelate bonds (Figure 1d) are formed between Ca^{2+} and alginate molecules. In the end, the pH of the solution is adjusted to about 7 to remove H^+ and, thus, ca-alginate beads are formed.

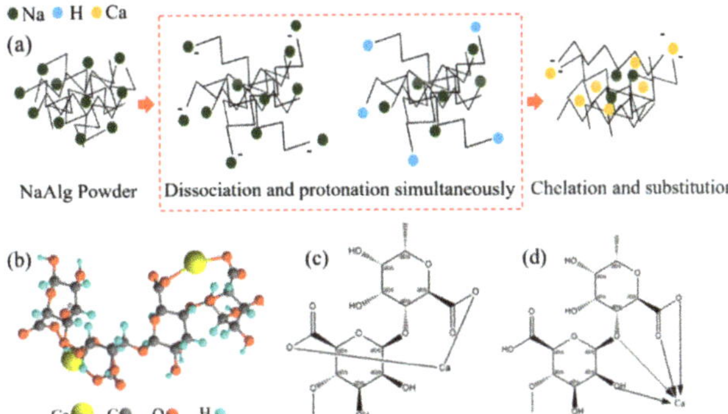

Figure 1. The preparation schematic diagram of PCA and the molecular structure of ca-alginate: (**a**) preparation schematic diagram of PCA, (**b**) the molecular chain of ca-alginate, (**c**) ionic bond, and (**d**) chelation.

Sodium alginate is protonated and dispersed when the pH value is less than the pKa value (≈ 3). The disperse phenomenon of sodium alginate powder in acid solutions with different pH values is different (the phenomenon shown in Table 1). The sodium alginate powder disperses better in an acid solution with a pH value of 0.01, but it disperses poorly and undergoes agglomeration in an acid solution with a higher pH value. The main reason for the agglomeration of sodium alginate in acid solution is the insufficient hydrogen ion content in the acid solution, which causes the molecular chain on the surface of sodium alginate to be protonated insufficiently and to agglomerate due to a lack of charge repulsion. The larger size of ca-alginate beads will be obtained due to the agglomeration. The large size is detrimental to the strength of cement-based materials [19]. In order to obtain the smaller size of ca-alginate beads, sodium alginate powder need to undergo better dispersion and less agglomeration. Strong acid with a pH value of about 0.01 was selected during preparation to obtain good dispersion. The sodium alginate powder should avoid contact with water before adding to the acid solution to avoid agglomeration and poor dispersion. Therefore, the preparation container was wetted with strong acid first.

Table 1. The disperse phenomenon of sodium alginate powder in acid solution with different pH values.

PH Value	Phenomenon
0.01	Good dispersion, no agglomeration
0.6	Good dispersion, a small amount of agglomeration
1.42	Poor dispersion, a lot of agglomeration
2.07	No dispersersion, agglomeration
2.55	No dispersersion, agglomeration
2.95	No dispersersion, agglomeration

The preparation steps of ca-alginate beads (PCA) were determined according to the preparation principle and the disperse phenomenon of sodium alginate in acid solutions with different pH values, as follows:

(1) Wetting the wall of the container with concentrated hydrochloric acid with a pH value of less than 0.01, and then adding 50 mL of the concentrated hydrochloric acid to the container;

(2) Turning on the magnetic stirrer and adding 50 g of sodium alginate powder to the acid while stirring slowly, so that the sodium alginate powders were protonated and dispersed better. The ion changes are shown in Figure 2b,c;

(3) Adding 50 g calcium chloride powder and stirring evenly, then adding deionized water to ensure the total volume of the solution in the reaction container was 2.25 L and the concentration of calcium chloride was 0.2 mol/L. The ion changes are shown in Figure 2d–f;

(4) Adjusting the pH value of the solution to about 7 with sodium hydroxide and stirring for 24 h to fully react the molecular chain of sodium alginate and calcium ions to form chelate bonds and ionic bonds. The ion changes are shown in Figure 2f,g;

(5) Washing and filtering the prepared ca-alginate beads with deionized water at least three times to remove excess ions;

(6) Putting the washed ca-alginate beads into a drying oven at 40 °C until the mass was constant. The dried ca-alginate beads (PCA) were obtained.

2.2.2. Preparation Method of Ca-Alginate Beads with a Lower Concentration of Sodium Alginate

The dropwise addition method was used to prepare ca-alginate beads (CA) with a lower concentration of sodium alginate [17]. The concentration of sodium alginate and the concentration of calcium chloride in this method were 2 wt% and 0.2 mol/L, respectively. The reaction system was stirred for 24 h to react fully between the molecular chain of sodium alginate and calcium ions to form chelate bonds and ionic bonds after the sodium alginate beads were dropped into the solution of calcium chloride with a syringe. After that, the ca-alginate beads were obtained. Then, the ca-alginate beads were washed and filtered with deionized water at least three times to remove excess ions. In the end, the washed ca-alginate beads were put into a drying oven at 40 °C until the mass was constant. The dried ca-alginate beads (CA) were obtained.

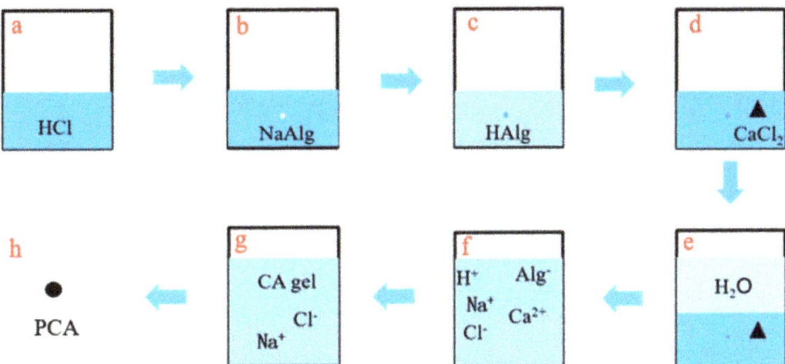

Figure 2. Ion changes in the preparation process of PCA.

2.3. Methods

2.3.1. Molecular Structures

The Nicolet is 5003190721 Fourier infrared spectrometer (FTIR-650) with a scanning resolution of 0.482 cm^{-1}, which was used to characterize the molecular structure of ca-alginate beads. The scanning range was 500 cm^{-1}–4000 cm^{-1}.

Here, X-ray photoelectron spectroscopy (XPS) was used to analyze the combination form of calcium ions and carboxyl groups. A 250Xi ESCLAB (Shanghai, China) photoelectron spectrometer with Al K$_\alpha$ rays was used. The pass energy of 100 eV and 20 eV were used for wide-range scanning spectra and narrow-range scanning spectra (high resolution) of C1$_s$ and O1$_s$, respectively. The XPS system software and instrument standard parameters were used for spectrum analysis and element composition quantification. The Origin mathematics software was used for high-resolution spectra to subtract linear background and split peaks with 80% Gaussian −20% Lorentzian mixed function. The C1$_s$ binding energy of saturated hydrocarbons, such as diffusion pump oil and other pollutants (284.8 eV), was used to correct the binding energy [20–23].

2.3.2. Morphology and Size

A DSX500 ultra-depth-of-field optical digital microscope produced by Japan's OLYM-PUS company (Shinjuku, Japan) was used to photograph the morphology of two kinds of ca-alginates beads. The BT-2001 laser particle size analyzer (Liaoning, China) was used to test the size distribution and average size of the two kinds of ca-alginates beads.

The morphology images were also obtained using a scanning electron microscope (SEM) produced by the TESCAN company (Brno, Czech Republic), with a VEGA3 tungsten filament at a magnification of 499 times and 1840 times in vacuum conditions at 30 kV. The samples for SEM were vacuum dried and coated with a thin layer of gold on the surface.

2.3.3. Pore Structure

Before the test, the prepared samples were dried in a 50 °C dryer for 24 h. The Auto-Pore IV9500 mercury porosimeter (Norcross, GA, USA, with a pressure range of 0 to 60,000 psi) was used to measure the pore size and pore distribution of the interconnected pores of PCA and CA. The powder dilatometer was selected for the test. Firstly, the drying sample (about 0.05 g) and the dilatometer assembly were weighed. The head of the dilatometer was sealed with resin. Then, the low-pressure and high-pressure tests were carried out successively. After measurement, the dilatometer was taken out and cleaned.

2.3.4. Swelling Kinetic Experiment

The cement filtrate was obtained by mixing 10 g of cement and 100 g of deionized water, stirring with a magnetic mixer for 3 h, and filtering with filter paper [11]. In order to simulate the self-healing environment in cement, the ca-alginate beads were first put into cement filtrate to adsorb and react with calcium ions, then dried and put into deionized water for swelling.

The swelling kinetics was measured with the tea bag method [24]. A small amount of dried ca-alginate beads and pre-moistened tea bags were weighed and recorded as m_c and m_f, respectively. They were put into cement filtrate or a healing environment. Then, they were taken out at regular intervals and wiped with filter paper to absorb redundant solutions on the surface of the tea bags. Their weight was recorded as m_t. The swelling rate $W(t)$ at t minutes was calculated according to Equation (1). Three samples with identical quality were measured for statistical analysis. The means were used for analyzing the swelling kinetics of ca-alginate beads.

$$W(t) = \frac{m_t - m_f - m_c}{m_c} \tag{1}$$

2.3.5. Statistical Analysis

The statistical analysis was performed on the characterization results of molecular structure, size, and swelling kinetics. The values were reported as the mean $\pm$ standard deviation.

3. Results and Discussion

3.1. Morphology and Size

Figure 3 shows the morphology and cross section of PCA and CA. The magnification was represented by the orange bars. Here, CA is a spherical particle with a smooth surface and looser structure (shown in Figure 3b,d), while PCA is an approximate ellipsoid with a rough surface and denser structure (shown in Figure 3a,c). That is because the sodium alginate used for PCA is not dissolved, and the molecular chain is not stretched. Therefore, the content of molecules per unit volume of PCA is higher and the density of molecules inside PCA is greater [17].

The average size of PCA is 89.90 μm. The average size of CA is 287.64 μm, which is nearly 3.2 times that of PCA (shown in Table 2). The size of PCA is reduced by 197.74 μm, which is significantly smaller than that of CA. Therefore, the size of PCA is more beneficial to cement-based materials [19].

Table 2. The diameters of two kinds of ca-alginate beads.

Prepare Method	Maximum Diameter/μm	Minimum Diameter/μm	Average Diameter/μm
PCA	392.90	0.50	89.90 $\pm$ 4.61
CA	457.16	149.05	287.64 $\pm$ 6.90

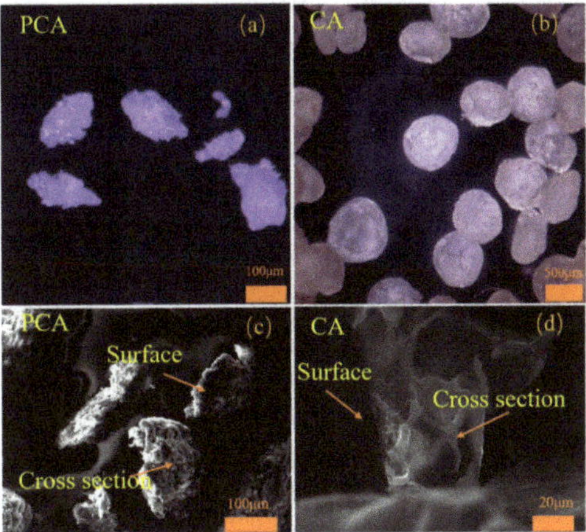

Figure 3. The morphology of PCA and CA: (**a,c**) the morphologies of PCA, (**b,d**) the morphologies of CA.

3.2. Molecular Structure

3.2.1. Molecular Structure of the Surface of Ca-Alginate Beads

The XPS spectrum was processed to characterize the molecular structures in the surfaces of sodium alginate and ca-alginate beads. Figure 4a shows the high-resolution scan spectra of O1s and the peak splitting results of sodium alginates. Two peaks appear in the peak splitting results of the high-resolution scan spectra of O1s. The lower binding energy peak (531.16 eV) corresponds to the ionic bond between the oxygen of carboxyl and the sodium ions. The carboxyl in sodium alginate is all combined with sodium ions by an ionic bond. The area fraction of oxygen of carboxyl in sodium alginate combined with sodium ions by the ionic bond is 33.01%, which is close to the theoretical calculation value of 33.3% [20,23]. The calculating result indicated that the experimental data and the result obtained by the processing method are reliable.

Two peaks also are found in the high-resolution scan spectra of O1s of CA and PCA (shown in Figure 4b,c). Among them, the area fractions of the lower binding energy peaks (531.57 eV, 531.47 eV) corresponding to the oxygen of carboxyl combined with calcium ions by the ionic bond are 29.52% and 29.89%, respectively. This is because the area fraction of oxygen of carboxyl in sodium alginate combined with sodium ion by ionic bond is the total content of the carboxyl in the ca-alginate beads. The proportion of carboxyl in ca-alginate beads combined with calcium ion by the ionic bond is 88.65% (29.52/33.3) and 89.76% (29.89/33.3), respectively. Therefore, the content of carboxyl forming the chelate bond is 11.35% (1–88.65%) and 10.24% (1–89.76%), respectively. The XPS spectra of two other identical samples of CA and PCA were also obtained for statistical evaluation. The statistical results show that the contents of the chelate bond of CA and PCA are 12.86 ± 1.07% and 11.85 ± 1.15%, respectively.

The chelate bond is also formed in the molecular structure of PCA. Furthermore, the content of the chelate bond of PCA is smaller than that of CA, which proves that the content of the chelate bond in PCA under the same calcium salt concentration is decreased. The decreased chelation is caused by the denser structure of PCA, which hinders the diffusion and crosslinking of calcium ions [17].

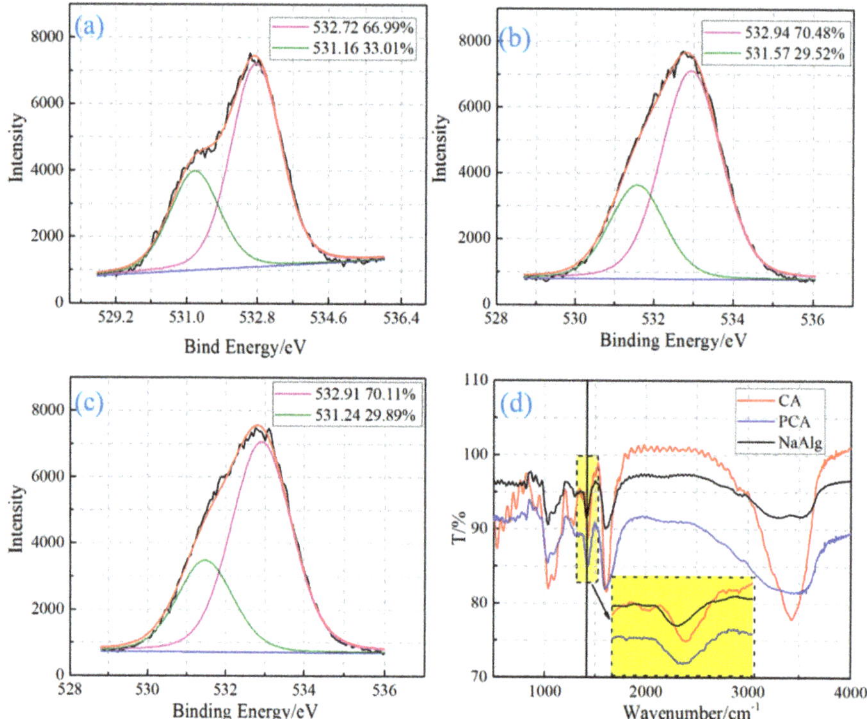

Figure 4. The results of XPS and FTIR: (**a**) the peaks splitting and fitting of O1s high-definition spectrogram of sodium alginate, (**b**) the peaks splitting and fitting of O1s high-definition spectrogram of CA, (**c**) the peaks splitting and fitting of O1s high-definition spectrogram of PCA, (**d**) the FTIRs of PCA, CA, and sodium alginate. Note: In Figure 4a–c, the black line represents the original data, the red line represents the curve after fitting, and the pink and green lines are the curves after peaking according to the fitted curve.

3.2.2. The Molecular Structure of Ca-Alginate Beads

The FTIRs of PCA, CA, and sodium alginate (NaAlg) are shown in Figure 4d. The stretching vibration peak of the O-H band ($\approx$3500 cm^{-1}) is the characteristic peak of intramolecular bonds, whose peak shape becomes stronger and sharper with the increase in the chelation bonds. However, the characteristic peak of the intermolecular bonds of the O-H ($\approx$3250 cm^{-1}) has little change with the increase in the chelation bonds [25]. The peak position ($\approx$3500 cm^{-1}) of PCA does not increase significantly compared with that of NaAlg, while that of CA increases significantly. The result proves that the content of the chelation bond of PCA is smaller under the same concentration of calcium salt.

The symmetrical stretching vibration peak ($\approx$1420 cm^{-1}) of the carboxyl group shifts to a high wave number when the carboxyl group of sodium alginate is combined with calcium ion by an ionic bond. The more calcium ion content, the more the shift occurs [25]. The symmetrical stretching vibration peak (1422 cm^{-1}) of the carboxyl group of PCA has a smaller shift, while the stretching vibration peak (1429 cm^{-1}) of the carboxyl group of CA has a larger shift. The change means that fewer ionic bonds are formed by the combination of the carboxyl group and calcium ion in PCA under the same concentration of calcium salt.

The decreased content of the chelation bond and ionic bond proves that calcium ions diffuse slowly inside the PCA.

3.3. Pore Structure

The differential distribution curves of the pore structures of PCA and CA are shown in Figure 5. The parameter $dv/d(\log(D))$ represents the derivative of the pore volume with respect to the log of the hole diameter. The pore size of the connected pores inside the PCA is significantly reduced. In addition, the pore size of PCA is significantly smaller compared with that of CA (shown in Figure 3c,d). Therefore, PCA has a higher molecular density according to the conclusion obtained by Stewart et al., who believe that the pore size inside the high molecular density ca-alginate beads is reduced [17].

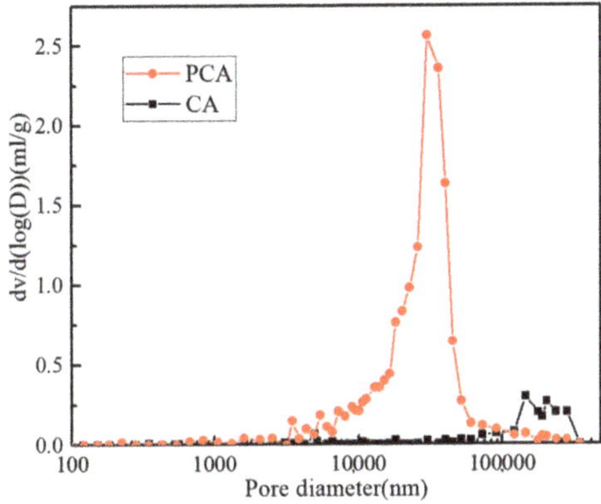

Figure 5. Pore diameter distributions of two kinds of ca-alginate beads.

3.4. Swelling Kinetic in Cement Filtrate and the Healing Environment

3.4.1. Swelling Kinetic in Cement Filtrate

Figure 6a shows the swelling processes of PCA and CA in the cement filtrate. The two kinds of ca-alginate beads swell rapidly in the first 5 min. Then, the swelling rate increases slowly due to the restriction of the molecular chain. It is interesting that the swelling rate starts to decrease when it reaches the maximum value. The decrease can be explained by other researcher's conclusions that ca-alginate beads adsorb calcium ions in cement filtrate, which leads to the decrease in crosslink density and, thus, the decrease in swelling rate [11]. In the end, the swelling balance is reached due to the balance between the elastic contraction force and the electrostatic repulsion force. The decrease in the swelling rate of PCA due to the adsorption of calcium ions is only 3.6%, while CA reaches 26.8%. The phenomenon means that the adsorption of PCA on calcium ions in the cement filtrate is indeed reduced. The decrease can be explained by the increased resistance and friction for calcium ions into PCA due to the smaller pore size inside PCA and the higher molecule density [17].

The ca-alginate beads also need to have a certain swelling ability to satisfy the internal curing of cement-based materials. Therefore, the diffusion behavior of water in ca-alginate beads is explored according to the swelling kinetics of ca-alginate beads in cement filtrate. The first 60% swelling data of the equilibrium swelling rate is fitted to determine the diffusion mechanism of water with the empirical formulas given by Ritger and Peppas, as follows:

$$\frac{W(t)}{W_{eq}} = kt^n \tag{2}$$

$$\ln\left(\frac{W(t)}{W_{eq}}\right) = n\ln t + \ln k \tag{3}$$

where W_{eq} is the equilibrium swelling rate; k is a constant of proportionality, which is related to the network structure; n determines the diffusion mechanism of water molecules.

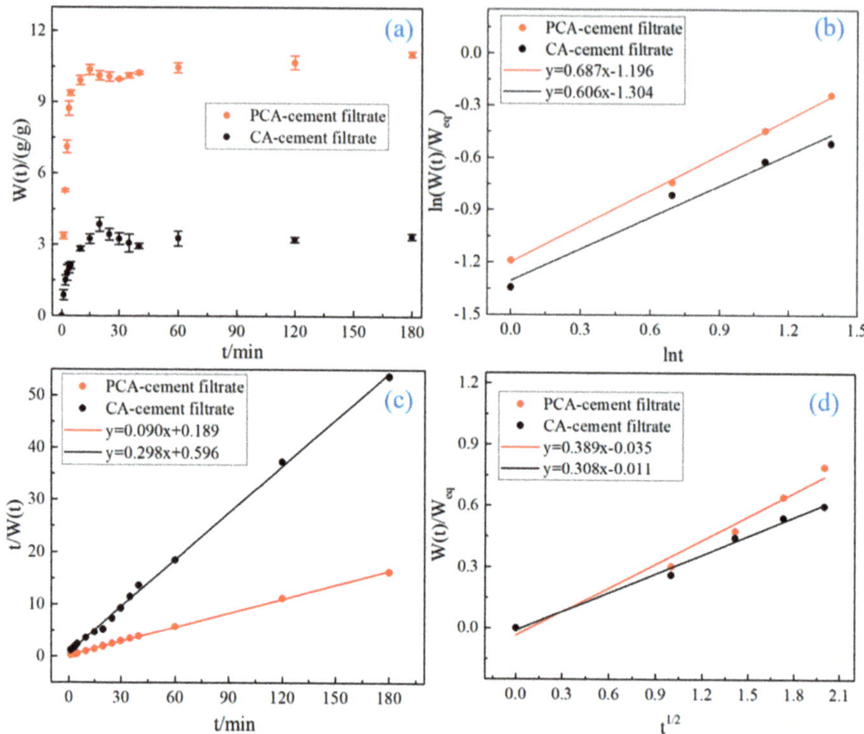

Figure 6. The swelling behaviors of PCA and CA in cement filtrate: (**a**) the swelling rate of two kinds of ca-alginate beads in cement filtrate over time, (**b**) The curve of $\ln (W(t)/W_{eq})$ over $\ln t$, and (**c**) the curve of $t/W(t)$ over t, (**d**) the curve of $W(t)/W_{eq}$ over $t^{1/2}$.

The $\ln (W(t)/W_{eq})$ to $\ln t$ curves of PCA and CA in cement filtrate are shown in Figure 6b. The relevant parameters n and k are obtained by linear fitting (shown in Table 3). The n values of PCA and CA in cement filtrate are both greater than 0.5, which means that the swellings of PCA and CA belong to non-Fickian swelling, and the diffusion rate of water is equivalent to the relaxation rate of ca-alginate beads. The similar n values indicate that the restriction of the molecular chain in the initial swelling of PCA is similar to that of CA.

Table 3. The swelling parameters of two kinds of ca-alginate beads in cement filtrate, deionized water, and a healing environment.

	W_{eq} (g/g)	W_{eq}^{*} (g/g)	K_0 ((g/g)/min)	R^2	n	k	R^2	D (cm²/s)	R^2
PCA-cement filtrate	11.06	11.11	0.043	0.999	0.687	0.302	0.998	2.06×10^{-6}	0.974
PCA-deionized water	12.61	12.5	0.023	0.999	0.663	0.290	0.997	1.7×10^{-6}	0.979
PCA-healing environment	12.50	12.66	0.032	0.999	0.599	0.325	0.996	1.95×10^{-6}	0.993
CA-cement filtrate	3.36	3.36	0.149	0.998	0.606	0.271	0.965	1.53×10^{-5}	0.988
CA-deionized water	3.98	4.00	0.098	0.999	0.682	0.248	0.992	1.63×10^{-5}	0.971
CA- healing environment	3.07	3.125	0.129	0.999	0.704	0.234	0.900	6.38×10^{-6}	0.922

The secondary kinetic model established by Schott, as shown in Equation (4), is used to determine the swelling rate and equilibrium swelling rate for the entire swelling process of PCA and CA. Equation (5) is obtained by the integral of Equation (4) according to the initial conditions $t = 0$, $W_t - W_0 = 0$, and is as follows:

$$\frac{\mathrm{d}(W(t))}{\mathrm{d}t} = K_s\left(W_{eq} - W(t)\right) \tag{4}$$

$$\frac{t}{W(t)} = A + Bt \tag{5}$$

where B is the reciprocal of the equilibrium swelling rate ($B = 1/W_{eq}^*$); A is the reciprocal of the initial swelling rate K_0 ($A = 1/(K_0 \, W_{eq}^2)$).

The curves of $t/W(t)$ to t are shown in Figure 6c. The parameters W_{eq}^* and K_0 are obtained through linear fitting and calculation (shown in Table 3). The fitting has a high correlation. The calculated equilibrium swelling rates are consistent with the experimental results, which means that the secondary kinetic model established by Schott is suitable for evaluating the dynamic swelling process of PCA and CA. The diffusion behaviors of water in relation to PCA and CA in cement filtrate are similar. The equilibrium swelling rate of PCA is 11.11 g/g, which is significantly greater than that of CA (3.36 g/g). The higher equilibrium swelling rate is caused by lower elastic contraction force brought on by the structure of the ca-alginate beads with less chelation and higher electrostatic repulsion, caused by their higher molecular density.

Stewart proved that the diffusion behaviors of water and calcium ions are similar [17]. The swelling data of the first 60% of the equilibrium swelling rate is adopted to determine the diffusion coefficients of water in PCA and CA with Equation (6), as follows [26,27]:

$$\frac{W(t)}{W_{eq}} = 4\left[\frac{Dt}{\pi D_0^2}\right]^{1/2} \tag{6}$$

where D is the diffusion coefficient.

The curves of $W(t)/W_{eq}$ to $t^{1/2}$ are shown in Figure 6d. The D values are obtained by linear fitting and calculation (shown in Table 3). In the cement filtrate, the diffusion coefficient of water in PCA is smaller than that of CA because the internal molecular structure of PCA is relatively denser and the pore size of PCA is smaller. The diffusion coefficient of water proves that the diffusion coefficient of calcium ions in PCA is decreased.

Although the diffusion rate of water is less than that of CA, the equilibrium swelling rate of PCA in the cement filtrate increases significantly. The equilibrium swelling rate of PCA in the cement filtrate reaches 11.11 g/g, which can satisfy the needs of the internal curing of cement-based materials.

3.4.2. Swelling Kinetic in the Healing Environment

Figure 7 shows the swelling behavior of PCA and CA in deionized water and the healing environment. The swelling parameters of PCA and CA in deionized water and the healing environment are shown in Table 3. The equilibrium swelling rate of PCA in the healing environment did not decrease compared with the equilibrium swelling rate in deionized water, while the equilibrium swelling rate of CA decreased by 21.9%. The decrease in the equilibrium swelling rate of CA is caused by the crosslink between CA and calcium ions in the cement filtrate.

The equilibrium swelling rate of PCA is 12.66 g/g in the healing environment, while that of CA is 3.12 g/g. This is a result of the larger molecular density of PCA and less chelation. In addition, the diffusion coefficient of water in the healing environment is still less than CA. However, the time for PCA to reach equilibrium swelling is similar to that of CA. That is probably because PCA, with its smaller size and rougher surface, has a larger specific surface area. Therefore, PCA can reach the swelling balance in the healing environment quickly, which is more conducive to the rapid closure of cracks of cement-based materials.

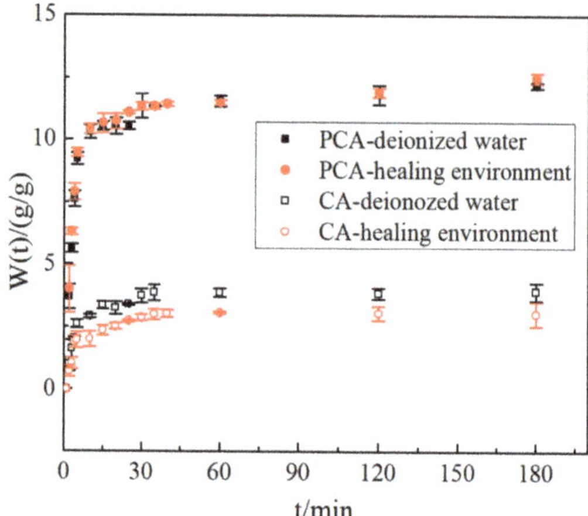

Figure 7. The swelling rate of two kinds of ca-alginate beads in cement filtrate over time.

4. Conclusions

This paper proposed the mechanism and steps for preparing environmentally friendly ca-alginate beads for self-healing cement-based materials based on protonation theory. In addition, the molecular structure and the swelling rates in cement filtrate and the healing environment of ca-alginate beads prepared by the proposed method were characterized. The following conclusions can be obtained based on the above experimental results:

1. The preparation mechanism and the preparation steps of ca-alginate beads with low adsorption to calcium ions are designed based on the protonation theory. The sodium alginate powder is directly protonated and dispersed in an acid solution, where it reacts with calcium salt solution with a pH value of 7 to form ca-alginate beads in the proposed method;

2. The chelating bonds and ionic bonds have been formed in the ca-alginate beads prepared by the proposed method. The ca-alginate beads prepared by the proposed method have higher molecular density, less chelate bond content, a rougher surface, and smaller pore size than the ca-alginate beads prepared by the dropwise addition method. Furthermore, the size prepared by the proposed method is reduced by 197.74 μm;

3. The adsorption to calcium ions of the ca-alginate beads prepared by the proposed method in the cement filtrate is reduced. The diffusion coefficient of water is also reduced. However, the equilibrium swelling rate is increased due to the increased molecular content and reduced chelation, which can satisfy the need for the internal curing of cement-based materials;

4. The equilibrium swelling rate of the ca-alginate beads prepared by dropwise addition method in the healing environment is reduced by 21.9% compared with that in deionized water due to the adsorption of calcium ions. Compared with ca-alginate beads prepared by dropwise addition method, the equilibrium swelling rate of ca-alginate beads prepared by the proposed method in the healing environment does not significantly decrease compared with that in deionized water. In addition, the equilibrium swelling rate can be quickly reached due to its large specific surface area, which is more conducive to the rapid closure of cracks of cement-based materials.

Author Contributions: X.Z.: Investigation, Software, Validation, Formal analysis, Data curation, Writing-original draft, Conceptualization, Methodology. Y.D.: Resources, Project administration, Funding acquisition. All authors have read and agreed to the published version of the manuscript.

Funding: This research is supported by China Postdoctoral Science Foundation with Grant Number (2022M710044).

Institutional Review Board Statement: Not applicable.

Informed Consent Statement: Not applicable.

Conflicts of Interest: The authors declare no conflict of interest.

References

1. Álvarez-Pinazo, G.; Cuesta, A.; García-Maté, M.; Santacruz, I.; Losilla, E.R.; De la Torre, A.G.; Aranda, M.A.G. Rietveld quantitative phase analysis of Yeelimite-containing cements. *Cem. Concr. Res.* **2012**, *42*, 960–971. [CrossRef]
2. Qin, L.; Gao, X.; Su, A.; Li, Q. Effect of carbonation curing on sulfate resistance of cement-coal gangue paste. *J. Clean. Prod.* **2021**, *278*, 123897. [CrossRef]
3. Wang, J.Y.; De Belie, N.; Verstraete, W. Diatomaceous earth as a protective vehicle for bacteria applied for self-healing concrete. *J. Ind. Microbiol. Biotechnol.* **2012**, *39*, 567–577. [CrossRef] [PubMed]
4. Chen, Y.; Gao, J.; Tang, L.; Li, X. Resistance of concrete against combined attack of chloride and sulfate under drying–wetting cycles. *Constr. Build. Mater.* **2016**, *106*, 650–658. [CrossRef]
5. Yang, Y.; Lepech, M.D.; Yang, E.; Li, V.C. Autogenous healing of engineered cementitious composites under wet–dry cycles. *Cem. Concr. Res.* **2009**, *39*, 382–390. [CrossRef]
6. Lee, H.X.D.; Wong, H.; Buenfeld, N.R. Self-sealing of cracks in concrete using superabsorbent polymers. *Cem. Concr. Res.* **2016**, *79*, 194–208. [CrossRef]
7. Snoeck, D.; Schaubroeck, D.; Dubruel, P.; De Belie, D. Effect of high amounts of superabsorbent polymers and additional water on the workability, microstructure and strength of mortars with a water-to-cement ratio of 0.50. *Constr. Build. Mater.* **2014**, *72*, 148–157. [CrossRef]
8. Jensen, O.M.; Hansen, P.F. Water-entrained cement-based materials: I. Principles and theoretical background. *Cem. Concr. Res.* **2001**, *31*, 647–654. [CrossRef]
9. Jensen, O.M.; Hansen, P.F. Water-entrained cement-based materials: II. Experimental observations. *Cem. Concr. Res.* **2002**, *32*, 973–978. [CrossRef]
10. Justs, J.; Wyrzykowski, M.; Bajare, D.; Lura, P. Internal curing by superabsorbent polymers in ultra-high performance concrete. *Cem. Concr. Res.* **2015**, *76*, 82–90. [CrossRef]
11. Mignon, A.; Snoeck, D.; D'Halluin, K.; Balcaen, L.; Vanhaecke, F.; Dubruel, P.; Van Vlierberghe, S.; De Belie, N. Alginate biopolymers: Counteracting the impact of superabsorbent polymers on mortar strength. *Constr. Build. Mater.* **2016**, *110*, 169–174. [CrossRef]
12. Engbert, A.; Gruber, S.; Plank, J. The effect of alginates on the hydration of calcium aluminate cement. *Carbohydr. Polym.* **2020**, *236*, 116038. [CrossRef] [PubMed]
13. Hu, M.; Guo, J.; Du, J.; Liu, Z.; Li, P.; Ren, X.; Feng, Y. Development of Ca^{2+}-based, ion-responsive superabsorbent hydrogel for cement applications: Self-healing and compressive strength. *J. Colloid Interface Sci.* **2019**, *538*, 397–403. [CrossRef] [PubMed]
14. Li, X.; Shen, Q.; Su, Y.; Tian, F.; Zhao, Y.; Wang, D. Structure-function relationship of ca-alginate hydrogels: A novel crystal-forming engineering. *Cryst. Growth Des.* **2009**, *9*, 3470–3476. [CrossRef]
15. Ma, Y.; Feng, Q. Alginate hydrogel-mediated crystallization of calcium carbonate. *J. Solid State Chem.* **2011**, *184*, 1008–1015. [CrossRef]
16. Liu, H.; Liu, F.; Ma, Y.; Douglas Goff, H.; Zhong, F. Versatile preparation of spherically and mechanically controllable liquid-core-shell alginate-based bead through interfacial gelation. *Carbohydr. Polym.* **2020**, *236*, 115980. [CrossRef]
17. Stewart, T.J.; Yau, J.H.; Allen, M.M.; Brabander, D.J.; Flynn, N.T. Impacts of ca-alginate density on equilibrium and kinetics of lead (II) sorption onto hydrogel beads. *Colloid Polym. Sci.* **2009**, *287*, 1033–1040. [CrossRef]
18. Cao, Y.; Shen, X.; Chen, Y.; Guo, J.; Chen, Q.; Jiang, X. Ph-induced self-assembly and capsules of sodium alginate. *Biomacromolecules* **2005**, *6*, 2189. [CrossRef]
19. Ding, H.; Zhang, L.; Zhang, P. Factors influencing strength of super absorbent polymer (SAP) concrete. *Tianjin Univ.* **2017**, *23*, 245–257. [CrossRef]
20. Chen, J.P.; Hong, L.; Wu, S.; Wang, L. Elucidation of interactions between metal ions and ca alginate-based ion-exchange resin by spectroscopic analysis and modeling simulation. *Langmuir* **2002**, *18*, 9413–9421. [CrossRef]
21. Dambies, L.; Guimon, C.; Yiacoumi, S.; Guibal, E. Characterization of metal ion interactions with chitosan by X-ray photoelectron spectroscopy. *Colloids Surf. A* **2001**, *177*, 203–214. [CrossRef]
22. Figueira, M.M.; Volesky, B.; Mathieu, H.J. Instrumental analysis study of iron species biosorption by sargassum biomass. *Environ. Sci. Technol.* **1999**, *33*, 1840–1846. [CrossRef]
23. Moulder, J.F.; Chastain, J.; King, R.C. Handbook of x-ray photoelectron spectroscopy: A reference book of standard spectra for identification and interpretation of XPS data. *Chem. Phys. Lett.* **1992**, *220*, 7–10. [CrossRef]
24. Snoeck, D.; Schroefl, C.; Mechtcherine, V. Recommendation of RILEM TC 260-RSC: Testing sorption by superabsorbent polymers (SAP) prior to implementation in cement-based materials. *Mater. Struct.* **2018**, *51*, 116. [CrossRef]

25. Sartori, C.; Finch, D.S.; Ralph, B.; Gilding, K. Determination of the cation content of alginate thin films by FTIR spectroscopy. *Polymer* **1997**, *38*, 43–51. [CrossRef]
26. Brazel, C.S.; Peppas, N.A. Modeling of drug release from Swellable polymers. *Eur. J. Pharm. Biopharm.* **2000**, *49*, 47–58. [CrossRef]
27. Okten, N.S.; Canakci, C.C.; Orakdogen, N. Hertzian elasticity and triggered swelling kinetics of poly (amino ester)-based gel beads with controlled hydrophilicity and functionality: A mild and convenient synthesis via dropwise freezing into cryogenic liquid. *Eur. Polym. J.* **2019**, *114*, 176–188. [CrossRef]

materials

Article

Analytical and Numerical Modeling of the Pullout Behavior between High-Strength Stainless Steel Wire Mesh and ECC

Xuyan Zou [1], Yawen Liu [2], Juntao Zhu [2,*], Ke Li [2,*] and Jinglong Cao [2]

[1] Department of Civil Engineering, Zhengzhou Institute of Technology, Zhengzhou 450044, China
[2] Department of Civil Engineering, Zhengzhou University, Zhengzhou 450001, China
* Correspondence: juntaozhu@zzu.edu.cn (J.Z.); irwinlike@163.com (K.L.); Tel.: +86-136-7493-9616 (J.Z.);
   +86-185-3094-8869 (K.L.)

Abstract: Bond behavior is a key factor in the engineering application of composite material. This study focuses on the constitutive model of the bond behavior between high-strength stainless steel strand mesh and Engineered Cementitious Composites (ECC). In this paper, the effects of strand diameter, bond length and transverse steel strand spacing on bond behavior were studied based on 51 direct pullout tests. Experimental results showed that the high-strength stainless steel strand mesh provided specimens an excellent ductility. Based on the experimental data, the existing bond–slip model was revised using the theory of damage mechanics, which fully considered the influence of the steel strand diameter on the initial tangent stiffness of the bond–slip curve. The results of the model verification analysis show that errors are within 10% for most parameters of the bond–slip model proposed, especially in the ascending section, the errors are within 5%, indicating that the calculated results using the revised model are in good agreement with the test results. In addition, the revised model was applied to the finite element analysis by using the software ABAQUS to simulate the pullout test, in which the spring-2 nonlinear spring element was used to stimulate the bond behavior between steel strand meshes and ECC. The simulation results show that the numerical analysis fits the experimental result well, which further verifies the accuracy of the model and the feasibility and applicability of the numerical analysis method.

Keywords: bond–slip relationship; engineering cementitious composites; finite element model; high-strength stainless steel wire mesh

Citation: Zou, X.; Liu, Y.; Zhu, J.; Li, K.; Cao, J. Analytical and Numerical Modeling of the Pullout Behavior between High-Strength Stainless Steel Wire Mesh and ECC. *Materials* **2022**, *15*, 5649. https://doi.org/10.3390/ma15165649

Academic Editor: F. Pacheco Torgal

Received: 23 July 2022
Accepted: 14 August 2022
Published: 17 August 2022

Publisher's Note: MDPI stays neutral with regard to jurisdictional claims in published maps and institutional affiliations.

1. Introduction

Engineered Cementitious Composites (ECC) is a novel cement-based material [1–3]. ECC is manufactured based on the designed theory of micromechanics and fracture mechanics and is composed mainly of cement, fine-graded aggregate, admixture, and is reinforced with short fibers [4]. As a high-performance cementitious composite, the hardened composite possesses high strength, excellent ductility, and significant strain hardening behavior [5,6].

There have been studies on the applications of ECC in engineering practices [7]. In some research, ECC was used to connect beam and column, without a transverse steel bar, in which it sustained substantial shear distortions under cyclic loads [8]. Mechtcherine [9] introduced the novel material to strengthen and repair buildings and infrastructure made of concrete and masonry. He demonstrated that the strengthening of structures with ECC layers increased their resistance to dynamic, energetic loading as with earthquakes, impact, or exposures. Some scholars [10] used ECC for the flexural repair of concrete structures with significant steel corrosion and achieved an excellent reinforcement effect, which proved that an ECC patch can fully recover the load-carrying capacity of the corroded steel rebar during the hardening of the rebar.

However, as a cement-based material, ECC cannot completely replace the steel bar to bear the load in a certain direction. Researchers have proposed their use in combination with reinforcement materials for numerous structural applications [11,12]. To develop a good performance of the composite structure, the assessment of bond behavior between reinforcement materials and ECC matrix is a key aspect [13].

Xu et al. [14] proposed a bond–slip model of steel bat-concrete where the parameters, including bar diameter, anchor length, concrete strength, and cover thickness, were considered comprehensively. A modified model [15], developed by Lundgren, can be used to predict splitting failures and the loss of bond and is certified by conducting a finite element model. Some researchers have also studied the influence of concrete age on bond strength [16,17]. The results demonstrated that bond strength between concrete and steel re-bar was affected by concrete age significantly and decreased with increasing curing age [18]. Based on these, Shen et al. [19] have further studied the bond behavior between steel bars and high-strength concrete at different ages and conducted a prediction model for the bond stress-slip relationship considering the effect of concrete age and concrete strength.

The bond behavior of steel bar and ECC is a hot research topic currently [20–23]. Hou [24] studied the effect of corrosion on the bond and found that corrosion scarcely weakened the bond toughness between ECC and corroded rebar below a 15% corrosion ratio. The influence of high temperature has also been investigated [25,26]. Test results proved that excellent bonding performance was built between the rebar and ECC, even at temperatures up to 800 °C.

Lee et al. [20] proposed an analytical model for the bond–slip relationship of steel bar-ECC, in which a nested iteration procedure was employed for embedded reinforcement under pull-out forces, and the effect of embedment length on the failure mode of reinforcement was studied. A more comprehensive constitutive model conducted by Zhou [27] can be adopted to predict the bond–slip behaviors of steel bar-ECC with FRP confinement, and used as a reference for application in practical engineering.

In the study of the bond behavior of FRP bars embedded in cement matrix, the researchers [28,29] found that the bond strength decreased with the increase of embedment length, and smaller diameter bars developed higher bond strengths than larger ones. A local bond–slip relationship model and the method for the determination of the parameters of the model was proposed [30]. The method can be applied to take into account different embedment lengths. Other studies were conducted to expose the effects of bonding by fiber [31,32]. The bond behavior of FRP-ECC was affected by fiber type, fiber volume fraction. PVA and hook end steel fibers were able to increase the bond behavior because they can resist and control the interfacial crack initiation, growth, and propagation.

The model describing bond–slip relationships between textiles and cementitious matrix was reported initially by Naaman et al. [33] and has been developed by other scholars [34]. Banholzer [35] proposed a model with iteration procedures that introduced the slip distribution law and boundary conditions. In further research [36], the transverse elements improved not only the bonding behavior but also the toughness of composite materials. Jiang [37] described the bond–slip curve in one continuous function that can be used to derive the bond force and proposed an equivalent method to evaluate the effect of weft yarns.

Steel strand has high strength, good fire resistance, and good economic, compared with reinforcement and textile. Currently, a combination of steel wire and cement-based material is popular in the rehabilitation of RC structures [38]. Kim [39] reported an experiment using steel wire mesh and permeable polymer mortar to retrofitted RC bridge columns and he suggested a satisfactory increase in load carrying capacity and improving substantially in the hysteretic behaviors of the column. Steel wires combined with ECC were used to improve the bearing capacity and durability of existing RC beams, in which the crack development in the concrete tension zone was delayed [40]. To develop a good performance of the composite structure, the assessment of bond behavior between steel strand and ECC matrix is a key aspect, but few relevant studies reported on this issue.

Compared with the aforementioned reinforced materials, the high-strength stainless steel strand has a large difference in both apparent morphology and mechanical properties.

Whether this type of bond property can be correctly expressed by existing models still needs further investigation. As shown in Figure 1, a series of pull-out experiments were conducted and the effect of parameters including the bar diameter, the relative embedded length, and the transverse steel strand spacing were investigated. The bond performance and the characteristics of bond–slip curve were discussed on the basis of the experimental results. Afterward, on the basis of the bond–slip model previously developed, a constitutive model was proposed for stimulating the bond–slip relationship between steel strand and ECC, which considered additionally the effect of the strand diameter on the initial tangential stiffness and the damage inside ECC substrate. Furthermore, the finite element was developed to verify the validity of the bond–slip model.

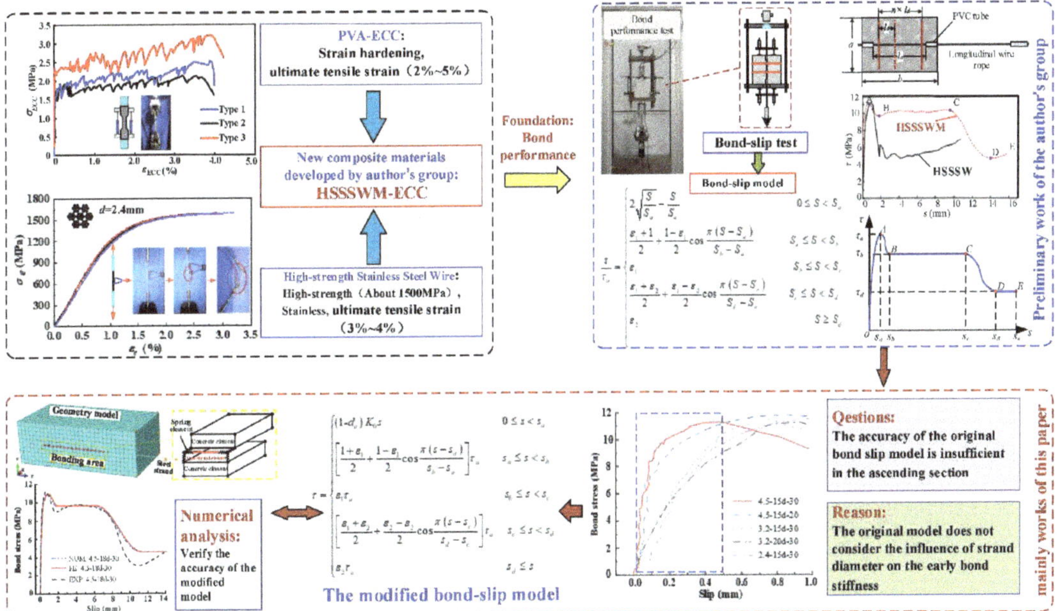

Figure 1. The workflow diagram.

2. Experimental Methods and Results

2.1. Specimen Design

In this paper, a total of 51 sets of pullout tests were conducted, which were divided into 17 groups according to different experimental parameters. Each group had three identical specimens. The parameters considered in the experiment were the strand diameter d (2.4 mm, 3.2 mm, and 4.5 mm), the relative embedded length l_a (15d, 18d, 20d, 22d, 25d, and 28d), and the transverse steel strand spacing l_d (20 mm, 30 mm and 40 mm).

As shown in Figure 2, the specimen size is $a \times b \times c$, where a indicates the specimen width (150 mm), b indicates the specimen length (100 mm, 150 mm, and 170 mm), and c indicates the specimen thickness (27 mm, 37 mm, and 50 mm). The specimens were cast in cuboid molds with a longitudinal steel strand embedded horizontally, $d/2$ mm below the central axis. Transverse steel wires are located at $d/2$ below the center of the specimen, and on the left and right end, extend 2–3 mm from the template. The longitudinal steel wires at the ends of the specimen are sheathed in PVC pipes to prevent the bonding between two materials, which can avoid the local failure of ends resulting from the stress concentration.

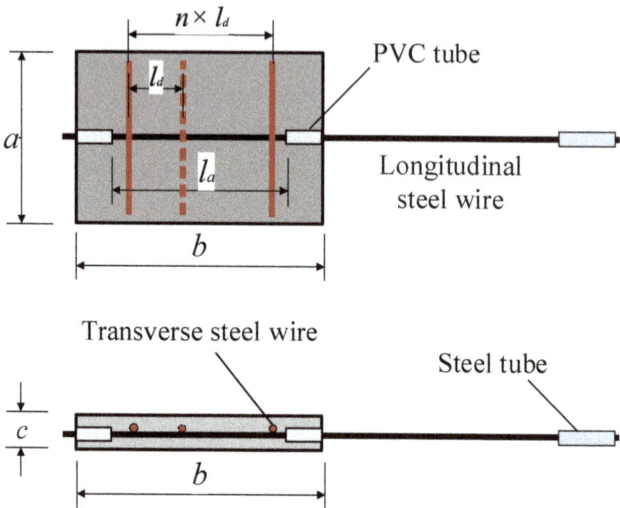

Figure 2. The diagram of specimens.

The standard specimens were manufactured according to the following steps: (1) fixtures were used to fix the steel strand in a predetermined position and make it in a tight state; (2) the mutually contacting parts of the steel strand in two directions were tied together by steel wires; (3) polyurethane foam sealing agent was applied to all the gaps in the template to avoid cement leakage; (4) ECC could be cast. In addition, a number of cubic specimens with sizes of 70.7 mm × 70.7 mm × 70.7 mm and 280 mm × 40 mm × 15 mm were also cast for compression and tension tests respectively. In the material properties test of ECC, the compressive strength is 32.45 MPa and the specimen exhibits a tensile strain-hardening behavior up to approximately 2.2% strain, with a tensile strength close to 2.83 Mpa and a cracking strength close to 1.89 MPa.

The specimens were numbered in the form A-B-C-D, where A denotes the diameter of the steel strand, B denotes the relative anchorage length, C denotes the spacing between transverse steel strands, and D represents the order of the specimen in the group. For example, a specimen named 4.5-15d-30-1 denotes that the diameter of the steel strand is 4.5 mm, the relative anchorage length is 15d, the spacing between transverse steel strands is 30 mm, and it is the first specimen in the group. Additional details on the experimental program are available in the literature [41].

2.2. Test Results

The setup for the pullout tests is shown in Figure 3. The AB segment is the free end, the CD is the loading end, and the BC segment is the actual anchoring segment of the longitudinal steel strand in the ECC. The loads were applied to the specimens by a 100 kN capacity servo-hydraulic testing machine, and were measured by the testing machine directly. Besides, four linear variable differential transducers (LVDTs) were installed to measure the slip of the steel strand relative to the cement.

Test results are given in Table 1, which includes the failure modes, the ultimate pullout force T_a, the ultimate bond stress τ_a, and the corresponding displacement s_a. The average bond stress and relative slip between anchorage steel strand and ECC were calculated as follows:

$$\tau = \frac{F}{\pi d l_a} \tag{1}$$

$$s = \frac{s_C + s_A}{2}, \tag{2}$$

where τ and s denote the average bond stress and slip respectively, F indicates the applied load, d indicates the steel strand diameter, l_a indicates the anchorage length. s_A denotes the slip at the free end and it can be ignored because the deformation of the AB segment steel strand is extremely small, and therefore the displacement at point A is regarded as point B. s_C denotes the slip at the loading end. Because the deformation of the *CD* segment cannot be ignored, s_C is calculated as follows:

$$s_{CD} = \frac{Fl_{CD}}{E_s A_s} \tag{3}$$

$$s_C = s_D - s_{CD}, \tag{4}$$

where l_{CD} is the *CD* segment length, E_S and A_S are the elastic modulus and the measured area of the steel strand, s_{CD} is the elastic deformation of the *CD* segment, and s_C and s_D are the displacements at point C and point D, respectively.

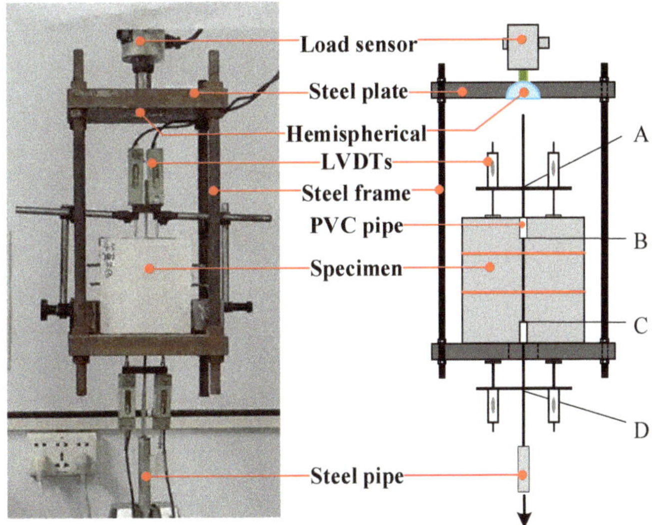

Figure 3. Schematic diagram of loading device.

Table 1. Summary of results from steel strand–ECC unidirectional pullout tests.

Group No.	Specimen No.	Specimen Size (mm × mm × mm)	T_a/kN	τ_a/MPa	s_a/mm	Failure Mode
	4.5-15d-0-1		10.74	11.32	0.91	P
A1	4.5-15d-0-2	150 × 150 × 50	10.92	11.37	0.92	P
	4.5-15d-0-3		10.80	11.28	0.91	P
	4.5-15d-20-1		11.14	11.68	0.82	P
A2	4.5-15d-20-2	150 × 150 ×50	10.65	11.18	0.61	P
	4.5-15d-20-3		10.74	11.53	0.69	P
	4.5-15d-30-1		11.10	11.63	0.72	P
A3	4.5-15d-30-2	150 × 150 × 50	10.65	11.17	0.75	P
	4.5-15d-30-3		10.63	11.14	0.72	P
	4.5-15d-40-1		10.38	10.89	0.76	P
A4	4.5-15d-40-2	150 × 150 × 50	10.14	10.85	0.87	P
	4.5-15d-40-3		10.64	11.16	0.91	P

Table 1. *Cont.*

Group No.	Specimen No.	Specimen Size (mm × mm × mm)	T_a/kN	τ_a/MPa	s_a/mm	Failure Mode
	4.5-18d-30-1		12.11	11.37	0.75	P
B1	4.5-18d-30-2	150 × 150 × 50	12.84	11.22	0.74	P
	4.5-18d-30-3		12.75	11.14	2.08	P
	4.5-20d-30-1		12.81	10.07	1.39	P
B2	4.5-20d-30-2	150 × 150 × 50	13.49	10.61	0.78	P
	4.5-20d-30-3		13.56	10.99	0.75	P
	4.5-22d-30-1		14.11	10.09	0.82	P
B3	4.5-22d-30-2	150 × 150 × 50	14.21	10.16	0.65	P
	4.5-22d-30-3		14.89	10.64	0.96	P
	4.5-25d-30-1		14.93	9.44	0.98	P
B4	4.5-25d-30-2	150 × 170 × 50	16.24	10.09	0.98	R
	4.5-25d-30-3		15.03	10.22	1.01	P
	4.5-28d-30-1		16.18	9.09	1.47	R
B5	4.5-28d-30-2	150 × 170 × 50	16.34	9.18	1.55	R
	4.5-28d-30-3		16.26	9.13	1.63	R
	3.2-15d-40-1		5.49	11.38	0.78	P
C1	3.2-15d-40-2	150 × 100 × 37	5.46	11.31	0.72	P
	3.2-15d-40-3		5.45	11.30	0.8	P
	3.2-18d-30-1		6.47	11.17	0.84	P
C2	3.2-18d-30-2	150 × 100 × 37	6.39	11.05	0.77	P
	3.2-18d-30-3		5.37	9.28	0.58	P
	3.2-20d-30-1		7.08	11.11	0.93	P
C3	3.2-20d-30-2	150 × 100 × 37	7.59	11.79	1.04	R
	3.2-20d-30-3		7.07	11.09	0.92	P
	3.2-22d-30-1		7.56	10.75	1.02	R
C4	3.2-22d-30-2	150 × 100 × 37	7.77	11.01	1.07	R
	3.2-22d-30-3		7.86	11.17	1.05	R
	2.4-15d-30-1		3.19	11.72	1.11	P
D1	2.4-15d-30-2	150 × 100 × 37	3.08	11.22	0.85	P
	2.4-15d-30-3		3.04	11.03	0.93	P
	2.4-18d-30-1		3.73	11.25	0.87	P
D2	2.4-18d-30-2	150 × 100 × 37	3.67	11.08	0.82	P
	2.4-18d-30-3		3.72	11.21	1.10	P
	2.4-20d-30-1		3.94	10.88	0.91	P
D3	2.4-20d-30-2	150 × 100 × 37	3.97	10.97	1.12	P
	2.4-20d-30-3		4.05	11.19	1.26	P
	2.4-22d-30-1		4.37	10.94	1.16	R
D4	2.4-22d-30-2	150 × 100 × 37	4.34	10.87	1.15	R
	2.4-22d-30-3		4.37	10.94	1.16	R

2.3. Analysis of the Pullout Process

As shown in Figure 4, the representative load-displacement curves are plotted according to the test results, and exhibit the following characteristics; (1) the ascending segment of the curve is very steep; (2) the slip at the free end was lower than the load end, which indicates that the pullout force starts from the load end and develops towards the free end; (3) with the join of transverse steel wires, the curves of those specimens show obvious ductility when the load is reduced to 80% of the maximum. Therefore, bond failure is transformed into ductile failure from brittle failure at the interface of the strand and ECC.

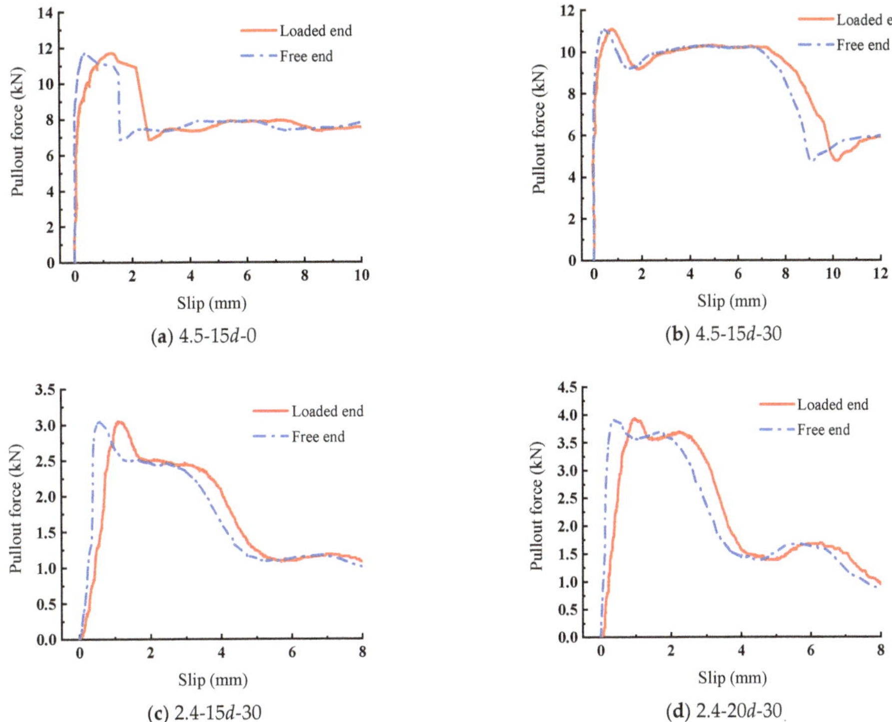

Figure 4. Representative pullout forces-slip curves.

3. Bond–Slip Model

3.1. Bond–Slip Curve

The representative bond–slip curves are shown in Figure 5, divided into five stages: upward stage, first descending stage, ductile strengthening stage, second descending stage, and residual stage.

(1) Upward stage (OA): In the initial stage, the bond stress develops rapidly while the slip increases very limited. The bonding force is mainly provided by chemical adhesive and mechanical interlocking. As the free end begins to slip, the chemical adhesive completely disappears, but the actions of mechanical interaction and friction resistance are gradually obvious;

(2) First descending stage (AB): After reaching the peak stress, the bond stress begins to decrease because of the rapid decrease of mechanical interaction force, and the increase of friction resistance is very limited. The role of the transverse steel wires is to reduce the decrease rate of the curve and thus help ECC bear the stress;

(3) Ductile strengthening stage (BC): After being destroyed to a certain degree, the traverse steel strand contributes to retain the mechanical interaction. At the same time, the friction resistance is constantly increasing. Under the combined action of the two forces, the bond stress remains stable, macroscopically;

(4) Second descending stage (CD): In this stage, the bond–slip curve shows a more obvious descending trend. This is because the action of mechanical interlocking and friction are both weakened gradually and the role of horizontal steel wires is also weakened further;

(5) Residual stage (DF): At last, the bond–slip curve tends to be flat, and the bonding force is provided only by friction resistance. Finally, the longitudinal steel strand is pulled out.

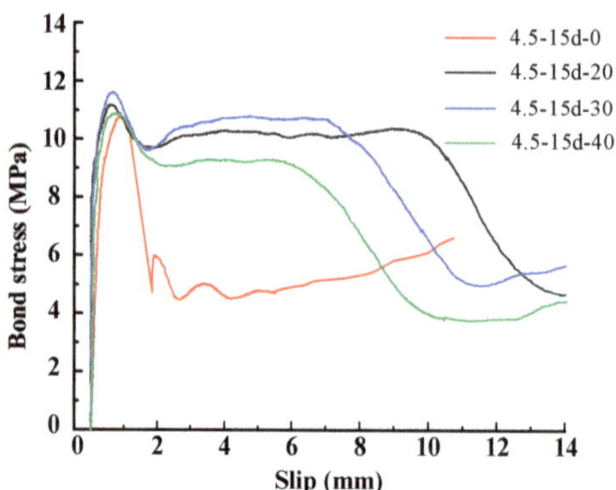

Figure 5. Representative bond stress-slip curves.

3.2. Development of an Analytical Bond-Slip Model

A suitable bond–slip constitutive relationship is the key to further research and the foundation for engineering application. As a novel civil engineering material, ECC has significant differences from conventional concrete, and there are few studies on the bond–slip relationship between ECC and steel strand. In the existing model [42,43], neither of them can describe such a distinctive ductile stage. Based on the test results and existing models, a new bond–slip constitutive model is proposed for the steel strand–ECC. The mathematical expression could be written as Equation (5), and the corresponding typical bond–slip model is plotted in Figure 6.

$$\tau = \begin{cases} (1-d_e)K_0 s & 0 \leq s < s_a \\ \left[\frac{1+\varepsilon_1}{2} + \frac{1-\varepsilon_1}{2}\cos\frac{\pi(s-s_a)}{s_b-s_a}\right]\tau_a & s_a \leq s < s_b \\ \varepsilon_1\tau_a & s_b \leq s < s_c \\ \left[\frac{\varepsilon_1+\varepsilon_2}{2} + \frac{\varepsilon_2-\varepsilon_2}{2}\cos\frac{\pi(s-s_c)}{s_d-s_c}\right]\tau_a & s_c \leq s < s_d \\ \varepsilon_2\tau_a & s_d \leq s \end{cases}, \tag{5}$$

where d_e denotes the damage evolution parameter; K_0 denotes the initial stiffness of the curve; τ_a, τ_b, τ_c and τ_d (unit: MPa) indicate the bond stress of points A, B, C and D, respectively; s_a, s_b, s_c and s_d (unit: mm) indicate slip value of points A, B, C and D, respectively; ε_1 and ε_2 denote dimensionless coefficients, which make it more concise to express the formula.

3.3. Prediction of the Model Parameters

Figure 7 shows the bond–slip curves in the upward stage. The curve is significantly affected by the strand diameter, hence when formulating the curve, it is reasonable to adopt a function involving diameter. Based on damage mechanics, a model is adopted to describe the bond–slip by Mazars J et al. [42]. It is taken here to plot the upward stage of bond–slip behavior between ECC and steel strand mesh. The model includes the influence of various factors having clear physical meanings, not just the diameter. The formulation is defined in Equation (6) where d_e denotes the damage evolution parameter, and K_0 denotes the initial stiffness of the curve.

$$\tau = (1-d_e)K_0 s \quad 0 \leq s < s_a \tag{6}$$

$$K_0 = \frac{\tau}{s} = \frac{0.8\alpha_0 E}{d} \tag{7}$$

$$d_e = 1 - \frac{\rho_e n}{n - 1 + (s/s_a)^n} \tag{8}$$

$$\rho_e = \frac{\tau_a}{K_0 s_a} \tag{9}$$

$$n = \frac{K_0}{K_0 s_a - \tau_a}. \tag{10}$$

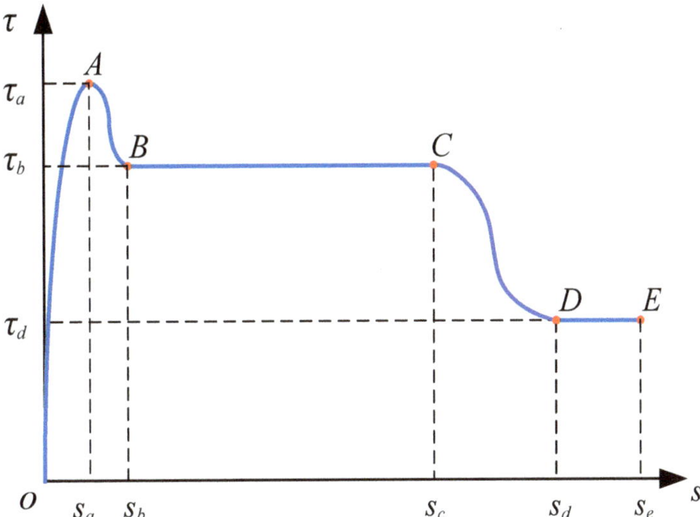

Figure 6. The typical bond–slip model for steel strand mesh and ECC.

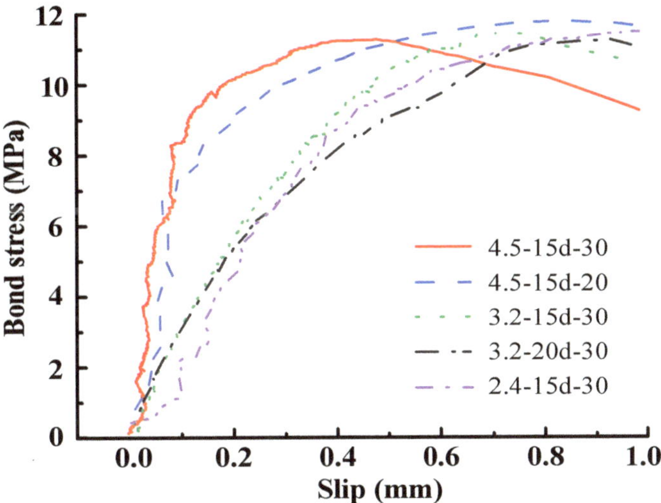

Figure 7. Bond slip curves in upward stage.

Zhou et al. [27] have explained the calculation method of K_0 in detail, and the final formulation could be written as Equation (7). However, there is no introduction to the calculation method of α_0, which is used to denote the instantaneous deformation depth. Fitting with the experimental data, α_0 is significantly affected by the strand diameter. It can be deduced as Equation (11) where $d_{3.2}$ indicates that the steel strand diameter is 3.2 mm. It is noticeable that the effect of steel strand diameter often has a limit value, and fitting of the test results denotes that $\alpha_0 \geq 0.03$ is more appropriate for steel strand–ECC. The comparison between measured and predicted bond–slip curves in the upward stage is shown in Figure 8, which proves that the proposed model fits well with test data.

$$\alpha_0 = \begin{cases} \left[\ln\left(-0.04(d/d_{3.2})^2 + 0.295(d/d_{3.2}) + 0.852\right)\right]^2 \\ \geq 0.03 \end{cases}. \tag{11}$$

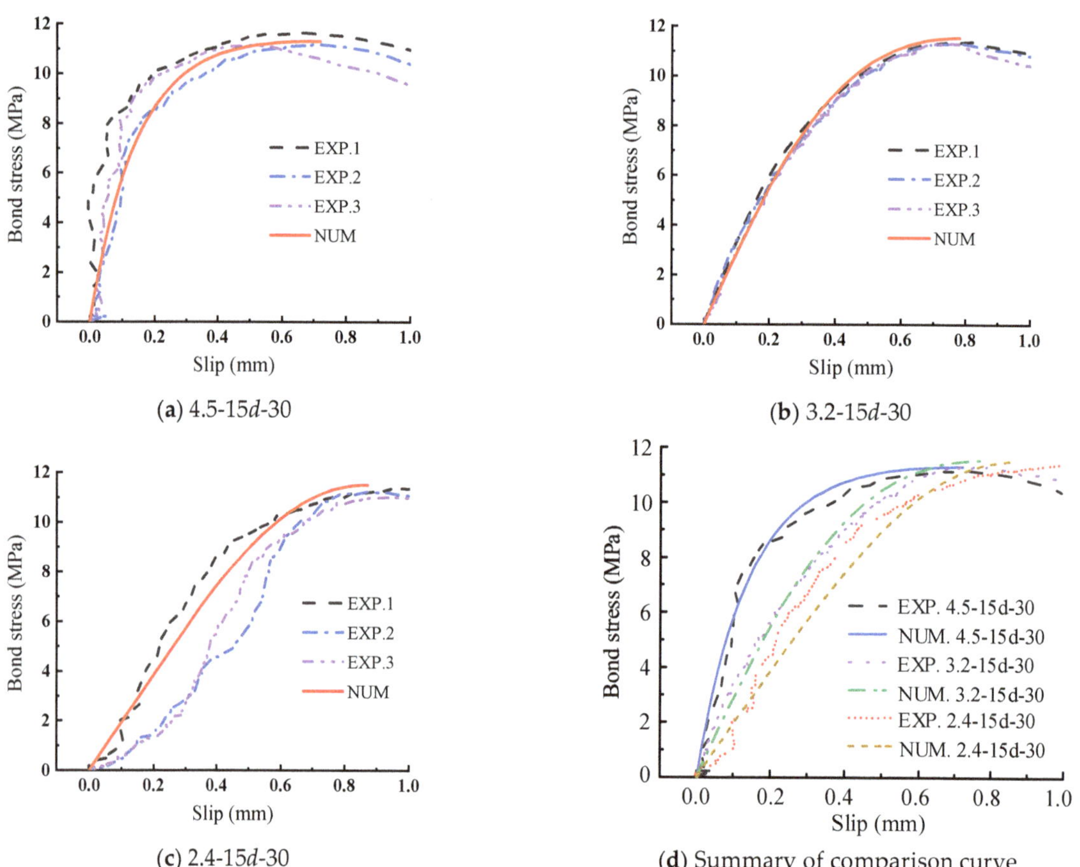

Figure 8. Comparison between measured and predicted bond slip curves in upward stage.

In Equation (5), s_a denotes the slip value corresponding to the peak value of bond strength, which is reduced due to part of the tensile stress is transferred to the transverse steel strand in the ECC, and consequently the shear deformation of the ECC becomes smaller. In the upward stage, where no obvious shear failure takes place, the relative slippage between the two materials is reduced due to the presence of transverse steel wires. The formulation of s_a could be written as Equation (12):

$$s_a = (0.002 \times d + 0.657) \times \left(0.018 \times \frac{l_a}{d} + 0.741\right) \times \left(-1.781 \times \frac{d}{l_d} + 1.402\right), \tag{12}$$

where d denotes the steel strand diameter, mm; l_a denotes the embedded length, mm; l_d denotes the transverse steel strand spacing, mm.

The peak value of bond strength τ_a can be obtained by solving the simultaneous equations of Equations (6) and (12), but when calculating τ_a only, this method is cumbersome, so it is necessary to establish a concise formulation. Based on the regression analysis of the experimental data in this paper, the following equation can be deduced:

$$\tau_a = \left(-0.06 \times \frac{d}{d_{3.2}} + 1.242\right) \times \left(-0.019 \times \frac{l_a}{d} + 5.828\right) \times \left(0.204 \times \frac{d}{l_d} + 0.578\right) \times f_{et}, \quad (13)$$

where d, $d_{3.2}$, l_a and l_b have already been introduced in the previous part of the paper, f_{et} denotes the tensile strength of ECC, f_{et} is 2.83 MPa.

The endpoint of the first descending stage τ_b, as shown in Figure 6, is written as $\varepsilon_1 \tau_a$. When $s > s_a$, without the intervention of the transverse steel strand, the bond force will be reduced to 0.5–0.6 times τ_a. The addition of the transverse steel strand makes ε_1 at least greater than 0.5. Based on the pull-out test result, the formula is shown as Equation (14), and the corresponding slip is shown as Equation (15):

$$\varepsilon_1 = 0.5 + \left(-0.037 \times \frac{l_a}{d} + 6.034\right) \times \left(0.057 \times \frac{d}{l_d} + 0.062\right), \quad (14)$$

$$s_b = \left(0.079 \times \frac{d}{d_{3.2}} + 0.111\right) \times \left(0.02 \times \frac{l_a}{d} + 2.61\right) \times (0.011 \times l_d + 2.394). \quad (15)$$

The length of ductile strengthening stage s_{c-b} is mainly related to the steel strand diameter d, the embedded length l_a, and the transverse steel strand spacing l_d. Fitting with the experimental data, it is given in Equation (16).

$$s_c - s_b = \left(0.188 \times \frac{d}{d_{3.2}} - 0.069\right) \times \left(-0.243 \times \frac{l_a}{d} + 9.26\right) \times (-0.216 \times l_d + 10.958). \quad (16)$$

In the same way as τ_b, τ_d is written as $\varepsilon_2 \tau_a$, and the formula is shown as Equation (17). The corresponding slip value s_b, which is ideally expressed as $s_b = g[\exp(d/d_{3.2}), l_a/d, l_d]$, because it is significantly affected by the steel strand diameter. The formula is shown as Equation (18).

$$\varepsilon_2 = \left(0.138 \times \frac{d}{d_{3.2}} + 0.072\right) \times \left(0.013 \times \frac{l_a}{d} + 0.323\right) \times \left(-0.1 \times \frac{l_a}{l_d} + 3.1\right), \quad (17)$$

$$s_d = e^{\left(1.531 \times \frac{d}{d_{3.2}} + 0.257\right)} \times \left(-0.013 \times \frac{l_a}{d} + 0.941\right) \times (-0.013 \times l_d + 1.814). \quad (18)$$

3.4. Model Verification

The prediction method of the model parameter introduced in Section 3.3 is used to predict the experiment curve. The comparison results of the prediction and experiment values are all gathered in Table 2, and the mean and coefficient of variation (C.V.) of the ratio of tested and predicted values were also listed. Figure 13 compares bond–slip curves measured in this test and predicted by the proposed model that considers different parameters including steel strand diameter, bond length, and transverse steel strand spacing. As shown in Table 2 and Figure 13, the curves predicted by the proposed model are always in good agreement with the test curves, demonstrating the satisfied accuracy of the proposed model in this paper.

Table 2. Comparison between measured and predicted parameters.

Specimen	$\tau_{a,EXP}$	$\tau_{a,NUM}$	$\tau_{a,EXP}/\tau_{a,NUM}$	$s_{a,EXP}$	$s_{a,NUM}$	$s_{a,EXP}/s_{a,NUM}$	$\varepsilon_{1,EXP}$	$\varepsilon_{1,NUM}$	$\varepsilon_{1,EXP}/\varepsilon_{1,NUM}$	$s_{b,EXP}$	$s_{b,NUM}$	$s_{b,EXP}/s_{b,NUM}$	$s_{c-b,EXP}$	$s_{c-b,NUM}$	$s_{c-b,EXP}/s_{c-b,NUM}$	$\varepsilon_{2,EXP}$	$\varepsilon_{2,NUM}$	$\varepsilon_{2,EXP}/\varepsilon_{2,NUM}$	$s_{d,EXP}$	$s_{d,NUM}$	$s_{d,EXP}/s_{d,NUM}$
4.5-15d-20	11.46	11.33	1.012	0.71	0.67	1.05	0.89	0.91	0.97	1.77	1.69	1.046	7.08	7.28	0.973	0.45	0.38	1.182	12.99	12.91	1.01
4.5-15d-30	11.31	11.05	1.024	0.73	0.76	0.96	0.88	0.89	1.00	1.81	1.76	1.030	4.56	4.91	0.928	0.33	0.40	0.833	12.13	11.83	1.03
4.5-15d-40	10.97	10.91	1.005	0.85	0.81	1.05	0.81	0.88	0.93	1.91	1.83	1.041	2.33	2.54	0.916	0.4	0.40	1.000	12.53	10.75	1.17
4.5-18d-30	11.30	10.94	1.033	0.75	0.80	0.93	0.91	0.88	1.03	1.80	1.80	0.999	4.92	4.28	1.150	0.35	0.42	0.835	10.43	11.21	0.93
4.5-20d-30	10.45	10.86	0.962	0.77	0.83	0.92	0.88	0.87	1.00	1.78	1.82	0.977	3.62	3.85	0.940	0.47	0.43	1.082	11.12	10.80	1.03
4.5-25d-30	9.77	10.67	0.915	1.00	0.90	1.11	0.83	0.86	0.96	1.85	1.88	0.981	2.59	2.79	0.929	0.33	0.47	0.702	9.97	9.77	1.02
3.2-15d-30	11.33	11.12	1.019	0.77	0.81	0.94	0.89	0.87	1.02	1.34	1.51	0.890	3.48	2.99	1.164	0.37	0.32	1.157	6.18	6.35	0.97
3.2-18d-30	11.11	11.01	1.009	0.81	0.86	0.94	0.92	0.87	1.06	1.43	1.54	0.930	2.60	2.60	0.997	0.33	0.34	0.970	5.43	6.02	0.90
3.2-20d-30	11.10	10.93	1.016	0.93	0.89	1.04	0.89	0.86	1.03	1.55	1.56	0.995	2.36	2.35	1.004	0.34	0.35	0.962	5.11	5.80	0.88
2.4-15d-30	11.21	11.16	1.005	0.81	0.84	0.97	0.79	0.87	0.91	1.35	1.35	1.000	1.46	1.81	0.806	0.27	0.27	0.997	4.60	4.33	1.06
2.4-20d-30	11.01	10.97	1.004	1.02	0.92	1.12	0.93	0.85	1.09	1.56	1.40	1.115	1.16	1.42	0.815	0.39	0.30	1.296	6.12	3.95	1.55
AVE			1.000			1.001			1.001			1.004			1.001			1.003			1.000
COV			0.033			0.073			0.054			0.060			0.094			0.038			0.083

4. Finite Element Modeling

Finite element analysis has gradually developed into an extremely crucial research tool that has been used to supplement and expand experiments by many scholars in engineering, biology, medicine, and aeronautic fields specimen in the group. Additional details on the experimental program are available in the literature [44,45]. It can solve several difficulties of experiments, such as long experimental periods, discrete results, and limited research parameters. Data that are difficult to measure or phenomena that cannot be observed also can be simulated by finite element. In this study, the general-purpose finite element package ABAQUS was used to check the effectiveness of the experimentally assessed relationship in simulating the bond behavior of ECC and steel strand by establishing a three-dimensional finite element. The steel stand and ECC are modeled using solid and beam elements, respectively. A nonlinear spring element is used to simulate the bond between steel strand and ECC, which is commonly adopted in literature for such a type of analysis.

4.1. Finite Element Model Geometry

Three-dimensional (3D_ full-scale models were created to simulate the pullout test, and these models are taken non-linear finite element analysis in order to better understand the bond behavior between ECC and steel strand mesh. A typical numerical model and the corresponding boundary conditions are plotted in Figure 9. The steel strand and ECC are modeled separately. ECC is established as a solid element, and the element type is C3D8R, the first-order, reduced integration hexahedral continuum element. Steel strand is established as a wire element due to the big aspect ratio, and the element type is B31, the 2-node linear beam in space.

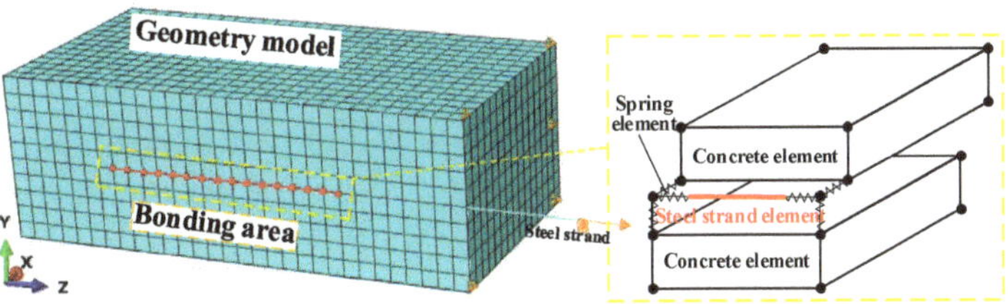

Figure 9. Typical finite element model.

The spring element is used to describe the force-displacement relationship between two nodes, and simulate the bond behavior between two materials, macroscopically. In ABAQUS, there are three types of spring elements: Spring-1, Spring-2, and Spring-A. Spring-2 element, all of which are imaginary mechanical models with only stiffness and no actual geometric dimensions and mass. Spring-2 element is adopted to simulate the bond behaviors between steel stand and ECC and it is a nonlinear spring element, including transverse stiffness and tangential stiffness. The methods to determine stiffness values are discussed in Section 4.2.

As shown in Figure 9, the identical constraint conditions of the experimental specimens were setup in the numerical model. The vertical displacement was confined to the face of concrete at the loaded end, and the other faces of the ECC specimen were left free of displacement constraints. The loading condition was a displacement to be put to the bottom node of the steel strand in the pullout direction.

4.2. Material Models, Properties, and Parameters

In ABAQUS, there is no standard method to simulate the material properties of ECC. As a cement-based composite material, ECC can be simulated by the same method as concrete, based on the reason that they share similar characteristics. Under low-pressure stress, the models adopted to simulate concrete properties include Concrete Smeared Cracking model, Concrete Brittle Cracking model, and Concrete Damage Plasticity model. In this paper, the Concrete Damage Plasticity (CDP) model is used to simulate the mechanical behavior of the ECC, and CDP is characterized by three sections: plasticity, compressive behavior, and tensile behavior. For the plasticity section, there are five parameters be defined: the dilation angle (ψ), eccentricity parameter (ε), the ratio of the biaxial compressive strength to the uniaxial compressive strength (f_{b0}/f_{c0}), the ratio of the second invariant of the deviatoric stress tensor in tensile meridians to that in compressive meridians (K_C), and Viscosity Coefficient (VC). These parameters are obtained by reference to available literatures and contrast trial calculation results. A set of data is accepted based on the numerical results that match the experimental or theoretical results with a minimum average error. The final values selected are listed in Table 3.

Table 3. Plasticity of ECC input data.

ψ	ε	f_{b0}/f_{c0}	K_C	VC
36	0.1	1.05	0.667	0

In the material property test, the elastic modulus and Poisson's ratio of ECC are 14.5 GPa and 0.255, respectively. For the compressive behavior of ECC, the uniaxial compressive stress-strain curve is shown in Figure 10a and can be expressed by:

$$\frac{\sigma}{\sigma_{cu}} = \begin{cases} 1.1\frac{\varepsilon}{\varepsilon_{cu}} + 0.5(\frac{\varepsilon}{\varepsilon_{cu}})^5 - 0.6(\frac{\varepsilon}{\varepsilon_{cu}})^6 & 0 \leq \frac{\varepsilon}{\varepsilon_u} < 1 \\[2ex] \dfrac{0.15(\frac{\varepsilon}{\varepsilon_{cu}})^2}{1-2\frac{\varepsilon}{\varepsilon_{cu}}+1.15(\frac{\varepsilon}{\varepsilon_{cu}})^2} & \frac{\varepsilon}{\varepsilon_{cu}} \geq 1 \end{cases} \tag{19}$$

where σ and ε denote the compressive strain and stress of ECC, respectively; σ_{cu} and ε_{cu} denote the maximal compressive strength (peak point of the curve) and strain of ECC, respectively; σ_{cu} is 32.45 MPa, and ε_{cu} is 0.022.

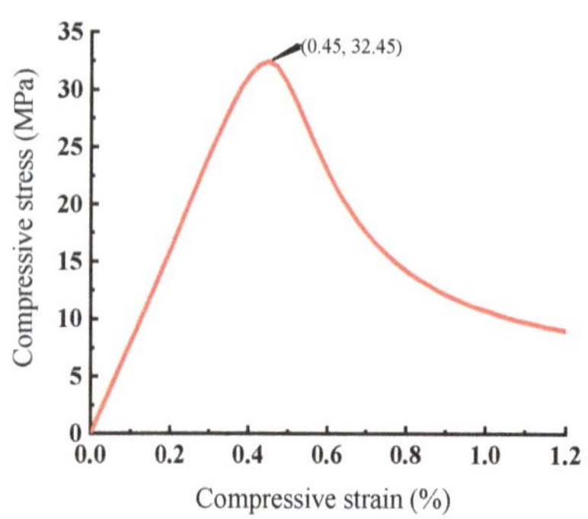

(**a**) Under uniaxial compression.

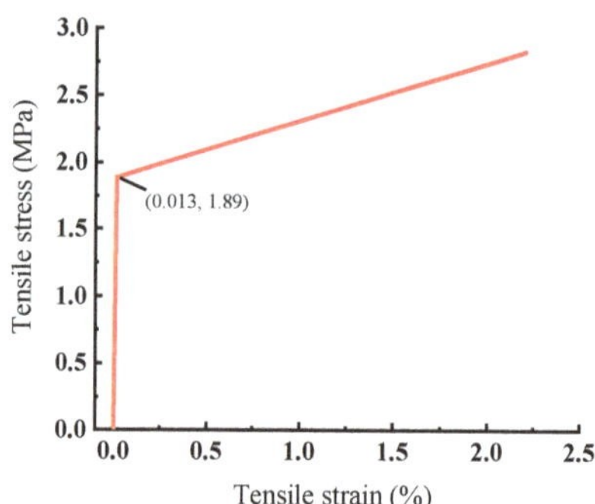

(**b**) Under uniaxial tension.

Figure 10. Constitutive relationships of ECC.

The uniaxial tensile stress-strain curve of ECC is shown in Figure 10b and can be expressed by the following equation:

$$\sigma = \begin{cases} E_{c0}\varepsilon & \varepsilon \leq \varepsilon_{tc} \\[1ex] (0.31\frac{\varepsilon}{\varepsilon_{tu}} + 0.69)\sigma_{tu} & \varepsilon_{tc} < \varepsilon \leq \varepsilon_{tu} \end{cases} \tag{20}$$

where E_{c0} denotes the elastic modulus of ECC, MPa; σ and ε denote the tensile stress and strain of ECC, respectively; σ_{tu} and ε_{tu} denote the maximal tensile strength (peak point of the curve) and corresponding strain, respectively, σ_{tc} and ε_{tc} denote the nominal tensile cracking stress and strain of ECC; σ_{tc} is 2.83 MPa, and ε_{tc} is 1.85×10^{-4}; σ_{tu} is 3.86 MPa, and ε_{tu} is 0.0271.

Longitudinal reinforcement was modeled using the elasticity plasticity model with isotropic strain hardening. The measured tensile properties are reported in Table 4. For steel strand under uniaxial tensile loading, the typical constitutive relationship is shown in Figure 11, which can be expressed as Equation (21), according to different steel strand diameters.

$$\sigma = \begin{cases} E_s\varepsilon_s & \varepsilon \leq \varepsilon_y \\ a(\frac{\varepsilon}{\varepsilon_u})^3 + b(\frac{\varepsilon}{\varepsilon_u})^2 + c\frac{\varepsilon}{\varepsilon_u} & \varepsilon_y < \varepsilon \leq \varepsilon_u \end{cases} \tag{21}$$

where E_s denotes the elastic modulus of steel strand; ε_y denotes the strain of steel strand, the plasticity beginning; σ and ε denote the tensile stress and strain of steel strand, respectively; σ_u and ε_u denote the maximal tensile strength and corresponding strain, respectively; a, b and c are dimensionless coefficients. For all these parameters, their values are obtained by fitting the test results and listed in Table 4.

Table 4. List of constitutive relationship parameters.

d/mm	E_s/GPa	ε_y	ε_u	a	b	c
2.4	130	0.0074	0.0307	1.33	−3.66	3.33
3.2	97	0.0098	0.0408	1.45	−3.52	3.25
4.5	108	0.0076	0.0378	0.90	−2.78	2.90

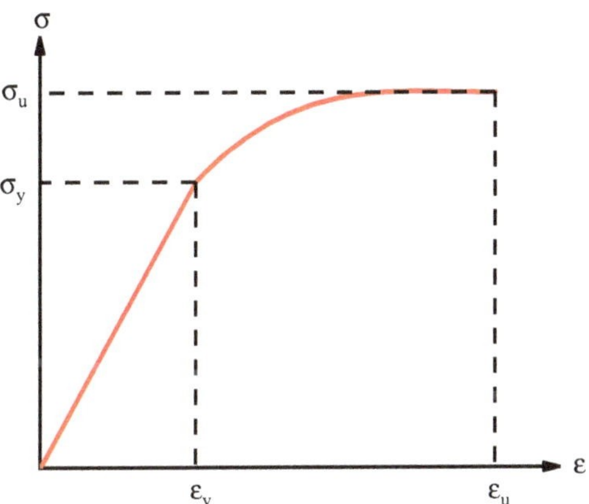

Figure 11. Typical constitutive relationships of steel strand.

The Spring-2 elements are established on the nodes where the ECC and steel strand coincide. The number and spacing of the non-linear spring elements are determined by the bond length of the steel strand. It is worth noting that to ensure the nodes of the steel strand element coincide with the nodes of ECC element, by selecting the appropriate element length. In the FE model, spring elements are established in normal and tangential directions, and are used to simulate the squeeze between the steel strand and ECC and the slip between the two materials, respectively. The stiffness of normal spring is taken as a constant value, which is the same as the elastic modulus of ECC, which is 14.5 GPa. The stiffness of the tangential spring is obtained through the conversion of the bond–slip relationship obtained through experiments, which can be expressed as follows:

$$K_t = \frac{d\tau}{ds} \times \pi d l_a. \tag{22}$$

According to the correspondence between τ and s in Equation (5), the tangential stiffness K_t is obtained by combining Equations (5) and (22). Thus, the constitutive relation between spring element force and node displacement difference as expressed in Equation (23) can be determined.

$$\begin{bmatrix} F_t \\ F_{v1} \\ F_{v2} \end{bmatrix} = \begin{bmatrix} K_t & 0 & 0 \\ 0 & K_{v1} & 0 \\ 0 & 0 & K_{v2} \end{bmatrix} \begin{bmatrix} \Delta u_t \\ \Delta u_{v1} \\ \Delta u_{v2} \end{bmatrix}, \tag{23}$$

where K_t, K_{v1} and K_{v2} are tangential spring stiffness and two normal spring stiffness, F_t, F_{v1} and F_{v2} denote the corresponding spring forces, Δu_t, Δu_{v1} and Δu_{v2} denote the corresponding displacement differences, respectively.

4.3. Comparison between Numerical and Experimental Results

As shown in Figure 12, the curve of the load-displacement relationship predicted by the proposed FE model is compared with the corresponding experimental results. It can be observed that the computed result fits the experimental result quite well when the curve is in the upward stage. In the latter part of a curve, the difference between the two curves is mainly due to the manufacturing error of the specimens. The comparison of the bond–slip curve is presented in Figure 13, in which the experimental curve, numerical curve, and stimulated curve are all plotted. In all comparison figures, the little difference is acceptable in highly nonlinear ranges of behavior; therefore, it proves that the developed model is accurate.

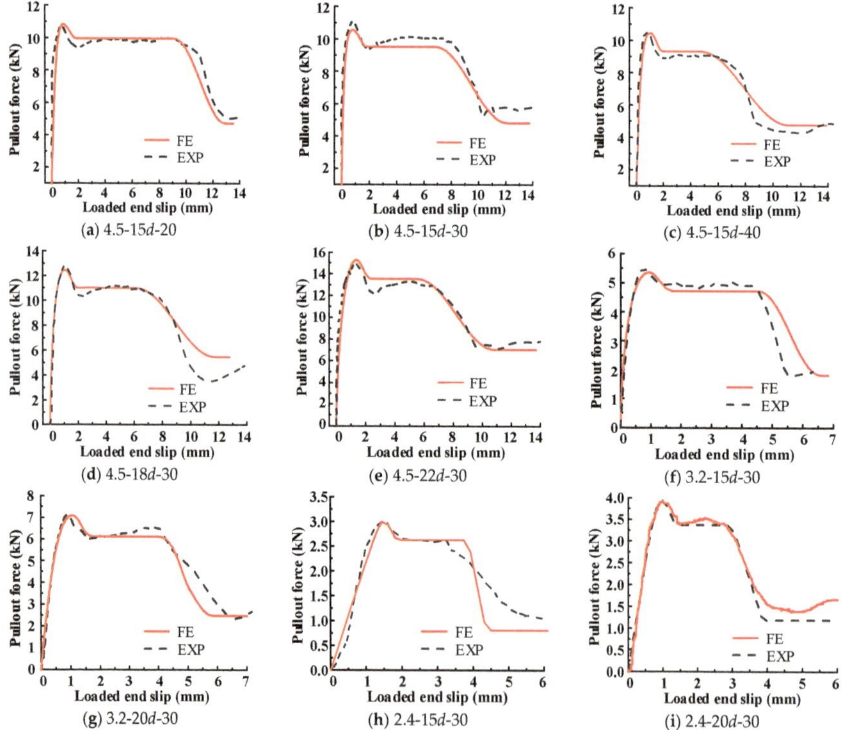

Figure 12. Comparison of force slip curves between simulation and test.

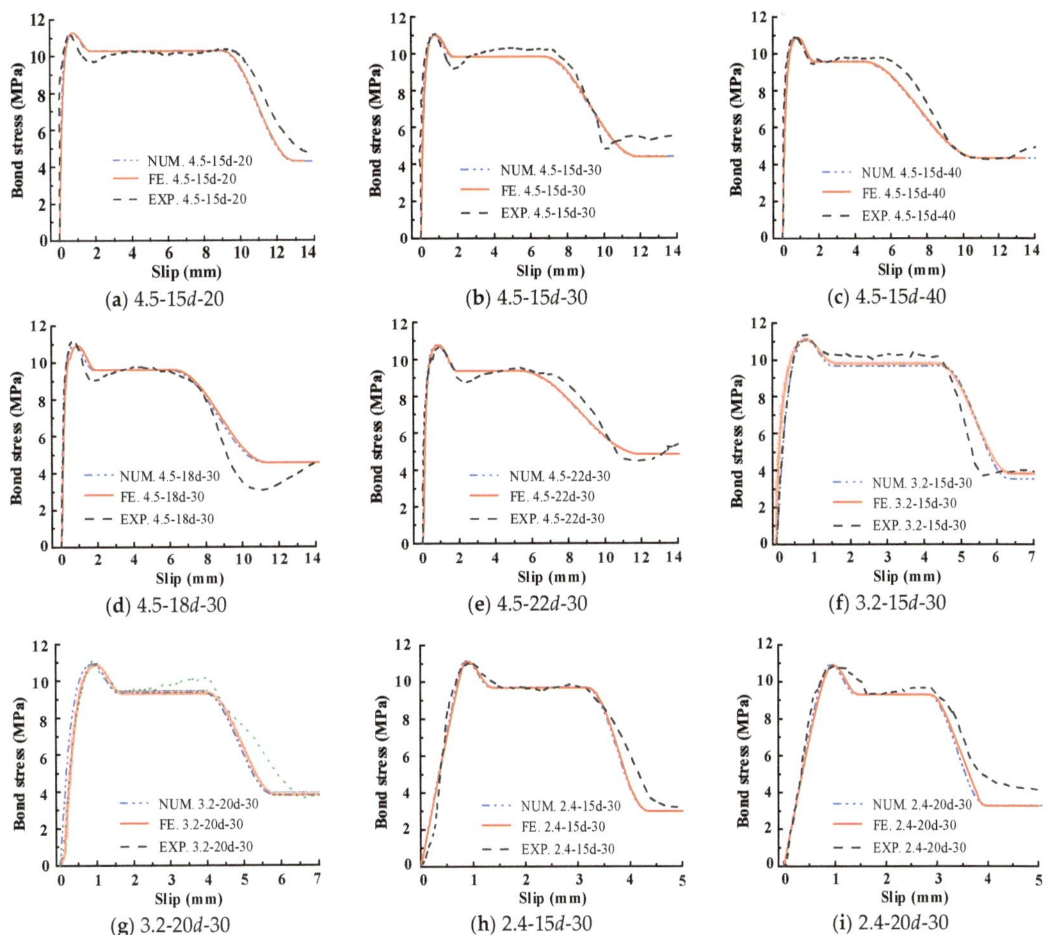

Figure 13. Comparison of bond slip curves of numerical calculation, simulation and test.

However, this model is still based on the average bond–slip relationship. As shown in Figure 4, there is a difference between the curves at the free end and loaded end. The study of local bond–slip relations may be carried out subsequently, and the entire bond–slip process will be carried out to analyze the real situation of the interface.

5. Conclusions

In this paper, experiments on seventeen groups of specimens with or without the addition of traverse steel stand were conducted to investigate the bond behavior of the steel strand–ECC system. For specimens with the traverse steel strand, the bond–slip curve showed an obvious ductility stage because the destruction speed of mechanical interlocking was reduced significantly with the part of stress in ECC borne by the traverse steel strand.

Based on the test data, a constitutive formula was proposed to predict the bond–slip relationship curve between the steel strand and ECC. The errors between the prediction and experiment values are within 10% for most parameters of the bond–slip model proposed, indicating the accuracy of the model. In addition, by introducing non-linear springs to model the bond behavior at the strand–ECC interface, the finite element models were established to stimulate the pull-out tests. In view of the results, the finite element model was able to simulate the bond behavior, which helps to develop a further understanding of the bar–concrete interaction during the pullout.

Author Contributions: Data curation, X.Z., Y.L. and K.L.; Formal analysis, X.Z. and Y.L.; Investigation, J.Z. and J.C.; Methodology, K.L.; Software, J.C.; Writing—original draft, Y.L.; Writing—review & editing, J.Z. All authors have read and agreed to the published version of the manuscript.

Funding: This work was financially supported by the National Natural Science Foundations of China (No. 51879243); China Postdoctoral Science Foundation (No. 2020M672236); Ministry of housing and urban rural development science and technology program (No. 2019-K-059).

Institutional Review Board Statement: Not applicable.

Informed Consent Statement: Not applicable.

Data Availability Statement: The data presented in this study are available on request from the corresponding author.

Conflicts of Interest: The authors declare no conflict of interest.

References

1. Sun, M.; Chen, Y.; Zhu, J.; Sun, T.; Shui, Z.; Ling, G.; Zhong, H.; Zheng, Y. Effect of Modified Polyvinyl Alcohol Fibers on the Mechanical Behavior of Engineered Cementitious Composites. *Materials* **2019**, *12*, 37. [CrossRef] [PubMed]
2. Zhong, J.; Shi, J.; Shen, J.; Zhou, G.; Wang, Z. Investigation on the Failure Behavior of Engineered Cementitious Composites under Freeze-Thaw Cycles. *Materials* **2019**, *12*, 1808. [CrossRef] [PubMed]
3. Zhang, D.; Yu, J.; Wu, H.; Jaworska, B.; Li, V.C. Discontinuous micro-fibers as intrinsic reinforcement for ductile Engineered Cementitious Composites (ECC). *Compos. Part B Eng.* **2020**, *184*, 107741. [CrossRef]
4. Wang, Y.; Zhang, Z.; Yu, J.; Xiao, J.; Xu, Q. Using Green Supplementary Materials to Achieve More Ductile ECC. *Materials* **2019**, *12*, 858. [CrossRef] [PubMed]
5. Yu, J.; Yao, J.; Lin, X.; Li, H.; Lam, J.Y.K.; Leung, C.K.Y.; Sham, I.M.L.; Shih, K. Tensile performance of sustainable Strain-Hardening Cementitious Composites with hybrid PVA and recycled PET fibers. *Cem. Concr. Res.* **2018**, *107*, 110–123. [CrossRef]
6. Yang, Z.; Du, Y.; Liang, Y.; Ke, X. Mechanical Behavior of Shape Memory Alloy Fibers Embedded in Engineered Cementitious Composite Matrix under Cyclic Pullout Loads. *Materials* **2022**, *15*, 4531. [CrossRef]
7. Zhu, J.; Xu, L.; Huang, B.; Weng, K.; Dai, J. Recent developments in Engineered/Strain-Hardening Cementitious Composites (ECC/SHCC) with high and ultra-high strength. *Constr. Build. Mater.* **2022**, *342*, 127956. [CrossRef]
8. Parra-Montesinos, G.J. High-performance fiber-reinforced cement composites: An Alternative for seismic design of structures. *ACI Struct. J.* **2005**, *102*, 668–675. [CrossRef]
9. Mechtcherine, V. Novel cement-based composites for the strengthening and repair of concrete structures. *Constr. Build. Mater.* **2013**, *41*, 365–373. [CrossRef]
10. Chen, Y.; Yu, J.; Leung, C.K.Y. Use of high strength Strain-Hardening Cementitious Composites for flexural repair of concrete structures with significant steel corrosion. *Constr. Build. Mater.* **2018**, *167*, 325–337. [CrossRef]
11. Du, Q.; Cai, C.; Lv, J.; Wu, J.; Pan, T.; Zhou, J. Experimental Investigation on the Mechanical Properties and Microstructure of Basalt Fiber Reinforced Engineered Cementitious Composite. *Materials* **2020**, *13*, 3796. [CrossRef] [PubMed]
12. Bandelt, M.J.; Frank, T.E.; Lepech, M.D.; Billington, S.L. Bond behavior and interface modeling of reinforced high-performance fiber-reinforced cementitious composites. *Cem. Concr. Compos.* **2017**, *83*, 188–201. [CrossRef]
13. Chen, Y.; Yu, J.; Younas, H.; Leung, C.K.Y. Experimental and numerical investigation on bond between steel rebar and high-strength Strain-Hardening Cementitious Composite (SHCC) under direct tension. *Cem. Concr. Compos.* **2020**, *112*, 103666. [CrossRef]
14. Xu, Y.; Shen, W.; Wang, H. An experimental study of bond-anchorage properties of bars in concrete. *J. Build. Struct.* **1994**, *15*, 26–37.
15. Lundgren, K. Bond between ribbed bars and concrete. Part 1: Modified model. *Mag. Concr. Res.* **2005**, *57*, 371–382. [CrossRef]
16. Li, H.; Xu, Z.; Gomez, D.; Gai, P.; Wang, F.; Dyke, S.J. A modified fractional-order derivative zener model for rubber-like devices for structural control. *J. Eng. Mech.* **2022**, *148*, 04021119. [CrossRef]
17. Fu, X.; Chung, D.D.L. Decrease of the bond strength between steel Rebar and concrete with Increasing curing age. *Cem. Concr. Res.* **1998**, *28*, 167–169. [CrossRef]
18. Song, X.; Wu, Y.; Gu, X.; Chen, C. Bond behaviour of reinforcing steel bars in early age concrete. *Constr. Build. Mater.* **2015**, *94*, 209–217. [CrossRef]
19. Shen, D.; Shi, X.; Zhang, H.; Duan, X.; Jiang, G. Experimental study of early-age bond behavior between high strength concrete and steel bars using a pull-out test. *Constr. Build. Mater.* **2016**, *113*, 653–663. [CrossRef]
20. Lee, S.W.; Kang, S.; Tan, K.; Yang, E. Experimental and analytical investigation on bond-slip behaviour of deformed bars embedded in engineered cementitious composites. *Constr. Build. Mater.* **2016**, *127*, 494–503. [CrossRef]

21. Deng, M.; Pan, J.; Sun, H. Bond behavior of steel bar embedded in Engineered Cementitious Composites under pullout load. *Constr. Build. Mater.* **2018**, *168*, 705–714. [CrossRef]

22. Cai, J.; Pan, J.; Tan, J.; Li, X. Bond behaviours of deformed steel rebars in engineered cementitious composites (ECC) and concrete. *Constr. Build. Mater.* **2020**, *252*, 119082. [CrossRef]

23. Hou, L.; Xu, R.; Zang, Y.; Ouyang, F.; Chen, D.; Zhong, L. Bond behavior between reinforcement and ultra-high toughness cementitious composite in flexural members. *Eng. Struct.* **2020**, *210*, 110357. [CrossRef]

24. Hou, L.; Liu, H.; Xu, S.; Zhuang, N.; Chen, D. Effect of corrosion on bond behaviors of rebar embedded in ultra-high toughness cementitious composite. *Constr. Build. Mater.* **2017**, *138*, 141–150. [CrossRef]

25. Li, X.; Bao, Y.; Xue, N.; Chen, G. Bond strength of steel bars embedded in high-performance fiber-reinforced cementitious composite before and after exposure to elevated temperatures. *Fire Saf. J.* **2017**, *92*, 98–106. [CrossRef]

26. Deshpande, A.A.; Kumar, D.; Ranade, R. Temperature effects on the bond behavior between deformed steel reinforcing bars and hybrid fiber-reinforced strain-hardening cementitious composite. *Constr. Build. Mater.* **2020**, *233*, 117337. [CrossRef]

27. Zhou, Y.; Fu, H.; Li, P.; Zhao, D.; Sui, L.; Li, L. Bond behavior between steel bar and engineered cementitious composite (ECC) considering lateral FRP confinement: Test and modeling. *Compos. Struct.* **2019**, *226*, 111206. [CrossRef]

28. Sayed Ahmad, F.; Foret, G.; Le Roy, R. Bond between carbon fibre-reinforced polymer (CFRP) bars and ultra high performance fibre reinforced concrete (UHPFRC): Experimental study. *Constr. Build. Mater.* **2011**, *25*, 479–485. [CrossRef]

29. Wang, H.; Sun, X.; Peng, G.; Luo, Y.; Ying, Q. Experimental study on bond behaviour between BFRP bar and engineered cementitious composite. *Constr. Build. Mater.* **2015**, *95*, 448–456. [CrossRef]

30. Focacci, F.; Nanni, A.; Bakis, C.E. Local bond-slip relationship for FRP reinforcement in concrete. *J. Compos. Constr.* **2000**, *4*, 24–31. [CrossRef]

31. Won, J.P.; Park, C.G.; Kim, H.H.; Lee, S.W.; Jang, C.I. Effect of fibers on the bonds between FRP reinforcing bars and high-strength concrete. *Compos. Part B Eng.* **2008**, *39*, 747–755. [CrossRef]

32. Kim, B.; Doh, J.H.; Yi, C.K.; Lee, J.Y. Effects of structural fibers on bonding mechanism changes in interface between GFRP bar and concrete. *Compos. Part B Eng.* **2013**, *45*, 768–779. [CrossRef]

33. Naaman, A.E.; Alwan, J.M.; Najm, H.S. Fiber Pullout and Bond Slip. 1. Analytical Study. *J. Struct. Eng.* **1991**, *117*, 2760–2790. [CrossRef]

34. Zhou, Y.; Wu, Y. General model for constitutive relationships of concrete and its composite structures. *Compos. Struct.* **2012**, *94*, 580–592. [CrossRef]

35. Banholzer, B.; Brameshuber, W.; Jung, W. Analytical simulation of pull-out tests—the direct problem. *Cem. Concr. Composites.* **2005**, *27*, 93–101. [CrossRef]

36. Dalalbashi, A.; Ghiassi, B.; Oliveira, D.V.; Freitas, A. Fiber-to-mortar bond behavior in TRM composites: Effect of embedded length and fiber configuration. *Compos. Part B Eng.* **2018**, *152*, 43–57. [CrossRef]

37. Jiang, J.; Jiang, C.; Li, B.; Feng, P. Bond behavior of basalt textile meshes in ultra-high ductility cementitious composites. *Compos. Part B Eng.* **2019**, *174*, 107022. [CrossRef]

38. Kim, S.Y.; Yang, K.H.; Byun, H.Y.; Ashour, A.F. Tests of reinforced concrete beams strengthened with wire rope units. *Eng. Struct.* **2007**, *29*, 2711–2722. [CrossRef]

39. Kim, S.H.; Kim, D.K. Seismic retrofit of rectangular RC bridge columns using wire mesh wrap casing. *KSCE J. Civ. Eng.* **2011**, *15*, 1227–1236. [CrossRef]

40. Yuan, F.; Chen, M.; Pan, J. Flexural strengthening of reinforced concrete beams with high-strength steel wire and engineered cementitious composites. *Constr. Build. Mater.* **2020**, *254*, 119284. [CrossRef]

41. Zhu, J.; Zhang, K.; Wang, X.; Li, K.; Zou, X.; Feng, H. Bond-Slip Performance between High-Strength Steel Wire Rope Meshes and Engineered Cementitious Composites. *J. Mater. Civ. Eng.* **2022**, *34*, 04022048. [CrossRef]

42. Mazars, J.; Pijaudiercabot, G. Continuum damage theory—application to concrete. *J. Eng. Mech.* **1989**, *115*, 345–365. [CrossRef]

43. Chu, S.H.; Kwan, A.K.H. A new bond model for reinforcing bars in steel fibre reinforced concrete. *Cem. Concr. Compos.* **2019**, *104*, 103405. [CrossRef]

44. Jamari, J.; Ammarullah, M.I.; Santoso, G.; Sugiharto, S.; Supriyono, T.; Prakoso, A.T.; Basri, H.; van der Heide, E. Computational Contact Pressure Prediction of CoCrMo, SS 316L and Ti6Al4V Femoral Head against UHMWPE Acetabular Cup under Gait Cycle. *J. Funct. Biomater.* **2022**, *13*, 64. [CrossRef]

45. Arulanandam, P.M.; Sivasubramnaian, M.V.; Chellapandian, M.; Murali, G.; Vatin, N.I. Analytical and Numerical Investigation of the Behavior of Engineered Cementitious Composite Members under Shear Loads. *Materials* **2022**, *15*, 4640. [CrossRef] [PubMed]

MDPI

St. Alban-Anlage 66

4052 Basel

Switzerland

www.mdpi.com

Materials Editorial Office

E-mail: materials@mdpi.com

www.mdpi.com/journal/materials